AF443418

Kolyma
Yenesei
Amur
Dniester
Volga
Danube
Dnieper
Rioni
Hwang Ho
Tokyo Bay
Tokuyama Bay
Minamata Bay
Nile
Bramaputra
Yangtze-kiang
Ganges
Si-kiang
Indus
Tapti
Godavari
Cochin Backwater
Irrawaddy
Vellar estuary
Salween
Mekong
Donreay
Ythan estuary
Tees estuary
Windscale
Sör Fjord
Ide Fjord
Clyde
Humber
Saltkälle Fjord
Zambesi
Shannon
Alde
Göta River
Blackwater
Valen
Thames
Waal
Ravenglass
Elbe
Liverpool Bay
Ems
Cardigan Bay
Wadden Sea
Solent
Mersey
Rhine
Severn
Scheldt
Murray
La Hague
Mause
Port Phillip Bay
Avon
Loire
Seine
Derwent
Beaulieu
Charente
Rhone
Po
Gironde
Var
Bay

Chemistry and Biogeochemistry of Estuaries

Chemistry and Biogeochemistry of Estuaries

Edited by

Eric Olausson and Ingemar Cato
Marine Geological Laboratory
University of Göteborg
Sweden

A Wiley–Interscience Publication

JOHN WILEY & SONS
Chichester · New York · Brisbane · Toronto

Copyright © 1980 by John Wiley & Sons Ltd.

All rights reserved.

No part of this book may be reproduced by any means, nor transmitted, nor translated into a machine language without the written permission of the publisher.

British Library Cataloguing in Publication Data:

Chemistry and biogeochemistry of estuaries.
  1. Estuarine oceanography
  2. Chemical oceanography
  I. Olausson, Eric   II. Cato, Ingemar
  551.4′609   GC97   79-41211

  ISBN 0-471-27679-0

Filmset in Northern Ireland at The Universities Press (Belfast) Ltd.
Printed by Pitman Press, Bath.

Contributing Authors

SIMON R. ASTON — Department of Environmental sciences, University of Lancaster, Lancaster, LA 1 4 YQ, England

KEN F. BOWDEN — Department of Oceanography, Bedford Street North, P.O. Box 147, Liverpool, L 69 3 BX, England

STIG R. CARLBERG — National Board of Fisheries, Hydrographic Division, Box 2566, S-403 17 Göteborg, Sweden

INGEMAR CATO — Marine Geological Laboratory, University of Göteborg, Box 33031, S-400 33, Göteborg, Sweden

JAN C. DUINKER — Netherlands Institute for Sea Research, Postbox 59, Den Burg, Texel, The Netherlands

DAVID DYRSSEN — Department of Analytical and Marine Chemistry, Chalmers University of Technology and University of Göteborg, Fack, S-402 20, Göteborg, Sweden

RHODES W. FAIRBRIDGE — Department of Geological Sciences, Schermerhorn Hall, Columbia University, New York, NY 100 27, USA

ULRICH FÖRSTNER — Institut für Sedimentforchung, University of Heidelberg, Postbox 103020, D-6900 Heidelberg 1, West Germany

ERIC OLAUSSON — Marine Geological Laboratory, University of Göteborg, Box 33031, S-400 33 Göteborg, Sweden

INGEMAR OLSSON — National Board of Fisheries, Box 2565, S-403 17 Göteborg, Sweden

HENK POSTMA — Netherlands Institute for Sea Research, Postbox 59, Den Burg, Texel, The Netherlands

BOB J. PRESLEY — Department of Oceanography, College of Geosciences, Texas A and M. University, College Station, Texas 77843, USA

LARS REUTERGÅRDH — National Swedish Environment Protection

Board, Special Analytical Laboratory, University of Stockholm, Wallenberg Laboratory, S-196 91, Stockholm, Sweden

RUTGER ROSENBERG   *National Board of Fisheries Institute of Marine Research, S-453 00 Lysekil, Sweden*

JOHN H. TREFRY   *Department of Oceanography, College of Geosciences, Texas A and M University, College Station, Texas 77483, USA*

MARGARETA WEDBORG   *Department of Analytical and Marine Chemistry, Chalmers University of Technology and University of Göteborg, Fack, S-402 20 Göteborg, Sweden*

WIM J. WOLFF   *Research Institute for Nature Management, Netherlands Institute for Sea Research, Postbox 59, Den Burg, Texel, The Netherlands*

Contents

Preface

The term estuary comes from the Latin adjective aestuarium which means tidal. In this book estuary is defined (by Fairbridge) as an inlet of the sea reaching into a river valley as far as the upper limit of the tidal rise. An estuary is a rather complex buffer zone between fresh and salt water environments where the salinity changes diurnally with the tide. These factors and the two opposing current systems have a strong influence on the biota as well as on the sedimentation and the physico–chemical processes in the estuary.

Every estuary is unique. There are, however, some general trends which make it possible to give outlines of the estuarine environments, the circulation and various processes and interactions going on in estuaries. The growing up of large cities nearby estuaries have in many cases caused environmental disturbances, particularly due to the discharge of domestic and industrial wastes. These problems as well as the problems of estuaries as traps for pollutants are dealt with in some chapters. The estuarine areas are of great interest also from many other aspects too, for example, its economic importance and from social and human aspects as well, but these problems are out of the scope of this book.

This book is intended as a monograph for a wide range of scientists in environmental sciences, e.g. geographers, geologists, biologists, oceanographers, and chemists. The focus is placed on the chemistry and biogeochemistry of estuaries but outlines are also given of physical factors, biotic transport and the biological processes necessary for a better understanding of the chemistry of estuaries. Human influences are also paid attention to.

Each chapter is written by a specialist in his field and their contributions treat the whole subject. However, each chapter can also be read separately as an introduction to the branch in question. For that reason a certain overlap in the text has been allowed.

The editors are most grateful to the authors for their collaboration. We also wish to thank John Wiley and Sons and particularly Dr Howard A.

Jones for their efficiency and cooperation which has much lightened the task of preparing this book.

Göteborg April, 1979

ERIC OLAUSSON
INGEMAR CATO

Chemistry and Biogeochemistry of Estuaries
Edited by E. Olausson and I. Cato
Copyright © 1980 by John Wiley & Sons Ltd.

R. W. FAIRBRIDGE
Department of Geological Sciences,
Columbia University, New York

1

The Estuary: Its Definition and Geodynamic Cycle

1 DEFINITION, CLASSIFICATION, AND LITERATURE REVIEW

1.1 Introduction

An estuary is, by common usage, a place where a river meets an inlet of the sea. It can be defined more precisely in physiographic or geomorphological terms as a river valley that is open to the ocean (e.g. Elie de Beaumont,

1845: in Schwartz, 1973); by implication, it is commonly a former or 'drowned' river valley into which the ocean now penetrates and which may or may not have some form of sediment bar at the entrance. In the case of a delta, the river may have built its own estuary by progradation. McGee (1890: reprint in Schwartz, 1973) remarked how 'the stream-carved configuration · · · passes into the sea · · · ' Dependent upon and following logically from this definition of the physical framework comes a hydrologic or ecological corollary: an estuary is a semi-enclosed arm of the sea merging with a river valley that is *influenced by tides and by the mixing of fresh and sea water.* Some confusion has been caused in the past because the corollary has sometimes been taken as the basic definition (as by Pritchard; in Lauff, 1967), which leads to false assumptions: not all areas of salt and fresh water mixing are estuaries (see Caspers: in Lauff, 1967)—in regions of high fluvial discharge, the mixing may extend to the edge of the continental shelf.

Like all geomorphic features of the earth's surface, an estuary is a dynamically evolving landform. On the scale of geologic time (10^2–10^6 yr), estuaries are ephemeral, like lakes. Eventually they fill up with sediment and, as actual geomorphic–hydrologic features, they become extinct or 'fossil'. The geomorphologist sees the estuary in the Davisian model of youth–maturity–old age. In contrast, the hydrodynamicist (as well as the hydrologist and the student of contemporary sedimentology) sees the estuary as an essentially steady-state model; that is to say, within an annual or decadal framework, the estuary tends towards an equilibrium state. In this chapter, only the long-term geodynamic aspect will be considered.

Every estuary is unique. Every estuary is subject to differing physical constraints and every estuary is therefore evolving at different rates. If one thinks of each estuary as having been recorded throughout its evolution on cinematic film, each picture frame documenting a new step in its growth, it is evident that, where different rates are involved, at any given time when one stops the film, all of the estuaries will be in different stages. In the present geological cycle no estuary in its present form is much older than 10,000 years (the boundary of the Pleistocene and Holocene epochs). During these last 10,000 years each estuary has advanced to a certain point, as now observed in the late twentieth century A.D. It can be predicted, assuming we have all the evidence furnished by the more rapidly evolving estuaries, precisely in what manner the more slowly developing examples will evolve. This information would be of inestimable value to conservationists, engineers and others dealing with the interactions of man with this extremely sensitive part of the coastal complex. Indeed, coastal zone management is today one of the crucial problem areas in the competitive field of economic development *versus* nature protection, and the estuarine division is certainly the most complex of all coastal environments. A useful reference is *The Coastline* (edited by Barnes, 1977; it includes an excellent summary on estuaries by Nelson-Smith).

A cautious assumption, noted above, required the prior knowledge of existing estuaries. That is regrettably a still largely incomplete picture, although world-wide attention is now available and several symposium type volumes have appeared during the last decade or so.

1.2 Literature review

Several societies and some new journals are more or less dedicated to the field. In Britain, there is the Estuarine and Brackish-Water Sciences Association, and in the United States, the American Society of Limnology and Oceanography, and also the Estuarine Research Federation. A major journal *Estuarine and Coastal Marine Science* (edited by Flemming, Gross, and Naylor) began publication in 1972 with bimonthly issues. Recently yet another journal, *Estuaries*, began to appear, replacing *Chesapeake Science.*

The most comprehensive introductory volume is certainly that entitled simply *Estuaries* (edited by Lauff, 1967). Evidence of significant steps forward is contained in another symposium: *Environmental Framework of Coastal Plain Estuaries* (edited by Nelson, 1972), which is more restrictive, dealing mainly with the Atlantic coasts of North and South America, but most importantly introducing the systems analytic approach. Specifically, Odum and Copeland (*l. op. cit.*) proposed a 'functional' classification of coastal systems, intended for ecological use, but equally applicable to geomorphic analysis inasmuch as biologic factors are just as much involved as physical (Figure 1). Energy sources are provided principally by: *fluid*

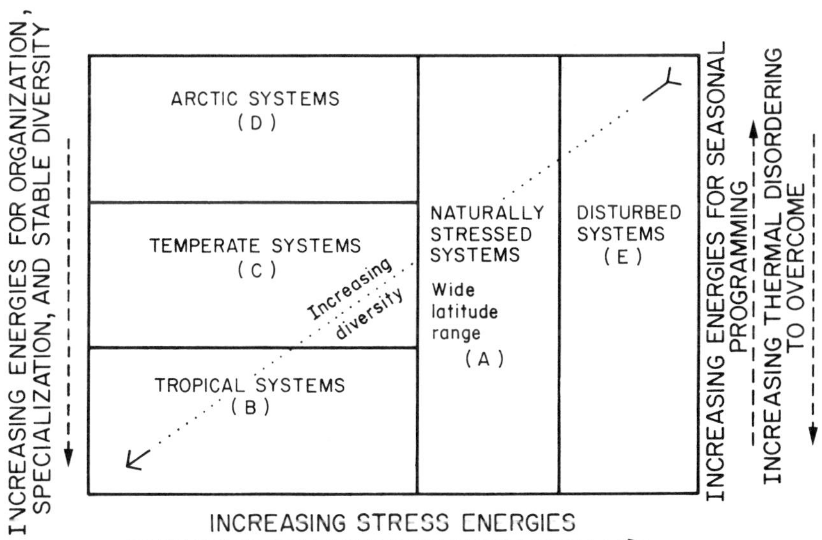

Figure 1. Systems analysis diagram appropriate to ecological and geomorphological classifications of estuaries (from Odum and Copeland, in Nelson, 1972, p. 24).

mechanics (ebb and flow of tides, river flow, coastal currents and wind), *light* (photosynthesis and thus plant growth, in day–night and annual cycles), and *organic fuels* (local and external supply for decay systems and metabolism). The energy drain involves *stress*, such as friction in the mechanical systems and biologic restrictions in the ecological systems; great stress, such as introduced by extremes of temperature and salinity cause mortality or require defensive adaptation on the part of organisms. Man-made stress introduces *disturbed systems*. All stress systems vary over a latitudinal-climatic spectrum: Tropical—Temperate—Arctic (Polar), with intermediate refinements. Vertical ecologic zonation, related to tide and season, intro-duces another overall modifier. A somewhat similar field is treated, but with a more European representation, in a text book edited by Barnes and Green (1972). Stress within the various systems received a special sym-posium treatment in Wiley (1976).

The state, or stage of development of an estuary can be analysed by an integration of studies based mainly on three disciplines: biological studies (e.g. population studies, faunal and floral), hydrologic studies (physical and chemical), and geological studies (geomorphic, sedimentologic, and strati-graphic or historical). Each of these areas are reviewed or touched on in the two volumes noted above.

A specific treatment of *Estuarine Chemistry* (edited by Burton and Liss, 1976) presents a useful collection. Because of the critical role of salinity in flocculating river-borne clays, the chemical studies may also be critical for sedimentology. Pritchard (1952) proposed that estuaries should be categorized as *positive* where sea water is diluted by fresh water (run off plus direct precipitation), although subsequently (e.g. in Lauff, 1967) he simply assumes that this is the normal state that requires no special identification. His *inverse* or *negative* estuary in which evaporation exceeds runoff and precipitation and sea water mixes with more saline water; this is, of course, seasonally important in tropical latitudes, but has been relatively little studied. Situations showing a relative balance between evaporation and freshwater supply Pritchard defined as *neutral*. It should be noted, however, that these labels identify seasonal states and are not year-round categories.

The fluid dynamics of the estuarine and related nearshore environments have received close attention in recent years from coastal engineers (Ippen, 1976; McDowell 1977). They include such matters as wave theory, tidal analysis, hurricane surges and various structural engineering problems. Interesting treatments of the use of vegetation in coastal engineering and the effects of estuarine dredging are included in *Estuarine Research* (vol. 2, edited by Cronin, 1975). A Russian collection on coastal dynamics is edited by Longinov (translation: 1969).

Numerous works are available on coastal ecological and biological prob-lems. A useful volume, *The Biology of Estuaries and Coastal Waters* (Perkins, 1974) reviews the field particularly from the point of view of human interactions: the benthos, pollution, waste disposal and management. The chemistry and biology, especially of the food chain or 'web' are treated in *Estuarine Research* (vol. 1, edited by Cronin, 1975).

The sedimentology of estuaries has not yet been accorded a volume to itself although numerous entries in the recent *Encyclopedia of Sedimentology* (edited by Fairbridge and Bourgeois, 1978) deal with specific aspects. *The Encyclopedia of Geomorphology* deals systematically with the morphology (edited by Fairbridge, 1968). A valuable series of sediment articles was included in the symposium volumes edited by Lauff (1967), Nelson (1972), Barnes and Green (1972), Wiley (1976), and by Burton and Liss (1976). Many of the classic papers by van Straaten, Kuenen, Reineck, Evans, and others are contained in the reprint volume edited by Klein (1976).

It has been shown by Rusnak (in: Lauff, 1967, p. 180) that the hydrological approach to the estuary, proposed by Pritchard (1952), applies also to a sedimentological classification. Estuaries receiving a stream discharge carrying a high load of sediment (either in traction or suspension) tend to fill rapidly in the manner of a prograding delta front: thus a *positive-fill estuary*. A low discharge situation may be countered by a high off-shore wave energy that brings off-shore or beach sediment into the estuary at every high tide: an *inverse-fill estuary*, such as was suspected by Häntzschel (1939) for the German Wattenmeer, and later confirmed by Hansen (1951) for the Danish Wadden Sea, by van Straaten (1950) for the Netherlands Wadden Sea and also in such areas as the Wash in Britain (Evans, 1965) and the U.S. East Coast (Meade, 1969). Many of the key papers are reprinted in Klein (1976).

Neutral estuaries, where for one reason or another there are extended periods of equilibrium, tend to develop anoxic (anaerobic) pockets where low pH and high organic concentrations favour the development of a euxinic or miniature Black Sea sediment environment. The deeper fjords of Norway (as shown by Strøm), British Columbia and elsewhere favour this situation; the strongly seasonal alternation of flow in the subtropics is inclined to favour the growth of semi-permanent sand bars that also create stagnant basins. Here Baas-Becking and Moore (1959) demonstrated FeS_2 mineralization (pyrite, etc). In the rivers of western France, Bourcart, and later Francis-Boeuf (1947) demonstrated the development of an unstable 'mud bung' that under the influence of high river discharge could be sluiced out at irregular intervals, creating remarkable alternations in conditions (Berthois and Barbier, 1954; Guilcher, 1963). In the case of some tropical or equatorial estuaries intense seasonal flow may bring a fresh water tongue out to the edge of the continental shelf, so that much of the finer sediment thus joins the turbidity current flows that transport it irreversibly into deep-sea basins. The surface water flow from the Amazon is sometimes drinkable up to 100 km offshore (Olson, 1970).

A field of inquiry that has been very little exploited so far is the Holocene stratigraphy of estuaries. It involves costly investigation by coring and radiocarbon dating, and so has been restricted to a few areas. The filling of some British estuaries is reviewed by Steers (in Lauff, 1967). Analysis of the Holocene history on the Gulf of Mexico by Rusnak (in Lauff, 1967) provides a basis for a preliminary estimation of accumulation rates. As will be discussed further below, this is an area urgently requiring systematic attention. Papers by Colquhoun et al. (in Nelson, 1972) on the South Carolina estuaries point the way.

1.3. The definition and classification of estuaries

The term estuary comes from the Latin substantive *aestus*, heat, boiling, tide; and specifically the adjective *aestuarium* means tidal. Thus the Oxford Dictionary defines it as 'the tidal mouth of a great river, where the tide meets the current'. Lyell is quoted as saying it is 'a term we confine to inlets entered both by rivers and tides of the sea'. Webster's Dictionary goes a little farther: '(a) a passage, as the mouth of a river or lake where the tide meets the river current; more commonly, an arm of the sea at the lower end of a river; a firth. (b) in physical geography, a drowned river mouth, caused by the sinking of land near the coast'. The French Larousse includes also an inlet only covered at high tide, e.g. the Bay of St. Michel, but as pointed out by Pritchard (1967) that is not an area of salt and freshwater mixing, there being no major river; such a definition would presumably include also the outer Bay of Fundy, which is certainly not estuarine, in contrast to the river channels leading into it (e.g. St. John, Passamaquoddy).

Contemporary literature tends to reflect the professional interests of the writers. Geomorphologists and physical geographers stick to the original definition, the upper limit of the estuary being the upper limit of tidal action, whereas the chemists prefer to define its upper limit as the innermost boundary of water mixing. In the latter sense, the definition proposed by Pritchard (1967) is as follows: 'An estuary is a semi-enclosed coastal body of water which has a free connection with the open sea and within which sea water is measurably diluted with fresh water derived from land drainage'. This excellently describes certain estuaries familiar to him, but it has totally lost the original, and critical, tidal and river qualifications.

A general review of estuaries is given in Fairbridge (1968). In non-tidal seas like the Mediterranean there are few estuaries, despite the presence of many rivers. Estuaries in semi-arid regions may not receive any freshwater for long periods; sometimes, as in southern California, Western Australia, and several parts of Africa, the estuary may become blocked by longshore sand drift, so that it is ephemerally isolated from the sea. In other regions the tidal limit, sometimes with a tidal bore, may reach 100 km or more above the limits of salt water intrusion. Pritchard's model is thus completely unrealistic for a globally acceptable definition.

Therefore it is proposed to define estuary (following Dionne, 1963) as follows: *An estuary is an inlet of the sea reaching into a river valley as far as the upper limit of tidal rise, usually being divisible into three sectors: (a) a marine or lower estuary, in free connection with the open sea; (b) a middle estuary, subject to strong salt and freshwater mixing; and (c) an upper or fluvial estuary, characterized by fresh water but subject to daily tidal action.* The limits between these sectors are variable, and subject to constant changes in the river discharge. Closely related to estuaries hydrologically, but different in terms of physical geography, are *sounds, lagoons,* and *deltas,* each of which may include or pass into an *estuary (sensu stricto).*

The bulk of the world's estuaries have only been in existence for the last 6,000 years, since when they have been progressively infilled, aided by a eustatically oscillating sea level and by sedimentation, furnished either from rivers, or by onshore and longshore drift. Where the fill has been principally from river-borne sediment, a *delta* grows at the expense of the estuary. At the present day it is the rivers with the smaller sediment discharges that have the open estuaries. Where long-shore drift is dominant, a *lagoon* or *sound* (a larger equivalent) is created due to the growth of *barrier islands* (Schwartz, 1973); for discussion of lagoons, see Fairbridge (1968) or Colombo (in Barnes, 1977). A lagoon tends to be oriented parallel to the coast, in contrast to the typical estuary, which is at right angles to it (LeBlanc and Hodgson, 1959: reprint in Schwartz, 1973, p. 67).

Deltas form best in seas of low tidal range and limited wave action (e.g. Mississippi, Fraser, Yukon, Orinoco, Nile, Po, Rhone, Danube) and here the estuaries fill most rapidly. Only streams with very low discharge rates still have open estuaries in these low-tide-range areas (e.g. in Texas). Estuaries are best preserved in seas of high tidal range (e.g. Seine, Thames, St. Lawrence, Hudson, Si-kiang), in regions of extreme glacial entrenchment (e.g. Norwegian fjords), in regions of crustal downwarp (e.g. Delaware Bay, Chesapeake Bay, Mobile Bay), and in semi-arid regions of minimum sediment supply.

From these examples it would seem to be clear that an estuary must relate to a pre-existing stream valley, and that, subject to various dynamic environmental factors, its existence is ephemeral, so that it will eventually fill.

The classification of estuaries has not hitherto been organized in any systematic way. In this chapter we shall make some attempt at organization, although this is to be sure a preliminary approach. Physiographic categories are summarized below, and these appear to fall into fairly distinctive genetic divisions, justification for which, it is hoped, will appear from the discussions of the various controlling factors.

The physiographic categories will be organized first in terms of (a) relative relief and (b) degree of blocking. In hydrodynamic terms, categories 1–2 are so-called 'disequilibrium estuaries', while 3–7 are mainly 'equilibrium' or 'constructional' estuaries. Category 3 is transitional. The disequilibrium situations are mainly due to the early Holocene sea-level rise where it has been off-set in some way by tectonics; thus estuaries in categories 1–2 (and some of 3) tend to look like artificial reservoirs created by dams. Category 5 is similar, but climatically controlled. Most of the others tend towards dynamic equilibrium.

The seven categories are as follows (Figure 2):

1. High relief estuary, U-shaped valley profile · · · *fjord*; related, but more subdued relief · · · *fjärd, firth, sea-loch*.
2. Moderate relief estuary, V-shaped profile, winding valley · · · *ria*; *aber* (western Britain); special limestone karst type · · · *calanque, cala*.
3. Low relief estuary, branching valleys, funnel-shaped plan · · · *coastal plain estuary* (open); flask-shaped, partly blocked by bar or barrier island (with lagoons or sounds) · · · *coast plain estuary* (with barrier).
4. Low relief estuary, L-shaped plan, lower course parallel to coast · · · *bar-built estuary*.
5. Low relief estuary, seasonally blocked by longshore drift and/or dunes, with/or without eolianite bar · · · *blind estuary*.
6. Delta-front estuary, ephemeral distributary · · · *deltaic estuary*; in interlobate embayment · · · *interdeltaic estuary*.
7. Compound estuary: flask-shaped ria backed by low plain · · · *tectonic estuary*.

Regional variations are recognized and these have been well summarized by Jennings and Bird (in Lauff, 1967, p. 121) with respect to one continent, Australia, but they can clearly be applied on a global basis. Slightly modified from Jennings and Bird, there are six of the 'dynamic environmental factors': 1. Fluvial hydrology; 2. Wave energy; 3. Tidal range; 4. Biologic sedimentary controls; 5. Sedimentology and mineralogy; 6. Geotectonics and Neotectonics. Each of these finds its place within a somewhat broader and more time-oriented organization, developed below.

This chapter will now turn to the problems relating more specifically to the dynamic geomorphology of estuaries. This aspect, that deals with the control of the physical form and growth of estuaries, is based on several independent variables: (1) regional history of sea level; (2) morpho-tectonic factors; (3) latitudinal-climatic factors; and (4) fresh-water discharge and sediment supply.

2 SEA LEVEL FACTORS

At any point on the earth's crust, the sea level is subject to tidal change (diurnal, fortnightly, and annual fluctuations) and to secular change that has

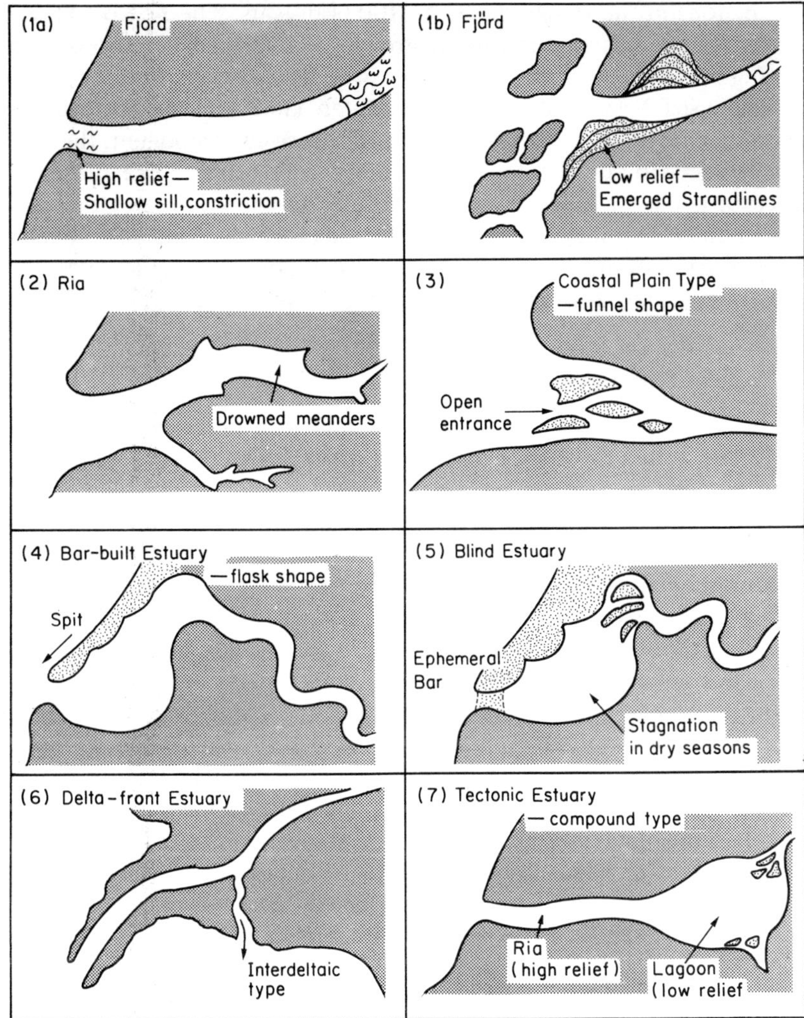

Figure 2. Sketch maps of basic estuarine physiographic types. Hydrodynamic characteristics are not considered here; discharge, tidal range, latitude (climate), and exposure all play important roles in modifying these examples, besides long-term secular processes such as tectonics and eustasy.

time functions in the range of 100 to 100,000 years. The first of these, the tidal component, affects the rate of growth of intertidal landforms such as salt marshes, coral reefs, and barrier islands. The average tide level constitutes mean sea level (MSL) and the related landforms or ecologic zones develop specific levels above or below it. The second component, the secular change, affects to some extent the shape of the estuary, the presence or absence of a bar or barrier, and contributes to the control of the rate of filling (mainly a climatic function). 'Secular change' means systematic changes of MSL of one year or more. Both components interact and need to be closely considered in appraising the regime of a delta.

2.1 The tidal component

The *tidal component* may involve fluctuations in water level from 0.1 to nearly 20 m, added to which exposure to storm waves ('fetch') and tsunamis (seismic sea waves) in certain places may lead to episodic increases in effective sea level height by values that range upwards from 2 m; a cyclonic low pressure system inversely affects sea level at approximately 10 mm per 1 mb atmospheric pressure, thus a 700 mb hurricane low will raise MSL by 3 m. Wave set-up under wind and current stress can sometimes triple this figure. Fluvial discharge also affects MSL, in some cases creating a *tidal bore* that reaches much farther upstream than does the salt-water mixing.

Estuaries in regions of low tidal range, e.g. the Baltic, Mediterranean and Black Seas, or near tidal amphidromes, e.g. Puerto Rico, tend to be more rapidly sedimented, because of the lack of ebb-tide flushing, and ecological stress such as would limit the growth halophytic swamps and mangrove. A large tidal range favours sediment removal from the main channels, and also encourages the intertidal fauna and flora to build up and out to a certain equilibrium boundary, which, once established, tends to remain constant for many centuries; this is known from comparison of air photographs over about 50 years and from ancient charts going back at least three centuries. Where exposure to a long fetch and extreme storminess is added to the high tidal factor, intertidal effects are extended, tending to severely limit the mud-living halophytes, but in contrast favouring the growth of coral reefs, which line tropical estuaries in low runoff areas (see Figures 4, 5 and 6).

2.2 Secular component

The secular component or slow change, affecting estuaries over periods in excess of one year, affects MSL by the actions of two independent variables. First there is the *eustatic variable*, that affects the volume of sea-water in the entire ocean (Fairbridge, 1961). The phenomenon of glacio-eustasy requires MSL to rise when the majority of world glaciers are melting and to sink when the glaciers are expanding. Sea-ice plays no role. There are other eustatic effects but none of them are as yet recordable on time scales shorter than 10^6 years, and need not concern us here.

Eustatic history over the last 2 million years (approximately the time of the Quaternary Ice Age) has involved a cyclic oscillation with a period of around 100,000 years and an amplitude of 100–150 m, modulated by shorter cycles of lower amplitude (e.g. around 42,000 years, 20,000 years, 8,000 years, 1,080 years, 360 years, and shorter wave lengths of ever-decreasing amplitude).

The maximum ice advance of the last glacial phase was in 17,000 B.P. ('Before Present'—before A.D. 1950), at which time MSL was approximately -135 m, and the shoreline in most parts of the world lay out near the edge of the continental shelf. Estuaries of that time correspond to the heads of the submarine canyons. No present-day estuary was then in existence, although many of the modern ones are now connected by shelf valleys or channels (e.g. Hudson Channel, seaward of New York) to the heads of those same submarine canyons. Most large shelf areas, e.g. the North Sea, the Sunda Shelf and the Sahul Shelf, were then dry land.

Progressive ice melting led to a rapid rise of sea level that brought the ocean in, for example, over the North Sea and into the Baltic, so that salt water reached Stockholm soon after 10,000 BP. By convention (Figure 3), the Pleistocene/Holocene boundary is taken as 10,000 C-14 years BP (Fairbridge, 1968, p. 525; Mörner, 1976). It reached into the Gulf of St. Lawrence even earlier (c. 13,000 B.P.) and the bones of fossil whale are found as far inland as the eastern borders of Michigan; at the same time the Hudson valley was flooded so as to join the St. Laurence basin ('Champlain Sea'). The Hudson Bay was flooded a little before 8,000 BP, the rise of water *under* the remnant ice cover probably played a major role in causing a catastrophic break-up and its rapid dissipation by means of icebergs. At this stage MSL was rising at over 10 mm/yr, thus drowning all coastlines and now most of the world's estuaries were beginning to take shape.

At 6,000 BP the MSL eustatic rise reached its present average level. This was the turning point in Holocene history. It set the clock for the infilling cycle of most of the world's present-day estuaries. Many of these estuaries had earlier histories, but traces of that evolution had largely been effaced by erosion during each successive glacial phase. We may mention, perhaps, the especially interesting early history of the Nile. Around 5 million yr BP the Mediterranean had become blocked, and evaporation lowered MSL by many hundreds of metres. The Nile, its largest feeder, cut its bed down to keep pace more or less with the falling base level. A canyon was cut back nearly 1,000 km to Aswan. Marine fossils of Pliocene age were found there in the high dam foundation work. This constitutes the longest known 'fossil' estuary. It has subsequently become filled by alluvial sediments and as an

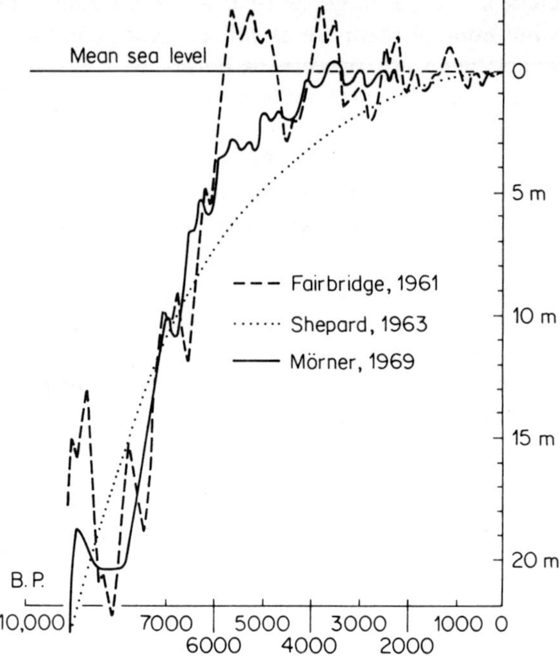

Figure 3. Curves of mean sea level over the last 10,000 years, as interpreted by different specialists. It may be noted that strong regional differences are indeed observed, and that some of the oscillations are still not certain, but new radiocarbon dates during the last decade seem to support the fluctuating model.

estuary has now ceased to exist. Many of the world's great deltas, e.g. the Mississippi, went through estuarine phases during the early Pleistocene interglacial phases, but have progressively prograded their former troughs so that the estuarine deposits are now buried by fluvial deposits.

From the year 6,000 BP until today many estuaries have reached their climax state, that is to say their dynamic equilibrium. They become progressively filled, partially or completely, by prograding fluvial sediments, which advance in the manner of a delta front. It is said that '*the delta is the enemy of the estuary*'; in other words, as the delta prograds it simply fills up the space of the estuary until it is entirely occupied, after which (like the Mississippi delta, for example) it continues to head seawards until it is near the edge of the continental shelf.

Since 6,000 BP sea level has fallen about 3 m on the average, although because of geodetic and tectonic effects it varies in different parts of the world. This 6,000 year period has also been marked by minor sea level fluctuations within the range of the smaller cycles mentioned above, notably 1,080 yr (an astronomic cycle that controls complex geodetic and climatic interactions). This periodicity dominated MSL behaviour until about 2,500 BP, subsequent to which a 360 year cycle has been prominent. Since AD 1890 world sea level has shown a secular rise, averaging about 1 mm/year. A minor pulsation of 45 yr is detected throughout the last 8,000 yr (Hillaire-Marcel and Fairbridge, 1978).

The second variable that affects the secular component of sea level change is any vertical change in the height of the earth's crust. Any youthful structural change of this sort is classified as 'neotectonic' and the ones that can be measured instrumentally as contemporary phenomena are known as 'recent vertical crustal movements' (RVCM). Rapid RVCMs are recorded of two independent types and in two quite distinct provinces on the earth's crust:

(a) *Isostatic reactions*

These occur in former glacier-loaded crustal areas, notably the northern half of North America, Scandinavia, the British Isles, Spitsbergen, Patagonia, etc. Isostatic uplift corresponding approximately to the original ice-thickness by a ratio of 1:3 (the density ratio of ice to lithosphere) is observed; the reaction, as the ice melted, was very rapid but central areas in the Hudson Bay and the Baltic are still rising by as much as 10 mm/yr. In the high uplift areas, the amplitude of isostatic uplift now exceeds that of eustatic uplift, so that the shoreline is progressively moving seaward. In such areas estuaries are becoming shallower and drying out, e.g. much of Stockholm, and parts of Edinburgh, Quebec City and Montreal are built on emerged estuaries.

Peripheral to the isostatic uplift areas are the so-called 'marginal bulges' where the crust bowed up slightly when the ice-sheet areas were loaded; those peripheral bulges have been recovering during the last several thousand years and undergoing slow subsidence, e.g. most of the U.S. East Coast, the southern Baltic, southern North Sea and southern Britain are affected by this slow crustal downwarp, so that estuaries here tend to be constantly revived, i.e. deepened. This deepening is locally offset by vigorous sedimentation.

(b) *Plate tectonic reactions*

Modern plate tectonic research shows that there are three types of crust in the world where unusual heat flow, high seismicity and marked gravity anomalies indicate exceptional stress conditions. These are at '*leading edge*' *plate margins* (i.e. subducting margins, associated with island arc or 'Pacific-type' or cordilleran continental margins), at *'trailing edge' plate margins* (i.e. rifted margins, associated with 'Atlantic type' coasts, major grabens, and strike-slip lineaments), and in *mid-plate disturbances* (associated with domes, often volcanic, rifts and graben structures). RVCMs, determined by geodetic relevelling over periods of twenty years or more disclose that some midplate areas usually with a diameter of the order of 100–500 km are rising or sinking at rates of 1–5 mm/yr, possibly in a fluctuating manner. because motions in excess of a few hundred years cannot be maintained without creating more lasting geomorphic features; e.g. the Cape Fear Arch on the eastern U.S. coast is a positive feature of this sort, and the Delaware Embayment is its negative equivalent. Plate margin vertical movements are even more rapid, and tend to be maintained in the same sense for long periods; thus the central Caucasus are rising at over 15 mm/yr according to RVCM relevelling, while some coral reefs in New Guinea, Timor and Taiwan show sustained upward displacements amounting to several hundred meters in several hundred thousand years (thus averaging 1 mm/yr over protracted periods, probably, however in brief but rapid steps).

3 MORPHO-TECTONIC FACTORS

The geological history of a coast affects not only its vertical position and geomorphic evolution, but sheds light on the lithologic nature of the rock material in which the estuary is sculpted, its reaction to weathering process, and thus also to the development of relief; this last point applies more particularly to the micro-relief, as distinct from the macro-relief, defined by the plate tectonic history and discussed above. Structural history, in detail, as well as those broad patterns, plays an important role in explaining the precise shape and form of each individual estuary.

Structurally defined, estuaries fall into two major categories: tectonically positive coasts, usually marked by hard rocks and a long-term history of net emergence (not necessarily active today); and tectonically negative coasts, marked by soft sedimentary material and a long-term history of net submergence (not necessarily active today). Some special cases exist, e.g. on limestone coasts, which will be noted below. To avoid confusion with Pritchard's 'positive' and 'negative' (or 'inverse') hydrologic regimes, one should avoid referring to these structural categories in the same terms, since they are unrelated. 'Emergence' and 'Submergence' are also terms liable to confusion because of the complexity of eustatic history. The safest procedure seems to be to categorize them as 'hard-rock estuaries' and 'soft-rock estuaries'; the rock is an observeable attribute and is not an inferred criterion; how subjective and relative terms like 'hard' and 'soft' can be quantitatively defined is not, however, a straightforward procedure, because the attribute varies in response to the energy input; a rock of intermediate hardness on a coast exposed to heavy wave attack may behave as a rather soft material whereas the same rock on the inner shore of an estuary may behave as a hard rock. With these provisos, 'hard' and 'soft' still seem to offer the simplest approach.

3.1 Hard-rock coasts

Similar in some ways to fjords, the hard rock estuaries are marked by steep relief, rocky outcrops and, at least in their outer parts, by considerable depths of water. (This attribute makes them of commercial interest as supertanker harbours and for naval facilities.) The upstream parts are characterized by a winding river course and overlapping spurs, features that are generally flattened or effaced in the case of fjords.

The most widely observed estuary-type in this class is the *ria*, as defined by von Richthofen (1886; see also Pannekoek, 1966; Fairbridge, 1968), a name associated with type examples along the northwestern coasts of Spain and Portugal, Brittany, S.W. England (Devon and Cornwall), S.W. Ireland, several parts of the Mediterranean, S.E. China, and S.E. Australia. The local term *aber* is used in Wales. The initial uplift of these ria-coasts took place long ago, in connection with the heat flow associated with continental rifting, in most cases within the range of about 250–50 million years ago. Following rifting, considerable vertical relief triggered the erosion of deep superposed river gorges in the near-coastal regions. Subsequently there followed a long period (at least 30 million years) of slow cooling and submergence. The postglacial submergence of 10,000–6,000 BP reoccupied earlier estuaries, the inner shores of which may not have changed much during the last 2 million years, e.g. the rias of S. W. Corsica which are rather precisely outlined by emerged beach deposits and erosional shore features, in parallel ribbons, ranging from Pliocene to Holocene. Estuaries of such great antiquity that are still 'living', are restricted to those valleys that are unblocked and are not fed by large drainage areas.

Some plate separations were attended by complex block-faulting and strike-slip structures. In some cases this has led to *compound rias*. One such is San Francisco Bay, where the Golden Gate begins like a ria, but in the 'back-bay' passes into a subsidence area due to graben development and downwarping; the same is true of the Tagus estuary at Lisbon.

Another interesting special case is that created by the parallel folds of 'Jura' or 'Appalachian' type. This is the *Dalmatian-type coast* (also as defined by von Richthofen, 1886), where the estuaries are known as *canali*. In this example, another plate-boundary affect, the river systems developed a superposed drainage crossing the fold trends, but subsequent streams stripped out softer formations in the axes of the folds. Slow crustal cooling and subsidence then drowned the coast, as in the case of normal rias, and the late Quaternary drowning has amplified this effect. Paradoxically the canali are typical of a submerged coastline, yet its basic geomorphic preparation required uplift.

Yet another special case is that introduced by limestone terrain, where collapsed karst tunnels result in dolinas and poljes; where invaded by the sea as in Majorca and other parts of the western Mediterranean, they are known as *calanques* or *cala*.

3.2 Soft-rock coasts

Associated with tectonically negative regions, are marked by such as downwarps of the earth's crust, fault-bounded grabens (rift structures) and by sedimentary infilling. The sediments in question are usually unconsolidated or weakly cemented or compacted, hence the term 'soft-rock', although customarily used by geologists for anything from a moldable clay to a relatively resistant shale. Downwarps of the crust, by their very nature, tend to be the loci of most of the world's greatest rivers. The post-glacial eustatic rise has drowned those great river mouths (such as in the Rhine–Maas–Schelde complex) or even entire lower valley systems (as in the Delaware Bay and Chesapeake Bay areas). Because of the soft nature of the headland rock types, that are easily eroded, vigorous longshore drift adds to the sediment discharged by the rivers. Thus, the building of bay-mouth spits, bars, and barrier islands is particularly facilitated in such regions.

The great rivers in question debouch either onto 'Atlantic-type' (rifted and downwarped) coasts, or onto 'Pacific-type' ('cordilleran', or subducted) coasts, or onto Marginal sea ('back-arc basin') coasts. The rifted borders of the first of these categories tend to have a zig-zag form corresponding to the intersections of a succession of triple points, following the theory of Burke. The 'failed arm' at each triple point is initially a graben that essentially locks in the locus of each major river and its delta. The Niger delta is cited as the classic example.

Rivers debouching onto the Pacific-type coasts are generally short-run, but with high discharge regimes, fed by the heavier precipitation engendered by the high cordilleran relief, but with limited catchment areas. Neither large deltas nor large estuaries occur, except where complicated by youthful graben structures, as in San Francisco Bay, Fraser River, Puget Sound or the Colorado delta on the Gulf of California.

In the case of Marginal Seas ('back-arc basins' in plate tectonic terminology) there are two contrasting shores. The outer side corresponds to an island arc, which may be quite youthful (e.g. Tonga Islands, Nicobar Islands) and therefore riverless, or mature (e.g. Japan, Sumatra, New Zealand, New Guinea) and possesses rivers of cordilleran type. The inner side impinges on ancient continents and is commonly the site of major river mouths (e.g. Mekong, Yangtze-Kiang, Yellow River, Amur River). Back-arc basins that form wedges terminating in cordilleran belts, such as the Andaman Sea, tend to be fed by streams that run parallel to the cordillera (e.g. the Salween and Irrawaddy). Where the back-arc basin has been truncated from the continent by rifting, as in the Tasman Sea, the youthful uplift along the fault scarps has caused most of the river systems to be deflected away from the basin in question; thus most of the large rivers of eastern Australia rise almost within sight of the Pacific but head away from it, to the SW or the NW. Only those easterly streams that have been augmented by river piracy, having captured parts of the original westerly directed headwaters, now have large catchment basins.

4 LATITUDINAL–CLIMATIC FACTORS

The surface of the globe is divisible into subparallel climatic zones that are more or less duplicated symmetrically in the two hemispheres. The latitudinal zones tend to be skewed in the Temperate Westerly belt by the continental east coast effect, and in the Tropical belt by the arid west coast effect. Furthermore, the massing of continental areas in the northern hemisphere is asymmetrically contrasted by a circumpolar waterway in the southern hemisphere. In spite of all the exceptions, however, some broad generalizations are helpful.

4.1 Polar and sub-polar zones

These include parts of the Temperate Zone formerly Polar. Land ice during the last glaciation stage (that began to retreat at 17,000 BP) reached as far as about latitude 45° in both hemispheres. Valley glaciers or ice-tongues peripheral to the great ice sheets descended through coastal valleys to scour them out, creating deep U-shaped cross-sections and overdeepened longitudinal sections (i.e. rising seaward to a threshold or rim). Postglacial eustatic rise drowned these valleys, but inasmuch as they were near the outer periphery of the ice sheet they were not subject to the great isostatic uplift of the interior situations. The geomorphic result is the classical *fjord*, the typical high-latitude estuary. Because of the overdeepening by glacier-scour, some fjords are extremely deep, over 1200 m in Norway and Greenland, but sometimes with quite shallow sills, so that miniature Black Sea (euxinic) conditions persist. Only in interior ice-sheet positions (as mentioned earlier) are the fjords now filled or totally emerged. Early Holocene shorelines, however, are traceable along the inner parts of many fjords showing a landward crustal rise, the gradient of which steepens as one goes back in time, beach-by-beach. Each of these old beach lines itself represents a time of eustatic still-stand and was separated from the next lower beach by an episode of rapid eustatic fall, preceeding the next fluctuation. These beaches in several parts of the world have been precisely dated by various mutually confirming methods that pin-point their dates to within one century or so (Hillaire-Marcel and Fairbridge, 1978). The sea-level stands in question are partially eustatic, but partly also related to climatic dynamics and to tidal effects. The ocean-facing beaches may often be traced into present-day estuaries where they merge with thalassostatic fluvial terraces, and with stepped fluvioglacial delta series.

Fjord coasts in the northern hemisphere are well known in Norway, Spitsbergen, Iceland, Greenland, Arctic Canada, Alaska, and British Columbia. In the southern hemisphere they are seen in New Zealand, Patagonia, the Antarctic Peninsula (Graham Land, etc.), and some of the Subantarctic Islands (e.g. Kerguelan).

Closely related to fjords are the *fjärds*, typically developed in southern Sweden, where the relief is less and lacks the peripheral gradients to the deep ocean (such as found in western Norway, Patagonia, and New Zealand). A discussion of the technical definition of the fjärd, along with other geomorphic forms mentioned here, may be found in Fairbridge (1968), but briefly it is a drowned arm of the sea, of glacial origin, but without the overdeepening and U-shaped profile of fjords. The examples of southern Sweden are modelled in hard Precambrian crystalline rocks; others in Denmark, e.g. Limfjord, are more gently contoured, being scoured in soft Tertiary and Quaternary deposits. (Note that in Danish 'fjord' is part of the common language meaning a protected arm of the sea, without geologic overtones; similarly 'fjärd' in Swedish. The terms when exported and used internationally thus acquire a technical precision.) In Scotland the term *firth* or *sea-loch* has a comparable meaning and is equally well applied to an estuary of glacial origin that lacks the U-shaped profile of the fjord. The fjärd coast is typical of eastern Canada, Maine, and Connecticut; also of southern British Columbia.

A feature of many middle and low latitude estuaries is the sandy bay-mouth bar or barrier (Schwartz, 1973). It is absent from the high latitude estuaries in sculpted hard-rock. Glacial action there has commonly stripped away the majority of loose soils and weathering debris (saprolite), so that waves beat ineffectively against hard-rock coastlines and lack the necessary detrital material for barrier building. In some of the intermediate (45–55° N) latitudes of Denmark, Britain, southeastern New England, and the Puget Sound region of Washington state, there is plentiful morainal debris that is reworked by wave action to generate bars for glacially scoured estuaries.

As a generalization for the Polar and Subpolar Zone, one may observe a dimensional gradation; the largest and deepest examples correspond to the higher latitude ice-sheet peripheries. In lower latitudes (now Temperate), the features in general are of reduced relief, and affected progressively less and less by isostatic crustal movements as one moves equatorwards.

4.2 Westerly-temperate zone

This is located approx. 55°–30° N and S. The climatic belt affects the coasts of the majority of the heavily populated industrial regions of the world, and is therefore subject to the most intense studies. Scientists should, however, beware of extrapolating from the familiar environments of eastern North America and northwestern Europe. Generalizations developed for this zone do not necessarily operate so universally in the high or low latitudes.

The primary geomorphological characteristic of the temperate belt is its vegetative cover. Apart from the areas where it is systematically cleared for agriculture and urban development, there is a green blanket over it that effectively prevents erosion, except in high relief areas. Inasmuch as most of the mountainous areas within this belt were glaciated during the Quaternary cold cycles, their upper slopes have been stripped bare of soil and their lower slopes cloaked with morainal debris. Thus even modest uplifts tend to furnish abnormally large sediment loads.

The equatorwards margins of the westerly-temperate zones often display a 'mediterranean' subtype that shows a transition into the semi-arid sub-tropics. In contrast to the year-round precipitation in the typical areas, this transition zone has a distinct dry (summer) season that results in the desiccation of grasslands and thus favours intense erosion, either by wind or with the first rains of winter before the ephemeral vegetation cover is reestablished. A special case is the limestone terrain within this belt which is liable to present totally bare and arid rock surfaces, even though subsurface water is plentiful in springs.

4.3 Tropical and equatorial zone

This is located approx. 30° N to 30° S. This belt embraces the largest fraction of the earth's total area, the equatorial circumference being a little over 40,000 km. Yet curiously it is the least known in terms of both coastal geomorphology and of estuarine studies in general.

Because of its higher air and water temperatures, and the absence of frost, biological factors play a greatly amplified role. They operate in several ways. Most importantly, biochemical weathering of bedrock is enormously acceler-ated in the bacteria-rich, warm-wet soils of the equatorial zone and wet-tropics. Since during interglacial maxima the monsoonal-type summer rains extended much farther poleward (at least 1,000 km) than today, these soils are now exposed to rapid mechanical erosion within the semiarid zones during their brief seasons of heavy precipitation. In this way a vast supply of loose weathered material is made available for redistribution by rivers and waves so that estuarine barriers tend to grow rapidly and accumulate to considerable dimensions.

On the mud banks and sand bars, mangrove (notably *Rhizophora*) readily takes root and then plays a major role in blocking longshore drift and tidal currents, causing additional build-up of coastal mudflats; aided by small eustatic fluctuations in sea level (even at the frequency of the 11–22 yr sunspot cycle: see Choubert in Fairbridge, 1966) the mangrove belt has in places grown many km in width during the last 6,000 years.

Within the trade-wind belts (approx. 30°–10° N and S), the subtropical high-pressure zones, extensive deserts develop. These are characterized either by no precipitation and therefore no rivers, or by endorheic (internal) drainage systems, that may terminate in lake basins or dissipate into the desert. In either case there are no exorheic outlets and no estuaries. Along the semi-arid coastal sectors nevertheless there are well developed river valleys that testify to formerly moister climates, whereas today the streams are either seasonal or quite ephemeral. The formerly important estuaries, or lagoon-type embayments are today frequently occupied by sabkhas ('seb-khas' in French-speaking regions), where algal mat sedimentation and evaporites are progressively filling in the basins (Fairbridge, 1968, p. 969). Sabkhas play the same role in the filling of tropical estuaries and lagoon embayments as salt marshes do in the temperate zones.

A very general feature of the semi-arid coasts is the incidence of *eolian-ites*, which are Quaternary calcareous dune sands that became lithified during wetter climatic cycles. Belts of them extend as reefs well offshore in places as well as above present sea level on the coastal plains. The dune building played a role of blocking the pre-existing stream valleys, so that many of the smaller or moderate sized streams are partly barred by hard eolianite reefs. Ephemeral streams develop seasonal estuaries, that revert to stagnant pools when longshore drift, migrating along the eolianite reefs seals off their entrances, when they become 'blind estuaries' (see Figure 2, item 5). This condition is widespread in Western Australia, southern California, Mexico, parts of southwestern Africa, southern Arabia and the Morocco–Mauritania coast of West Africa.

An exceptional situation exists on the Red Sea where plate tectonic rifting is relatively youthful (initiated about 30 million years ago), the rifted sides of the graben still presenting a marked relief. This happened when the area was closer to the equator than now and received heavy precipitation; deep gorges were cut, but as the lithosphere has cooled, the rifted walls are gradually subsiding, thus drowning the river valleys and creating rias. Quaternary eustasy has amplified this condition. These valleys in many instances hardly receive any precipitation today and accordingly remain as open estuaries, inactive as stream outlets and in part blocked by coral growth.

Their name in Arabic is *sherm* (Schmidt, 1923; and discussion in: Fairbridge, 1968, p. 327).

In situations that are somewhat removed from strong freshwater discharge, coral reefs flourish in warm, well-lit, shallow water. Their position is usually established by the level of a eustatic still-stand, either one prior to the last glacial (c. 125,000 B.P.), or by some fluctuations during the early or mid-Holocene (e.g. around 9,500, 7,500 and 6,000 BP). They do not grow up randomly from the floors of deep, sediment-filled lagoons, but require starting points, that is to say footholds established during eustatic low stages on some pre-existing projections. Coral reefs, particularly barrier reefs, tend to play a role with respect to estuaries comparable to that of sandy barriers on the non-coral coasts. They are, however, much more locked in position and can scarcely migrate by longshore erosion and drift. Their growth is absolutely inhibited by the fresh water jet that emerges from even small rivers at flood times. Thus there are regular breaks or 'passes' in all barrier reefs, except off semi-arid coasts (e.g. NE Queensland). Debris produced by wave actions tends to be washed over the barrier or patch reef to accumulate on the lee side permitting some expansion of the reef into relatively deep water. But once a reef-bounded estuary is established, it tends to become stabilized, only to be modified by the rates of stream-based sediment discharge and the trapping agency of the mangrove. Reef-bound estuaries in northwestern Australia, backed by a semi-arid hinterland, have growing reefs along the sides of the estuaries in their outer sectors, especially in areas of high tidal range, setting up powerful tidal currents that keep the estuaries sluiced clear; only the inner parts of the reefs become mud-covered and then populated by mangrove, as are also the delta regions.

River mouths with very large discharges in tropical-equatorial regions tend to be completely free of reefs. Thus, for example, the 1,500 km-long Great Barrier Reefs of Queensland are situated far offshore, whereas fringing reefs are limited to headlands. Reef-bound estuaries are restricted to the drier climatic areas. The barrier reefs continue around the SE end of New Guinea and the Louisiade Islands for another 1,500 km, but onshore reefs are scarce; even the barrier is interrupted for about 100 km off the deltaic mouth of the Fly River which carriers a high discharge with a heavy load of suspended mud at almost all seasons.

Coral reefs are not universal in the tropical and equatorial seas. While their outer (poleward) limits are close to 30° N and S, they are controlled by a minimum average water temperature for the coldest month of about 15–17 °C. Thus, the semi-arid tropical west coasts, subject to cold currents, originating in reefless high latitudes, i.e. currents that set *towards* the equator are not only unfavourable ecologically, they are also without coral spawn (planulae) that are essential for propagation and colonization. In short, high discharge of freshwater and mud, low water temperature, both ambient and with respect to origin of planulae, exercise insuperable ecologic constraints over the growth of reef corals. In this way the high discharge waters of the Atlantic, particularly off Africa and South America, have large estuaries and are almost coral-free, likewise the waters around the Indian Peninsula and China with their monsoon regimes; the western shores of the Americas, both North and South, are not so much affected by discharge, but by waters of cold origin. This leaves the East Indies, West Indies, East Africa, Australia, and Oceania as the coral-rich seas. In these regions the closeness of the coral growth to the estuary mouth is largely dependent upon the river discharge. Those estuaries with a semi-arid hinterland are much more influenced by the reef constrictions than those with equatorial or strong monsoonal rains. In the latter category, a rather mature Pacific island like Fiji has a large estuary, the Rewa, where there are only offshore reefs.

An additional factor, noted already in the Temperate Zone (especially Mediterranean), is the incidence of limestone coasts, far more prevalent in the Tropical and Equatorial zones, inasmuch as ancient coral reefs are now present in emerged structural positions. This lithology creates a littoral karst terrain that is essentially riverless. Rainfall tends to infiltrate into such terrain, joining a subsurface drainage system and to re-emerge either offshore, or in localized resurgent springs; in either case, this limestone terrain tends to protect growing reefs from local mud or fresh-water contamination. Some estuaries of the limestone ria type ('calanque') are known in equatorial latitudes, e.g. the sungeis of Aru.

5 FRESHWATER DISCHARGE AND SEDIMENT SUPPLY

If it is assumed that the principal estuarine drowning process ceased about 6,000 BP and if the delta is truly the 'enemy of the estuary', then the principal constraint will be the presence or absence of a barrier of some sort. All other factors being equal, it can be reasoned that the stage of sedimentation and of infilling reached so far depends upon (a) the discharge and sediment supply of the rivers relative to the size of the estuarine basin and (b) the exposure to storms and offshore sediment supply. Although the experiment does not appear to have been attempted hitherto, it would seem that a general quantitative solution to this problem is feasible. Many of the necessary values are already available, although some would require more detailed study before the calculations could be undertaken.

Stream discharge and sediment supply can be organized to recognize two regional controls of the variables: the latitudinal-climatic category and (within each) the large-scale morpho-tectonic category. Sediment supply by streams, is usually quoted in terms of (a) clastic solids transported by traction (bed load) plus turbulence (suspended load), together representing the mechanical denudation products of the drainage basin; and (b) dissolved material transported in solution, representing the chemical denudation of the basin. The combined value $(a+b)$ equals the total (E), the denudational (erosional) lowering of the hinterland. The values a and b are determined by hydrologic monitoring, standardized to a value for time (t) of one average year. On the other hand the value for E can also be approached, to obtain a long-term average figure extending over lengthy geologic periods. Two figures for E can thus be obtained:

E_h (hydrologically derived historical annual average) and E_g (geomorphically derived data to provide a mean figure averaged over geological time, possibly a multi-million year annual average).

The difference between E_h and E_g will then reflect the relative disequilibrium of the resent regime, which may be explained in terms of (a) human interference (accelerated runoff, accelerated erosion, pollution, damming and river control leading to sediment traps); (b) recent climate change (subtropical rivers in particular undergo dramatic variations on both short and long term bases); (c) recent vertical crustal movements and neotectonics of longer duration (related to short and long term variations in the earth's tectonic energy supply, such as caused by changes in global spin-rate, glaciation/deglacial loading). In most recent studies concerning the trends attributable to these disequilibrium factors, the results indicate late Holocene accelerations, so that usually $E_h > E_g$. An exceptional situation exists in the Caspian Sea, a closed (endorheic) basin, where the damming of the Volga and general human use of the normal discharge to the Caspian is at the present time causing an annual deficit, which with evaporation is leading to a very rapid drop in water level. Estuaries in the Caspian are therefore decreasing in size and deltas are rapidly prograding, but the expansion of the latter is not due to increased sediment load, but to lowered water level.

Values for E_g are of variable quality depending upon stratigraphic and geomorphic information. For example in a semi-stable platform situation, such as a Precambrian shield, the distribution of small erosional remnants of, say 100 million yr-old shallow marine Cretaceous formations on hill tops or plateaus suggests that the elevation difference between these outliers and the sediment plains represents the total denudational loss in the interim; there is, however, a possibility that the Cretaceous capping was further covered by an Eocene formation, now totally removed, so that the denudational time frame should have been 50×10^6 yr, not 100×10^6 yr. In denudation chronologies of mountain belts, results are even more difficult to refine, in part, because of repeated burials and pulsatory uplifts.

Estimates of the values for E_h and E_g may be compared with calculations for the global rates of marine sedimentation, extending from the shelf to the deep sea. Like the values for continental erosion, those for marine sedimentation can also be based on short-term (penecontemporary) or long-term (greater than million-year) averages.

24   *R. W. Fairbridge*

Studies of long-term sea-level behaviour agree that high-sea level (*thalassocratic*) phases correspond to sluggish runoff, and to both low erosion and low sedimentation rates (Figure 4); the same phases will favour globally maritime and mild climates, universal vegetative cover and a very low ratio of mechanical to chemical denudation (Fairbridge, 1967). This creates a soil-weathering regime known as 'biostatic' according to the terminology of Erhart (1973). Although high sea level favours estuary expansion, the same eustatic control decreases the mechanical load of stream transport and decreases the infilling rate (Figure 5). Paleogeographic studies disclose that estuarine facies expand enormously in the stratigraphic record during thalassocratic phases.

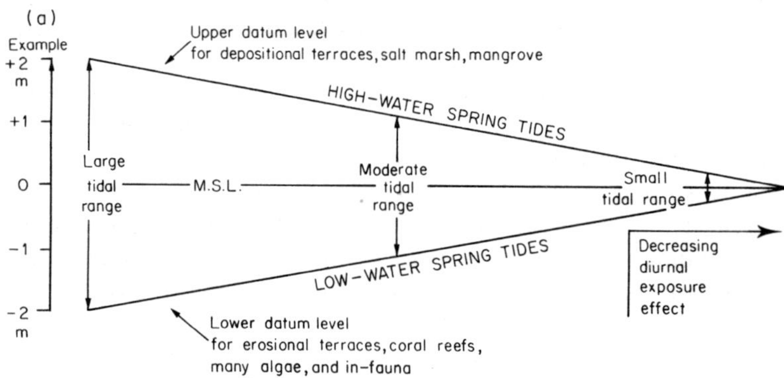

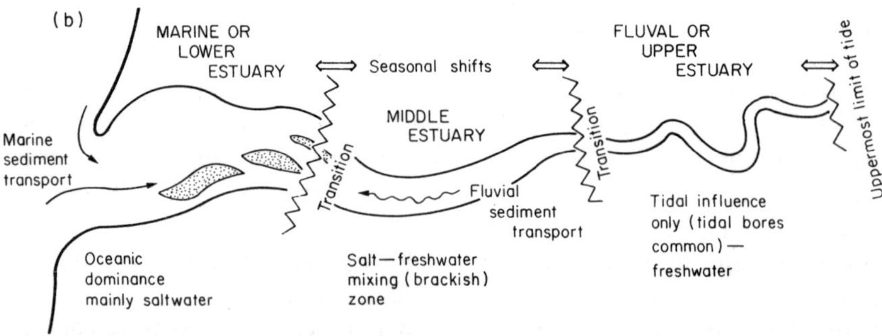

Figure 4. (a) Effect of high and low tidal range on estuarine sedimentation, ecologic and geomorphic features. M.S.L. = mean sea level. Actual geodetic height of M.S.L. tends to rise progressively towards the upper estuary. (b) Idealized map of a typical estuary showing three divisions, lower, middle, and upper; the boundaries are transition zones that shift according to season, weather, and tides.

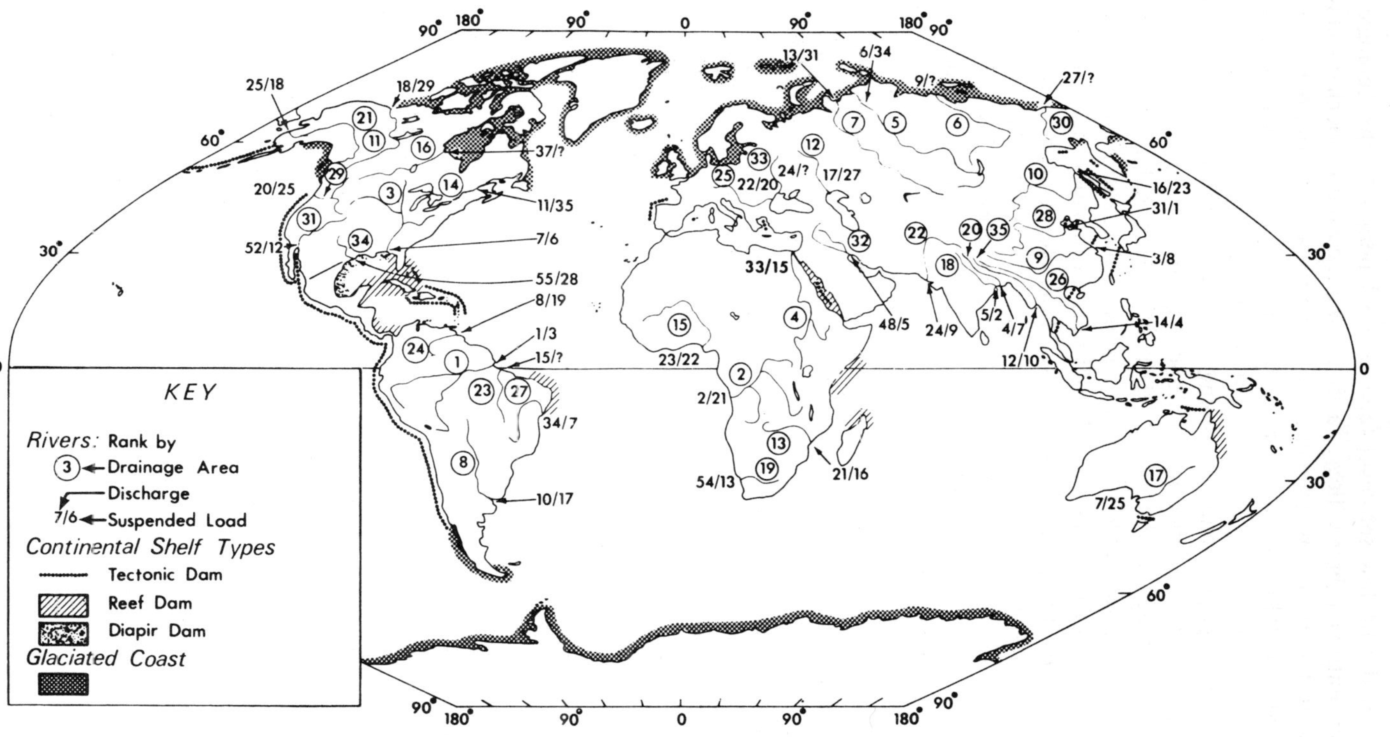

Figure 5. Estuary filling relates to sediment brought down by rivers. World rivers are here classified in order of rank by drainage areas and by the discharge/suspended load. Effects are modified by the structural and physiographic characteristics of the continental shelf (Walker, 1975).

In contrast, during low sea-level (epeirocratic) phases of the geological past, streams entrench their valleys, vastly raising the ratio of mechanical to dissolved load (Figure 5). The increase in the relative area of continents then raises the continentality:oceanicity ratio, favouring climatic extremes, the growth of deserts and even ice-ages. This is the *epeirocratic* regime, with continental relief high. Climates, becoming more arid, lead to the creation of soil duricrusts, notably silicified laterite, and vegetation become sparse; this is Erhart's 'rhexistatic' condition (1973). It may be noted that long-term sea-level changes are essentially tectono-eustatic (related to plate tectonics: rifting, sea-floor spreading rates and collisions). Glaciation adds another component in the form of glacio-eustasy; thus, for example, a 200 m tectono-eustatic fall may then be augmented by a 100 m glacio-eustatic pulse. The latter tends to be short-term and rapid, in terms of geologic time.

It is the transition phase from a low eustatic stand to a high one, followed by a stable high stand, that favours estuarine filling (the contemporary state of affairs). The opposite phase results in a removal or sluicing-out of the estuarine fill of a previous cycle. The erosional material resulting from the eustatic lowering is transported seaward, but the shoreline now corresponds to the edge of the continental shelf, so the sediment is immediately entrained in turbidity currents which transport it into the abyssal plains (see, for example, the Amazon transport during the last glacial phase: Damuth and Fairbridge, 1970). This is a one-way way, irreversible process on time scales of less than 10^8 yr. Fresh sediment can only come from the land, but, as sea-level rises, there will also be wave-reworking of coastal plain sediments now covered by waters of the continental shelf.

The problem of *residence-time* for sediments within the estuary can now be considered. Goldberg (1971) has calculated, from the total global marine sedimentation rate, that 1.8×10^{10} tons of sediment (as suspended solids) pass into the ocean annually. Airborne material is two orders of magnitude less; and iceberg transport is likewise not a major factor. Of this 1.8×10^{10} tons, an appreciable fraction must 'rest' for some time in the estuarine environment where it will become involved in the local biological and physical processes, and will interact with the various biochemical agents found there. In the case of estuaries in tectonically positive or relatively stable areas, this residence time is likely to be brief, ranging for perhaps one week up to 10,000 yr (the average duration of an interglacial phase). Only in subsiding estuaries will the sediments find a semi-permanent resting phase within their original estuarine framework.

A brief consideration of some selected rivers will now be offered, considered in terms of latitudinal-climatic zone and morphotectonic province. Figures were obtained from Strakhov (1967). Mechanical denudation (erosion) is expressed as E_m and chemical as E_c. The ratio of E_m/E_c is expressed as R.

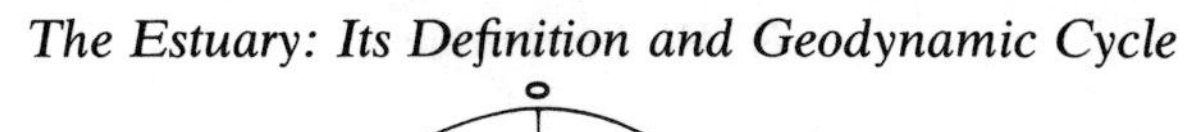

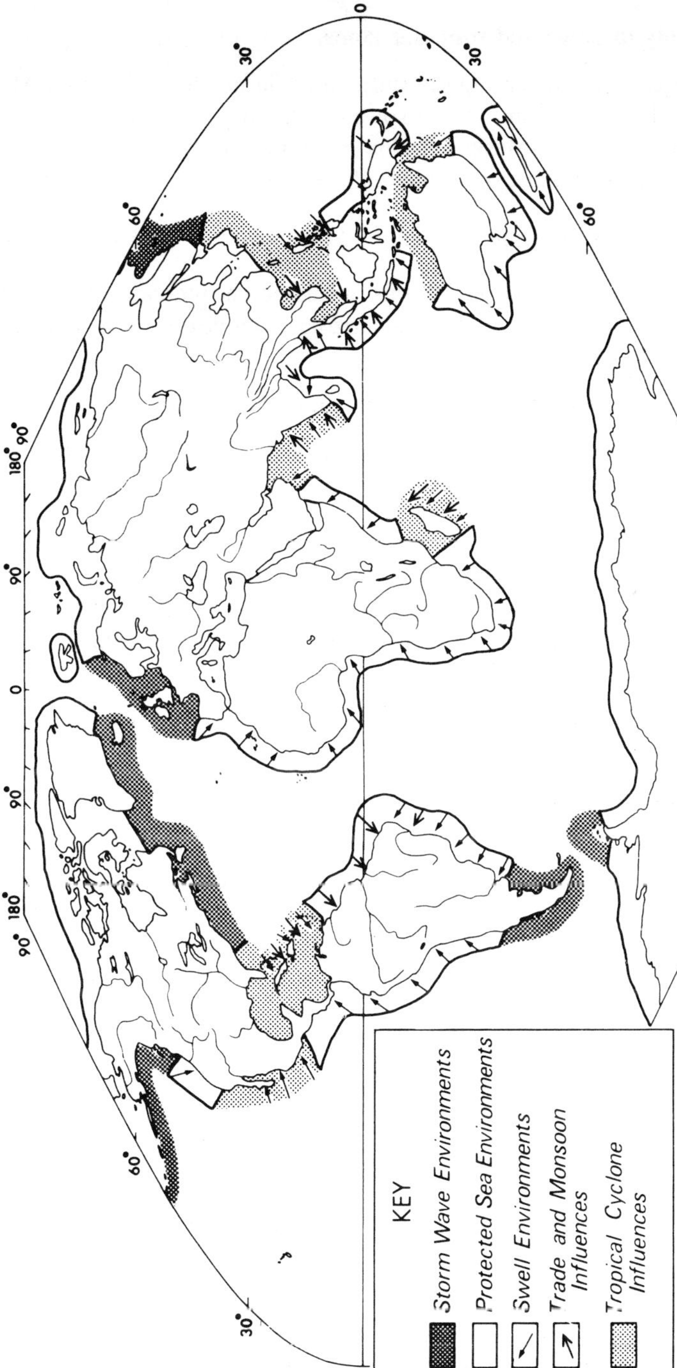

Figure 6. Estuary blocking relates largely to wave action, that is dependent in turn on fetch, swell and tidal range (Walker, 1975).

5.1 The supply in polar and subpolar zones

Several major rivers drain high latitude lands in northern Canada, Alaska, and Siberia. There are none in the southern hemisphere. Examples:

The *Yukon* (Canada, Alaska) drains a mountainous basin in youthfully revived Mesozoic–Cenozoic orogenic belts, extensively glaciated, furnishing:

$$E_m = 103 \text{ tons/km}^2; \quad E_c = 22 \text{ tons/km}^2. \quad R = 4.7.$$

The *Amur* (eastern Siberia) drains a dissected plateau of youthfully revived Precambrian and early Palaeozoic complexes, not extensively glaciated, furnishing:

$$E_m = 28 \text{ tons/km}^2; \quad E_c = 10 \text{ tons/km}^2. \quad R = 2.8.$$

The *Kolyma* (northeastern Siberia) drains a deeply dissected platform of Precambrian complexes with some Phanerozoic overlap, still extensively affected by permafrost furnishing:

$$E_m = 7 \text{ tons/km}^2; \quad E_c = 5 \text{ tons/km}^2. \quad R = 1.4.$$

The *Yenisei* (western Siberia) drains the Paleozoic Altai Mts and low-relief Precambrian shield with marginal Phanerozoic basins, affected extensively by permafrost, furnishing:

$$E_m = 4 \text{ tons/km}^2; \quad E_c = 11 \text{ tons/km}^2. \quad R = 0.36.$$

5.2 The supply in temperate-westerly zones

Numerous major rivers in these zones are known in both hemispheres, but human interference (agriculture and engineering) reaches a maximum in this belt. Examples:

The *Brioni* (southwestern Georgia, USSR), a very short-run mountain river that drains the southern Caucasus; the mountains are now rising at over 15 mm/yr, and the river debouches into the Black Sea, where the estuary is sinking at 2 mm/yr.

$$E_m = 2,000 \text{ tons/km}^2; \quad E_c = 209 \text{ tons/km}^2. \quad R = 9.8.$$

The *Mississippi* (U.S.) drains mainly a Paleozoic sediment-covered platform, very extensively covered by solifluction material, morainic debris (tills), outwash, and loess from Quaternary glaciation, furnishing:

$$E_m = 118 \text{ tons/km}^2; \quad E_c = 28 \text{ tons/km}^2. \quad R = 4.2.$$

La Plata-Parana (Argentina, Brazil, Paraguay, Bolivia) rises partly in the Andean foothills and also drains a low-relief stable platform and basin area covered by undeformed Phanerozoic sediments; minor component of

Quaternary moraines, loess and dunes. Furnishes:

$$E_m = 75 \text{ tons/km}^2; \qquad E_c = 18 \text{ tons/km}^2. \qquad R = 4.1.$$

The *Dniester* (southwestern Ukraine, USSR) rises in the Carpathian Mountains, of moderate relief, youthfully folded and drains a low-relief basin on the margin of the Russian Platform in Ukraine; extensive loess. Furnishes:

$$E_m = 31 \text{ tons/km}^2; \qquad E_c = 39 \text{ tons/km}^2. \qquad R = 0.8.$$

The *Volga* (central Russia, USSR) drains the Russian Platform, a low-relief Paleozoic area; Quaternary-morainic areas and extensive loess. Furnishes:

$$E_m = 18 \text{ tons/km}^2; \qquad E_c = 32 \text{ tons/km}^2. \qquad R = 0.56.$$

The *Dnieper* (southern Russia; USSR) drains the drier parts of the Russian Platform; few Quaternary moraines but extensive loess. Furnishes:

$$E_m = 4 \text{ tons/km}^2; \qquad E = 17 \text{ tons/km}^2. \qquad R = 0.24$$

5.3 The supply in tropical and equatorial zones

The zones span the equatorial zone (with heavy precipitation all the year-round) to the strongly seasonal (monsoonal) summer rainfall savanna belts and eventually to the riverless hyperarid desert belts; in the latter there are no streams except highly ephemeral flows, or rivers that rise elsewhere (e.g. Nile, Niger, Diamantina).

The *Amazon* (Peru and Brazil) drains the eastern slopes of the Andes and its major tributaries drain the largely sandstone plateaus of the Guyana and Brazilian shields. Some of the largest shield tributaries (e.g. the Xingu) contribute almost nothing. The collective load is:

$$E_m = 60 \text{ tons/km}^2; \qquad E_c = 13 \text{ tons/km}^2. \qquad R = 4.6.$$

5.4 The sediment fill

Some generalizations can now be presented, relating the climate, fluvial sediment supply and potential rate of estuarine filling.

In the high latitude situations it is evident that forest cover and permafrost severely limit low-relief basins, but a high relief headwaters such as the Yukon, with its plentiful 'prepared' sediment, loose and easily transported morainic and solifluction debris contributes to a heavy sediment load. As a result, the Yukon has no estuary today, but rather a very rapidly prograding delta, favoured by a shallow offshore platform, the Bering Shelf. In contrast the Amur, Kolyma, and Yenisei all have long unfilled estuaries. None face excessively stormy oceans with shoreward sediment transport. All are protected much of the year by sea ice. All are subjected to strong spring (melt) floods and much of the upstream load is deposited in middle courses on their flood plains. The ratio of solids to dissolved load (R) decreases with relief. Increase in dissolved load must also be understood as an increase in potential (inorganic) nutrients for the estuarine biota.

For the temperate latitudes the same conditions hold true, except that there is no permafrost, and in many cases there are considerably augmented supplies of 'prepared' morainic debris, of reworked solifluction cover, or of wind-blown loess (with silt size particles, in part soluble because of its periglacial mechanical origin). The large supplies of dissolved material from the loess-covered steppes of U.S. and Russia are notable: this also applies to the Hwang-Ho and Yangtze-Kiang, Chinese rivers, not discussed here in detail.

The true equatorial streams in low relief areas furnish almost no solids or dissolved loads because of the massive plant cover and deeply leached soils (see Gibbs, in Wiley, 1976; and in Nelson, 1972); only where the vegetation is artificially cleared does the stream receive any major contribution. Those rivers that have their headwaters in the high mountains are, of course, quite different and become modified mountain rivers. Such mountain streams will quite commonly have a sediment load of 2,000 tons/km^2 and a dissolved load of 200–300 tons/km^2, thus with an E_m/E_c ratio (R) of up to 10. Progressive dilution downstream greatly reduces these values except in the relatively restricted examples of actively subducting plate margins or recent orogenic coasts. The example quoted, the Rioni River (ancient Phasis R., of Colchis) has a back-filled estuary, aided by rapid mountain uplift and basin subsidence.

A feature of some equatorial and tropical zone rivers is the 'interior delta'. This phenomenon is encountered where the system has its headwaters in the equatorial or high monsoon rainfall latitudes and then flows through a semi-arid sector. This type of regime is best known in the case of the Niger, but also occurs on the Nile (El Sudd), the Zambesi (Okavango Swamps), the Murray (Riverina District). In each case the bulk of the sediment load is thus trapped within the swampy interior delta, so that only the lower course furnishes sediment to the marine delta and its estuary.

In general, one may assume that a long-course, low gradient river delivers only fine-grained sediment to the estuary; the fine silt grade consists mainly of finely ground quartz and stable heavy minerals which are unreactive in most estuaries, but the finest material, the clay size grade, consists of clay minerals which are important as chemical reagents, playing a role as chelating compounds, as well as a having a biological role as bacterial substrates.

Sands and gravels reach the estuary only in three situations: from short-run, high gradient rivers; from climatically and tectonically stressed conditions (or man-induced analogs); or from residual reworking. An example of the latter can be seen in the semi-arid rivers of North and West Africa,

Arabia, and Western Australia where quartz-rich dune sands are either blown or washed into the estuary or carried down in the ephemeral stream bed loads. During the Quaternary cold periods the subtropics, tropics and even some equatorial zones become almost universally desiccated. Desert sands extended much farther towards the equator than they do today; for example much of the Congo (Zaire) Basin is underlain by desert sands. Quartz is extraordinarily stable in most soils and thus these Quaternary sands are subject to reworking by streams and revived by wind following man-made interference (overgrazing and so on). Another type of inherited stress product is the widespread availability of coarse boulders (erratics, till) and outwash sands (sandur) left by the last glaciation. In otherwise low energy situations, the ready presence of this inherited material may create an estuarine fill that is quite out of equilibrium with its expected regime.

6 SUMMARY

An estuary, defined as a river valley open to the ocean, is variably affected by tides and saltwater mixing. There are estuaries in non-tidal seas and in exceptional cases pseudoestuaries can occur in non-marine environments as in Lake Baikal. Thus the basic definition must rest on physiographic or geomorphological criteria. Hydrologic considerations follow as corollaries to the physical setting.

The estuary is an ephemeral feature in long-term, geologic history and must be regarded as a dynamically evolving land-form that will go through a life cycle from valley creation (by fluvial or ice erosion), followed by the drowning phase (eustatic rise of sea level, with/or without vertical crustal motion), and ending with the progressive infilling (mainly deltaic, but some offshore sedimentation, aided by eustatic drop in sea level with/or without vertical crustal motion). Completely filled, 'fossil' estuaries can be studied as end-members of this evolution. A Davisian 'youth–maturity–old age' terminology could be applied to it. Because of the powerful role of inheritance in geology, some estuaries go through repeated histories in more or less the same structural position over periods of 10^3 yr up to 10^7 yr.

The vast majority of the world's estuaries today were created by the postglacial rise of sea level that stabilized at 6,000 BP. Vertical crustal motions are generally slower than eustatic, so that although many estuaries show some tectonic modification, almost all are dominated by this world-wide eustatic control. It 'set the time clock ticking', being the chronometer that measures and regulates the progressive infilling of the estuary. Thus sea level behaviour is seen as the fundamental constraint of estuaries.

Estuaries are categorized into three latitudinal–climatic groups that are influenced by (a) the climatic–geomorphic history of their setting during the Quaternary period, essentially the last 2 million yr and more especially the

                    R. W. Fairbridge

Table of World's Major Rivers and Estuary Types.

	Length km	Basin area × 1000 km²	Discharge m³/sec	Discharge/ area ratio l/sec. km²	Estuary D = delta F = funnel B = barrel
North America					
Mississippi-Missouri	6,400	3,250	18,000	5.5	D
St. Laurence (system)	3,100	1,380	8,700	7.0	F
South America					
Amazon (system)	6,300	7,050	200,000	28.3	D/F
La Plata (system)	4,630	4,350	27,700	6.5	F
Orinoco	2,120	900	12,600	14	D/F
Europe					
Volga	3,700	1,380	8,300	5.6	D
Danube	2,850	817	6,400	7.8	D
Rhine	1,360	224	2,500	11.2	D/F
Seine	780	79	500	6.3	F
Po	680	75	1,500	21.4	D
Thames	340	16	80	5.3	F
Asia					
Ob-Irtysh	5,200	2,430	11,400	3.8	F
Yangtze-Kiang	5,000	1,770	30,000	17.6	F
Indus	3,180	960	6,100	7.6	D
Irrawaddy	2,000	410	11,000	25.5	D
Yenisei	4,250	2,600	17,600	6.7	F
Lena	4,400	2,425	15,700	6.5	D
Amur	4,225	1,970	11,000	5.8	F
Hwang-Ho (Yellow R)	4,800	1,260	—	—	D
Brahmaputra-Ganges	2,980	935	17,700 (com-bined)	19.0	D/F
Africa					
Congo	4,370	3,690	41,400	11.0	F
Nile-Kagera	6,700	2,800	1,600	0.6	D/B
Niger	4,070	1,500	14,300	9.5	D
Zambesi	2,820	1,330	—	—	D
Australia					
Murray-Darling	1,630	910	400	0.4	B

last glacial cycle of 100,000 yr; and (b) their setting in terms of plate tectonics, which largely define vertical crustal motions other than those introduced by glacial isostasy. The three climatic categories are the polar and subpolar, the westerly-temperate zones, and the tropical and equatorial zones. Each can then be subdivided in terms of vertical crustal motion. Uplift coasts are generally expressed as hard-rock, high-relief sectors, and downwarp coasts as soft-rock, low-relief settings.

The rate of estuary filling is dependent on two variables: (a) the rates of sediment supply from fluvial discharge on the one hand, and (b) the effectiveness or otherwise of wave-induced onshore and long-shore drift, littoral sediment availability and organic sediment builders in forming a barrier that obstructs the estuary mouth and favours the accumulation of the fluvial sediment supply as well as more storm-carried offshore material (compare Figures 5 and 6).

Both of these variables are highly dependent upon the latitudinal–climatic controls. Thus glacially gouged estuaries in high latitudes are generally open because of low sediment supply, whereas those in the northern temperate belt are liable to rapid blocking by longshore drift barriers, due to the availability of vast quantities of glacially prepared debris. Correspondingly in low, semi-arid latitudes, the estuary is liable to blocking because of weak stream discharge and large supplies of inherited desert sand or by growth of coral reefs. In the high-discharge, equatorial estuaries reef growth is inhibited, but mangrove colonization stabilizes the estuarine muds.

It must be stressed that hydrologic studies of estuaries today are far more advanced than are geomorphic and stratigraphic studies, which require sophisticated drilling and dating equipment. Accurate prediction of future estuary behaviour must, however, await accurate analyses of estuarine histories. And because of the multiple variables, every estuary is unique.

Acknowledgements

For critical and constructive reading of the manuscript, grateful acknowledgement is due to Donald Swift (Miami), M. Grant Gross and Hanna Bremer (Köln).

REFERENCES

Baas Becking, L. G. M., and Moore, D. (1959). The relation between iron and organic matter in sediments, *J. Sedim. Petr.*, **29**, 454–458.

Barnes, R. S. K. (ed.) (1977). *The Coastline*, 356 pp., Wiley, New York.

Barnes, R. S. K., and Green, M. (eds.) (1972). *The Estuarine Environment*, Applied Science, London.

Berthois, L., and Barbier, M. (1954). Recherches sur la sédimentation en Loire, Bull. C.O.E.C., **6**, 387–397.

Bird, E. C. F. (1968). *Coasts*, M.I.T. Press, Cambridge, Mass., 246 pp.

Burton, J. D., and Liss, P. S. (eds.) (1976). *Estuarine Chemistry*, Academic Press, 229 pp., London and New York.

Cronin, L. E. (ed.) (1975). *Estuarine Research*, 2 Vols., 788 pp., and 587 pp., Academic Press, New York.

Damuth, J. E., and Fairbridge, R. W. (1970). Equatorial Atlantic deep-sea arkosic sands and ice-age aridity in tropical South America, *Geol. Soc. Am. Bull.*, **81**, 189–206.

Dionne, J. C. (1963). Towards a more adequate definition of the St. Lawrence estuary, *Zeitschr. f. Geomorph.*, **7** (1), 36–44.

Dyer, K. R. (1973). *Estuaries: a Physical Introduction*, 140 pp., Wiley, London.

Erhart, H. (1973). *Itinéraires Géochimiques et Cycle Géologique de Silicium*, 217 pp., Doin, Paris.

Evans, G. (1965). Intertidal flat sediments and their environments of deposition in the Wash, *Quart. J. Geol. Soc., London*, **121**, 209–241.

Fairbridge, R. W. (1961). Eustatic changes in sea-level. In: *Physics and Chemistry of the Earth*, **4**, 99–185.

Fairbridge, R. W. (1967). Carbonate rocks and paleoclimatology in the biogeochemical history of the planet, in *Carbonate Rocks* (Eds. G. V., Chilingar *et al.*), pp. 399–432, Elsevier, Amsterdam.

Fairbridge, R. W. (ed.) (1968). *The Encyclopedia of Geomorphology*, 1295 pp., Reinhold, New York (reprint by D. H. R./Academic Press).

Francis-Boeuf, C. (1947). Recherches sur le milieu fluviomarin et les dépôts d'éstuaire, *Ann. Inst. Oceanogr.*, N.S (paris), **23**, 149–344, (Thèse, Paris).

Goldberg, E. D. (1971). Atmospheric dust, the sedimentary cycle and man, *Comments Earth Sci. Geophys.*, **1**, 117–132.

Guilcher, A. (1963). Estuaries, deltas, shelf, slope, in *The Sea* (Eds. M. N. Hill), pp. 620–654, vol. **3**, Interscience, New York.

Hansen, D. V., and Rattray, M.jr. (1966). New dimensions in estuary classification. *Limnolog. Oceanogr.*, **11**, 319–331.

Hansen, K. (1951). Preliminary report on the sediments of the Danish Wadden Sea, *Medd. Dansk Geol. For.*, **12** (1), 26 p.

Häntzschel, W. (1939). Tidal flat deposits (Wattenschlick), in *Recent marine Sediments* (Ed. P. D. Trask), pp. 195–206, S.E.P.M., sp. publ. **1**, Tulsa.

Hillaire-Marcel, C., and Fairbridge, R. W. (1978). Isostasy and eustasy of Hudson Bay, *Geology*, **8**, 117–122.

Inman, D. L., and Nordstrom, C. E. (1971). On the tectonic and morphologic classification of coasts, *J. Geol.*, **79**, 1–21.

Ippen, A. T. (ed.) (1976). *Estuary and Coastline Hydrodynamics*, McGraw-Hill Book co., New York.

Kidson, C. (1963). The growth of sand and shingle spits across estuaries, *Zeitschr. f. Geomorph.*, **7**, 1–22.

Klein, G. de V. (ed) (1976). *Holocene Tidal Sedimentation*, Benchmark Papers (v. 30), Stroudsburg, PA.: Dowden, Hutchinson, and Ross, 423 pp.

Lauff, G. H. (ed.) (1967). *Estuaries*, Am. Assoc. Adv. Sci., publ. 83, 757 p., Washington, D.C.

Longinov, V. V. (ed.) (1969). Dynamics and morphology of sea coasts, *Jerusalem*: *Israel Progr. Sci. Transl.*, TT 68-50355.

McDowell, D. M. (1977). *Hydraulic Behaviour of Estuaries*, Wiley, New York, 292 pp.

Meade, R. H. (1969). Landward transport of bottom sediments in estuaries of the Atlantic Coastal Plain, *J. Sedim. Petr.*, **39**, 222–234.

Mörner, N. A. (1969). The late Quaternary history of the Kattegatt and the Swedish west coast, *Sveriges Geol. Undersökn.* ser. C, **640**, 487 pp.

Mörner, N. A. (ed.) (1976). The Pleistocene/Holocene boundary: a proposed boundary stratotype in Gothenburg, Sweden. *Boreas*, **5**, 193–275.

Nelson, B. W. (ed.) (1972). Environmental framework of coastal plain estuaries, *Geol. Soc. Amer.*, mem. **133**, 619 pp.

Officer, C. B. (Chm.), (1977). *Estuaries, Geophysics, and the Environment*, National Academy of Sciences, Washington, D.C., 127 pp.

Olson, R. E. (1970). *A Geography of Water*, Brown, Dubuque, Iowa.

Pannekoek, A. J. (1966). The ria problem, *Tids. Kon. Nederl. Aardr. Genootsch.*, **83** (3), 289–297.

Perkins, E. J. (1974). *The Biology of Estuaries and Coastal Waters.* Academic Press, New York, 678 pp.

Pritchard, D. W. (1952). Salinity distribution and circulation in the Chesapeake Bay estuarine system, *J. Marine Res.*, **11,** 106–123.

Schmidt, W. (1923). Die Scherms an der Rotmeerküste von el-Hedschas, *Petermanns Mitt.*, **69,** 118–121. (see also Rathjens and von Wissmann, 1933, *ibid.*, **79,** 183–187).

Schubel, J. R. (ed.) (1971). The Estuarine Environment: Estuaries and Estuarine Sedimentation, *Amer. Geol. Inst.*, Washington, D.C., 324 pp. (Short Course Lecture Notes).

Schwartz, M. L. (1973). *Barrier Islands*, Academic Press; Dowden, Hutchinson and Ross. Benchmark, **9,** 451 pp.

Shepard, F. P. (1963). Thirty-five thousands years of sea-level. In: *Essays in marine geology in honor of K.O. Emery.* Univ. of Southern California Press, Los Angeles, 1–10.

Shepard, F. P., and Wanless, H. R. (1971). *Our Changing Coastlines*, McGraw-Hill, New York, 579 pp.

Steers, J. A. (1964). *The Coastline of England and Wales*, Cambr. Univ. Press, Cambridge, 750 pp (2nd ed.).

Strakhov, N. M. (1967–70). *Principles of Lithogenesis.* Oliver and Boyd, Edinburgh, 3 rols.: 245 pp., 609 pp., 577 pp.

Turekian, K. K. (1971). Rivers, tributaries and estuaries, in: *Impingement of Man on the Oceans*, John Wiley and Sons, New York, 9–73.

Van Straaten, L. M. J. U. (1950). Environment of formation and facies of the Wadden Sea sediments, *Tijdschr. Kon. Nederl. Aardr. Gen.*, **67,** 354–368.

Von Richthofen, F. (1886). *Führer für Forschungstreisende*, Janecke, Hannover, 734 pp.

Walker, H. J. (1975). Coastal morphology, *Soil Science*, **119,** 3–19.

Wiley, M. (ed.) (1976). *Estuarine Processes*, Academic Press, New York, 2 vols., 588 pp., 444 pp.

Chemistry and Biogeochemistry of Estuaries
Edited by E. Olausson and I. Cato
Copyright © 1980 by John Wiley & Sons Ltd.

K. F. BOWDEN

Department of Oceanography,
University of Liverpool

2

Physical Factors: Salinity, Temperature, Circulation, and Mixing Processes

1 INTRODUCTION

The definition of an estuary here used is that given by Fairbridge (this volume), stating that an estuary is 'an inlet of the sea reaching into a river valley as far as the upper limit of tidal rise'. This definition differs from that given by Pritchard (1952) in taking the tidal influence rather than the presence of a measurable concentration of sea water as defining the upper limit of an estuary. In some cases, as for example where river water enters the estuary over a weir, there is little difference between these two definitions, whereas in other estuaries there is a long upper reach in which the tidal rise and fall is appreciable but the water remains purely river water, derived from land drainage. The extreme case is found in the Amazon, where the volume of river flow is so great that no appreciable amount of sea water penetrates the estuary at all, although the tidal influence extends upstream to a distance of 735 km from the mouth. The mixing of river water and sea water takes place entirely in the ocean, in a plume spreading as far as 1,000 km seaward from the river mouth (Gibbs, 1960).

The treatment in this chapter is concerned primarily with conditions in the lower part of an estuary where river water and sea water both contribute to the distribution of salinity and to the dynamics of the flow. There are two basic features of estuaries which are important from a dynamical point of view. The first is that there is a horizontal gradient of density, increasing from the point or points of river inflow towards the sea. The second is the presence, in most estuaries, of tidal currents with periodic flood and ebb between the sea and the estuary. The density gradient acting alone would cause the river water to flow out as a surface layer over an intruding wedge of sea water with a limited amount of entrainment. The action of the tidal currents, on the other hand, is to introduce a considerable amount of kinetic energy, some of which is used to mix together the river water and sea water and so modify the circulation. It is the interaction between river flow and tidal currents which, in combination with the topography of the estuary, gives rise to a range of types of circulation.

Three broad categories of estuary circulation may be identified: highly stratified, as in the *salt wedge estuary* or in many fjords; *partially mixed estuary*, in which the vertical density gradient is significant and inhibits vertical mixing to some extent, and *well-mixed estuaries*, in which the vertical variation of salinity is very small. A rough indication of the characteristics of an estuary is given by the ratio of the river discharge during a tidal cycle to the tidal prism, which is the volume of water entering the estuary from the sea during the flood tide. If this ratio is of order 1, a salt wedge, or arrested flow, condition normally exists. If it is of order 0.1 the estuary is likely to be partially mixed and when it is of order 0.01 the estuary is probably well mixed. These figures should not be taken as more than an order of magnitude indication, however. As discussed in section 2.2, it is doubtful whether a single parameter can indicate adequately the overall characteristics of the flow.

Not only the overall intensity of tidal action in an estuary, as represented by the tidal prism, but also the distribution of tidal elevations and currents within an estuary are important factors in determining the circulation and mixing properties. Tidal movements in an estuary are generated by the tidal wave entering from the sea, the direct effects of the tide-generating forces of the moon and sun on the estuary being negligible in most cases. The hydrodynamic treatment of tides is a highly developed subject which will not be considered here. A summarized treatment has been given by Officer (1976) and the extensive literature on the subject was reviewed by Dronkers (1969). It will be assumed here that the distribution of tidal elevations and currents in an estuary is known, either from observations or from the use of a hydraulic or mathematical model of the area. Tidal models usually give the vertically averaged current and involve the assumption that the water is of uniform density. These restrictions are adequate for tidal purposes but not for calculating the estuarine circulation, in which density gradients and the variation of velocity along a vertical are essential features.

2 MAIN FEATURES OF CIRCULATION AND MIXING

2.1 Types of circulation

The main features of the circulation and salinity distribution, each averaged over a tidal period, are represented diagrammatically in Figures 1, 2, and 3 for three broad categories of estuary. In the salt wedge estuary, shown in Figure 1, the river flow dominates the circulation almost completely. The salt water extends as a wedge into the river and, in the absence of friction, the interface would remain horizontal and extend up-river to where the bed was at sea level. In practice, due to a small amount of friction between the layers, the interface slopes downwards slightly in the upstream direction. As a result of the Coriolis force, arising from the earth's rotation, the interface also slopes downwards to the right, in the northern hemisphere, looking towards the sea. The best known example of a salt wedge estuary is the Mississippi, where the wedge extends for more than 300 km upstream in the main channel at low river discharges but is pushed completely out of the estuary at very high discharges.

If the velocity of the seaward moving layer of river water exceeds a certain value, internal waves formed at the interface tend to break, causing an entrainment of salt water into the upper layer as in Figure 2. The salinity of the water in this upper layer therefore increases and its volume is also

 K. F. Bowden

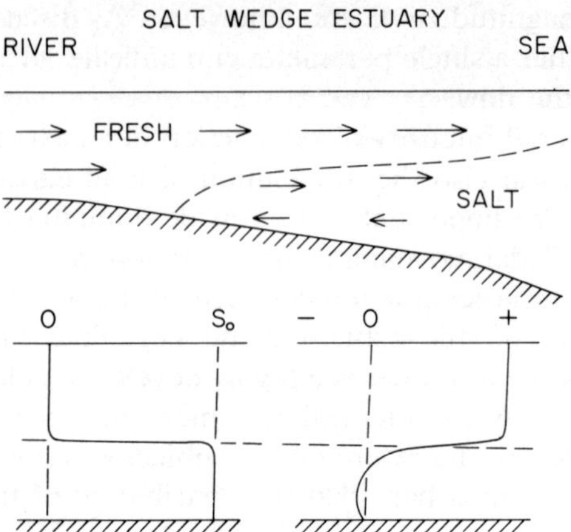

Figure 1. Salt wedge estuary. Above: section along the estuary. Below: typical salinity and velocity profiles.

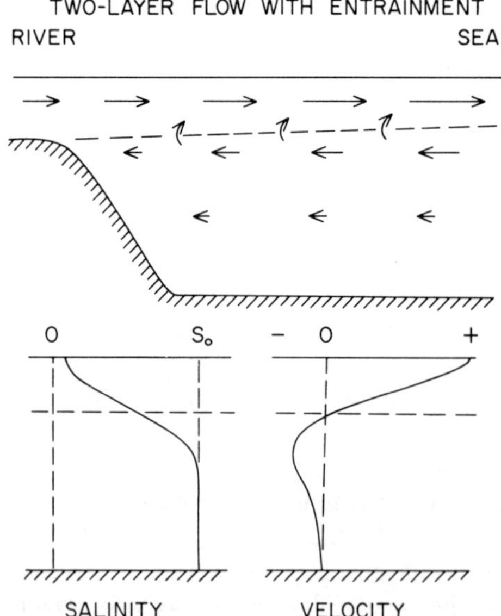

Figure 2. Two-layer flow with entrainment. Above: section along the estuary. Below: typical salinity and velocity profiles.

increased as it moves towards the sea. This usually results in a higher velocity of flow rather than an increase in depth of the layer. There is no significant entrapment of fresh water downwards, so that the salinity of the deeper layer is almost unchanged but there is a slow movement of water upstream to compensate for the loss by entrainment. If the estuary is deep, the upstream movement becomes negligible at greater depths. This type of circulation is typical of the inner basins of many fjords.

In some cases disturbances at the interface cause it to become broadened into a mixing zone, known as the halocline, in which turbulent processes allow a transfer of the fresher water downwards as well as of salt water upwards. The halocline is thus an intermediate layer of large salinity gradient separating the upper and lower layers in which the gradients are very small. This type of salinity structure was described by Tully (1958) in the fjord-type Alberni Inlet, in British Columbia, and is also found in a number of coastal plain estuaries as an intermediate stage between the salt wedge and partially mixed types.

In the partially mixed estuary, vertical mixing due to tidal currents extends throughout the depth and the halocline becomes replaced by a gradual increase in salinity from surface to bottom, as indicated in Figure 3. There is still a two-layered flow, with the surface of no motion, which separates the seaward flowing upper layer from the upstream flow of the lower layer, usually occurring a little above mid-depth. In this type of circulation the volumes of seaward flowing water in the upper layer and of

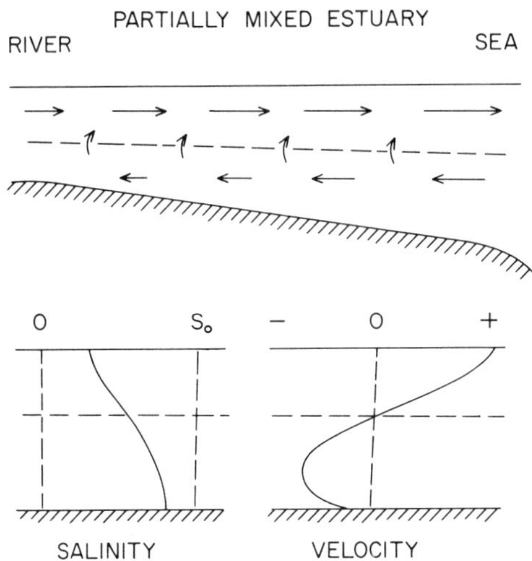

Figure 3. Partially mixed estuary. Above: section along the estuary. Below: typical salinity and velocity profiles.

upstream flow in the lower layer both increase in the downstream direction, with an upward flow of water at mid-depth to provide continuity. The volume of flow in each layer is typically an order of magnitude greater than that of the river discharge but an order of magnitude less than that of the oscillatory tidal currents. Thus if the volume rate of river discharge is R, the average rate of estuarine flow might be $10R$ seaward in the upper layer, and $9R$ upstream in the lower layer while the tidal flow might have an amplitude of $100R$. Naturally there is a wide variation in these ratios, as well as in the vertical range of salinity, from one estuary to another or for a given estuary under different conditions of river discharge.

When the influence of tidal mixing relative to that of the river flow is very strong, the estuary becomes well-mixed, with very little variation in salinity or mean current with depth. A convenient definition of a well mixed estuary is that the mean velocity is seaward at all depths, although its magnitude may decrease from surface to bottom. In this type of estuary the upstream transport of salt, needed to compensate for the seaward transport of salt by the mean flow, is brought about by diffusive processes rather than by advection although, as discussed later, the term 'diffusive' may cover a number of different physical processes.

2.2 Stratification–circulation diagram

A continuous spectrum of estuary classification is provided by the stratification–circulation diagram, introduced by Hansen and Rattray (1966) and illustrated in Figure 4. In this diagram the ordinate is $\delta S/S_0$, where δS is the difference in salinity between surface and bottom and S_0 is the depth-mean salinity, both averaged over a tidal cycle. The abscissa is u_s/U_f, where u_s is the surface velocity, averaged over a tidal cycle, and U_f is the discharge velocity, i.e. the rate of river discharge R divided by the cross-sectional area A.

The broad categories mentioned above are represented by distinctive areas on the diagram. Type 1 is the well mixed estuary in which the mean flow is seaward at all depths and the upstream transfer of salt is by diffusive processes. Type 2 is the partially mixed estuary in which the net flow reverses at depth and both advective and diffusive processes contribute to the upstream flux of salt. In Type 3 the differential flow in the two layers becomes predominant, with advection accounting for more than 99% of the salt transfer. Type 4 is the salt wedge estuary, in which the stratification is still greater and the fresh water flows out over an almost stationary deep layer with very little interaction. As seen in the Figure, Types 1, 2, and 3 are each subdivided, rather arbitrarily, according to the degree of salinity stratification. The (a) subdivision corresponds to $\delta S/S_0 < 0.1$ and (b) to $\delta S/S_0 > 0.1$. Fjord estuaries are generally of Type 3(b).

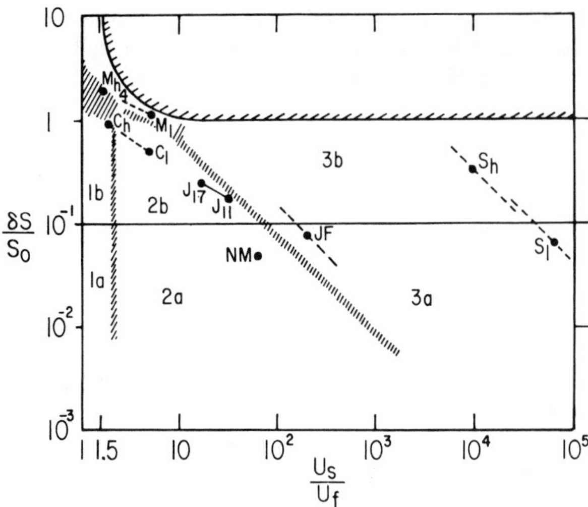

Figure 4. Stratification–circulation diagram. Code to stations: M, Mississippi River mouth; C, Columbia River estuary; J, James River estuary; NM, Narrows of the Mersey estuary; JF, Strait of Juan de Fuca; S, Silver Bay. Subscripts h and l refer to high and low river discharge; numbers indicate distance (in miles) from the mouth of the James River estuary. (Reproduced by permission of the American Society of Limnology and Oceanography from D. V. Hansen and M. Rattray, 1966, *Limnology and Oceanography*, **11**, 319–326, Fig. 2.)

If observations of salinity and currents have been made in a particular estuary and the river discharge is also known, the values of $\delta S/S_0$ and u_s/U_f can be calculated and the point corresponding to a given cross-section may be plotted on the diagram, enabling inferences to be drawn about the circulation and mixing characteristics of the estuary. Points corresponding to different cross-sections, for the same river discharge, may be joined up to form a line. A change in river discharge, in general, causes the points to move to different positions and may, if the change is considerable, indicate a change in the estuary type. Since the stratification–circulation diagram was introduced by Hansen and Rattray, a number of other investigators have used it to plot their results and compare conditions in different estuaries. As an example, Bowden and Gilligan (1971) plotted data for four sections in the Narrows of the Mersey estuary, England, for a series of values of river discharge in this way. This estuary was essentially of Type 2(a), although at the higher discharges the points at some sections moved into the Type 2(b) area.

In order to be able to use the diagram in a predictive capacity, Hansen and Rattray suggested that the quantities $\delta S/S_0$ and u_s/U_f could be related to two bulk parameters characterizing an estuary. They proposed the parameters P and F_m defined by:

$$P = U_f/U_t, \qquad F_m = U_f/U_d$$

where U_t is the r.m.s. tidal current speed and U_d is a densimetric velocity defined by:

$$U_d = (gD\Delta\rho/\rho)^{1/2}$$

where g is the acceleration due to gravity, D the depth of water, $\Delta\rho$ is the density difference between river water entering at the head of the estuary and sea water at the mouth and ρ is the mean density. F_m is thus a densimetric Froude number. Using data available from six estuaries, Hansen and Rattray derived empirical equations relating P and F_m to the theoretical parameters used in an analysis on which the stratification–circulation diagram was based. Lines of constant P and F_m were then plotted on the diagram, indicating that u_s/U_f depends only on F_m while $\delta S/S_0$ depends on both P and F_m. If the river discharge, tidal current velocities, mean depth of the estuary and density difference between fresh water and the salt water at the mouth are known, P and F_m may be calculated and then the values of u_s and $\delta S/S_0$ estimated from the diagram.

Other parameters have been used by different authors to characterize the circulation properties of an estuary. Thus Fischer (1972) defined an 'estuarine Richardson number' by

$$Ri_E = g\frac{\Delta\rho}{\rho}\frac{Q_f}{bU_t^3}$$

where Q_f is the river discharge and b the width of the estuary, the other symbols being as before. Larger values of Ri_E correspond to increasingly stratified estuaries. Ri_E is related to the bulk parameters of Hansen and Rattray by

$$Ri_E = \frac{P^3}{F_m^2}$$

By plotting lines of constant Ri_E on the Hansen and Rattray diagram, Fischer (1976) showed that $\delta S/S_0$ depends primarily on Ri_E, which suggests that F_m and Ri_E may be a convenient pair of parameters for representing the type of circulation and stratification. Figure 5, reproduced from Fischer (1976), shows the diagram drawn in this way.

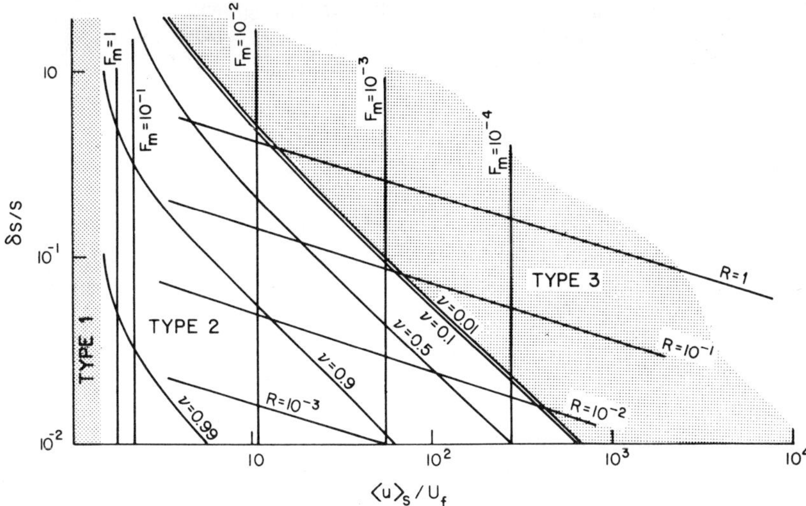

Figure 5. Stratification–circulation diagram, showing lines of constant diffusive fraction ν, densimetric Froude number F_m and estuary Richardson number R. (Reproduced, with permission, from H. B. Fischer, 1976, *Annual Review of Fluid Mechanics*, **8**, 107–133, Figure 2. © 1976 by Annual Reviews Inc.)

2.3 Lateral circulation

In the discussion above, the flow along the estuary and its variation with depth has been considered, but no account taken of variations across the estuary and the secondary circulations which may develop. Where the depth varies across a cross-section the upstream flow in the lower layer tends to be concentrated in the deeper part of the channel while the downstream flow may extend from surface to bottom in the shallower parts. This is illustrated by the profiles for four stations across Southampton Water shown in Figure 6, from Dyer (1977). If the cross-section changes longitudinally in area or form, corresponding changes will occur in the pattern of flow.

In the flow of a meandering river it is well known that the axis of maximum current swings towards the outside of the bend, with a corresponding increase in scour and deepening of the bed. The water level rises towards the outside and a secondary circulation is set up, with a downflow of water on the outside and an upflow on the inside of the bend. In a salt wedge estuary, as described by Dyer (1977) the upper layer of fresh water behaves in a similar way to river flow at a bend. In a partially mixed estuary, however, the strong coupling between the upper and lower layers may lead to the direction of the secondary circulation being reversed, with upward flow over the deeper water on the outside of the bend. This is because the higher salinity water tends to be concentrated in the deeper part of the

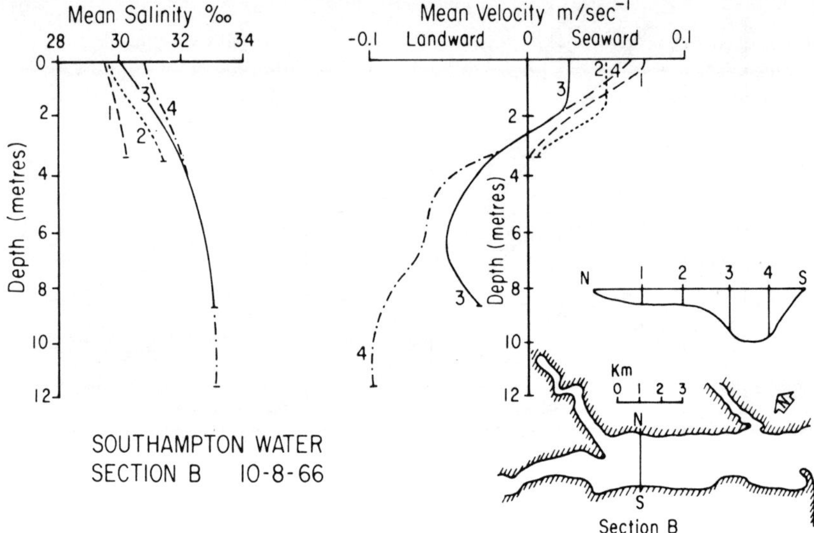

Figure 6. Variation in mean slainity and velocity profiles across an estuary: four stations across Southampton Water. (Reprinted from K. R. Dyer, 1977, in *Estuaries, Geophysics, and the Environment,* National Academy of Sciences, Washington, D.C. pp. 22–29, Figure 2.2, with the permission of the National Academy of Sciences, Washington, D.C.)

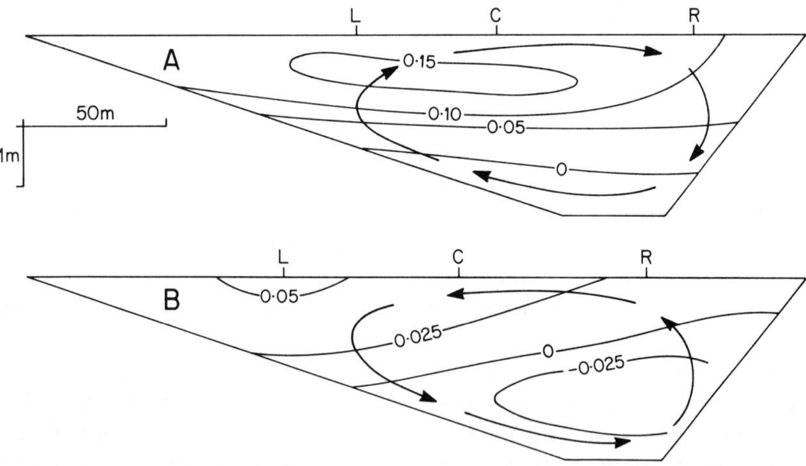

Figure 7. Secondary flow on a cross-section of a left hand bend of the Vellar estuary. (a) during high river flow, when of salt wedge type: (b) during low river flow, when partially mixed. Contours are the longitudinal velocity in $m\,s^{-1}$, positive downstream. (Reproduced by permission of the Colston Research Society from K. R. Dyer, 1978, Colston Research Symposium: Tidal Energy and Estuary Management.)

channel, with the isohalines rising towards the outer bank. This change in the secondary circulation has been observed in the Vellar estuary, India, as shown in Figure 7, which changes from the salt wedge to the partially mixed type when the river discharge decreases.

In a wide, shallow estuary with strong tidal mixing the conditions will tend to be vertically homogeneous but the maximum flood current often takes a different position in the cross-section to the maximum ebb. The Gironde estuary, France, is an example of this type, with strong cross-stream transports of water and salt at certain stages of the tide.

2.4 Temperature effects

The distribution of temperature in an estuary is dependent on the temperatures of the incoming river and sea water, the mixing processes and also on the exchange of heat through the surface while the water is passing through the estuary. In most estuaries the river water is colder than the sea water in winter and warmer in summer, due to the greater heat capacity of the sea and its slower response to the heating and cooling processes. In these cases the effect of temperature is to increase the density difference between river and sea water in summer and reduce the difference in winter. The main effect of heat exchange through the sea surface is to increase the tendency to density stratification in summer. A thermocline will tend to form at the depth of the halocline, thus increasing the density gradient. In winter the net loss of heat through the surface will tend to produce convective mixing in the surface layer and so reduce the density gradient due to salinity. On the whole the dynamical effects of temperatures changes are much smaller than those of salinity in estuaries, unlike the conditions in many areas of the open sea.

3 QUANTITATIVE TREATMENT OF CIRCULATION AND SALINITY DISTRIBUTION

3.1 Theoretical framework

Following the descriptive account of the various types of estuary circulation given in the previous section, the methods which have been developed to provide quantitative estimates of the currents and salinity distribution will be discussed fairly briefly in this section. Further details may be found in the books by Dyer (1973) and Officer (1976), which also give fuller references to the original papers.

A theoretical framework is provided by the equations of continuity of volume of water and mass of salt and the momentum equation, relating the

                                    K. F. Bowden

change of momentum of an element of water to the forces acting on it. These forces include gravity, the pressure gradients produced by the slope of the free surface and the horizontal gradient of density, and shearing stresses arising from turbulence in the water and friction at the estuary bed. It is usually assumed that the internal shearing stresses may be related to the gradients of mean velocity by coefficients of eddy viscosity. The turbulent stresses are typically several orders of magnitude greater than those due to molecular viscosity, which can usually be neglected.

The pressure gradient in the water depends on the density, which itself is a function of salinity, temperature, and pressure, as described in textbooks of physical oceanography (e.g. Neumann and Pierson, 1966). In an estuary the variation of density with pressure may usually be neglected and the effect of variations in temperature is often negligible compared with that of salinity. In that case it is adequate to take the density as a linear function of salinity, i.e.

$$\rho = \rho_0(1 + aS)$$

where ρ_0 is the density of fresh water at the appropriate temperature, S is the salinity in parts per thousand (‰) and a is a constant, numerically equal to 7.8×10^{-4}, to a close enough approximation.

The partially mixed estuary will be dealt with in some detail as considerable progress has been made in the theoretical study of this case. Rectangular coordinate axes will be taken with the x-axis along the estuary, positive seawards, y across the estuary and z vertically downwards. The corresponding components of velocity at a point will be denoted by u, v, and w. Pressure will be denoted by p and shearing stresses by terms such as τ_{zx}, which denotes the stress per unit area, perpendicular to the z-axis, acting in the x-direction.

From a comprehensive set of observations in the James River, flowing into Chesapeake Bay, Pritchard (1954, 1956) demonstrated that the most important terms in the salt balance equation were those due to the mean horizontal flow along the estuary and to vertical turbulent diffusion. In the momentum equation the most important terms were those due to the horizontal pressure gradient and to vertical shear in the horizontal velocity component u. Denoting the coefficient of eddy viscosity in vertical shear by N_z, the significant shear stress term is given by:

$$\tau_{zx} = -\rho N_z \frac{\partial u}{\partial z}$$

N_z is analogous to the kinematic coefficient of molecular viscosity, but is usually greater by a factor of 10^2 to 10^3. Similarly, denoting the coefficient of eddy diffusion in the vertical direction by K_z, we have:

$$\text{Vertical turbulent flux of salt per unit area} = -\rho K_z \frac{\partial S}{\partial z}.$$

Observations of the type made by Pritchard enable estimates to be made of the numerical values of N_z and K_z.

3.2 Solution for a partially mixed estuary

A theoretical treatment of the mean circulation and salinity distribution, averaged over a tidal period, was given for a partially mixed estuary by Rattray and Hansen (1962), retaining only the terms which Pritchard had shown to be significant. The estuary was assumed to be straight and narrow, of rectangular cross-section with uniform width but variable depth and the coefficients of eddy viscosity N_z and eddy diffusion K_z were assumed constant. The solution gave the velocity profile, indicating a seaward flow in the upper layer and an upstream below it, and the salinity profile with the salinity increasing from surface to bottom.

In a subsequent paper, Hansen and Rattray (1965) extended their analysis to include a longitudinal eddy diffusion term, represented by $-\rho K_x \, \partial S/\partial x$, in the salt balance equation. In this case the upstream flux of salt, required to balance the downstream flux by the depth-mean flow, included a diffusive component as well as the advective component arising from the variation of current and salinity with depth. The quantity ν, expressing the ratio of the diffusive flux to the advective flux, was an important parameter in the solution. By comparison with observed distributions of velocity and salinity, Hansen and Rattray estimated that ν ranged from 0.1 in the James River, through 0.5 in the Narrows of the Mersey estuary, to 0.8–0.9 for the Columbia River. The solution given in their 1965 paper was the basis of the classification scheme, described above, introduced by Hansen and Rattray (1966).

An alternative derivation of the Rattray and Hansen (1962) solution for the central part of a partially mixed estuary was given by Officer (1976, 1977). He extended the treatment to the case of a finite velocity at the bottom with τ_b, the frictional stress per unit area at the bed, related to the bottom velocity u_b by the equation:

$$\tau_b = \rho k \, | \, u_b \, | \, u_b = -\rho N_z \left(\frac{\partial u}{\partial z} \right)_b \tag{1}$$

where k is the bottom friction coefficient. In practice, u_b may be taken as the velocity at 1 m above the bed. It then follows that the surface velocity u_s

                     K. F. Bowden

is given by:

$$u_s = \frac{gh^2}{6N_z}\left(i - \frac{1}{4}\frac{\lambda h}{\rho}\right) + u_0 \qquad (2)$$

and the bottom velocity u_b by:

$$u_b = -\frac{gh^2}{3N_z}\left(i - \frac{3}{8}\frac{\lambda h}{\rho}\right) + u_0 \qquad (3)$$

where h is the depth of water, λ is the horizontal density gradient $\partial\rho/\partial x$, u_0 is the river discharge velocity and i is the surface slope, $\partial\zeta/\partial x$, where ζ is the elevation. The slope i may be found from the subsidiary equation

$$u_0 = \frac{gi}{3N_z}h^2 - \frac{g\lambda}{8\rho N_z}h^3 - \left[\frac{gh}{k}\left(\frac{1}{2}\frac{\lambda h}{\rho} - i\right)\right]^{1/2} \qquad (4)$$

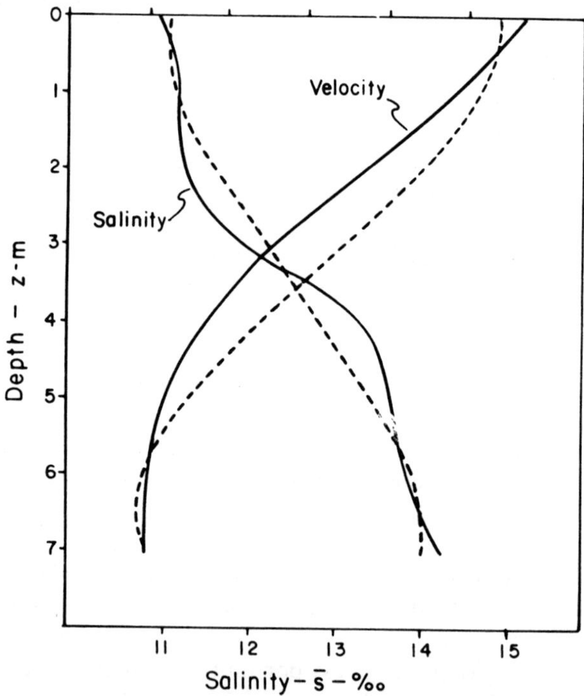

Figure 8. Typical profiles of mean velocity and salinity for the James River estuary. Observations ———; theory ‑‑‑. (Reprinted from C. B. Officer, 1977, in *Estuaries, Geophysics, and the Environment,* National Academy of Sciences, Washington, D.C., pp. 13–21, Figure 1.1, with the permission of the National Academy of Sciences, Washington, D.C.)

The total salinity difference between surface and bottom is given by:

$$\delta S \equiv S_b - S_s = \frac{1}{20} \frac{h^2}{K_z} \frac{\partial S}{\partial x} (3u_s - u_b - 2u_0) \tag{5}$$

Equations for the profiles of velocity and salinity with depth were also given by Officer but are not reproduced here. Figure 8 shows a comparison of typical velocity and salinity profiles observed in the James River (Pritchard, 1954) and those calculated by Officer (1977). There are appreciable deviations at mid-depth, in the region of the halocline, which are probably due to the eddy coefficients N_z and K_z being somewhat lower there than above or below, whereas the theory assumes them to be independent of depth. Similar comparisons of calculated and observed profiles were given for other estuaries, with a comparable degree of agreement. It should be remembered, however, that the values of N_z and K_z were chosen to give the best fit in each case and were not assigned independently.

The above equations may be used in conjunction with field data to determine values of the parameters i, k, N_z and K_z from the observations of u_s, u_b, u_0, $\partial S/\partial x$, h and δS. This was done by Officer (1977) for six selected estuaries and his results are shown in Table 1. Two of the areas, Long Island

Table 1. Properties of some representative estuaries.

(a) Observed values

Location	$S_b - S_s$ (‰)	$u_s - u_0$ (cm s^{-1})	$u_b - u_0$ (cm s^{-1})	u_0 (cm s^{-1})	$\partial S/\partial x$ (‰ km^{-1})	h (m)
Long Island Sound	0.8	22	−5		0.035	31
Mersey Estuary	0.9	12	−9		0.24	20
Southampton Water	1.5	8	−6		0.22	12
James Estuary	3.0	12	−10		0.59	7
Columbia River Estuary	15.0	29	−27	43	1.41	10.6
Vellar Estuary	23.5	9	−4		2.01	2.7

(b) Computed values

Location	N_z (cm^2 s^{-1})	K_z (cm^2 s^{-1})	k $\times 10^{-3}$	i cm/km	K_z/N_z
Long Island Sound	8.1	9.5	11	0.03	1.2
Mersey Estuary	41	25	4.5	0.16	0.6
Southampton Water	12	3.3	3.3	0.12	0.3
James Estuary	4.4	2.2	0.8	0.14	0.5
Columbia River Estuary	17	8	0.7	0.54	0.5
Vellar Estuary	0.9	0.1	6.0	0.17	0.1

(Reprinted from C. B. Officer, 1977, *Estuaries, Geophysics, and the Environment*, National Academy of Sciences, Washington, D.C., pp. 13–21, Tables 1.1 and 1.2, with permission of the National Academy of Sciences, Washington, D.C.)

Sound and the Mersey estuary, may be described as well mixed to weakly stratified, Southampton Water and the James River as partially stratified while the Columbia River and the Vellar estuary, on the east coast of India, are strongly stratified. These Tables are useful in indicating the range of values of velocity and salinity differences which occur in the field and the corresponding range in the parameters.

3.3 Fjord circulation

A similar approach may be made to the circulation in the inner basin of a fjord-type estuary, treating the velocity and salinity as continuous functions of depth. Because of the tendency for the circulation to be concentrated in the surface layer, however, it is no longer possible to neglect the field acceleration terms, $u\partial u/\partial x$ and $w\partial u/\partial z$, in the equation of motion. On the other hand, since the surface layer, in which the net movement is seaward, is typically a small fraction of the total depth in a fjord, the velocity may be assumed to become negligible at great depth. Dynamical treatments of this case have been given by Rattray (1967) and Winter (1973), who have been able to reproduce the form of the velocity and salinity profiles. An example, showing a comparison of observed and calculated velocity and salinity profiles for Knight Inlet, B.C., is reproduced in Figure 9.

As in the examples of partially mixed estuaries, the values of the eddy coefficients were chosen to provide the best fit with the observations. This theoretical treatment does not apply in the vicinity of a sill, which is often a region of considerable mixing between the outflowing and inflowing layers, particularly if the tidal currents are strong. The circulation in a fjord is also sensitive to changes in the wind regime: an up-inlet wind may reverse the surface flow and cause significant changes in the deeper water, as discussed in more detail later.

3.4 Salt wedge circulation

A salt wedge is likely to develop in a relatively shallow estuary if a high river discharge is combined with weak tidal currents. The interface between the outflowing river water and the salt wedge remains sharp and is bounded at its upstream edge by the estuary bed. The two parameters F_m and P, defined above, usually have values $F_m > 10^{-2}$, $P > 10^{-1}$ in this case. Upward flux of salt through the interface is very slow and has a negligible dynamical effect. Horizontal gradients of salinity within the two layers are small and the circulation is driven by the pressure gradient arising from the slopes of the free surface and the interface. Several theoretical studies of the salt wedge circulation have been made using a two-layer model with an interfacial stress, depending on a coefficient which cannot be specified directly but

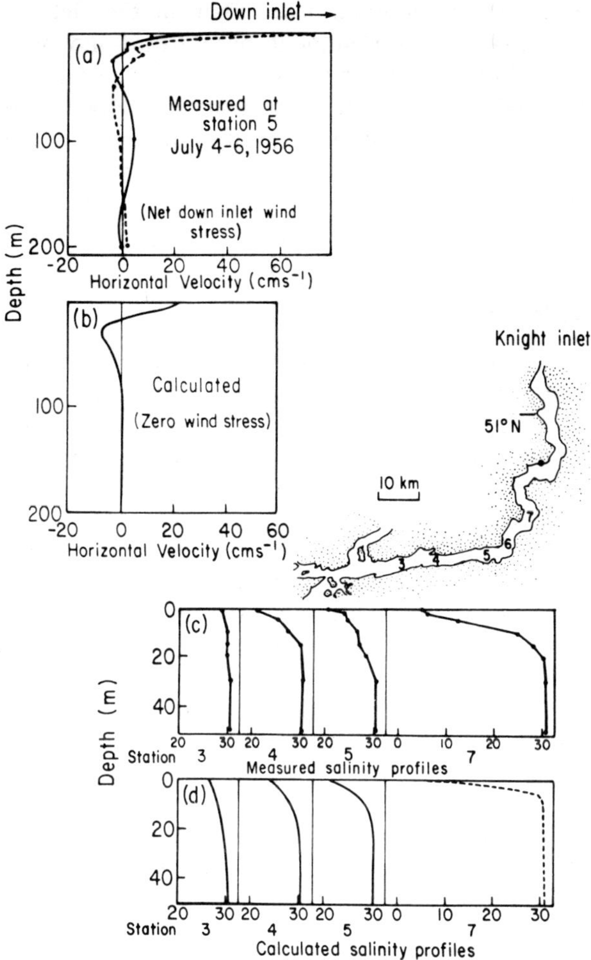

Figure 9. Comparison of (a) measured and (b) calculated horizontal velocity profiles for Station 5 in Knight Inlet. Comparison of (c) salinity measurements acquired at the four stations indicated in the map of Knight Inlet with (d) calculated depth profiles of salinity. (Reproduced with permission from D. F. Winter, 1973, *Estuar. Coastal Mar. Sci.*, **1**, 387–400, Figure 6. Copyright by Academic Press Inc. (London) Ltd.)

has to be chosen to fit observations, either in the field or in model experiments. A more recent treatment by Rattray and Mitsuda (1974) assumed a laminar boundary layer to exist between the salt wedge and the upper layer. The authors were able to calculate the velocity profile through the wedge and the stress distribution, as well as the shape of the wedge and its length in given steady flow conditions. Reasonably good agreement was found with observed conditions in the Mississippi and other salt wedge estuaries.

3.5 Plumes and fronts

Plumes of low salinity water spreading horizontally over the underlying sea water are of common occurrence where a stratified estuary enters a wider channel or the open sea and are particularly pronounced when they occur in association with a salt wedge estuary. Plumes are characteristically very shallow, that extending from the mouth of the Mississippi being only about 1 m deep and 10 km wide. As the river water spreads, sea water is entrained from below so that there is an increase in the salinity and volume of water in the plume. A dynamical treatment leading to a method of predicting changes in the breadth, thickness, and other properties of a spreading plume was given by Wright and Coleman (1971) and related to observed conditions at the mouth of the Mississippi.

Fronts tend to develop at the lateral boundaries of a plume. A study of the plume formed by the Connecticut River flowing into Long Island Sound and the associated fronts was made by Garvine (1974) and Garvine and Monk (1974). A front is characterized by the isohalines rising steeply to the surface, with a sharp horizontal gradient in salinity. The front itself tends to advance over the lower layer but at a lower velocity than that of the water in

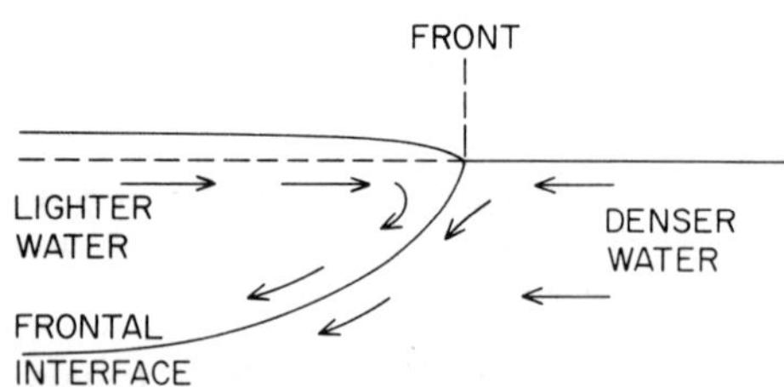

Figure 10. Diagram of the velocity structure at a front. The arrows indicate velocities relative to the moving front. (Based on, with permission of the American Meteorological Society, R. W. Garvine, 1974, *J. Phys. Oceanogr.*, **4**, 557–569, Figure 1.)

the surface layer. The front is, in fact, a zone of convergence with surface water sinking along the interface between the layers. In addition to having a sharp transition in salinity, a front is often marked by a change in colour and turbidity of the water and by a line of foam or debris, concentrated by the convergence. A schematic diagram of the currents in the neighbourhood of a front, based on Garvine (1974) is shown in Figure 10.

4 MIXING PROCESSES

4.1 Treatment of mixing in estuaries

An overall view of the mixing properties of an estuary is provided by the methods which provide an estimate of the flushing time of the estuary as a whole or of major divisions of it. The simplest of the more detailed treatments is the steady state one-dimensional approach, in which the distribution of properties along the length of the estuary is considered but averages are taken over each cross-section. Such a treatment is clearly justified for a well mixed estuary and it can be applied formally to a partially stratified one, but in that case it may obscure significant details of the vertical or lateral distribution. More attention has been given, up to the present, to longitudinal dispersion and its representation by one-dimensional models (analytical, numerical or hydraulic) than to any other formulation. The longitudinal dispersion, discussed in more detail below, is almost invariably influenced by variations in the vertical or lateral directions and their effects are parametrized in the one-dimensional treatment by including them in an effective coefficient of dispersion.

A two-dimensional treatment, involving averaging in either the vertical or transverse direction, is the next advance in complexity. In relatively narrow estuaries, lateral averaging can be used to eliminate transverse variations but the vertical current shear and salinity stratification is represented. In this way the vertical and longitudinal dispersion of a pollutant, released at a certain depth in a given cross-section, may be studied. As a half-way step to the laterally-averaged two-dimensional model, a two-layer model may be used. This is a natural choice if the surface layer is separated by a fairly sharp halocline from the layer below. If the salinity gradient is more gradual, the vertical distribution may still be simplified to a two-layer model by suitably averaging the velocity and salinity through each layer and introducing appropriate parameters to represent exchange of momentum and matter across the interface. In a broad estuary, especially where there is a considerable transverse variation in depth, the lateral variations in current and salinity may be more significant than those in the vertical. In that case a vertically averaged two-dimensional representation, taking into account the lateral variation of properties, may be more appropriate. In some estuaries the vertical and lateral variations appear to be of comparable significance for mixing purposes and a complete three-dimensional treatment is then desirable.

4.2 Flushing time

In the simplest estimate of flushing time, fresh water is used as a tracer and it is assumed that fresh water is being removed by flushing at the same rate as it is being added by river discharge. The flushing time t_1 is then given by:

$$t_1 = F/R \tag{6}$$

where R is the rate of influx of fresh water and F is the total volume of fresh water accumulated in the estuary. If S is the salinity at any point within the estuary and S_0 is the salinity of the sea water outside which is available for mixing, the freshwater content at that point is given by:

$$f = \frac{S_0 - S}{S_0} \tag{7}$$

To determine F, the estuary is divided into a suitable number of elements of volume δV and the appropriate value of f assigned to each element. The total fresh water content is then given by:

$$F = \sum f \delta V \tag{8}$$

where the summation is carried out over the total volume V.

4.3 Water and salt budget method

An alternative approach is to treat the estuary as a simple basin, with river inflow at its head and a two-layer exchange between the estuary and the sea, as in Figure 11. In the Figure, there is an outflow of water of lower salinity S_1 above an inflow of water of salinity S_2. The corresponding rates of transport of water are T_1 and T_2. The same treatment and notation can be used if the inflow from the sea and the outflow to it take place through two openings separated horizontally from one another.

Assuming a steady state and no change in water levels in the estuary, the equations expressing the conservation of volume of water and mass of salt lead to the following equations for the rates of inflowing and outflowing transport:

$$\text{Outflow:} \quad T_1 = T_2 + R = \frac{S_2 R}{S_2 - S_1} \tag{9}$$

$$\text{Inflow:} \quad T_2 = \frac{S_1 R}{S_2 - S_1} \tag{10}$$

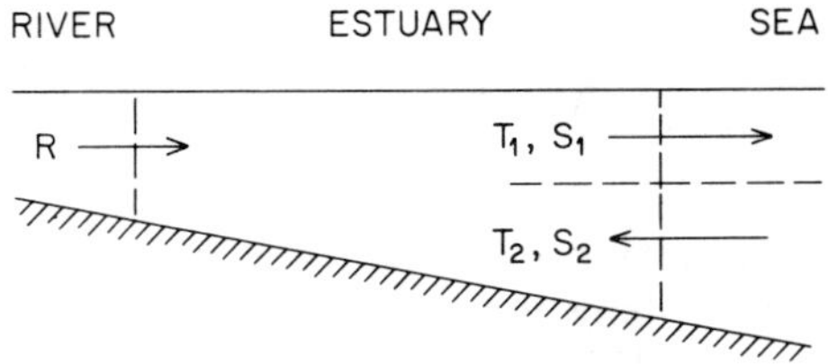

Figure 11. Two-layer exchange of water
between an estuary and the sea.

If the cross-sectional areas of the two layers are known, the mean velocities
of flow may be calculated. The flushing time is then given by:

$$t_2 = \frac{V}{T_1} = \frac{V(S_2 - S_1)}{S_2 R} \tag{11}$$

where V is, as before, the total volume of the estuary.

If changes occur in sea level and in the volumes of fresh water and sea
water within the estuary, the equations for T_1 and T_2 have to be modified by
the inclusion of additional terms (see Waldichuk, 1957).

The water and salt budget method does not take into account any
diffusive transport of salt across the boundary sections, so that the rates of
advective transport may be overestimated and the estimate of t_2 from
equation (11) reduced if horizontal diffusion is significant. The reduced
value of t_2 may still be a reasonably good estimate of the actual flushing
time, however, since the diffusive processes would act to increase the rate of
flushing. In the method described in Section 4.2, leading to the estimate t_1,
the processes by which fresh water is removed from the estuary are not
considered explicitly. The effect of diffusive transport is taken into account
implicitly in that it would affect the distribution of salinity within the estuary
and hence that of the fresh water content f.

4.4 Tidal prism methods

Where tidal movements appear to be the main mixing agency an attempt
may be made to compute the rate of exchange from data on the tides and
the physical dimensions alone. In the tidal prism method, now known to be
unreliable in its simplest form, it is assumed that all the water entering on
the flood tide becomes completely mixed with the water in the estuary and
the ebb flow consists of the mixed water. It is assumed further that none of
the mixed water will return on the following flood, which will consist of new
sea water. In that case a fraction $P/(V_0 + P)$ of the estuary water will be
replaced on each tide, where P is the tidal prism, i.e. the intertidal volume,
and V_0 is the volume of water in the estuary at low water. The flushing time

will then be

$$t_3 = \frac{V_0 + P}{P} \text{ in tidal periods} \tag{12}$$

In some estuaries the mixing each tide may be much less than assumed above and the actual flushing time may be of the order of 10 times that estimated from equation (12).

A refinement of the method was introduced by Ketchum (1951) who divided the estuary into a number of segments, the length of each being equal to the tidal excursion, and assumed that complete mixing occurred within each segment. Dyer and Taylor (1973) have given a modification of Ketchum's method, adopting a rather more general approach. The estuary is segmented in a similar way but an efficiency parameter is introduced to allow for the mixing on each tide being less than complete. The authors applied the method to the Thames estuary, England, and were able to predict the salinity distributions at high and low water reasonably well, taking the efficiency factor as 0.9.

Although the tidal segmentation method is a drastic simplification of the actual mixing processes, it is sometimes able to provide useful estimates of the dispersion and flushing of pollutants in a fairly well mixed estuary. It has the advantage of being based on a minimum amount of observational data and so may be of value in a preliminary study.

4.5 Advection–diffusion equation

Water movements by themselves cause an advective transport of properties and lead to dispersion when the current velocity varies with position but turbulent diffusion is also needed in order to produce mixing. The nature of turbulent diffusion can be demonstrated by considering the instantaneous value of a velocity component u as the combination of a mean value $\bar{u}$ and a turbulent deviation u', i.e.

$$u = \bar{u} + u'$$

The mean value $\bar{u}$ is determined by averaging over a suitable interval of space or time. The instantaneous concentration C of any property may be separated into a mean and a turbulent component in a similar way, i.e.

$$C = \bar{C} + C'$$

It may be shown (e.g. by Okubo, 1971) that the equation for the mean concentration $\bar{C}$ involves terms $\overline{u'C'}$, $\overline{v'C'}$ and $\overline{w'C'}$ as well as the mean velocity components $\bar{u}$, $\bar{v}$, $\bar{w}$ and the mean concentration $\bar{C}$. While the distribution of mean velocity may be assumed to be known, the covariance functions such as $\overline{u'C'}$ are usually unknown. It would be possible to calculate them if the statistical properties of the turbulent fields of velocity and concentration were completely specified but usually this is not the case. The practice usually adopted is to relate the covariance functions to gradients of the mean concentration by introducing coefficients of eddy diffusion, K_x, K_y and K_z, where K_x is defined by:

$$\overline{u'C'} = -K_x \frac{\partial \bar{C}}{\partial x} \tag{13}$$

with similar definitions for K_y and K_z. The coefficients K_x, K_y and K_z may be functions of position and time and may take different values for different properties, e.g. for heat and salt. Molecular diffusion has been neglected in this formulation, as in most problems involving dispersion in estuaries or the sea, its effects are several orders of magnitude smaller than those of turbulent diffusion. Values of K_z are typically of the order 10^{-1} to $10 \, \text{cm}^2 \, \text{s}^{-1}$ and those of K_x and K_y of the order 10^3 to $10^5 \, \text{cm}^2 \, \text{s}^{-1}$, whereas the molecular diffusivity for salt is about $2 \times 10^{-5} \, \text{cm}^2 \, \text{s}^{-1}$ and that for heat about $1.4 \times 10^{-3} \, \text{s}^{-1}$.

With coordinate axes taken as in Section 3.1, the equation for the distribution of the mean concentration $\bar{C}$ of any property may be written:

$$\frac{\partial \bar{C}}{\partial t} + \bar{u}\frac{\partial \bar{C}}{\partial x} + \bar{v}\frac{\partial \bar{C}}{\partial y} + \bar{w}\frac{\partial \bar{C}}{\partial z} = \frac{\partial}{\partial x}\left(K_x\frac{\partial \bar{C}}{\partial x}\right) + \frac{\partial}{\partial y}\left(K_y\frac{\partial \bar{C}}{\partial y}\right) + \frac{\partial}{\partial z}\left(K_z\frac{\partial \bar{C}}{\partial z}\right)$$
$$+ F(\bar{C}, x, y, z, t) \tag{14}$$

where $F(\bar{C}, x, y, z, t)$ is a term representing the rate of production, consumption or decay of the property C. For a conservative property, such as salinity, this term is zero.

Equation (14) is the advection–diffusion equation which, either complete or in a simplified form, provides the basis of all mathematical models for computing the distribution of salt or of chemical or biochemical properties in an estuary. The use of the equation requires the current distribution to be known, either from observations or from hydraulic or mathematical modelling. The main problems then arising in the use of equation (14) are in assigning appropriate values to the eddy diffusion coefficients K_x, K_y, and K_z. As introduced above, the coefficients are entirely empirical, based on the idea that, by analogy with molecular diffusion, the flux of material should be proportional to the concentration gradient. A physical basis exists for relating the eddy coefficients to the characteristics of the turbulence in the form:

$$K_x = |\bar{u}'| 1_x \tag{15}$$

where $|\bar{u}'|$ is a mean turbulent velocity (irrespective of sign) and 1_x is a 'mixing length', as introduced by Prandtl, analogous to the mean free path in molecular motion. The length 1_x and the corresponding lengths 1_y and 1_z

K. F. Bowden

may be identified roughly with characteristic linear dimensions of the eddies producing turbulent transport in the x, y, and z directions.

4.6 Longitudinal dispersion

In general it is better to maintain the distinction between advective and diffusive terms when applying equation (14). In some cases, however, the two processes interact in such a way that, with limited observational data available, it is more convenient to treat them together. An example of this is the 'shear effect', the name given to dispersion in the direction of a current, arising from interaction between shear in the current and turbulent diffusion in the direction of the velocity gradient. The effect is essentially that described by Taylor (1954) for turbulent flow in a pipe and extended to an open channel by Elder (1959). Under certain assumptions, the flux of material due to the shear effect is found to be proportional to the longitudinal gradient of concentration $\partial \overline{C}/\partial x$, so that an effective eddy diffusivity, or dispersion coefficient, K'_x may be defined. An oscillatory flow, such as a tidal current, can also give rise to longitudinal dispersion by the shear effect under suitable conditions (Bowden, 1965; Okubo, 1967).

The dispersion of salt, or of a pollutant, along an estuary is the resultant effect of several processes acting together, including the shear effect mentioned above. If $u(x, y, z, t)$ and $S(x, y, z, t)$ are the values of velocity and salinity respectively at the point (x, y, z) at time t, then the instantaneous rate of transport of salt in the x-direction across a complete vertical cross-section is given by:

$$F_x = \overline{AuS} \tag{16}$$

where A is the cross-sectional area and the bar denotes averaging over the cross-section. The average rate of transport over a complete tidal period may be represented by:

$$\langle F_x \rangle = \langle \overline{AuS} \rangle \tag{17}$$

where the brackets denote averaging with respect to time over a tidal period. By separating both u and S into a number of components which allow for the variation with vertical and lateral position in the cross-section and with time, it may be shown (e.g. by Fischer, 1972) that the mean flux $\langle F_x \rangle$ is made up of contributions arising from:

(1) advection of salt by the mean flow;
(2) correlated variations (with time) in the cross-sectionally averaged values of velocity and salinity;
(3) shear effect associated with vertical variations in tidally-averaged velocity and salinity;

(4) similar effect to (3) but arising from lateral shear;
(5) shear effect associated with time-varying shear in the vertical direction;
(6) similar effect to (5) but arising from lateral shear;
(7) variations of cross-sectional area during the tidal period;
(8) turbulent diffusion in the longitudinal direction, which is frequently smaller in magnitude than any of the preceding terms.

In the one-dimensional representation of estuary circulation, only term (1) is considered independently and the remaining terms (2)–(8) are all included in the dispersion term so that

$$\langle F_x \rangle = A_0 \left\{ u_0 S_0 - D_x \frac{\partial S_0}{\partial x} \right\} \tag{18}$$

where A_0 is the tidally-averaged cross-sectional area, u_0 and S_0 are the mean flow and mean salinity averaged over the cross-section and tidal period and D_x is the coefficient of longitudinal dispersion.

In the steady state $\langle F_x \rangle = 0$ and equation (18) becomes that given by Stommel (1953) for computing the dispersion coefficient at various sections of an estuary from the river discharge Q and the longitudinal distribution of salinity, i.e.

$$D_x = \frac{QS_0}{A} \left/ \frac{\partial S_0}{\partial x} \right. \tag{19}$$

since $Q = Au_0$. The values of D_x as a function of x can then be used to calculate the rate of dispersion along the estuary of a pollutant introduced at a given section in conditions of similar river flow. In his 1953 paper, Stommel applied the method to the Severn estuary. Since then the method has been very widely used and it is still the basis of many one-dimensional models used in estuary management.

Allowing for changes with time and for the presence of a source or decay term in the case of a non-conservative constituent, the one-dimensional dispersion equation for the concentration C is written:

$$\frac{\partial}{\partial t}(AC) + \frac{\partial}{\partial x}(AuC) = \frac{\partial}{\partial x}\left(D_x A \frac{\partial C}{\partial x}\right) + F(C, x, t) \tag{20}$$

where the cross-sectional area A, as well as u, C, D_x and the source/decay function F are all regarded as functions of the longitudinal coordinate x. In this form equation (20) may be used to calculate changes in concentration within a tidal period, taking u as the value of the tidal velocity plus river flow at a given time. The equation may be simplified if one is concerned only with slow changes in concentration, averaged over a tidal period. An alternative approach is the 'slack-tide' approximation, in which the distribution is calculated for times of high water slack or low water slack, when the tidal velocity may be set equal to zero and u becomes the river flow velocity u_0. A discussion of the various methods of solution was given by Harleman and Thatcher (1974).

Determinations of the coefficient D_x for a number of estuaries have been summarised by Officer (1976, 1977). They range from 10^5 to $10^7 \, \mathrm{cm^2 \, s^{-1}}$, the values for a given estuary usually, although not invariably, increasing in the downstream direction and being greater at higher river discharges. In view of the diversity and complexity of the processes contributing to D_x as defined in this way, it is not surprising that empirical determinations of D_x in actual estuaries have led to a wide range of values which cannot readily be related to bulk properties of the estuary.

4.7 Vertical turbulent mixing

A knowledge of the rate of vertical mixing is needed in estimating the dispersion of a pollutant discharged at a given depth and in practice this usually involves assigning a suitable value to the coefficient K_z. Examples were given above in Table 2 where values of K_z, assumed constant with depth, in the range 0.1 to $25 \, \mathrm{cm^2 \, s^{-1}}$ were needed to give reasonable agreement between the observed and calculated salinity distribution in selected estuaries. If K_z is constant, a pollutant discharged into the water at mid-depth will be effectively mixed throughout the depth on a time scale of the order of h^2/K_z, where h is the depth of water. In general, however, the coefficient K_z varies along a vertical. In homogeneous flow K_z may be expected to be approximately equal to the eddy viscosity N_z and reach its maximum value at mid-depth.

The presence of a stable density gradient tends to reduce both coefficients but K_z to a greater extent than N_z. The effect of stratification may be expressed in terms of a suitable parameter such as the local Richardson number Ri, defined by:

$$Ri = \frac{g}{\rho}\frac{\partial \rho}{\partial z} \bigg/ \left(\frac{\partial u}{\partial z}\right)^2 \qquad (21)$$

where $\partial \rho/\partial z$ and $\partial u/\partial z$ are the gradients of density and velocity at a particular depth. There are often difficulties in determining these gradients with sufficient precision and it is then possible, as an alternative, to take an overall value of Ri, based on larger scale features of the flow. In a tidal current an overall Richardson number may be defined by:

$$Ri_0 = \frac{gH\Delta\rho}{\rho U^2} \qquad (22)$$

where U is the depth-mean velocity, H the total depth and $\Delta\rho$ the total increase in density from surface to bottom. Ri_0 as defined in this way is still a function of time, varying with U and H during a tidal period.

Various empirical equations have been derived for the values of the coefficients N_z and K_z in neutral conditions and their dependence on Ri in the presence of stratification. Several of these were discussed by Bowden and Hamilton (1975) and Blumberg (1977) in connection with the choice of parameters to use in numerical models of estuarine circulation. From a comparison of the various equations with the experimental data, Officer (1976) deduced that the dependence of N_z and K_z on Ri was represented fairly well by the simple equations:

$$N_z = A_0(1+Ri)^{-1} \tag{23}$$

$$K_z = A_0(1+Ri)^{-2} \tag{24}$$

where $N_z = K_z = A_0$ for $Ri = 0$.

4.8 Dispersion in two-layer flow

As the stratification in an estuary increases, the density gradient tends to become concentrated in a thin layer at mid-depth with relatively well mixed layers above and below, extending to the surface and bottom respectively. In this case a two-layer model may be an adequate representation of circulation and mixing in the estuary. The exchange of momentum and material across the interface is then parametized by coefficients of interfacial friction and exchange. A model of this type described by Pritchard (1969) is the two-layer analogue of the segmented prism models referred to earlier. Reference should be made to Pritchard's paper or to a summary of the method by Dyer (1973) for further details.

4.9 Lateral mixing and topographic effects

Turbulent diffusion across an estuary may be characterized by an eddy diffusion coefficient K_y which is larger, sometimes by two or three orders of magnitude, than the vertical coefficient K_z. The time scale for effective mixing across the estuary is of the order of B^2/K_y, where B is the breadth, analogous to the vertical mixing time scale h^2/K_z. Although B is usually large compared with h, the mixing time scale across the estuary is often comparable with, or even shorter than, that in the vertical because of the much higher value of K_y compared with K_z. If a transverse circulation is present, this will enhance the rate of lateral mixing by a shear effect arising from the interaction of the vertical shear $\partial u/\partial z$ with vertical diffusion.

The contribution of lateral mixing to longitudinal dispersion by the shear effect has been referred to earlier. There are several other ways in which topographic features of an estuary of irregular width may increase the longitudinal dispersion, as discussed by Fischer (1976). The process known as 'trapping' occurs where there is a series of shoreline irregularities, in the form of basins or embayments, along the sides of an estuary, in which portions of contaminated water flowing past may be temporarily trapped. Later they are released gradually into the flow after the main body of the contaminant has passed, thus increasing the dispersion along the estuary. This process was treated analytically by Okubo (1973) who derived an equation for the effective longitudinal diffusivity in terms of the geometry of the system and an exchange rate between the 'traps' and the main channel. 'Tidal pumping' is the name given to the effect of a tidal residual flow which varies across the estuary, as in the formation of flood and ebb channels. An example, cited by Fischer (1976), is the net circulation in San Francisco Bay, a complex estuarine area, as computed by a two-dimensional numerical model.

4.10 Complex mixing problems

The above discussion has been concerned with the distribution of a property which may have sources or sinks within the estuary, or be subject to chemical or radioactive decay, but is not dependent on other varying properties. In problems involving water-quality the concentration of one quantity usually interacts with that of other quantities, by a series of chemical or biochemical reactions. The simplest case is that of the decay or oxidation of a sewage effluent and its effect on the dissolved oxygen in the water. In such cases, the concentration of each constituent is governed by an equation similar to (14) or to its one-dimensional analogue, equation (20), but with additional terms on the right hand side expressing its dependence on the concentration of other constituents. The system is thus represented by a set of coupled differential equations. In simplified examples, analytical solutions may be found (Officer, 1976), but in applying the method to real estuaries it is usually necessary to formulate a model and use numerical methods. Further references are given by O'Connor *et al.* (1977) and Harleman (1977).

5 VARIABLE CONDITIONS AND INTERMITTENT EVENTS

5.1 Wind effects

In the description of estuary circulation given above the emphasis has been on the mean circulation pattern obtained by averaging over one or more complete tidal periods. It has been tacitly assumed that the mean circulation determined in this way is steady with time or changes only slowly as, for example, with a seasonal variation in river discharge. In fact the river flow may change rapidly, on a time scale of a few days which is short compared with that needed for the estuary to become adapted to a new steady state. Another source of short term changes in circulation, which is often more important, is the effect of wind stress on the water surface.

The stress of the wind on the sea surface, τ_s, may be expressed as

$$\tau_s = C_D \rho_a W^2 \tag{23}$$

where W is the wind speed, measured at a standard height, ρ_a is the density of the air and C_D is a drag coefficient dependent on the height at which the wind is measured, the roughness of the water surface and the stability of the air in the first few metres above the surface. A typical value of C_D is 1.3×10^{-3} for W measured at a height of 10 m, so that stresses of a few dyne cm^{-2} are common at moderate wind speeds.

The initial effect of the wind stress is to produce a surface flow of water approximately in the direction of the wind, so that it is the component of wind along the estuary which is most effective in modifying the circulation. In the open sea the surface speed is usually about 2 per cent of the wind speed but in an estuary the ratio may be higher, although the layer of water affected is often shallower. A wind blowing up an estuary frequently generates an upstream surface flow which more than compensates for the normal seaward movement, giving a net surface flow upstream.

Due to the constraint of the estuary boundaries and the stratification of the water, where it is present, secondary effects are soon developed. The upstream surface flow produced by an up-estuary wind, for example, tends to raise the water level at the head of the estuary, setting up a pressure gradient which reinforces the downstream flow at an intermediate depth. In the bottom layer the normal upstream flow is likely to persist so that a three-layer flow pattern is established.

The effects of wind on circulation in coastal plain estuaries were recognized in the earlier observations and were included in the theoretical models of Rattray and Hansen (1962). More recently, a year-long study of the Potomac estuary by Elliott (1978) showed that there were fluctuations in the circulation pattern, on a time scale of 2 to 5 days, largely attributable to the wind. The expected pattern of seaward surface flow and landward bottom flow, although the most commonly observed, occurred for only 43% of the time. The reverse of this circulation was present for 21% of the time. Total inflow, or storage, occurred for 22% of the time and the remainder was accounted for by total discharge or three-layer circulations. The storage and discharge periods appeared to be caused by the interaction of the Potamac estuary with Chesapeake Bay and the effect of wind on the coastal area at its outer boundary.

In fjords the effects of varying winds on the currents and density structure are frequently very marked. A detailed study of effects in Alberni Inlet, British Columbia, was made by Farmer (1972) who developed a dynamical description which explained many of the observed features. A strong up-inlet wind reversed the flow in a thin surface layer, producing an up-inlet flow of up to $50\,\mathrm{cm\,s^{-1}}$. This was accompanied by a sudden thickening of the low salinity surface layer at the head, which took several days to relax to its normal thickness. The disturbance was communicated to positions lower down the inlet by the action of internal waves.

5.2 Renewal of deep water in fjords

Where the deep part of a fjord or inlet is separated from the sea by a sill, the deep water does not usually take part in a steady circulation but is replaced intermittently when sufficiently dense sea water is able to flow in over the sill and sink to the bottom layer. The renewal process is thus determined partly by the distribution of density in the water outside the sill and partly by changing density of the deep fjord water. In some cases renewal may occur annually, in conjunction with a seasonal cycle in density, the actual onset of renewal possibly being triggered by wind effects. In other cases the renewal may not occur every year, but depend on year-to-year variability in the conditions. Renewal processes in the Nordfjord, Norway, which is divided by a second sill into an inner and outer basin, were described by Saelen (1967). The outer basin is replenished with oceanic water every summer but in the inner basin renewal took place only in the years 1932, 1940, and 1947 during the period 1931–1954. In the intervening years the oxygen content fell steadily from 90% to less than 50% of saturation. Gade (1973) has treated the replacement in such cases as a stochastic process. In some cases, a number of renewal events may occur each year; the density of the sea water source, fresh water input, wind effects, and the strength of tidal currents over the sill all playing some part in the process.

5.3 Variations within the tidal period

The vertical profile of salinity, as well as the velocity profile, often varies considerably within a tidal period. In the Narrows of the Mersey estuary, illustrated in Figure 12, as in many other partially mixed estuaries, the high current speed and high turbulent intensity during the flood tide enhance the vertical mixing, which may be almost complete. At high water slack tide the reduction in turbulent mixing enables a stable stratification to develop. This is broken down to some extent during the ebb flow, but not entirely, and a more pronounced stratification, amounting to several parts per thousand in

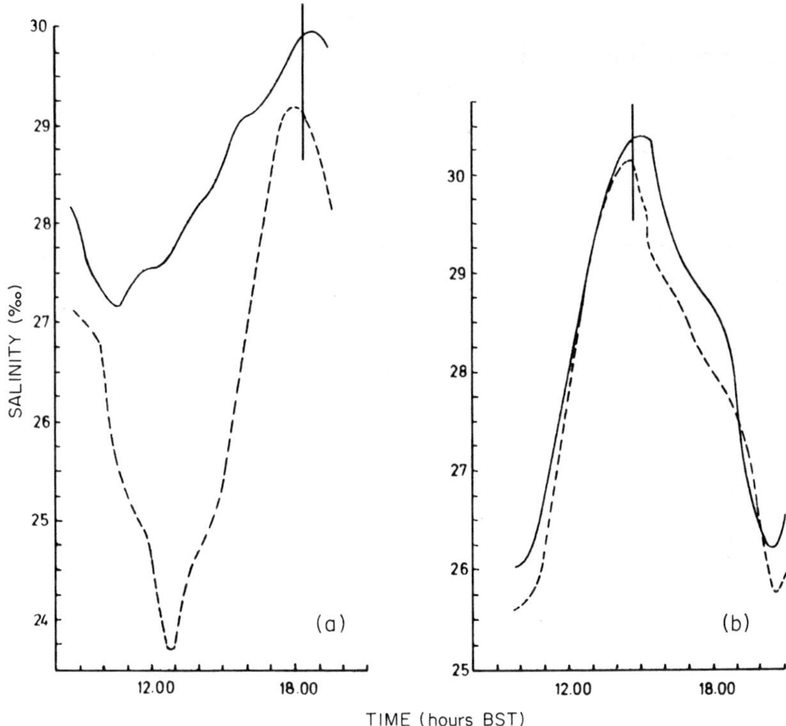

Figure 12. Salinity variation during a tidal period at a station in the Mersey estuary ――― 1 m below surface, ——— 1 m above bottom: (a) 28 August, 1963: tidal range 3.8 m, river discharge 60 m^3 s^{-1}; (b) 17 October, 1962: tidal range 9.1 m, river discharge 24 m^3 s^{-1}. The vertical line indicates the time of high water. (Reproduced by permission of the Royal Astronomical Society from K. F. Bowden and S. H. Sharaf El Din, 1966, *Geophys. J. R. Astr. Soc.*, **10**, 383–399, Figure 7.)

salinity, is produced near low water. The degree of stratification developing within a tidal period is dependent on the amplitude of the tidal current and tends to be greater at neaps than at spring tides, as well as tending to increase with increased river flow.

The variation in the rate of vertical mixing needs to be taken into account in considering problems such as the initial dispersion of a pollutant during the first few hours after release. The resultant effect of mixing and advection over a complete tidal period may be estimated incorrectly if only the average distributions of current and salinity are considered. Reference was made in Section 4 to some numerical model investigations in which changes during the tidal period were taken into account.

                    K. F. Bowden

6 SUMMARY

The circulation in an estuary is largely determined by the interaction of the river discharge, the tidal currents and the geometrical configuration of the estuary. The various types of circulation which can occur, ranging from salt wedge estuaries and fjords to partially mixed and well mixed coastal plain estuaries, are described. Typical distributions of current and salinity are related to parameters of the estuary and, in particular, to a stratification–circulation diagram. Methods by which quantitative estimates of these distributions can be made in certain cases are described. The mixing properties of an estuary arise from the combined effects of currents and turbulent diffusion. Various methods of assessing these effects range from simple estimates of the flushing time to the use of the advection–diffusion equation. Finally, changes due to variable conditions and intermittent events are considered. Wind effects may modify considerably or even reverse the normal estuarine flow for periods of several days. On a longer time scale are intermittent events such as the renewal of deep water in fjords.

REFERENCES

Blumberg, A. F. (1977). Numerical model of estuarine circulation. *J. Hydraul. Div., ASCE*, **103**, (HY3), 295–310.

Bowden, K. F. (1965). Horizontal mixing in the sea due to a shearing current. *J. Fluid Mech.*, **21**, 83–95.

Bowden, K. F., and Gilligan, R. M. (1971). Characteristic features of estuarine circulation as represented in the Mersey estuary. *Limnol. Oceanogr.*, **16**, 490–502.

Bowden, K. F., and Hamilton, P (1975). Some experiments with a numerical model of circulation and mixing in a tidal estuary. *Estuar. Coastal Mar. Sci.*, **3**, 281–301.

Bowden, K. F., and Sharaf El Din, S. H. (1966). Circulation, salinity, and river discharge in the Mersey estuary. *Geophys. J. R. Astr. Soc.*, **10**, 383–399.

Dyer, K. R. (1973). *Estuaries: a Physical Introduction*. John Wiley & Sons, London, 140 pp.

Dyer, K. R. (1977). Lateral circulation effects in estuaries. In: *Estuaries, Geophysics and the Environment*, pp. 22–29. National Academy of Sciences, Washington, D. C.

Dyer, K. R., and Taylor, P. A. (1973). A simple segmented prism model of tidal mixing in well-mixed estuaries. *Estuar. Coastal Mar. Sci.*, **1**, 411–418.

Dronkers, J. J. (1969). Tidal computations for rivers, coastal areas and seas. *J. Hydraul. Div., ASCE*, **95**, (HY1), 29–77.

Elder, J. W. (1959). The dispersion of marked fluid in turbulent shear flow. *J. Fluid Mech.*, **5**, 544–560.

Elliott, A. J. (1978). Observations of the meteorologically induced circulation in the Potomac estuary. *Estuar. Coastal Mar. Sci.*, **6**, 285–299.

Farmer, D. M. (1972). The influence of wind on the surface waters of Alberni Inlet. *Ph.D. Thesis*, Institute of Oceanography, University of British Columbia.

Fischer, H. B. (1972). Mass transport mechanisms in partially stratified estuaries. *J. Fluid Mech.*, **53**, 671–687.

Fischer, H. B. (1976). Mixing and dispersion in estuaries. *Ann. Rev. Fluid Mech.*, **8,** 107–133.

Gade, H. G. (1973). Deep water exchanges in a sill fjord; a stochastic process. *J. Phys. Oceanogr.*, **3,** 213–219.

Garvine, R. W. (1974). Dynamics of small-scale oceanic fronts. *J. Phys. Oceanogr.*, **4,** 557–569.

Garvine, R. W., and Monk, J. D. (1974). Frontal structure of a river plume. *J. Geophys. Res.*, **79,** 2251–2259.

Gibbs, R. J. (1970). Circulation in the Amazon River estuary and adjacent Atlantic Ocean. *J. Mar. Res.*, **28,** 113–123.

Hansen, D. V., and Rattray, M. (1965). Gravitational circulation in straits and estuaries. *J. Mar. Res.*, **23,** 104–122.

Hansen, D. V., and Rattray, M. (1966). New dimensions in estuary classification. *Limnol. Oceanogr.*, **11,** 319–326.

Harleman, D. R. F. (1977). Real-time models for salinity and water-quality analysis in estuaries. In: *Estuaries, Geophysics and the Environment*, pp. 84–93. National Academy of Sciences, Washington, D. C.

Harleman, D. R. F., and Thatcher, M. L. (1974). Longitudinal dispersion and unsteady salinity intrusion in estuaries. *La Houille Blanche*, No. 1/2, 25–33.

Ketchum, B. H. (1951). The exchange of fresh water and salt waters in tidal estuaries. *J. Mar. Res.*, **10,** 18–38.

Neumann, G., and Pierson, W. J. (1966). *Principles of Physical Oceanography.* Prentice-Hall, Englewoods Cliffs, N. J., 545 pp.

O'Connor, D. J., Thomann, R. V., and DiToro, D. M. (1977). Water-quality analyses of estuarine systems. In: *Estuaries, Geophysics and the Environment*, pp. 71–83. National Academy of Sciences, Washington D. C.

Officer, C. B. (1976). *Physical Oceanography of Estuaries (and Associated Coastal Waters).* John Wiley & Sons, New York, 465 pp.

Officer, C. B. (1977). Longitudinal circulation and mixing relations in estuaries. In: *Estuaries, Geophysics and the Environment*, pp. 13–21. National Academy of Sciences, Washington, D. C.

Okubo, A. (1967). The effect of shear in an oscillatory current on horizontal diffusion from an instantaneous source. *Internat. J. Oceanol. Limnol.*, **1,** 194–204.

Okubo, A. (1971). Horizontal and vertical mixing in the sea. In: *Impingement of Man on the Oceans*, (Hood, D. W., ed.) pp. 89–168, John Wiley & Sons, New York.

Okubo, A. (1973). Effect of shoreline irregularities on streamwise dispersion in estuaries and other embayments. *Netherlands J. Sea Res.*, **6,** 213–224.

Pritchard, D. W. (1952). Estuarine hydrography. *Adv. Geophysics*, **1,** 243–280.

Pritchard, D. W. (1954). A study of the salt balance in a coastal plain estuary. *J. Mar. Res.*, **13,** 133–144.

Pritchard, D. W. (1956). The dynamic structure of a coastal plain estuary. *J. Mar. Res.*, **15,** 33–42.

Pritchard, D. W. (1969). Dispersion and flushing of pollutants in estuaries. *J. Hydraul. Div.*, ASCE, **95** (HY1), 115–124.

Rattray, M. (1967). Some aspects of the dynamics of circulation in fjords. In: *Estuaries* (Lauff, G. H., ed.). Publ. 83, pp. 52–62. Amer. Assoc. Adv. Sci., Washington, D. C.

Rattray, M., and Hansen, D. V. (1962). A similarity solution for circulation in an estuary. *J. Mar. Res.*, **20,** 121–133.

Rattray, M., and Mitsuda, E. (1974). Theoretical analysis of conditions in a salt wedge. *Estuar. Coastal Mar. Sci.*, **2,** 375–394.

Saelen, O. H. (1967). Some features of the hydrography of Norwegian fjords. In: *Estuaries* (Lauff, G. H., ed.) Publ. 83, pp. 63–70. Amer. Assoc. Adv. Sci., Washington, D. C.

Stommel, H. (1953). Computation of pollution in a vertically mixed estuary. *Sewage and Industrial Wastes*, **25**, 1065–1071.

Taylor, G. I. (1954). The dispersion of matter in turbulent flow through a pipe. *Proc. Roy. Soc.*, **A219**, 186–203.

Tully, J. P. (1958). On structure, entrainment and transport in estuarine embayments. *J. Mar. Res.*, **17**, 523–525.

Waldichuk, M. (1957). Physical oceanography of the Strait of Georgia, British Columbia. *J. Fish. Res. Bd. Canada*, **14**, 321–486.

Winter, D. F. (1973). A similarity solution for steady-state gravitational circulation in fjords. *Estuar. Coastal Mar. Sci.*, **1**, 387–400.

Wright, L. D., and Coleman, J. M. (1971). Effluent expansion and interfacial mixing in the presence of a salt wedge, Mississippi River delta. *J. Geophys. Res.*, **76**, 8649–8661.

Chemistry and Biogeochemistry of Estuaries
Edited by E. Olausson and I. Cato
Copyright © 1980 by John Wiley & Sons Ltd.

D. DYRSSEN and M. WEDBORG

*Department of Analytical and Marine Chemistry,
Chalmers University of Technology
and University of Göteborg, Göteborg.*

3

Major and Minor Elements, Chemical Speciation in Estuarine Waters

1. INTRODUCTION

The composition of river waters varies to a great extent, depending on the main source of dissolved salts. The two major sources are sea salts added to the river with precipitation, and weathering of rock by rain containing carbon dioxide, the latter process being the quantitatively more important one (Burton, 1976). Dissolved constituents and particulate matter are brought to the estuary, where the river mixes with sea water. As a contrast to river waters, the composition of sea water is not subject to variations, at least not as far as the major constituents are concerned. In general, the major constituents of river water are calcium and carbonate, while sodium, magnesium, chloride, and sulphate form the major part of dissolved sea water constituents (see Table 1).

This chapter will deal with the chemical forms of the constituents of river and sea water and the chemical aspects of the estuarine mixing of these

Table 1. Concentrations (in moles/l., M) of the main ionic constituents of standard sea water, high and low alkalinity river water.

Constituent	SSW	HARW	LARW
Carbonate alkalinity (A_c)	0.0024	0.0014	0.0001
log K_{p12} (25 °C)	−1.51	−1.47	−1.47
log K_{12} (25 °C)	5.86	6.34	6.35
log K_1 (25 °C)	8.95	10.30	10.33
log K_{12}/K_1	−3.09	−3.96	−3.98
$[HCO_3^-]_t$	0.00186	0.00136	0.0001
$[CO_3^{2-}]_t$	0.000275	0.000018	9.4×10^{-8}
pH_t	8.12	8.43	7.30
Sodium	0.47932	0.00030	0.00030
Potassium	0.01045	0.000065	0.000065
Magnesium	0.05439	0.00020	0.000030
Calcium	0.01053	0.00057	0.000087
Chloride	0.55862	0.00020	0.00020
Sulphate	0.02889	0.00015	0.00015
Ionic strength	0.7	0.004	0.001
Chlorinity ‰	19.374	0.0071	0.0071
Salinity ‰	35	0.144	0.041

waters. Knowledge of the oxidation states and chemical forms of dissolved species is an essential link in understanding biogeochemical processes and, increasingly important, can be necessary when trying to anticipate the effects of environmental pollution.

The mixing of river water with sea water is a drastic process. The increase of ionic strength from approx. 0 to 0.7 M, together with the change of composition, causes removal of some constituents by flocculation and change of dissolved chemical forms of others ('dissolved' is operationally defined, see Section 7). Furthermore, biological processes in the estuary may also affect constituent forms (cf. Morris *et al.*, 1978). If a certain component present in river and sea water is conservative, i.e. changes total concentration only as a result of physical mixing, its total concentration will have a straight line relationship to salinity (or, more properly, chlorinity), either increasing or decreasing as the salinity increases, depending on whether the concentration is larger in sea water or in river water. However, if the constituent is removed from (added to) the water by some process, the result obtained when plotting total concentration *vs.* salinity will show a negative (positive) deviation from the theoretical dilution line. Thus measurement of total concentration as function of salinity will, ideally, reveal whether a component behaves conservatively or not. In practice, the results are often difficult to interpret for a number of reasons. This technique, as well as the practical problems associated with the measurements and conclusions, have been treated by Liss (1976).

Sholkovitz (1976) and Eckert and Sholkovitz (1976) used another approach to investigate estuarine flocculation. Sholkovitz (1976) collected and filtered samples from Scottish rivers and sea water, which were mixed at varying salinities in the laboratory. The flocculants formed were collected and analysed.

These two approaches both give as a result that iron and aluminium behave non-conservatively on mixing. In addition, Sholkovitz (1976) found that manganese, phosphorus, organic carbon and humic substances are flocculated. His experiments showed that flocculation was completed between 0 and 15–20‰ salinity. Practically no additional removal was observed at higher salinities. Sholkovitz (1976) concluded that dissolved organic matter plays an important role in controlling the non-conservative behaviour of inorganic constituents. The mechanism of flocculation was studied by Eckert and Sholkovitz (1976). They suggested that flocculation results from both electrostatic and chemical interactions of the major sea salts with river borne colloidal humic substances.

In order to make calculations practically feasible we shall use average concentrations of dissolved river water constituents. It is necessary, however, to distinguish between high and low alkalinity river water (abbreviated HARW and LARW, respectively; see Table 1). The general outline of the

                          D. Dyrssen and M. Wedborg

calculations is as follows:

1. The speciation of the major components of sea water (abbreviated SW), HARW, and LARW is calculated (Section 4). The free concentrations of inorganic ligands are of special importance for later calculations on trace metals.
2. The results are used to calculate the free concentration of inorganic ligands in estuarine waters as functions of salinity (Section 5).
3. The ligand concentrations are used to calculate what inorganic forms of trace metals will dominate in the different waters (Section 6).
4. The stability of inorganic trace metal forms is compared to that of organic forms. Because of the low concentrations of organic ligands we have assumed that only ion exchange reactions in organic polymers with locally enhanced ligand concentration will be of importance (Section 7).
5. Elements that may change oxidation state when transferred from an oxidizing to a reducing environment (e.g. stagnant estuarine bottom waters) are considered. These calculations require prior calculation of pH and redox potential, *Eh*. In Section 8, the calculations are shown and theoretical values of *Eh* are compared to experimental results. Formation of sulphide complexes is also considered.

The models used for speciation calculations in this chapter are all equilibrium models, i.e. the natural waters considered are looked upon as systems where chemical equilibrium prevails. This approach is based on knowledge of the chemical nature of species in solution provided by fundamental solution chemistry. Thus it is based on well established chemical principles, and it is further useful since it renders possible quantitative calculations. However, the following points should be remembered when applying equilibrium models: If there are kinetic barriers to a reaction, equilibrium concentrations of the species involved may never be attained. Thus ammonium is practically not oxidized to nitrate by inorganic reactions. Biological activity (e.g. nitrifying bacteria) may catalyze such slow reactions so that they approach equilibrium. Stability constants of important complexes are sometimes not well known, or not known at all, or they have been determined under favourable laboratory conditions in media different from natural waters in composition, all of which introduces uncertainties into the calculations.

In order to perform the calculations outlined above, the following information must be available:

— total concentrations of all constituents of interest
— stability constants for all possibly important complexes. Formally, the electron is regarded as a ligand, allowing calculation of the 'free concentration' of electrons and use of the pE-concept, which is analogous to the pH-concept (see Section 8).

If necessary approximations are made in a chemically correct way, the calculations can often be performed manually, using a simple desk calculator. In order to improve the results of the calculations and minimize the time needed, it is advisable to use an appropriate computer program (e.g. HALTAFALL, see Section 2) to carry out the successive approximations.

2. SOME FUNDAMENTAL CONCEPTS IN THE DETERMINATION AND USE OF STABILITY CONSTANTS

The purpose of this section is to provide some fundamental information which is necessary in order to understand the calculations and conclusions presented in this chapter. First of all, since terminology concerning stability constants is not standardized, the terminology applied in this chapter is defined. The principles of calculating free concentrations of different species using total concentrations and stability constants are discussed. Finally, the important medium dependence of stability constants is dealt with.

2.1 Definitions

The stability constants are defined according to Sillén and Martell (1964, 1971). Thus the stability constant for the stepwise addition of a ligand, L, to a metal ion, M

$$ML_{n-1} + L \rightleftharpoons ML_n$$

is denoted

$$K_n = [ML_n]/[ML_{n-1}][L]$$

For $n = 0$, $K_n = K_0$ is set equal to 1.

Another way of looking at the complex formation leads to the definition of the cumulative stability constant:

$$M + nL \rightleftharpoons ML_n; \quad \beta_n = [ML_n]/[M][L]^n$$

where

$$\beta_0 = K_0 = 1 \text{ and } \beta_1 = K_1$$

These two ways of defining stability constants are equivalent mathematical formal ways of describing the complex formation. They give information about the strength of complexes but not about the mechanism of complex formation. Consequently, it is a practical matter which one of them should be used for a calculation.

Since there is no formal reason to distinguish between H^+ and other central ions, the consecutive constant of an acid–base equilibrium is defined analogously to the 'K'-constant above:

$$H^+ + H_{n-1}L \rightleftharpoons H_nL; \quad K_{1n} = [H_nL]/[H^+][H_{n-1}L]$$

Comparing the 'K'-constants shows that the first index denotes the number of ligands, and the second, the number of central ions. When the second index equals 1 it is always omitted. The same index rules apply to the 'beta'-constant, which is only used for metal–ligand equilibria. The constant K_{p12} for the carbonate system, used in sections 3 and 4 is defined

$$CO_2(\text{atm}) \rightleftharpoons CO_2(\text{tot, aq}); \quad K_{p12} = [CO_2]_t/[CO_2]_{\text{atm}}$$

where

$$[CO_2]_t = [CO_2] + [H_2CO_3]$$

and $[CO_2]_{\text{atm}}$ is the partial pressure of carbon dioxide in the atmosphere.

2.2 Principles of calculation

The chemical system consisting of the metal ion M and the ligand L in solution is completely described by the mass balance equations and the stability constants of the complexes formed:

$$M_t = [M] + [ML] + \cdots + [ML_n] = \sum_0^n [ML_i] = \sum_0^n \beta_i [M][L]^i = \sum_0^n \beta_i ml^i$$

$$L_t = [L] + [ML] + \cdots + n[ML_n] = [L] + \sum_1^n i\beta_i [M][L]^i$$

$$= l + \sum_1^n i\beta_i ml^i$$

For the sake of simplicity, possible hydrolysis reactions of M and protonation reactions of L have been omitted. In real systems, however, such complexes must also be considered. They are often important.

It is obvious that the mass balance equations above form a system of two equations and two unknowns, m and l. Thus the system can be solved for m and l, at least in principle. Unfortunately, the equations are almost always of a higher degree and are therefore difficult or impossible to solve in practice. If, on the other hand, the free concentrations, m and l, were known, the calculation of the concentrations of all complexes would be trivial: $[ML_n] = \beta_n ml^n$.

In reality, total concentrations and perhaps one free concentration, such as pH, are known and a computer program will be of great help, or even necessary, to perform the calculation. Such a program essentially performs the equivalent to approximate hand calculations, only much faster and with negligible errors. Examples of such programs are the HALTAFALL program, worked out by Sillén and his coworkers (cf. Ingri *et al.*, 1967 and Dyrssen *et al.*, 1968) and the SOLGASWATER program (Eriksson, 1979).

The latter program works faster than HALTAFALL and is more reliable for calculations on multiphase systems. In this work, we have used the HALTAFALL program.

Constituents of natural waters appear in widely different total concentrations. One consequence, which is important to realize, is that a micro or trace constituent will only be able to bind a very small part of a major constituent. As an example, the concentration of $HgCl_n$ may be neglected in the expression for Cl_t but certainly not in the expression for Hg_t. Another example is magnesium and fluoride. mgF^+ constitutes less than 0.1% of Mg_t but almost 50% of F_t in sea water. Many good approximations can be made on the basis of the differences in total concentrations of the constituents. Once this fact is realized it is quite obvious that it is sufficient to use a computer program in order to calculate the speciation of the major constituents. The free ligand concentrations are then conveniently used to calculate the percentage distribution of the trace metal from

$$[ML_n]/M_t = \beta_n l^n \Big/ \sum_0^n \beta_i l^i$$

where l denotes all important ligands.

2.3 Choice of medium. The concept of conditional constants

Frequently, stability constants are determined at very low ionic strengths and the thermodynamic constant at infinite dilution is evaluated from extrapolation. For river waters, where the ionic strength is generally below 0.01 M, these constants may be used, but estuarine and sea water are too far from the pure medium, water, for these constants to be applied. Addition of electrolytes to water results in interactions between the ions and the water (hydration) and among the ions themselves. Such an ionic medium obviously offers an environment different from pure water. The interaction forces involved are generally divided into two groups:

— non specific interactions, depending on ionic strength only, i.e. the number of ions
   and
— specific interactions or complex formation reactions depending on the nature of the ions

In the absence of complex forming medium ions, Davies' formula

$$\log f = -0.5\, z^2 \left(\frac{\sqrt{I}}{1+\sqrt{I}} - 0.3\, I \right)$$

$z =$ ionic charge

$f =$ activity coefficient

$I =$ ionic strength

can be used to account for non specific interaction.
In practice, it can be used for mono- and divalent ions up to $I = 1$ M.

However, most ionic media have at least some complex forming tendencies, which naturally makes it difficult to use constants valid for pure water to calculate stability constants valid at higher ionic strengths. Instead, the constants are determined in the medium for which they are intended to apply (the ionic medium approach). Such constants are often referred to as medium dependent or conditional constants (for a detailed outline, see Ringbom, 1963). The carbonate system in sea water can be used as an example: When carbon dioxide is dissolved in sea water it is hydrated and ionized. Furthermore, the carbonate and hydrogen carbonate ions form complexes with calcium, magnesium and, to some extent, sodium. The hydrogen ions form complexes with sulphate. Thus the following mass balance equations are obtained:

$$[CO_3^{2-}]_t = [CO_3^{2-}] + [MgCO_3] + [CaCO_3] (+[NaCO_3^-])$$
$$= [CO_3^{2-}](1 + K_{MgCO_3}[Mg^{2+}] + K_{CaCO_3}[Ca^{2+}])$$
$$[HCO_3^-]_t = [HCO_3^-] + [MgHCO_3^+] + [CaHCO_3^+]$$
$$= [HCO_3^-](1 + K_{MgHCO_3}[Mg^{2+}] + K_{CaHCO_3}[Ca^{2+}])$$
$$[H^+]_t = [H^+] + [HSO_4^-] = [H^+](1 + K_{HSO_4}[SO_4^{2-}])$$

All these complexes are included in the protonation constants K_1 and K_{12} which are conditional constants, valid only for the medium in which they are determined. These conditional constants are defined

$$K_1 = [HCO_3^-]_t/[H^+]_t[CO_3^{2-}]_t$$
$$= K'(1 + K_{MgHCO_3}[Mg^{2+}] + K_{CaHCO_3}[Ca^{2+}])$$
$$/(1 + K_{MgCO_3}[Mg^{2+}] + K_{CaCO_3}[Ca^{2+}])(1 + K_{HSO_4}[SO_4^{2-}])$$
$$K_{12} = [CO_2]_t/[H^+]_t[HCO_3^-]_t$$
$$= K''/(1 + K_{MgHCO_3}[Mg^{2+}] + K_{CaHCO_3}[Ca^{2+}])(1 + K_{HSO_4}[SO_4^{2-}])$$

where K' and K'' depend on ionic strength only, and K_{ML} denotes the K_1 constant. It is obvious from these definitions that K_1 and K_{12} will change if the free sulphate concentration or the free concentrations of calcium and magnesium are changed.

When studying weak complexes (small stability constants) or protonation of ligands, $NaClO_4$, $NaCl$, or $NaNO_3$ are often used as medium electrolytes. The reason for this choice is that ClO_4^-, Cl^-, and NO_3^- form very weak proton complexes. In addition metal ions form only weak perchlorate and nitrate complexes. Since sea salt mainly consists of sodium chloride, 0.7 M NaCl has often been used as reference medium, neglecting the weak complex formation with Na^+ or Cl^-. The fact that Na^+ is a medium constituent makes it impossible to determine complexes between Na^+ and L from a titration of L in NaCl with hydrochloric acid. Even if the studies of

Elgquist and Wedborg (1975, 1978, 1979) indicate complex formation between chloride and Mg^{2+}, Ca^{2+} we shall simplify our calculations by neglecting the formation of $MgCl^+$ and $CaCl^+$. This simplification will only slightly influence the 'free' concentrations of chloride and sulphate. In addition, many stability constants of magnesium and calcium complexes have been determined in chloride media neglecting chloride complexation.

3 MAJOR IONIC CONSTITUENTS OF SEA AND RIVER WATERS

According to Mackenzie and Garrels (1966), Burton (1976) and Liss (1976) the main constituents (>1 ppm) of river waters are sodium, potassium, magnesium, calcium, chloride, sulphate, hydrogen carbonate and silicon. Of these SiO_2 is only slightly ionized ($H^+ + SiO(OH)_3^- \rightleftharpoons Si(OH)_4$, $\log K_1 = 9.5$–9.9) at the pH of river water and may be neglected together with nitrate and fluoride in an approximate calculation of the ionic balance. It will be shown below (See section *Speciation of carbonate* below) that the formation of hydroxide and carbonate complexes will depend on the alkalinity of the river water. The high alkalinity water is representative of rivers entering into the Baltic proper and the value of 1.4 mM, which by no means is extreme, fits to the residual alkalinity found by Andersson and Dyrssen (1978):

$$A_t(\text{mmoles/kg}) = 0.0473\ Cl + 1.42$$

where Cl is the chlorinity of the Baltic water, which is a mixture of North Sea water and river water. The value of 0.1 mM is representative of low alkalinity lake and river waters in Western Scandinavia, and fits to data for the Idefjord on the border of Norway and Sweden (Danielsson and Dyrssen, 1975)

$$A_t = 0.07S + 0.10\ \text{mmoles/kg}$$

The concentrations in sea water were taken from Dyrssen and Wedborg (1974). The concentrations of the carbonate ions were calculated (see below Section 4.1) from the alkalinities and stability constants in Table 1. The constants were taken from Hansson (1973), Almgren *et al.* (1975) and Smith and Martell (1976). It was also assumed that the surface water was saturated with carbon dioxide at a mean partial pressure in the atmosphere of 330 ppm (330×10^{-6} atm). From Table 4 in the paper of Mackenzie and Garrels (1966) we conclude that the sodium and chloride concentrations in river water are approximately equal to 7 mg/l., which is 0.30 mmoles Na^+/l. and 0.20 mmoles Cl^-/l. The sulphate concentration varies between 10 and 20 mg/l. and we shall use a rough mean value of 0.15 mmoles/l. The mean weight ratio of sodium and potassium in river waters is 2.7, which corresponds to a mole ratio of 4.6 giving an approximate potassium concentration in river water of $0.3/4.6 = 0.065$ mmoles/l. The weight ratio of magnesium and calcium in river waters varies, but 0.21 may be taken as an approximate mean value. This corresponds to a mole ratio of 0.35. The ionic balance will give the following equation:

$$[HCO_3^-] + 2[CO_3^{2-}] + [Cl^-] + 2[SO_4^{2-}] = [Na^+] + [K^+] + 2[Mg^{2+}] + 2[Ca^{2+}]$$

or, using $[Mg^{2+}] = 0.35 \times [Ca^{2+}]$ and the definition of the carbonate alkalinity (see Section 4.1) together with concentrations given in Table 1

$$A_c + 0.0002 + 2 \times 0.00015 = 0.0003 + 0.000065$$
$$+ 2[Ca^{2+}] \times 1.35; \quad [Ca^{2+}] = (A_c + 0.000135)/2.7 \text{ moles/l.}$$

The concentrations of magnesium and calcium in Table 1 were calculated with this equation.

According to Liss (1976) the major ionic constituents will behave conservatively upon estuarine mixing. A chlorinity (Cl) titration with silver nitrate is thus a simple way of determining the degree of mixing. For a known value of Cl the concentration of another major constituent may be calculated with the following equation:

$$\frac{[X] - [X]_{RW}}{Cl - 0.0071} = \frac{[X]_{SW} - [X]_{RW}}{19.374 - 0.0071} = k$$

or

$$[X] = k \times Cl + [X]_{RW} - k \times 0.0071$$

If we apply this equation to calcium in the mixing of high alkalinity river water with standard sea water we obtain

$$[Ca] = 5.143 \times 10^{-4} \, Cl + 0.000566 \, \text{M}$$

If this equation is transferred from moles/l. to g/kg using a density of 1 kg/l. the coefficients will be

$$[Ca] = 0.0206 \, Cl + 0.0227 \text{ g/kg}$$

This equation agrees very well with the equation for the Baltic proper given by Andersson and Dyrssen (1978). The corresponding relation for the salinity will be

$$S = 1.8 \, Cl + 0.131$$

which indicates that the Knudsen relations $S = 1.805 \, Cl + 0.03$ fits better to an input of low alkalinity river water. This is confirmed by the following calculation with data from Table 1:

$$\frac{S - 0.041}{Cl - 0.0071} = \frac{35 - 0.041}{19.374 - 0.0071} = 1.805$$

or

$$S = 1.805 \, Cl + 0.028$$

4 INORGANIC LIGAND CONCENTRATIONS IN SEA AND RIVER WATER

4.1 Total carbonate ion concentrations

The main carbonate equilibrium in natural waters is the disportionation equilibrium of hydrogen carbonate:

$$2\,HCO_3^- \rightleftharpoons CO_2 + CO_3^{2-} + H_2O$$

Since

$$K_{12} = [CO_2]_t/[H^+]_t[HCO_3^-]_t$$

and

$$K_1 = [HCO_3^-]_t/[H^+]_t[CO_3^{2-}]_t$$

the constant for the disportionation equilibrium will be

$$K = K_{12}/K_1 = [CO_2]_t[CO_3^{2-}]_t/[HCO_3^-]_t^2$$

K is larger in seawater than in river water (see Table 1), since $\log K_1$ decreases more than $\log K_{12}$ as the ionic strength is increased and since magnesium and calcium form stronger complexes with CO_3^{2-} than with HCO_3^- (see p. 83; cf. the definition of the conditional constants on p. 77).

The concentration of carbon dioxide, which mainly consists of unhydrated $CO_2([CO_2]/[H_2CO_3] \approx 600)$, varies with the partial pressure of carbon dioxide in the surface atmosphere, the temperature of the water, the salinity, and the biological activity (photosynthesis, respiration, consumption and decay of phytoplankton). However, since most stability constants are given for 25 °C we shall calculate the concentrations of the carbonate ions for a mean atmospheric content of 330 ppm using the stability constants in Table 1 for 25 °C.

Thus

$$[CO_2]_t = 330 \times 10^{-6}\, K_{p12}$$

$$A_c = [HCO_3^-]_t + 2[CO_3^{2-}]_t$$

and

$$K = 330 \times 10^{-6}\, K_{p12}[CO_3^{2-}]_t/[HCO_3^-]_t^2$$

The values of $[HCO_3^-]_t$ and $[CO_3^{2-}]_t$ are given in Table 1. The $pH_t = -\log[H^+]_t$ is then calculated from the equations for K_{12} and $[CO_2]_t$ above

$$pH_t = \log K_{12} - \log K_{p12} + 3.48 + \log[HCO_3^-]_t$$

The index 't' stands for both total and titrated since all forms of carbonate (both 'free' carbonate ions and carbonate complexes) react with hydrochloric acid in an alkalinity titration. In a Gran evaluation of the alkalinity according to Dyrssen and Sillén (1967) and Hansson and Jagner (1973) pH_t

is determined from the excess of hydrochloric acid. Thus it will also include protons bound to sulphate

$$[H^+]_t = [H^+] + [HSO_4^-] = [H^+](1 + K_{HSO_4}[SO_4^{2-}])$$

For sea water Dyrssen and Hansson (1972–73) and Dyrssen and Wedborg (1974) obtained

$$pH_t = pH - 0.12$$

which agrees quite well with Bates and Culberson (1977). For river water $K_{HSO_4} \approx 100$ and $[SO_4^{2-}] = 0.00015$. Thus

$$[H^+]_t = [H^+](1 + 100 \times 0.00015)$$

and

$$pH_t = pH - 0.006$$

Speciation of carbonate

Trace metals such as uranium(VI), copper(II) and lead(II) will barely influence the speciation of a main constituent like hydrogen carbonate.

Table 2. some carbonate complex formation constants in river water ($I \sim 0$) and sea water (in parenthesis).

$\log[MCO_3]/[M][CO_3]$

Mg^{2+}	2.88	(1.5)
Ca^{2+}	3.15	(1.5)
UO_2^{2+}	~ 9.6	(~ 8.1)
Cu^{2+}	6.75	(~ 5.25)
Pb^{2+}	~ 5.9	(~ 4.4)

$\log[M(CO_3)_2]/[M][CO_3]^2$

UO_2^{2+}	15.6	(~ 14.1)
cu^{2+}	9.92	(~ 8.4)
Pb^{2+}	8.2	(~ 6.7)

$\log[M(CO_3)_3]/[M][CO_3]^3$

UO_2^{2+}	20.7	(~ 19.2)

$\log[MHCO_3]/[M][HCO_3]$

Mg^{2+}	0.95	(0)
Ca^{2+}	1.0	(0)
Mn^{2+}	1.8	(~ 0.8)
UO_2^{2+}	~ 3.9	(~ 2.9)
Cu^{2+}	~ 3.2	(~ 2.2)
Pb^{2+}	~ 3.2	(~ 2.2)

Magnesium and calcium may, however, form appreciable amounts of carbonate complexes. In addition pH shifts due to biological uptake of CO_2 (photosynthesis) and CO_3^{2-} (formation of zoo-plankton shells) and decomposition of biological matter will alter the speciation. For surface sea water (K_{p12} and the total concentrations are given in Table 1):

$$[CO_2]_t = 330 \times 10^{-6} K_{p12} = 0.0000102 \text{ M}$$

$$[HCO_3^-]_t = 0.00186 = [HCO_3^-] + [MgHCO_3^+] + [CaHCO_3^+]$$

$$= [HCO_3^-](1 + K_{MgHCO_3}[Mg^{2+}] + K_{CaHCO_3}[Ca^{2+}])$$

$$[CO_3^{2-}]_t = 0.000275 = [CO_3^{2-}] + [MgCO_3] + [CaCO_3]$$

$$= [CO_3^{2-}](1 + K_{MgCO_3}[Mg^{2+}] + K_{CaCO_3}[Ca^{2+}])$$

The 'free' ion concentrations of magnesium and calcium are calculated below (p. 87) as $[Mg^{2+}] = 0.0502$ M and $[Ca^{2+}] = 0.0086$ M. The speciation of surface sea water may then be calculated with the constants in Table 2.

For river water the 'free' concentrations of magnesium and calcium are practically equal to the total concentrations in Table 1. With constants in Table 2 for river water the same set of equations are used to calculate the carbonate speciation in both high and low alkalinity river water. The results of the calculations are summarized in Table 3.

From this table we may conclude that HCO_3^- is always the dominating species in river waters as well as sea water. Furthermore, as the ionic strength and magnesium concentration increase with estuarine mixing and

Table 3. Carbonate speciation. Logarithms of concentrations (in moles/l.) for standard sea water (SSW), high and low alkalinity river waters (HARW and LARW). Calculated at 25 °C and $p_{CO_2} = 330$ p.p.m. The percentage of total carbonate are given in parenthesis for values >0.1%.

	SSW	HARW	LARW
CO_2	−4.99	−4.95	−4.95
	(0.5%)	(0.8%)	(10%)
HCO_3^-	−2.76	−2.87	−4.00
	(82%)	(97%)	(90%)
$MgHCO_3^+$	−4.05	−5.62	−7.57
	(4.1%)	(0.2%)	
$CaHCO_3^+$	−4.82	−5.11	−7.06
	(0.7%)	(0.6%)	
CO_3^{2-}	−4.02	−5.04	−7.09
	(4.5%)	(0.7%)	
$MgCO_3$	−3.82	−5.85	−8.73
	(7.1%)	(0.1%)	
$CaCO_3$	−4.58	−5.13	−8.00
	(1.2%)	(0.5%)	

increasing salinity, the species $MgHCO_3^+$, CO_3^{2-} and $MgCO_3$ will constitute a more and more appreciable part of the total carbonate concentration. Carbon dioxide will always be a minor species except in low alkalinity waters and waters where considerable decay of organic matter takes place without free access to gas exchange with the atmosphere, e.g. in waters below the photozone and pore waters. The value of $-\log[Ca^{2+}][CO_3^{2-}]$ will be 6.09 for sea water, 8.28 for high alkalinity river water and 11.15 for low alkalinity river water. The solubility products (pK_s; $K_s = [Ca^{2+}][CO_3^{2-}]$ in the presence of solid $CaCO_3$) for calcite at 25 °C are 6.22 and 8.33 for sea and river water, respectively. Thus calcite should precipitate in surface sea water. In fact, except for some rare cases of 'whitings', practically all calcium carbonate found in the oceans is biogenic (mainly zooplankton shells). Calcium carbonate should dissolve readily in low alkalinity river waters.

4.2 The ionic product of natural waters. Hydroxide concentration

Dyrssen and Hansson (1972–73) showed that the titrated ionic product of water was considerably larger in sea water ($pK_w = 13.19$ at 25 °C) than in 0.7 M sodium chloride ($pK_w = 13.77$ at 25 °C). In river water the value of pK_w at 25 °C is slightly dependent on the ionic strength

$$pK_w = 14.00 - \sqrt{I}$$

or 13.94 for high alkalinity and 13.97 for low alkalinity river water. The main influence on the titrated ionic product of sea water, $[H^+]_t[OH^-]_t$, is the complexation of H^+ by SO_4^{2-} and OH^- by Mg^{2+}. Thus

$$[H^+]_t[OH^-]_t = ([H^+]+[HSO_4^-])([OH^-]+[MgOH^+])$$

$$= [H^+][OH^-](1+K_1[SO_4^{2-}])(1+K_1[Mg^{2+}])$$

Using $K_1[SO_4^{2-}] = 0.324(K_1 = 26.2)$ for HSO_4^- and $K_1[Mg^{2+}] = 1.85(K_1 = 36.9)$ for $MgOH^+$ according to Dyrssen and Hansson (1972–73) we obtain

$$\log(1+K_1[SO_4^{2-}]) = 0.122$$
$$\log(1+K_1[Mg^{2+}]) = 0.455$$
$$\log[H^+]_t[OH^-]_t + pK_w = 0.58$$

which fits to $13.77 - 13.19 = 0.58$. The $\log[OH^-]_t$ in sea water will be $pH_t - 13.19 = -5.07$ and $\log[OH^-] = -5.07 - 0.46 = -5.53$ or $pOH = 5.53$. For high alkalinity river water $pH_t = 8.43$ (see Table 1) and $pK_w = 13.94$. Thus

$$pH_t + p[OH^-]_t = 13.94 - \log(1+K_1[SO_4^{2-}]) - \log(1+K_1[Mg^{2+}])$$
$$K_1[SO_4^{2-}] = 100 \times 0.0015 = 0.015$$

and

$$K_1[Mg^{2+}] = 380 \times 0.0002 = 0.076$$

which gives

$$pH_t + p[OH^-]_t = 13.94 - 0.01 - 0.03 = 13.90$$

Then $\log[OH^-] = 8.43 - 13.90 - 0.03 = -5.50$. For low alkalinity river water there is practically no influence of sulphate and magnesium complexation on H^+ and OH^- and $\log[OH^-]$ will be $7.30 - 13.97 = -6.67$.

4.3 Chloride and bromide

Chloride forms very weak complexes with sodium, potassium magnesium, and calcium according to Smith and Martell (1976) and Elgquist and Wedborg (1975, 1978, 1979). Thus for calculation of trace metal complexation we may use $pCl = -\log Cl_t$. for sea water $pCl = 0.25$ and for river water $pCl = 3.7$.

Dyrssen and Wedborg (1974) showed that mercury may form bromide and mixed chloride and bromide complexes in sea water in spite of the low bromide concentration: $[Cl^-]_t/[Br^-]_t = 648$ or $pCl - pBr = -2.81$. This relation should approximately be valid for river water too since their chemical behaviour is similar. Thus $pBr = 3.06$ for sea water and 6.5 for river water.

4.4 Fluoride

The total concentration of fluoride in sea water is 1.32 mg/l. while the concentration in river waters may vary considerably. Burton (1976) and Liss (1976) use 0.1 mg/l. as an average value. The complex formation constants for magnesium and calcium vary with the concentration of sodium chloride according to Table 4.

For sea water we may use

$$K_1 = [MgF^+]/[Mg^{2+}][F^-] = 18.8$$

and

$$K_1 = [CaF^+]/[Ca^{2+}][F^-] = 4.22$$

Table 4. Fluoride complexation constants for magnesium and calcium at 25 °C and different concentrations of sodium chloride (I moles/l.) according to Elgquist (1970).

I	K_1 for MgF$^+$	K_1 for CaF$^+$
0	63	(13)
0.1	28.7	(6)
0.4	22.0	5.01
0.7	18.8	4.22
1	18.6	3.85

86   *D. Dyrssen and M. Wedborg*

The 'free' concentrations of magnesium and calcium are 0.0502 and 0.0086 M, respectively. Thus, neglecting the weak tendency for sodium to form fluoride complexes

$$F_t = [F^-] + [MgF^+] + [CaF^+] = 6.95 \times 10^{-5} \text{ M}$$

or

$$F_t = [F^-](1 + 18.8 \times 0.0502 + 4.22 \times 0.0086) = 1.980[F^-]$$

$$[F^-]/F_t = 0.505, \text{ i.e. } 50.5\% \text{ 'free' fluoride}$$

$$[MgF^+]/F_t = 18.8 \times 0.0502/1.98 = 0.477, \text{ i.e. } 47.7\% \text{ MgF}^+$$

$$[CaF^+]/F_t = 4.22 \times 0.0086/1.98 = 0.018, \text{ i.e. } 1.8\% \text{ CaF}^+$$

The fluoride ligand concentration in sea water will thus be

$$[F^-] = 0.505 \times 6.95 \times 10^{-5} \text{ M or } pF = -\log[F^-] = 4.45$$

For high alkalinity river water the corresponding calculation will give

$$F_t = 5.25 \times 10^{-6} \text{ M} = [F^-](1 + 63 \times 0.0002 + 13 \times 0.00057) = 1.02[F^-]$$

$$[F^-]/F_t = 0.980, \text{ i.e. } 98.0\% \text{ F}^-$$

$$[MgF^+]/F_t = 63 \times 0.0002/1.02 = 0.0124, \text{ i.e. } 1.2\% \text{ MgF}^+$$

$$[CaF^+]/F_t = 13 \times 0.00057/1.02 = 0.0073, \text{ i.e. } 0.7\% \text{ CaF}^+$$

Thus $pF = -\log[F^-] = 5.3$. The complexation of fluoride in low alkalinity river water will be still smaller. Thus for all river waters we use $pF = 5.3$ as a mean value of the fluoride concentration.

4.5 Sulphate

The complexation of sulphate in sea water is somewhat controversial since different methods for the determination of the stability constants do not give the same results (cf. Elgquist and Wedborg, 1974, 1975, 1978). Table 5 summarizes the constants for sodium, potassium, magnesium and calcium. Neglecting the formation of chloride complexes with magnesium and calcium in sea water we have chosen the following constants for $NaSO_4^-$, KSO_4^-, $MgSO_4$ and $CaSO_4$: 1.8, 2.5, 6.3, and 17.3 in sea water and 4.3, 6.2, 95, and 115 in river water. The following set of equations will solve the speciation of sulphate in sea water:

$$[SO_4]_t = 0.02889 = [SO_4^{2-}] + [NaSO_4^-] + [KSO_4^-] + [MgSO_4] + [CaSO_4]$$

$$[Na]_t = 0.47932 = [Na^+] + [NaSO_4^-]$$

$$[K]_t = 0.01045 = [K^+] + [KSO_4^-]$$

$$[Mg]_t = 0.05439 = [Mg^{2+}] + [MgSO_4] + [MgHCO_3^+] + [MgCO_3] + [MgF^+]$$

$$[Ca]_t = 0.01053 = [Ca^{2+}] + [CaSO_4] + [CaHCO_3^+] + [CaCO_3] + [CaF^+]$$

In these equations we may neglect the concentrations of MgF^+ and CaF^+ in

Table 5. Sulphate formation constants at 25 °C for the main metal ions in sea and river water.

Ion	Ionic strength	$\log K_1$
Na^+	0	0.70
	0.7 (SW)	0.09–0.30
	0.004 (RW)	0.64^a
K^+	0	0.85
	0.7 (SW)	0.40
	0.004 (RW)	0.79^a
Mg^{2+}	0	2.23
	0.7 (SW)	0.80–1.09
	0.004 (RW)	1.98^a
Ca^{2+}	0	2.31
	0.7 (SW)	1.22–1.56
	0.004 (RW)	2.06^a

[a] These constants have been estimated from the values at zero ionic strength taken from Smith and Martell (1976) using

$$\log K_1 - \sqrt{I} \text{ for } KSO_4^- \text{ and } NaSO_4^-$$

and

$$\log K_1 - 4\sqrt{I} \text{ for } MgSO_4 \text{ and } CaSO_4$$

comparison with the concentration of the other species and the total concentrations of magnesium and calcium.

Solving these equations with the HALTAFALL computer program (Section 2.2) gives the following results:

$$[SO_4^{2-}] = 0.0124 \text{ M } (42.8\%); \quad pSO_4 = 1.91$$

$$[NaSO_4^-] = 0.0204 \text{ M } (36.1\%)$$

$$[KSO_4^-] = 0.0003 \text{ M } (1.1\%)$$

$$[MgSO_4] = 0.0039 \text{ M } (13.6\%)$$

$$[CaSO_4] = 0.00185 \text{ M } (6.4\%)$$

$$[Na^+] = 0.4689 \text{ M} \qquad [Mg^{2+}] = 0.0502 \text{ M}$$

$$[K^+] = 0.0101 \text{ M} \qquad [Ca^{2+}] = 0.0086 \text{ M}$$

For river water

$$[NaSO_4^-]/[Na^+] \approx 4.3[SO_4]_t = 6.45 \times 10^{-4}$$

$$[KSO_4^-]/[K^+] \approx 6.2[SO_4]_t = 9.3 \times 10^{-4}$$

$$[MgSO_4]/[Mg^{2+}] \approx 95[SO_4]_t = 1.4 \times 10^{-2}$$

$$[CaSO_4]/[Ca^{2+}] \approx 115[SO_4^{2-}]_t = 1.7 \times 10^{-2}$$

Thus less than two percent of the metal ions form sulphate complexes and the 'free' sulphate will be practically equal to the total concentration, i.e. $[SO_4^{2-}] = 0.00015$ or $pSO_4 = 3.88$.

Table 6. Summary of some average ligand concentrations given as $-\log[L] = pL$ in standard sea water and high and low alkalinity river water.

	SSW	HARW	LARW
pF	4.46	5.3	5.3
pCl	0.25	3.7	3.7
pBr	3.06	6.5	6.5
pOH	5.53	5.50	6.67
$pHCO_3$	2.76	2.87	4.00
pCO_3	4.02	5.04	7.09
pSO_4	1.91	3.88	3.88

To summarize this section the ligand concentrations of fluoride, chloride, bromide, hydroxide, hydrogen carbonate, carbonate, and sulphate are calculated and given in Table 6 for sea water and high and low alkalinity river waters on the basis of the total concentrations of the major constituents in Table 1. In addition the speciation of total carbonate (see Table 3), fluoride and sulphate have been calculated on the basis of stability constants given in Tables 2, 4, and 5. The values of pH and pOH are calculated from the carbonate alkalinity and the assumption that surface waters are saturated with carbon dioxide.

5 INORGANIC LIGAND CONCENTRATION IN ESTUARINE WATERS

In order to calculate the ligand concentration in mixtures of sea and river water in an estuary one has to know the ionic composition of the river water and the dependence of the stability constants on the salinity. In general one may assume that the main constituents behave conservatively (cf. Liss, 1976). The decrease of the stability constants is most pronounced below 5 ‰ salinity (i.e. 0.1 M ionic strength) as may be seen from Figure 1. This figure also shows that the protonization of a trivalent anion (phosphate) is more dependent on the ionic strength than the protonization of a divalent (carbonate) or monovalent (hydrogen carbonate or sulphide) ion. Furthermore, the solubility product (pK_{so}) of barium sulphate will decrease to a larger extent than $\log K_1$ for the formation of MgF^+ and CaF^+. Deviations from Davies' formula (p. 77) will occur if Na^+ forms an ion pair (e.g. with PO_4^{3-}). Furthermore, it may be seen that the stability constants measured in sea water media (full-drawn curves) are lower than those measured in sodium chloride of the same ionic strength. This is due to complexation with the main constituents (e.g. CO_3^{2-} with Mg^{2+}). Ligand concentration shifts will be summarized in Figure 2 as $\log[L]$ ($[L]$ in moles/l., M) versus the salinity (in g/kg, ‰).

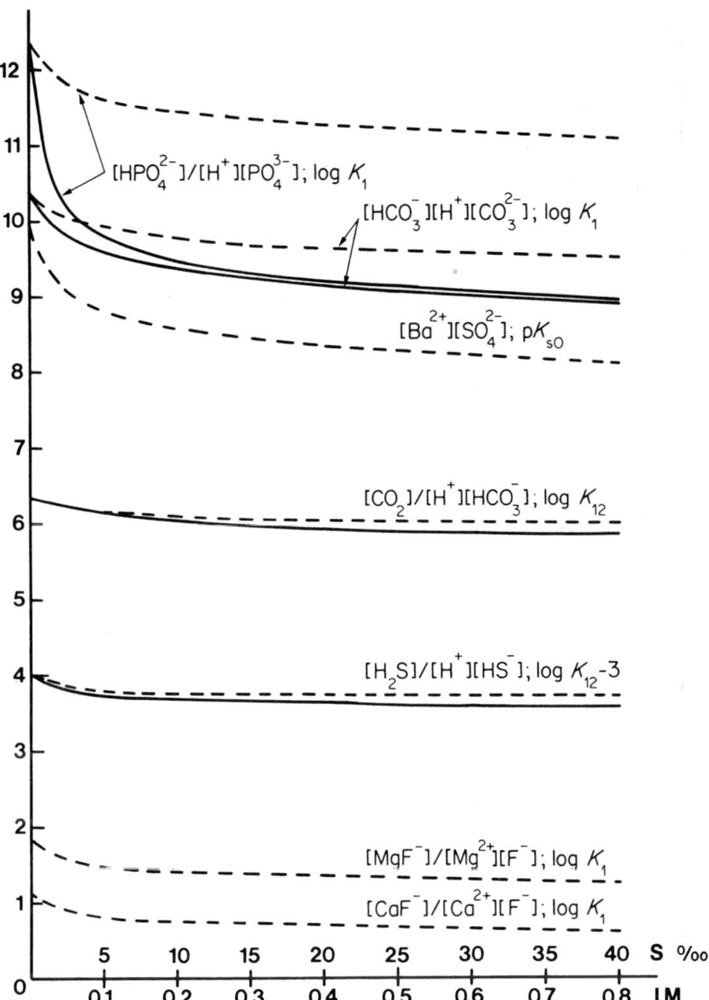

Figure 1. Conditional or medium stability constants as a function of the sea water salinity (full-drawn curves) or concentration of sodium chloride (dashed curves).

5.1 Chloride and bromide

Since we neglect complexation of chloride and bromide with the main cations their concentrations are simply given by the equations (cf. Table 1)

$$\frac{S - S_r}{[\text{Cl}^-] - 0.0002} = \frac{35 - S_r}{0.55862 - 0.0002}$$

                    D. Dyrssen and M. Wedborg

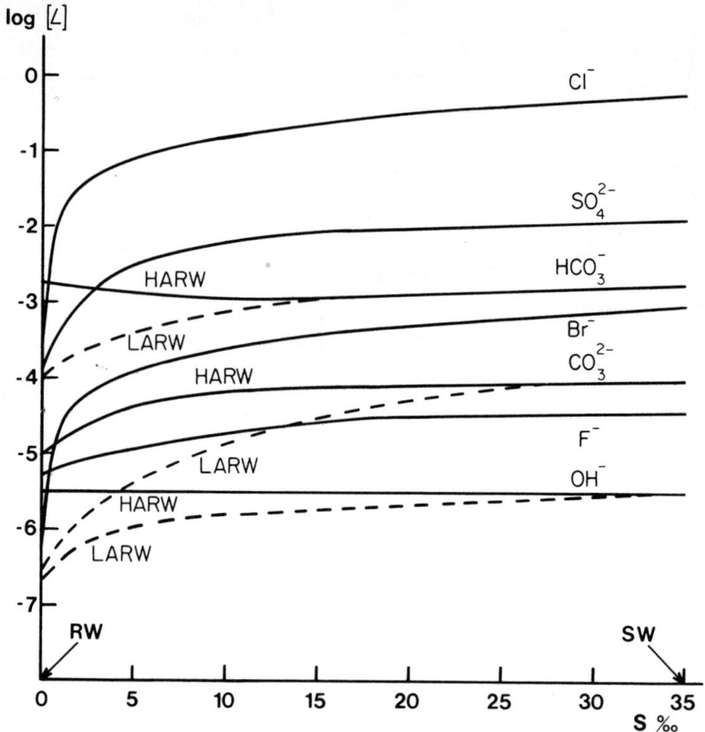

Figure 2. Concentrations of inorganic ligands in estuarine waters plotted as log [L] against the salinity. When separate the low alkalinity river water curves are dashed.

and

$$pCl - pBr = -2.81$$

where S_r is the salinity of the river water (0.144 for HARW and 0.041 for LARW).

In Figure 2 we have used a rough mean value for $S_r = 0.1‰$.

5.2 Fluoride and sulphate

The difference in pF for sea and river water is less than one unit and in order to draw the $\log[F^-]$ vs. S curve in Figure 2 it is sufficient to calculate the pF value for $S = 5‰$ which corresponds to the ionic strength of 0.1 M. The total concentration of fluoride may be calculated from

$$\frac{S - S_r}{F_t - 5.26 \times 10^{-6}} = \frac{35 - S_r}{(69.5 - 5.26) \times 10^{-6}}$$

For $S = 5‰$ and $S_r = 0.1‰$ $F_t = 14.3 \times 10^{-6}$ M Magnesium and calcium will bind fluoride and sulphate. Their total concentrations, which are needed for the calculation of the ligand concentrations, may be obtained from the corresponding equations using the total concentrations given in Table 1. Since Mg_t and Ca_t for river water will affect Mg_t and Ca_t in an estuarine mixture only to a minor extent we have chosen a mean of high and low alkalinity river waters. Thus for $S = 5‰$ and $S_r = 0.1‰$ we obtain

$$Mg_t = 0.00774 \text{ M}$$
$$Ca_t = 0.00176 \text{ M}$$
$$(SO_4)_t = 0.00419 \text{ M}$$

The approximate stability constants at $S = 5‰$ (0.1 M ionic strength) will be

$$28.7 \text{ for } MgF^+$$
$$6.0 \text{ for } CaF^+$$
$$44 \quad \text{for } MgSO_4$$
$$63 \quad \text{for } CaSO_4$$
$$3.8 \text{ for } NaSO_4^-$$
$$5.0 \text{ for } KSO_4^-$$

Using these constants and neglecting the small part ($<1\%$) of magnesium and calcium complexed by carbonate ions we must solve the following set of equations:

$$F_t = 0.0000143 = [F^-] + [MgF^+] + [CaF^+]$$
$$Na_t = 0.0676 = [Na^+] + [NaSO_4^-]$$
$$K_t = 0.00152 = [K^+] + [KSO_4^-]$$
$$Mg_t = 0.00774 = [Mg^{2+}] + [MgF^+] + [MgSO_4]$$
$$Ca_t = 0.00176 = [Ca^{2+}] + [CaF^+] + [CaSO_4]$$
$$(SO_4)_t = 0.00419 = [SO_4^{2-}] + [NaSO_4^-] + [KSO_4^-] + [MgSO_4] + [CaSO_4]$$

These equations are solved by the HALTAFALL computer program (p. 76). The following results were obtained:

$$[Mg^{2+}] = 0.00697 \text{ M} \qquad [SO_4^{2-}] = 25.18 \times 10^{-4} \text{ M}$$
$$[Ca^{2+}] = 0.00152 \text{ M} \qquad [NaSO_4^-] = 6.41 \times 10^{-4} \text{ M}$$
$$[F^-] = 1.18 \times 10^{-5} \text{ M} \qquad [KSO_4^-] = 0.189 \times 10^{-4} \text{ M}$$
$$[MgF^+] = 0.237 \times 10^{-5} \text{ M} \qquad [MgSO_4] = 7.72 \times 10^{-4} \text{ M}$$
$$[CaF^+] = 0.0108 \times 10^{-5} \text{ M} \qquad [CaSO_4] = 2.41 \times 10^{-4} \text{ M}$$

                     D. Dyrssen and M. Wedborg

5.3 Carbonate and hydroxide

In this case we have to treat high and low alkalinity waters separately. Again we shall calculate $[HCO_3^-]$, $[CO_3^{2-}]$ and $[OH^-]$ for a salinity of 5‰ (0.1 M ionic strength). The carbonate alkalinities will be

$$A_c = \frac{0.001}{35} S + 0.0014$$

for high alkalinity river water and

$$A_c = \frac{0.0023\,S}{35} + 0.0001$$

for low alkalinity river water. For $S = 5$‰ we obtain $A_c = 0.00154$ and 0.00043 M respectively. The following constants may be estimated from Figure 1:

$$\log K_{12} = 6.15 \text{ for hydrogen carbonate and}$$
$$\log K_1 = 9.57 \text{ for carbonate}$$

Furthermore we need the following constants for the calculations:

$$[MgOH^-]/[Mg^{2+}][OH^-] = 170$$
$$\log[H^+][OH^-] = -13.78$$
$$[HSO_4^-]/[H^+][SO_4^{2-}] = 35$$
$$[MgCO_3]/[Mg^{2+}][CO_3^{2-}] = 58$$
$$[CaCO_3]/[Ca^{2+}][CO_3^{2-}] = 107$$
$$[MgHCO_3^+]/[Mg^{2+}][HCO_3^-] = 4.1$$
$$[CaHCO_3^+]/[Ca^{2+}][HCO_3^-] = 4.6$$

Using K_{12} and K_1K_{12} together with $\log[CO_2] = -4.98$ we obtain

$$[H^+]_t[HCO_3^-]_t = 10^{-11.13}$$
$$[H^+]_t^2[CO_3^{2-}]_t = 10^{-20.70}$$

which together with

$$A_c = [HCO_3^-]_t + 2[CO_3^{2-}]_t$$

give $pH_t = 8.27$ and 7.75 for high and low alkalinity river water, respectively. $[H^+]$ may then be calculated from

$$H_t = [H^+](1 + K_1[SO_4^{2-}]) = [H^+](1 + 35 \times 0.00252)$$

or

$$pH_t = pH - 0.037$$

which gives $\log[OH^-] = -5.47$ and -5.99 for high and low alkalinity river water.

Using

$$[HCO_3^-]_t = [HCO_3^-] + [MgHCO_3^+] + [CaHCO_3^+]$$

and

$$[CO_3^{2-}]_t = [CO_3^{2-}] + [MgCO_3] + [CaCO_3]$$

the following results were obtained for $[Mg^{2+}] = 0.00697$ M and $[Ca^{2+}] = 0.00172$ and 0.00133 M for high and low alkalinity river water:

	HARW	LARW
$[HCO_3^-]_t$	0.00138	0.000417
$[CO_3^{2-}]_t$	0.000069	0.0000063
$\log[HCO_3^-]$	-2.88	-3.40
$\log[CO_3^{2-}]$	-4.37	-5.40

Thus from the above calculations of inorganic ligand concentrations in estuarine waters, the following conclusions can be drawn:

Conclusions. For a ligand like fluoride where the free fluoride concentration in sea water is only seven times higher than in river water $\log[L]$ will not change appreciably upon mixing river water with sea water. For ligands like chloride and bromide where the concentrations are almost 3000 times higher in sea water than in river water there will be drastic changes especially below 5‰ salinity as can be seen from Figure 2. The effect of the river water alkalinity will be important only for the 'free' concentrations of the HCO_3^-, CO_3^{2-} and OH^- ligands.

6 TRACE METAL BINDING TENDENCIES WITH INORGANIC LIGANDS

The binding tendency of a trace metal may be estimated from the value of $\beta_n[L]^n$ since (see Section 2.1)

$$[ML_n] = [M]\beta_n[L]^n$$

where $\beta_0 = 1$ for [M]. If the $\beta_n[L]^n$ for two monovalent ions (e.g. Cl^- and OH^-) have values close to each other, mixed complexes may be formed. If they have not been measured, the constants of mixed complexes may be estimated statistically (see Dyrssen and Wedborg, 1974), e.g. for the ligands L and A

$$\beta_{m,n} = \frac{N!}{m!\,n!} \sqrt[N]{{}^L\beta_N^m \, {}^A\beta_N^n}$$

where $N = m + n$ and ${}^L\beta_N$ and ${}^A\beta_N$ are the stability constants for the ligands

L and A. The stability constants are in general taken from a compilation by Smith and Martell (1976) with ionic medium corrections in accordance with this compilation and Figure 1.

For our calculations we have selected biologically important trace metals for which the stability constants are comparatively well known. Some important trace metals such as titanium, vanadium, chromium, molybdenum, and tin are not included because of their complicated chemistry in aqueous solutions. Reduced forms of some elements are treated separately in Section 8.

6.1 Hydroxide and carbonate

In a natural surface water system, which usually can be regarded as a weakly buffered solution of hydrogen carbonate, a trace metal may react in three ways with the hydrogen carbonate ion. Stability constants for carbo-

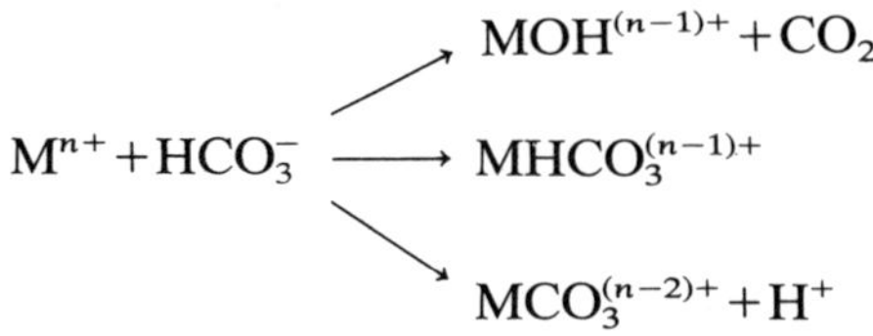

$$M^{n+} + HCO_3^- \longrightarrow \begin{cases} MOH^{(n-1)+} + CO_2 \\ MHCO_3^{(n-1)+} \\ MCO_3^{(n-2)+} + H^+ \end{cases}$$

nate complexes of trace metals have only been measured for UO_2^{2+}, Cu^{2+}, and Pb^{2+} (see Table 2), which indicates that hydrolysis of the trace metal usually has prevented an investigation of the complexation with HCO_3^- and CO_3^{2-}. Table 7 summarizes the values of $\log K_1[OH^-]$ for high and low alkalinity river waters. This table shows that the major cations in natural water (i.e. sodium, potassium, magnesium, and calcium) have small tendencies to hydrolyse in natural waters. This is also true for the monovalent trace metals lithium, rubidium, caesium, silver, and thallium. Some divalent trace metals have also a weak tendency to hydrolyse, namely manganese, iron, cobalt, nickel, and cadmium. The hydrolysis tendency of zinc is also rather weak. The values of $K_1[OH^-]$ for copper and lead seem to vary around 1 depending on the alkalinity of the river water. The following trace metals have a marked tendency to hydrolyse: beryllium, titanium, thorium, uranium (VI) (the uranyl ion), iron (III), cobalt (III), mercury (II), the methyl mercury ion, and aluminium. However, if the tendency to form complexes with other ligands (e.g. chloride, fluoride, carbonate) is more pronounced (i.e. has a larger value of $K_1[L]$) then hydroxide species may be of minor importance.

The tendency to hydrolyse in sea water will probably be less marked than in high alkalinity river water since K_1 is lower (see Figure 1) in sea water while $\log[OH^-]$ is practically the same (see Figure 2). As a rough guess we

Table 7. Values of $\log K_1$ at $I=0$ and $\log K_1[\text{OH}^-]$ for some metals in high and low alkalinity river waters (pOH $=5.5$ and 6.7).

Metal ion	$\log K_1$	$\log K_1[\text{OH}^-]$ HARW	LARW
Be^{2+}	8.6	3.1	1.9
Mg^{2+}	2.6	-2.9	-4.1
Ca^{2+}	1.3	-4.2	-5.4
TiO^{2+}	22.6	17.1	15.9
Th^{4+}	10.8	5.3	4.1
UO_2^{2+}	8.2	2.7	1.5
Mn^{2+}	3.4	-2.1	-3.3
Fe^{2+}	4.5	-1.0	-2.2
Co^{2+}	4.3	-1.2	-2.4
Ni^{2+}	4.1	-1.4	-2.6
Cu^{2+}	6.3	0.8	-0.4
Fe^{3+}	11.8	6.3	5.1
Co^{3+}	13.5	8.0	6.8
Ag^+	2.0	-3.5	-4.7
CH_3Hg^+	9.2	3.7	2.5
Zn^{2+}	5.0	-0.5	-1.7
Cd^{2+}	3.9	-1.6	-2.8
Hg^{2+}	10.6	5.1	3.9
Pb^{2+}	6.3	0.8	-0.4
Al^{3+}	9.0	3.5	2.3

will use $\log K_1 - 0.8$ for tri- and tetravalent ions and $\log K_1 - 0.5$ for divalent ions in sea water.

Since uranium (VI), copper (II), and lead (II) have a tendency to form both hydroxide and carbonate complexes we have calculated the values of $\log \beta_n[\text{L}]^n$ for these trace metals. The results are presented in Table 8. For uranium (VI) UO_2CO_3 will dominate in low alkalinity river water together with UO_2OH^+ (and probably $UO_2(OH)_2$) and $UO_2(CO_3)_2^{2-}$. In high alkalinity river water and in sea water the carbonate complexes will dominate, especially $UO_2(CO_3)_2^{2-}$ and $UO_2(CO_3)_3^{4-}$. Lead has a smaller tendency to form carbonate complexes than copper, but a greater tendency to form chloride complexes. In low alkalinity river water $Cu(OH)_2$ and Pb^{2+} will dominate together with Cu^{2+}, $CuCO_3$, $CuOH^+$ and $PbOH^+$. In high alkalinity river water $Cu(OH)_2$, $CuCO_3$, $PbCO_3$ and $PbOH^+$ will dominate together with $CuOH^+$, $CuHCO_3^+$, and $PbHCO_3^+$. In sea water $Cu(OH)_2$, $PbCl^+$ and $PbCl_2$ will dominate (cf. Zirino and Yamamoto, 1972; and Paulson, 1978) together with $CuCO_3$, $CuOH^+$, $Cu(CO_3)_2^{2-}$, $PbCO_3$ and $PbOH^+$.

In general the stability constants decrease for each step in the formation of the saturated complex. Quite frequently, however, the two first steps in

D. Dyrssen and M. Wedborg

Table 8. Values of $\log \beta_n [L]^n$ for hydroxide and carbonate complexes of the divalent uranyl, copper, and lead complexes. The values of pL are given in Table 6.

	UO_2^{2+}	Cu^{2+}	Pb^{2+}
Sea water			
M^{2+}	0	0	0
MOH^+	2.2	0.3	0.3
$M(OH)_2$	<2.2	0.2–1.7	−0.8
$MHCO_3^+$	0.1	−0.6	−0.6
MCO_3	4.1	1.2	0.4
$M(CO_3)_2^{2-}$	6.1	0.3	−1.4
$M(CO_3)_3^{4-}$	7.1		
MCl^+	−0.6	−0.4	0.9
MCl_2			1.0
MSO_4	−0.1	−1	−1
HARW			
M^{2+}	0	0	0
MOH^+	2.7	0.8	0.8
$M(OH)_2$	<2.7	1.0–2.6	0.1
$MHCO_3^+$	1.0	0.3	0.3
MCO_3	4.6	1.7	0.9
$M(CO_3)_2^{2-}$	5.5	−0.2	−1.9
$M(CO_3)_3^{4-}$	5.6		
MCl^+	−3.5	−3.3	−2.1
LARW			
M^{2+}	0	0	0
MOH^+	1.5	−0.4	−0.4
$M(OH)_2$	<1.5	−1–0.3	−2.2
$MHCO_3^+$	−0.1	−0.8	−0.8
MCO_3	2.5	−0.3	−1.2
$M(HCO_3)_2^{2-}$	1.4	−4.3	−6.0
$M(CO_3)_3^{4-}$	−0.6		
MCl^+	−3.5	−3.3	−2.1

the hydrolysis of a metal ion are close to each other, and in the case of $HgOH^+$ and $Hg(OH)_2$ $\log K_2 = 11$ is larger than $\log K_1 = 10$. This also seems to be the case for $CuOH^+$ and $Cu(OH)_2$ ($\log K_1 = 5.8$ and $\log K_2 = 7$ for sea water, although Paulson obtained a value of $\log K_2 = 5.5$). For trace metals with very high values of $\log K_1 [OH^-]$ like titanium (IV) (see Table 7) the saturated complex will dominate, i.e. $Ti(OH)_4$.

Metal ions have a tendency to form polynuclear cationic hydroxide complexes, e.g. $Cu_2(OH)_2^{2+}$, $Fe_2(OH)_2^{4+}$, $Al_3(OH)_4^{5+}$, as a prestep to the formation of colloids and precipitates of metal hydroxides by hydrolysis (e.g. $2Cu^{2+} + 2H_2O \rightarrow Cu_2(OH)_2^{2+} + 2H^+$). Mixed hydroxides with several different metal ions may also be formed. In addition hydroxides like FeOOH or

$Fe(OH)_3$ have a marked tendency to adsorb trace metals. Otherwise the concentration of trace metal ions is generally too low and the pH too high to allow the formation of highly charged polynuclear complexes in natural waters. In order to see how well the metal ion is neutralized in natural waters one may calculate the values of $\log \beta_n [OH^-]^n$ for the mononuclear complexes. This has been done for Fe^{3+}, Al^{3+}, and Th^{4+} in Table 9. The conclusions one can draw from this table are that $Fe(OH)_3$ dominates in both sea water (SW) and low alkalinity river water (LARW), but since $[Fe(OH)_2^+] > [Fe(OH)_4^-]$ in LARW and $[Fe(OH)_2^+] < [Fe(OH)_4^-]$ in SW this can be taken as an indication that colloidal $Fe(OH)_3$ may reverse its charge in an estuary from $+$ to $-$. In the case of aluminium, $Al(OH)_4^-$ will be the dominating species in sea water and $[Al(OH)_4^-]$ is always greater than $[Al(OH)_2^+]$. For thorium $Th(OH)_3^+$ and $Th(OH)_4$ are the dominating species in SW. In river water saturated with $Fe(OH)_3$ and $Al(OH)_3$ the concentrations can be calculated from $K_{s0}\beta_3$. For iron $[Fe(OH)_3] = 10^{-7}$ M $= 5.6$ μg/l. and for aluminium $[Al(OH)_3] = 10^{-6.5}$ M $= 8.5$ μg/l. These concentrations are considerably lower than average total concentrations of 670 μg Fe/l. and 650 μg Al/l. Thus only a small part of these trace metals can be in the dissolved form. In the case of thorium the total concentration of dissolved thorium in river water will be

$$Th_t = [Th(OH)_3^+] + [Th(OH)_4] = [Th^{4+}](10^{11.1} + 10^{11.3}) = [Th^{4+}] \times 3.25 \times 10^{11}$$

$$[Th^{4+}][OH^-]^4 = 10^{-44.7}$$

$$[OH^-] = 10^{-6.67}$$

Table 9. Formation constants and values of $\log \beta_n [OH^-]^n$ for hydroxide complexes of iron (III), aluminium and thorium in sea water and low alkalinity river water. Constants in parenthesis are rough estimates from data compiled by Smith and Martell (1976).

	Fe^{3+}		Al^{3+}		Th^{4+}	
	$I = 0$	$I = 0.5$	$I = 0$	$I = 0.5$	$I = 0$	$I = 0.5$
$\log \beta_1$	11.8	11.01	9.0	8.48[b]	10.8	9.6[a]
$\log \beta_2$	22.3	(21.2)	(18.7)	(17.6)[b]	21.1	19.9[a]
$\log \beta_3$	(31.8)	(30.5)	(27.0)	(25.7)[b]	(31.1)	(30)
$\log \beta_4$	34.4	(33.1)	33.0	(31.7)	(38)	(37)
$\log K_{s0}$	-38.8		-33.5		-44.7	
	SW	LARW	SW	LARW	SW	LARW
MOH	5.5	5.1	3.0	2.3	4.1	4.1
$M(OH)_2$	10.1	9.0	6.5	5.4	8.8	7.8
$M(OH)_3$	*13.9*	*11.8*	9.1	7.0	13.4	*11.1*
$M(OH)_4$	11.0	7.7	9.6	6.3	*15.0*	*11.3*

[a] $I = 1.0$ M
[b] $I = 0.1$ M

These equations give $Th_t = 10^{-6.51}$ M $= 72$ $\mu g/l$. of which 61.3% is $Th(OH)_4$ and 38.7% $Th(OH)_3^+$. The average concentration of thorium is <0.1 $\mu g/l$., which indicates that thorium is absorbed on particulate matter (e.g. colloidal $Fe(OH)_3$). Beneš and Steinnes (1974, 1975) showed that 26% of the thorium in river water could diffuse into a bag of Visking tubing with a mean pore diameter of 4.8 nm.

6.2 Chloride and bromide

According to Figure 2 there is a drastic increase in $\log[Cl^-]$ and $\log[Br^-]$ upon mixing river water with sea water. For metal ions that hydrolyse in river water ($\log K_{OH} > 7$) and form strong chloride complexes ($\log K_{Cl} > 3$) this could lead to a ligand exchange of OH^- for Cl^-. Table 10 summarizes the values of the first stability constant of chloride (K_1). The constants for both hydroxide and chloride have been grouped in different ranges in Table 11. From this table we may see that the group Mg^{2+}, Ca^{2+}, Mn^{2+}, Fe^{2+}, Co^{2+}, and Ni^{2+} will form only weak chloride complexes and not be hydrolysed to any great extent. In the next group the *uranyl ion* and *aluminium* have been discussed above and *beryllium* should most likely appear as $Be(OH)_2$. Chloride complexing can be neglected since $K_{OH}[OH^-] \gg K_{Cl}[Cl^-]$

Table 10. Stability constants for chloride complexes according to Smith and Martell (1976). *I* is given in moles/l. (M).

Metal ion	$\log K_1$	I
Be^{2+}	-0.3	0.7
Mg^{2+}	$-1.0\,(-0.4)$	3(0.7)
Ca^{2+}	0.08	0.7
Th^{4+}	1.4 (0.3)	0(0.5)
UO_2^{2+}	0.2 (−0.1)	0(1)
Mn^{2+}	0.04	1
Co^{2+}	-0.05	1
Ni^{2+}	0	1
Cu^{2+}	0.4(−0.06)	0(3)
Fe^{3+}	1.5(0.8)	0(3)
Co^{3+}	1.4	3
Ag^+	3.3	0
CH_3Hg^+	5.2	0.1
Zn^{2+}	0.5	0
Cd^{2+}	2.0	0
Hg^{2+}	6.7	0.5
Pb^{2+}	1.6	0

Table 11. Stability constants at $I = 0$ and $25\,°C$ for hydroxide and chloride complexes.

$\log K_{OH} \leq 4.5$		$6 < \log K_{OH} < 7$	$8 < \log K_{OH} < 10$	$\log K_{OH} > 10$	
Mg^{2+}, Ca^{2+} $Mn^{2+}, Fe^{2+}, Co^{2+}, Ni^{2+}$			Be^{2+}, UO_2^{2+} Al^{3+}	TiO^{2+} Ga^{3+}	$\log K_{Cl} < 0.25$
Zn^{2+}, Cd^{2+}	Cu^{2+}, Pb^{2+}			Th^{4+} Fe^{3+}, Co^{3+} In^{3+}	$0.25 < \log K_{Cl} < 3$
Ag^+		CH_3Hg^+		Hg^{2+} Tl^{3+}	$\log K_{Cl} > 3$

in both river and sea water. This is also true for *titanium* and *gallium* that should appear as $Ti(OH)_4$ and $Ga(OH)_3$ ($\log \beta_3 = 31.7$) and $Ga(OH)_4^-$ ($\log \beta_4 = 39.4$). Hydroxide complexes should also dominate for *thorium* (see above), *iron* (III) (see above), *cobalt* (III), and *indium* (for which $In(OH)_3$ dominates). The silver ion is only weakly hydrolysed and a calculation of chloride and bromide complexation shows that Ag^+ (70%) and $AgCl$ (30%) will dominate in river water while $AgCl_2^-$ (35%), $AgCl_3^{2-}$ (42%) and $AgCl_4^{3-}$ (23%) will dominate in sea water. In an estuary with a salinity of 5‰ $AgCl_2^-$ (78%) will dominate besides $AgCl$ (10%) and $AgCl_3^{2-}$ (12%). Chloride complexes of silver thus become dominating at very low salinities. *Copper* and *lead* were treated above (see Table 8) and it was found that for lead $PbCl^+$ and $PbCl_2$ may dominate in sea water. From Tables 7 and 10 it can be seen that *zinc* forms weaker chloride complexes than *cadmium*, but is more easily hydrolysed. Table 12 summarizes the approximate speciation percentages of these two trace metals. The carbonate complexation of zinc and cadmium is not known, but most likely they form somewhat stronger complexes than magnesium, but weaker than copper and lead; our guess is $\log K_1 = 1.8$ and 0.8 for HCO_3^- and $\log K_1 = 4.5$ and 3.0 for CO_3^{2-} in river and sea water, respectively. The results of the speciation calculations in Table 12 show that the complexation of Zn^{2+} and Cd^{2+} is not very strong except for cadmium in sea water where extensive formation of chloride complexes occur ($CdCl^+$, $CdCl_2$ and $CdCl_3^-$). Carbonate complexes appear in high alkalinity river water, where also $ZnOH^+$ and $Zn(OH)_2$ are formed. In sea water Zn^{2+} and $ZnCl^+$ dominate (40 and 22%, respectively). The stability constants are not very well known and it is easy to obtain quite different percentages as shown for sea water by Dyrssen and Wedborg (1974), Zirino and Yamamoto (1972) and Ahrland (1975). In spite of these differences the general tendency for inorganic complexing of zinc and

Table 12. The percentage of different inorganic zinc and cadmium complexes in sea water and high and low alkalinity river waters.

Complex	Zinc			Cadmium		
	SW	HARW	LARW	SW	HARW	LARW
M^{2+}	40	34	96	2	69	96
MOH^+	4	11	2	negl.	2	negl.
$M(OH)_2$	6	42	0.6	negl.	negl.	negl.
$MHCO_3^+$	0.4	3	0.6	negl.	5	0.5
MCO_3	4	11	0.2	0.2	22	negl.
MSO_4	5	0.4	1	0.3	1	1
MCl^+	22	negl.	negl.	41	1	2
MCl_2	6	negl.	negl.	36	negl.	negl.
MCl_3^-	negl.	negl.	negl.	20	negl.	negl.
$MOHCl$	12	negl.	negl.	0.1	negl.	negl.

cadmium in river and sea water may be seen from the percentages in Table 12.

Methyl mercury, mercury (II), and thallium (III)

These all form both strong hydroxide, chloride, and bromide complexes. In the case of methyl mercury CH_3HgCl will dominate in sea water and CH_3HgOH in river water as can be seen in Table 13. Table 14 shows that there will be drastic changes in the complexing of mercury (II) in an estuary. With increasing salinity the complexing will be shifted from $Hg(OH)_2$ in river water to $HgOHCl$ in low salinity waters. In sea water there will be a mixture of chloride and bromide complexes ($HgCl_4^{2-} > HgCl_3Br^{2-} \sim HgCl_3^- > HgCl_2Br^- \sim HgCl_2 \sim HgClBr$).

Table 13. The values of $\log K_1[L]$ and percentage (in parenthesis) complexation of methyl mercury in sea water and high and low alkalinity river water.

	SW	HARW	LARW
CH_3Hg^+	0 (negl.)	0 (negl.)	0 (negl.)
CH_3HgOH	3.7 (6%)	3.9 (99.5%)	2.7 (92%)
CH_3HgCl	4.9 (92%)	1.6 (0.5%)	1.6 (7%)
CH_3HgBr	3.3 (2%)	0.1 (negl.)	0.1 (negl.)

Table 14. Values of $\log \beta_n [L]^n$ for mercury (II) and thallium (III) in sea and river water.

Complex	Mercury (II)		Thallium (III)	
	SW	RW	SW	RW
M	0	0	0	0
MOH	4.5	3.9–5.1	7.3	6.7–7.9
$M(OH)_2$	10	*8.5–10.8*	14.2	13.1–15.4
$M(OH)_3$	negl.	negl.	*21.0*	18.7–22.2
$M(OH)_4$	negl.	negl.	*18.9*	14.3–19
MCl	6.5	3.6	6.5	4.0
MCl_2	*12.7*	6.6	11.3	6.1
MCl_3	*13.4*	3.8	13.7	5.4
MCl_4	*14.1*	1.4	15.3	3.5
MBr	5.9	3	5.8	3.4
MBr_2	11	4.9	10.3	5.1
MBr_3	10.2	0.7	11.8	3.6
MBr_4	8.8	−3.9	11.9	0.1
MClBr	*12.6*	6.5	11.1	5.9
MCl_2Br	*12.7*	3.2	13.5	5.3
MCL_3Br	*13.4*	0.7	15.1	3.3
MOHCl	11.7	7.7–8.9[a]	13.0	9.7–10.9
$M(OH)_2Cl$	negl.	negl.	*19.8*	14.1–16.4
$M(OH)_3Cl$	negl.	negl.	*18.6*	11.7–15.2
$M(OH)_2Cl_2$	negl.	negl.	17.7	8.6–10.9
$MOHCl_2$	negl.	negl.	17.0	8.9–10.1

[a] At a salinity of 2‰ $\log \beta_{11}[OH^-][Cl^-]$ will be 10.3–11 while $\log \beta_3[Cl^-]^3 \approx \log \beta_4[Cl^-]^4 \approx 10$

6.3 Fluoride

From the calculations of the 'free' ligand concentration in Section 4.4 it was seen that about 50% of the total fluoride in sea water was complexed by magnesium while practically all (98%) was 'free' in river water. Table 15 summarizes the tendency for metal ions to form fluoride complexes. In general the trace metal ions that form strong fluoride complexes also have a marked tendency to hydrolyse in natural waters. From Figure 2 and the adjacent text it can be seen that $\log[F^-] - \log[OH^-] \leq 1.37$ also in low alkalinity river water. Even for those elements that form strong fluoride complexes $\log K_1 = \log \beta_1$ for OH^- is at least two units greater than $\log K_1$ for F^- (see Table 16). The difference in $\log \beta_n$ increases rapidly with n. This implies that $Al(OH)_3$, $Al(OH)_4^-$, $Th(OH)_3^+$ and $Th(OH)_4$ will dominate over any fluoride complex. Table 17 confirms that conclusion.

  D. Dyrssen and M. Wedborg

Table 15. Tendency of metal cations to form fluoride complexes. The major cations are printed in italics.

Li^+, Na^+, K^+	Very weak tendency
Ba^{2+}	$\log K_1 < 0$ [a]
Ag^+	
Ca^{2+}, Sr^{2+}	Weak tendency
Mn^{2+}, Fe^{2+}, Co^{2+}, Ni^{2+}, Cu^{2+}	$0 < \log K_1 < 1$
Cd^{2+}	
Mg^{2+}	Moderate tendency
CH_3Hg^+, Hg^{2+}	$1 < \log K_1 < 2$
Zn^{2+}, Pb^{2+}	
Be^{2+}, UO_2^{2+}	Marked tendency
Fe^{3+}, Ga^{3+}, In^{3+}	$4 < \log K_1 < 6$
Al^{3+}, Th^{4+}	Strong tendency
	$\log K_1 > 6$

[a] $\log K_1 = [MF^{(n-1)+}]/[M^{n+}][F^-]$. For $M^{n+} = H^+$ $\log K_1 = 3$

6.4 Sulphate

'Free' divalent trace metal ions in sulphate containing natural waters will form sulphate complexes to a certain extent. Since the complexing takes place through ion pairing of the hydrated ions the stability constants for MSO_4 will be approximately 10 in sea water and 100 in river waters. Thus $K_1[SO_4^{2-}] = [MSO_4]/[M^{2+}] = \frac{1}{8}$ in sea water and $\frac{1}{70}$ in river water. This will

Table 16. Values of $\log K_1[L]$ for hydroxide and fluoride complexes of tracemetals in sea water and low alkalinity river water. The values of pF and pOH are given in Table 6.

	SW		LARW	
Metal ion	OH^-	F^-	OH^-	F^-
Cu^{2+}	0.3	−3.8	−0.4	−4.1
Cd^{2+}	−2.1	−4.0	−2.8	−4.3
CH_3Hg^+	3.2	−3.1	2.5	−3.6
Hg^{2+}	4.6	−3.4	3.9	−3.7
Zn^{2+}	−1.0	−3.7	−1.7	−4.2
Pb^{2+}	0.3	−3.0	−0.4	−3.3
Be^{2+}	2.6	0.3	1.9	0
UO_2^{2+}	2.2	−0.1	1.5	−0.4
Fe^{3+}	5.5	0.7	5.1	0.7
Ga^{3+}	5.9	0	5.5	0
In^{3+}	3.7	−0.7	3.3	−0.7
Al^{3+}	2.7	1.7	2.3	1.7
Th^{4+}	4.1	3.1	4.1	3.1

Table 17. Values of $\log \beta_n [L]^n$ for hydroxide and fluoride complexes of aluminium and thorium in low alkalinity river water.

	Aluminium		Thorium	
	OH^-	F^-	OH^-	F^-
$\log \beta_1 l$	2.3	1.7	4.1	3.1
$\log \beta_2 l^2$	5.4	2.0	7.8	4.5
$\log \beta_3 l^3$	7.0	0.8	11.1	3.9
$\log \beta_4 l^4$	6.3	−2.1	11.3	2.0
$\log \beta_5 l^5$	n.f.	−6.0	n.f.	n.f.
$\log \beta_6 l^6$	n.f.	−10.9	n.f.	n.f.

n.f. = not formed

concern divalent trace metal ions that do not have a marked tendency to hydrolyse or form chloride complexes (see Table 11), i.e. Mn^{2+}, Fe^{2+}, Co^{2+}, Ni^{2+} and Zn^{2+} and Cd^{2+} in river waters (Table 12).

Thus to summarize: Several important ions (Mn^{2+}, Fe^{2+}, Co^{2+} and Ni^{2+}) have a rather weak tendency to form inorganic complexes in sea and river waters. Zinc and cadmium also have rather weak tendencies to form inorganic complexes in estuarine water, but the percentage of free cadmium ion becomes quite small in sea water due to extensive complexation by chloride. Copper and lead have a moderate tendency to form complexes with almost all inorganic ligands which leads to a complex mixture. The calculated percentage speciation will vary with the slight variations in the choice of stability constants. The uranyl ion has a marked tendency to form carbonate complexes even in low alkalinity river waters. Titanium (IV), thorium, iron (III) and aluminium are extensively hydrolysed. Methyl mercury will form CH_3HgOH in river waters, but has a marked tendency to form CH_3HgCl at low salinities. Mercury will be hydrolysed in river waters (i.e. $Hg(OH)_2$), but will exchange the OH^- ligand for Cl^- and Br^- at rather low salinities. $HgCl_4^{2-}$, $HgCl_3Br^-$ and $HgCl_3^-$ will dominate in sea water. Fluoride cannot compete with hydroxide. Sulphate complexing occurs to some extent with Mn^{2+}, Fe^{2+}, Co^{2+}, Ni^{2+} and Zn^{2+} in sea water.

7 COMPLEXING BY ORGANIC LIGANDS IN ESTUARINE WATERS

In general there is more 'dissolved' organic matter than particulate organic matter in rivers, estuaries and coastal waters (cf. Head, 1976). The 'dissolved' organic matter may also consist of colloidal matter since a 0.45 μm pore size filter is often used to separate the two fractions. We shall use a rough value of 10 mg carbon/l. for the 'dissolved' organic matter and

5 mg/l. for the particulate organic matter. Part of the organic matter is lost in the estuary by coagulation of colloidal material (cf. Burton, 1976). The composition of the organic matter is not very well known. Part of the material (e.g. carbohydrates) has very low metal complexing ability. The concentration of individual, dissolved carboxylic and amino acids is usually low ($<10^{-6}$ M, cf. Williams, 1975). This concentration is too low to form any appreciable amounts of trace metal complexes (cf. Dyrssen and Wedborg, 1974). In river waters about 50% of the organic carbon may be humic substances. According to the model for fulvic and humic acids and insoluble humin material (cf. Gamble and Schnitzer, 1973, the links must, however, be covalent; hydrogen and metal ions may though have a cross-linking function) the polymers consist of ligands of the following type (protonated form):

COOH	OH	OH, OH
MW = 122	MW = 94	MW = 110
CHO, OH	OH, COOH	COOH, COOH
MW = 122	MW = 138	MW = 166

The mean number of carbon of these five ligands is 6.8 and the mean molecular weight 125. Thus 7.5 mg C/l. corresponds to a total ligand concentration of $0.0075/12 \times 6.8 = 9.2 \times 10^{-5}$ M. In a polymer with ten to several thousands of these ligand units there will certainly be very strong metal binding sites, which in part may be due to the high negative charge of the site by dissociation of several carboxylic acid groups and in part may be due to a favourable steric arrangement of the chelating groups. The complexing ability of the simple ligands (see Table 18) will, however, show the tendency for metal complexing by humic material. The concentration level of 9.2×10^{-5} M may be compared with the average river concentration (cf. Liss, 1976) of 650 μg Al/l. $= 2.4 \times 10^{-5}$ M, of 670 μg Fe/l. $= 1.2 \times 10^{-5}$ M and of 7 μg Cu/l. $= 1.1 \times 10^{-7}$ M. Aluminium and iron may therefore cover a considerable fraction of the metal-binding sites. The tendency of aluminium and iron (III) to form positively charged hydroxo polymers is, however, very pronounced below the isoelectric pH (cf. Ottewill and Holloway, 1975), and the aggregation with the negatively charged humic material may only involve the surface sites.

Table 18. Stability constants for acetate (1), phthalate (2), hydrogen phthalate (3), salicylate (4), hydrogen salicylate (5), tironate (6) and hydrogen tironate (7).

Metal ion	(1)	(2)	(3)	(4)	(5)	(6)	(7)
H^+	4.6	4.9	2.8	13.4	2.8	12.5	7.6
Be^{2+}	1.6	4.0		12.4		12.9	4.2
Mg^{2+}	0.5					6.9	2.0
Ca^{2+}	0.5	1.1		(4.5)	0.2	5.8	2.2
Th^{4+}	3.9	5.9			4.3		
UO_2^{2+}	2.4	4.4		12.1		(15.9)	6.4
Mn^{2+}	0.7					8.6	2.3
Fe^{2+}	0.7			6.6			
Co^{2+}	0.8	1.5				9.5	3.1
Ni^{2+}	0.8	1.7	0.7			10.0	3.0
Cu^{2+}	1.7	2.8	1.2	10.6		14.3	5.5
Fe^{3+}	3.2	(5.2)		16.4	4.4	20.4	9.7
Ag^+	0.4						
CH_3Hg^+	3.2						
Zn^{2+}	0.6	2.2		6.9		10.4	3.3
Cd^{2+}	1.2	2.5		5.6		9.1	
Hg^{2+}	5.6					19.9	
Pb^{2+}	2.1					(13.4)	
Al^{3+}	1.5	3.2		12.9		16.6	

The ligand concentration of fulvic and humic acids depends on the degree of polymerization. Inside a polymer the concentration may be comparable with an ion exchange resin, e.g. 1 to 10 equivalents per litre (cf. Gamble and Schnitzer, 1973), but as low molecular weight monomers the total concentration will approach 9.2×10^{-5} M. For each individual monomer the concentration is of course considerably lower. Trace metals may be bound to polymeric forms if they can compete favourably with protons and the major cations, and if they are not too extensively complexed by OH^-, Cl^- and CO_3^{2-} ions. The pH of the water and the concentration of major cations will thus be important for the trace metal content of humic material.

It may therefore be concluded that the concentration of individual organic ligands is in general too low to form any appreciable amounts of trace metal complexes. However, inside a polymer the ligand concentration may be quite high and ion exchange reactions may occur with adsorbed trace metal ions and inorganic complexes. The exchange reaction that thus takes place between a trace metal and humic material can be written as follows:

$$M^{n+} + H_x(Na, K)_y(Mg, Ca)_z L \rightleftharpoons MH_{x-s}(Na, K)_{y-t}(Mg, Ca)_{z-u} L$$
$$+ sH^+ + t(Na, K)^+ + u(Mg, Ca)^{2+}$$

where $s + t + u = n$, i.e. for iron (II) $n = 3$ and Fe^{3+} may thus replace three

                    D. Dyrssen and M. Wedborg

monovalent ions (H^+, Na^+, or K^+) or one monovalent ion and one divalent ion (Mg^{2+} or Ca^{2+}). The sorption of a trace metal will thus depend on the pH and the concentrations of major cations, which increase with the salinity. Trace metals originally present in the humic material or taken up in river water may therefore be desorbed in estuarine waters.

Stability constants for acetic acid derivatives are shown in Table 18.

8 REDOX FORMS OF COMMON ELEMENTS

Most estuarine waters are well aerated. Thus the partial pressure of oxygen in the atmosphere ($p = 0.21$ atm) should set a stable value of pE in surface waters:

$$0.5\ O_2(g) + 2H^+ + 2e^- \rightleftharpoons H_2O$$

$$K = [H^+]^{-2}[e^-]^{-2}p_{O_2}^{-0.5} = 10^{41.55}$$

For pH = 8, i.e. $[H^+] = 10^{-8}$ M

$$[e^-]^{-2} = 10^{41.55}10^{-16}0.21^{0.5}$$

$$-\log[e^-] = pE = 12.6$$

$$Eh = pE(RT(\ln 10)/F) = 12.6 \times 0.059155 = +0.746\ \text{V}$$

A platinum electrode will only show a potential of $+0.45$ V (cf. Breck, 1974 and Parsons, 1975), which most likely is due to the slow kinetics in the breaking of the O_2 bond and the formation of O_2 from two H_2O. Such slow reactions are not involved in the oxygen–peroxide equilibrium:

$$O_2(g) + 2H^+ + 2e^- \rightleftharpoons H_2O_2$$

$$K = [H_2O_2][H^+]^{-2}[e^-]^{-2}p_{O_2}^{-1} = 10^{23.5}$$

For $[H^+] = 10^{-8}$ M and $[H_2O_2] = 10^{-8}$ M

$$[e^-]^{-2} = 10^{23.5}10^{-16}10^{8}0.21$$

$$-\log[e^-] = pE = 7.4$$

$$Eh = +0.44\ \text{V}$$

This value is close to the value found in natural waters, which implies that the potential of the platinum electrode is set by the formation of a surface layer of hydrogen peroxide with a concentration of 10^{-8} M. The current (electron flow) is, however, very small which results in a low pE buffer capacity and considerable fluctuations in the potential readings. This situation can hardly be improved by using a membrane electrode with a $Fe(CN)_6^{3-}/Fe(CN)_6^{4-}$ buffer (cf. Breck, 1974).

8.1 Reducing conditions

Even if the general condition of estuarine waters is thus oxidizing the elements can be exposed to reducing conditions. The main reducing agent is organic matter which is subjected to bacterial decay also in the sediment surface layer. The decay and diagenesis (cf. Berner, 1971 and Murray *et al.*, 1978) that takes place below the surface layer by entrained water and downward diffusion of oxygen and sulphate and upward diffusion of reduction products is usually too slow to influence the estuarine water on an annual (seasonal) time scale. Thus old organic matter such as humic material most likely reacts too slowly, and sunken photosynthetic matter together with dead organisms, fecal products and organic waste should be the main oxygen consumers. In some special situations the estuarine water is stagnant and saline (e.g. in fjords and basins). The water then becomes depleted of its oxygen and nitrate within a week or two and the sulphate will be the electron acceptor for the bacterial decay of the organic matter. Only in recent years the oxidation rates per unit bottom area have been measured (cf. Rowe *et al.*, 1975, Almgren *et al.*, 1975, Granéli, 1977, Dyrssen and Hallberg, 1978, Balzer, 1978, and Jansson and Hagström, 1978). The carbohydrate part of the organic matter produces carbon dioxide and hydrogen sulphide according to the following reactions;

$$O_2 + CH_2O \rightarrow CO_2 + H_2O$$
$$SO_4^{2-} + 2CH_2O \rightarrow 2HCO_3^- + H_2S$$

The carbon dioxide produced by oxygen consumption will lower the pH and may dissolve some metal hydroxides and carbonates. The hydrogen sulphide may either rise as gas bubbles several metres in the estuarine water before it is chemically oxidized to a mixture of sulphur, thiosulphate, sulphite and sulphate or stay in the sediment as a metal sulphide (e.g. FeS). Quite frequently a layer of yellowish sulphur (contained in the microorganism Beggiatoa, cf. Goldhaber and Kaplan, 1974) can be observed on the sediment surface.

The protein part of the organic matter will primarily be hydrolysed and degraded to ammonium–ammonia, which subsequently can either be used as a nutrient or be nitrified by bacteria to nitrate. Gaseous nitrogen (N_2) and dinitrogen oxide (N_2O) may also be formed. Organic phosphate is usually hydrolyzed into soluble inorganic phosphate. Part of this phosphate may, however, be bound to the sediment as hydroxophosphates of calcium and iron (II, III).

The electrode potential of a sulphide containing (reduced) sediment may be estimated from the following equilibria assuming the presence of solid

                     D. Dyrssen and M. Wedborg

iron (II) sulphide and sulphur:

$$FeS(s) + H^+ \rightleftharpoons HS^- + Fe^{2+}; \qquad \log K = -3.1$$

$$S(s) + 2H^+ + 2e^- \rightleftharpoons H_2S; \qquad \log K = 4.8$$

$$H^+ + HS^- \rightleftharpoons H_2S; \qquad \log K = 7.0$$

These constants have been taken from Dyrssen and Hallberg (1978), Sillén and Martell (1964) and Almgren *et al.* (1976), respectively. The pH of oxygen depleted sediments is usually low, in the order of 7.5. We shall assume that the salinity of the estuarine bottom water is 20‰ and that 1% of this sulphate is in the form of H_2S and HS^-. Thus

$$[H^+] = 10^{-7.5} \text{ M}$$

$$[HS^-] + [H_2S^-] = 0.01 \times 0.02889 \times 20/35 = 1.65 \times 10^{-4} \text{ M}$$

$$= [HS^-](1 + 10^7[H^+]) = 1.316[HS^-]$$

$$[Fe^{2+}] = 10^{-3.1}[H^+]/[HS^-] = 10^{-6.7} \text{ M} = 11 \ \mu\text{gFe/l.}$$

$$[H_2S]/[H^+]^2[e^-]^2 = 10^{4.8}$$

$$2pE = 4.8 - 2pH - \log[H_2S] = -5.8$$

$$pE = -2.9; \qquad Eh = -2.9 \times 0.059155 = -0.171 \text{ V}$$

This value is close to the values found by Bågander and Niemistö (1978) for reducing sediment layers in the Baltic. Cf. also Whitfield (1969).

8.2 Redox forms

Many elements do not change their oxidation state during biological and geological cycling. This is true for most of the major elements. In Table 19

Table 19. Oxidized and reduced forms of elements in natural waters.

Oxidized forms	Reduced forms
O_2	H_2O
CO_2—HCO_3^-—CO_3^{2-}	CO, $(CH_2O)_n$, C, $(CH_2)_n$, CH_4
NO_3^-	NO_2^-, N_2O, N_2, $NHCO$ (protein), $NH_4^+ - NH_3$
SO_4^{2-}	H_2S—SH^-—$MS - MSH^+$, org-S
$H_2VO_4^-$—HVO_4^{2-}	VO^{2+}—$VO(OH)^+$, $V(OH)_2^+$
CrO_4^-	$Cr(OH)_3$
MnO_2	Mn^{2+}
$Fe(OH)_3$	Fe^{2+}
$Co(OH)_3$	Co^{2+}
$HAsO_4^{2-}$—$H_2AsO_4^-$	$As(OH)_3$
IO_3^-	I_2, I^-

we have summarized redox forms of elements that do appear in different forms in river and coastal waters. Oxygen, carbon, nitrogen, and sulphur are all involved in the decay or organic matter, which has been treated above. Vanadium is interesting as a blood pigment in tunicates (sea squirts, ascidians), which may be involved in the synthesis of their cellulose tunic. Cobalt occurs in the important vitamin B-12. Iodine is also biologically important and its different redox forms have been studied quite extensively by Wong and Brewer (1974, 1976, 1977) among others. Chromium and arsenic are common industrial pollutants in estuaries. Chromium (III) and cobalt (II) form very strong and inert complexes and are consequently converted into other forms very slowly. In an air-saturated estuarine water with $pE = 12.6$ and $pH = 8$ the thermodynamically stablest form may be calculated from the constants for the ligand e^- (redox equilibria) compiled by Sillén and Martell (1964). Since the constants often have been derived by a combination of thermodynamic data according to Latimer their actual value in an estuarine medium may differ by one or two orders of magnitude. This is not a serious limitation since the speciation calculation in nearly every case gives a clear-cut answer to which oxidation state is stablest under oxidizing and under reducing ($pE = -2.9$ and $pH = 7.5$) conditions.

8.3 Iodine

The iodide–iodine constant gives

$$[I^-]^2/[I_2][e^-]^2 = 10^{21.04}$$

Since Wong and Brewer (1974) have shown that $[I^-] \leq 10^{-6.7}$ M it follows that $[I_2] \leq 10^{-9.24}$ M for $pE = 12.6$ and still less at $pE = -2.9$. The disproportionation equilibrium between iodate-iodide and iodine is fairly well known. Thus

$$[I_2]^3/[IO_3^-][H^+]^6[I^-]^5 = 10^{46.55}$$

Combining these two equations gives

$$[I^-]/[IO_3^-][H^+]^6[e^-]^6 = 10^{109.67}$$

or

$$[I^-]/[IO_3^-] = 10^{-13.93}$$

in an oxidizing environment and

$$[I^-]/[IO_3^-] = 10^{82.07}$$

in a reducing environment. This reducing environment could be furnished inside an alga, which could then give off iodide to the water where slow oxidation to iodate takes place. The rates of these different processes will determine the actual values of $[IO_3^-]/[I^-]$, which may be determined analytically according to Wong and Brewer (1974, 1976, 1977).

8.4 Arsenic

The redox potential between arsenic (III) and (V) is set by

$$[As(OH)_3]/[H_3AsO_4][H^+]^2[e^-]^2 = 10^{18.9}$$

At $pH = 8$ $HAsO_4^{2-}$ is the dominant form of arsenic (V) and from Smith and Martell (1976) we obtain

$$[H_3AsO_4]/[H^+]^2[HAsO_4^{2-}] = 10^{9.2}$$

Combining these equations gives

$$[As(OH)_3]/[HAsO_4^{2-}][H^+]^4[e^-]^2 = 10^{28.1}$$

For oxidizing conditions

$$[As(OH)_3]/[HAsO_4^{2-}] = 10^{-29.1}$$

and for reducing conditions

$$[As(OH)_3]/[HAsO_4^{2-}] = 10^{3.9}$$

Hydrogen sulphide will reduce arsenic (V) and further stabilize arsenic (III) by the following equilibrium:

$$As(OH)_3 + 1.5H_2S \rightleftharpoons 0.5As_2S_3(s) + 3H_2O: \qquad \log K = 12.6.$$

8.5 Vanadium

The constants of the following equations for vanadium (III), (IV) and (V) are taken from Sillén and Martell (1964)

$$[VO^{2+}]/[VO_2^+][H^+]^2[e^-] = 10^{16.90}$$
$$[V^{3+}]/[VO^{2+}][H^+]^2[e^-] = 10^{6.66}$$

All three vanadium cations are hydrolysed at pH 7.5 and 8 since

$$[V^{3+}] = [V(OH)_2^+][H^+]^2 10^{6.2}$$
$$[VO^{2+}] = [VOOH^+][H^+] 10^{5.0}$$
$$[VO_2^+] = [H_2VO_4^-][H^+]^2 10^{7.3}$$

In an oxidizing environment $H_2VO_4^-$ (vanadium (V)) will be stabler than $VOOH^+$ (vanadium (IV)), which is stabler than $V(OH)_2^+$ (vanadium (III)). In a reducing environment $V(OH)_2^+$ seems to be slightly favoured.

8.6 Chromium

CrO_4^{2-} is the stablest form of chromium (VI) since $[HCrO_4^-]/[H^+]$ $[CrO_4^{2-}] = 10^{6.5}$ and dimerization to $Cr_2O_7^{2-}$ does not occur at trace concentrations. The redox potential is set by

$$[Cr(OH)_3]_{s,hydr}[OH^-]^5/[CrO_4^{2-}][e^-]^3 = 10^{-6.9}$$

or for $[OH^-] = 10^{-5.5}$ M

$$[Cr(OH)_3]_{s,hydr}/[CrO_4^{2-}][e^-]^3 = 10^{20.6}$$

Thus for oxidizing conditions chromium (VI) will be the stablest form since

$$[Cr(OH)_3]_{s,hydr}/[CrO_4^{2-}] = 10^{-17.2}$$

while for $pE = -2.9$

$$[Cr(OH)_3]_{s,hydr}/[CrO_4^{2-}] = 10^{29.3}$$

i.e. hydrated chromium (III) hydroxide should dominate.

8.7 Cobalt

The redox potential is set by

$$[Co^{2+}]/[Co^{3+}][e^-] = 10^{31}$$

which implies that cobalt (II) should be the stablest form. Cobalt (III) forms, however, very strong (and inert) complexes. For the ligand OH^- we may e.g. estimate that

$$[Co(OH)_3]/[Co^{3+}][OH^-]^3 = 10^{37.5}$$

or

$$[Co[OH)_3]/[Co^{3+}] = 10^{21}$$

for $pOH = 5.5$ and for $pE = 12.6$

$$[Co^{2+}]/[Co(OH)_3] = 10^{-2.6}$$

Thus in an oxidizing environment complexing may favour the inert three valent state of cobalt.

8.8 Manganese

Manganese may also exist in a three valent state besides the two and four valent forms. The equilibrium constants are

$$[Mn^{2+}]/[MnO_2]_s[H^+]^4[e^-]^2 = 10^{41}$$

$$[Mn(OH)_2]_s[OH^-]/[Mn(OH)_3]_s[e^-] = 10^{2.6}$$

The solubility product of $Mn(OH)_2$ gives

$$[Mn^{2+}][OH^-]^2/[Mn(OH)_2]_s = 10^{-13}$$

and by comparison with iron (III) and thorium we estimate

$$[Mn(OH)_3]/[Mn(OH)_3]_s \approx 10^{-7} \, \text{M}$$
$$[Mn(OH)_4]/[MnO_2]_s \approx 10^{-7} \, \text{M}$$

Then

$$[Mn^{2+}]/[Mn(OH)_4] = 10^{-9.2}$$
$$[Mn^{2+}]/[Mn(OH)_3] = 10^{0.5}$$

and manganese (IV) will dominate in oxidizing conditions. Manganese (II) will be the main oxidation state in reducing environments.

8.9 Iron

The redox potential of iron (II)–iron (III) is well known and

$$[Fe^{2+}]/[Fe^{3+}][e^-] = 10^{13}$$

The stable inorganic form of iron (III) was $Fe(OH)_3$ and

$$[Fe(OH)_3]/[Fe^{3+}][OH^-]^3 \approx 10^{31}$$

For

$$[OH^-] = 10^{-5.5} \, \text{M}$$
$$[Fe(OH)_3]/[Fe^{3+}] = 10^{14.5}$$

and

$$[Fe^{2+}]/[Fe(OH)_3] = 10^{-14.1}$$

Thus iron (III) hydroxide is the stablest form in air-saturated estuarine waters. In a reducing environment with $pE = -2.9$ and $pOH = 6.5$, Fe^{2+} is stabler since $[Fe^{2+}]/[Fe(OH)_3] = 10^{4.4}$. In a reducing sediment with the solid phases FeOOH and FeS we may calculate the constant for the following equilibrium:

$$FeOOH(s) + HS^- + e^- \rightleftharpoons FeS(s) + 2OH^-; \qquad \log K_s$$

from the constants for the following equilibria:

$$FeOOH(s) \rightleftharpoons Fe^{3+} + 3OH^-; \qquad \log K = -37.5$$
$$FeS(s) + H^+ \rightleftharpoons Fe^{2+} + HS^-; \qquad \log K = -3.1$$
$$Fe^{3+} + e^- \rightleftharpoons Fe^{2+}; \qquad \log K = 13$$
$$H_2O \rightleftharpoons H^+ + OH^-; \qquad \log K = -14$$
$$\log K_s = 2\log[OH^-] - \log[HS^-] - \log[e^-]$$
$$= -37.5 + 3.1 + 13 + 14 = -7.4$$

For $\log[OH^-] = -6.5$ and $\log[e^-] = +2.9$ we obtain $\log[HS^-] = 7.4 - 13 - 2.9 = -8.5$. Since $[HS^-]$ is considerably higher than $10^{-8.5} \text{M}$ in a reducing estuarine environment containing hydrogen sulphide the equilibrium will be shifted to the right until all iron (III) hydroxide is dissolved.

Manganese and iron will be introduced in divalent form in river water by weathering reactions (cf. Mackenzie and Garrels, 1966 and Boyle and Edmond, 1977),

$$(Fe, Mn)CO_3 + CO_2 + H_2O \rightarrow (Fe, Mn)^{2+} + 2HCO_3^-$$

and

$$(Fe, Mn)Al \; silicate + 2CO_2 + H_2O \rightarrow Al \; silicate + (Fe, Mn)^{2+}$$
$$+ 2HCO_3^- + Si(OH)_4$$

The oxidation of Fe^{2+} to FeOOH is a rather slow process (cf. Kester *et al.*, 1975) that is catalysed by ferrooxidans bacteria (cf. Zajic, 1969).

8.10 Sulphide complexing in anoxic basin water.

Anoxic basin water has often a low $pH \approx 7.5$ (i.e. $[H^+] = 10^{-7.5}$ M) because of the oxygen depletion. Furthermore, due to the oxidation of carbohydrate the hydrogen carbonate concentration may be rather high (≈ 0.003 M). The decomposition of proteins and organic phosphates will likewise lead to high concentrations of ammonium (≈ 0.00013 M) and phsophate (≈ 0.000015 M, see Almgren *et al.*, 1975). The hydrogen sulphide concentration will depend on the sulphate available and thus the salinity. In a stagnant fjord we obtained values as high as 0.0006 M. Using the approximate stability constants for dihydrogen sulphide in a cool, saline basin the concentration of hydrogen sulphide and sulphide are calculated from:

$$[HS^-]/[H^+][S^{2-}] = 10^{14}$$
$$[H_2S]/[H^+][HS^-] = 10^7$$
$$0.0006 = [H_2S] + [HS^-] + [S^{2-}]$$

or

$$0.0006 = [HS^-]10^{-0.5} + [HS^-] + [HS^-]10^{-6.5} = 1.316[HS^-];$$
$$[HS^-] \approx 10^{-3.3} \; M; \qquad [S^{2-}] \approx 10^{-9.8} \; M$$

Using these ligand concentrations and the solubility products compiled by Smith and Martell (1976) we may calculate $[M^{2+}]$ (cf. Table 20). For MnS and Tl$_2$S these values are so high that the sulphides should dissolve completely. Since these metal ions have rather weak tendencies to form complexes with OH^-, Cl^- and HS^- they may exist either in the form of Mn^{2+} and Tl^+ or as carbonate complexes. Using the solubility product of $MnCO_3$ ($= 10^{-9.3}$) and protonation constant of CO_3^{2-} ($= 10^{10.3}$) it is possible to calculate the constant for the following equilibrium (eq. const $= K$):

$$MnCO_3(s) + H^+ \rightleftharpoons Mn^{2+} + HCO_3^-; \qquad \log K = 10.3 - 9.3 = 1$$

D. Dyrssen and M. Wedborg

Table 20. Sulphide speciation calculations using the stability constants and solubility products in Smith and Martell (1976), pH $= 7.5$, pHS $= 3.3$ and pS $= 9.8$. The concentrations are given in log units.

Metal species	$\log[MH_xL_y]$	Remarks
Mn^{2+}	-3.7	$= 2\times10^{-4}$ M or $11000\ \mu g\ Mn/1$
Fe^{2+}	-8.3	$= 0.28\ \mu g\ Fe/1$
Co^{2+}	-15.8	
Ni^{2+}	-16.8	
Cu^{2+}	-26.3	
Cu^{+}	-19.4	comparable with Ag^+
Ag^{+}	-20.0	
AgHS	-10.0	$= 0.011\ \mu g\ Ag/1$
$Ag(HS)_2^-$	-9.4	$= 0.043\ \mu g\ Ag/1$
Tl^{+}	-5.7	$= 2\times10^{-6}$ M or $409\ \mu g\ Tl/1$
TlHS	-6.7	
Zn^{2+}	-14.6	
ZnHSOH	-5.9	$= 88\ \mu g\ Zn/1$
$ZnHS_2^-$	-6.3	$= 33\ \mu g\ Zn/1$
Cd^{2+}	-16	
$CdHS^+$	-11.7	
$Cd(HS)_2$	-8	$= 1\ \mu g\ Cd/1$
$Cd(HS)_3^-$	-9.4	
$Cd(HS)_4^{2-}$	-10.3	
Hg^{2+}	-41.2	
$Hg(HS)_2$	-10.1	
$HgHS_2^-$	-8.8	$= 0.3\ \mu g\ Hg/1$
HgS_2^{2-}	-9.6	
Pb^{2+}	-17.7	

Using $[H^+] = 10^{-7.5}$ M and $[HCO_3^-] = 0.003$ M, $[Mn^{2+}] = 1.05\times10^{-4}$ M or 105 μM. This shows that $MnCO_3(s)$ is stabler than $MnS(s)$, but since $[Mn]_t < 100\ \mu$M it is questionable whether rodochrosite ($MnCO_3$, see Murray *et al.*, 1978) is dissolved or not. If it is dissolved then the concentration of manganese may be regulated by ion exchange equilibria with clay minerals or humic material.

The ligands HS^- and S^{2-} do not only form solid sulphides, but also soluble complexes such as AgHS, CH_3HgS^-, ZnHSOH, $Cd(HS)_2$, and $HgHS_2^-$.

Table 21. Stability constants for $M(HS)_2$ and MHS_2^- estiminated from dithizone extraction constants (cf. Irving, 1977, p. 49).

Values in parenthesis are compiled by Smith and Martell (1976) and used for the interpolation together with $\log K_{ex}$. The concentrations of $M(HS)_2$ and MHS_2^- in the presence of solid MS are calculated with the values of $[M^{2+}]$, $[HS^-]$ and $[S^{2-}]$ in Table 20.

Metal ion	$\log\dfrac{[M(HS)_2]}{[M^{2+}][HS^-]^2}$	$\log[M(HS)_2]$	$\log\dfrac{[MHS_2^-]}{[M^{2+}][HS^-][S^{2-}]}$	$\log[MHS_2^-]$
Fe^{2+}	10.0	-4.9	16.0	-5.4
Co^{2+}	14.6	-7.8	20.8	-8.1
Ni^{2+}	12.5	-10.9	18.6	-11.3
Cu^{2+}	22.8	-10.1	29.5	-9.9
Zn^{2+}	15.2	-6	(21.4)	-6.3
Cd^{2+}	(14.6)	-8	20.8	-8.3
Hg^{2+}	(37.7)	-10.1	(45.5)	-8.8
Pb^{2+}	13.5	-10.8	19.6	-11.2

Some equilibrium constants have been determined (see Table 20), but other metal ions should form similar complexes. Using dithizonate (cf. Irving, 1977) as a model ligand we have tried to estimate the constants for other divalent ions in Table 21. Tables 20 and 21 show that complexing by HS^- and S^{2-} will give rise to far higher metal concentrations than the free ion concentrations set by the solubility product $[M^{2+}][S^{2-}]$. Further equilibrium studies are needed in order to make better calculations on anoxic basin and pore waters (cf. Lu and Chen, 1977).

9 SOME GENERAL CONCLUSIONS

The main object of speciation calculations and trace metal studies in estuarine waters is to evaluate the fate of trace metals at an important stage in their biogeochemical cycles. Man has increased the river input of trace metals considerably. Large-scale mining and intensive agriculture has greatly enhanced the input of both inorganic and organic particulate matter. These activities have also increased the effect of natural weathering of carbonates, silicates and sulphide ores by working up virgin parts of the earth's surface. The enormous use of iron, copper, zinc, and lead with world productions in the order of several millions of metric tons per year have led to considerable city outfalls by metal industries, corrosion, and the use of leaded fuels. If the heavy metals only entered the sediments, where the pore water concentration was set by the solubility, the environmental damage might be quite small. However, if the decay of organic matter sets free the trace metals in the surface sediments, which now and then are resuspended by tides and storms, it is quite possible that the heavy metals could damage the estuarine aquatic life. Both *in situ* studies (cf. Balzer, 1978 and Dyrssen and Hallberg, 1978) and laboratory experiments (cf. Lu and Chen, 1977; and Granéli, 1977) are needed to determine the release rates from estuarine sediments. So far these studies have been incomplete. One has measured the oxygen consumption only (Jansson and Hagström, 1978 and Granéli, 1977) or the trace metals (Lu and Chen, 1977). Usually pore water concentrations are measured (e.g. Murray *et al.*, 1978), but not surface release rates. A complete study should, however, not only measure oxygen consumption and release rates of decay products of organic matter (hydrogen carbonate, sulphide, phosphate, ammonium, amino acids and other dissolved organic compounds), but also the release of trace metals and formation of solid phases (sulphides, carbonates, silicates). Thus, a complete study involves a tremendous amount of analytical determinations, which of course should be combined with studies of the bottom organisms and bacteria. In many cases the equilibrium state is instantaneously attained and in cases where kinetic barriers impede the reactions, analytical data compared with speciation calculations show where these barriers appear. Thus one makes use of the wealth of studies in solution chemistry. In sections 3, 4, 5, 6, 7, 8, and 9 we have given the main results of our calculations of ligand concentrations and complex formation tendencies. Special attention has been drawn to the importance of the river water alkalinity and exchange reactions with protons and major cations in humic material. The chelating tendency of adjacent carboxylic and phenolic groups in humics has been demonstrated for model ligands. The importance of soluble sulphide complexes has been stressed in our treatment of reduced environments.

10 REFERENCES

Ahrland, S. (1975). Metal complexes present in seawater, pp. 219–244 in *The Nature of Seawater* edited by E. D. Goldberg, Dahlem Konferenzen, Berlin, 719 pp.

Almgren, T., Danielsson, L.-G., Dyrssen, D., Johansson, T., and Nyquist, G. (1975). Release of inorganic matter from sediments in a stagnant basin. *Thalassia Yugoslavica*, **11**, 19–29.

Almgren, T., Dyrssen, D., and Strandberg, M. (1975). Determination of pH on the moles per kg seawater scale (M_w). *Deep-Sea Research*, **22**, 635–646.

Almgren, T., Dyrssen, D., Elgquist, B., and Johansson, O. (1976). Dissociation of hydrogen sulphide in seawater and comparison of pH scales. *Mar. Chem.*, **4**, p. 289–297.

Andersson, L., and Dyrssen, D. (1978). Excess calcium and alkalinity in the Baltic and Southern Kattegatt. Submitted to *Deep-Sea Research*.

Balzer, W. (1978). Untersuchungen über Abbau organischer Materie und Nährstoff-Freisetzung am Boden der Kieler Bucht beim Übergang vom oxischen zum anoxischen Milieu. Reports Sonderforschungsbereich 95 Wechsellwirkung Meer–Meeresboden, Universität Kiel, Nr., 36 März 1978.

Bates, R. G. and Culberson, C. H. (1977). Hydrogen ions and the thermodynamic state of marine systems, p. 45–61 in *The Fate of Fossil Fuel CO$_2$ in the Oceans* edited by N. R. Anderson and A. Malahoff, Plenum Press, New York and London, 749 pp.

Beneš, P., and Steinnes, E. (1974). In situ dialysis for the determination of the state of trace elements in natural waters. *Water Research*, **8**, 947–953.

Beneš, P., and Steinnes, E. (1975). Migration forms of trace elements in natural fresh waters and the effect of the water storage. *Water Research*, **9**, 741–9.

Berner, R. A. (1971). *Principles of chemical sedimentology*, McGraw-Hill Book Company, 240 pp.

Boyle, E. A., and Edmond, J. M. (1977). The mechanism of iron removal in estuaries. *Geochim. Cosmochim. Acta*, **41**, 1313–1324.

Breck, W. G. (1974). Redox levels in the sea, pp. 153–179 in *The Sea* Vol. 5: Marine Chemistry edited by E. D. Goldberg, John Wiley & Sons, 895 pp.

Burton, J. D. (1976). Basic properties and processes in estuarine chemistry, p. 1–36, in *Estuarine Chemistry* edited by J. D. Burton and P. S. Liss, Academic Press, 299 pp.

Bågander, L. E., and Niemistö, L. (1978). An evaluation of the use of redox measurements for characterizing recent sediments. *Estuarine and Coastal Marine Science* **6**, 127–134.

Danielsson, L.-G., and Dyrssen, D. (1975). Chemical investigation of the Idefjord on July 24, 1975. Report on the chemistry of sea water XVI. Department of analytical chemistry, University of Göteborg.

Dyrssen, D., Jagner, D., and Wengelin, F. (1968). *Computer Calculation of Ionic Equilibria and Titration Procedures.* Almqvist & Wiksell, Stockholm, 250 pp.

Dyrssen, D., and Hallberg, R. (1979). Anoxic sediment reactions—a comparison between box experiments and a fjord investigation. *Chem. Geol.*, **24**, 151–159.

Dyrssen, D., and Hansson, I. (1972–73). Ionic medium effects in seawater—a comparison of acidity constants of carbonic acid and boric acid in sodium chloride and synthetic sea water. *Mar. Chem.*, **1**, 137–149.

Dyrssen, D., and Sillén, L. G. (1967). Alkalinity and total carbonate in sea water; a plea for p-T-independent data. *Tellus*, **19**, 113–121.

Dyrssen, D., and Wedborg, M. (1974). Equilibrium calculations of the speciation of elements in seawater, p. 181–195 in *The Sea*, Vol. 5: Marine Chemistry edited by E. D. Goldberg, John Wiley & Sons, 895 pp.

Eckert, J. M., and Sholkovitz, E. R. (1976). The flocculation of iron, aluminium, and humates from river water by electrolytes. *Geochim. Cosmochim. Acta*, **40**, 847–8.

Elgquist, B. (1970). Determination of the stability constants of MgF$^+$ and CaF$^+$ using a fluoride ion selective electrode. *J. inorg. nucl. Chem.* **32**, 937–944.

Elgquist, B., and Wedborg, M. (1974). Sulphate complexation in seawater. A relation between the stability constants of sodium sulphate in seawater from a determination of the stability of hydrogen sulphate and a calculation of the single ion activity coefficient of sulphate. *Mar. Chem.*, **2**, 1–15 (see also *Mar. Chem.*, **3**, 79–80).

Elgquist, B., and Wedborg, M. (1975). Stability of ion pairs from gypsum solubility. Degree of ion pair formation between the major constitutents of seawater. *Mar. Chem.*, **3**, 215–225.

Elgquist, B., and Wedborg, M. (1978). Stability constants of NaSO$_4^-$, MgSO$_4$, MgF$^+$, and MgCl ions pairs at the ionic strength of seawater by potentiometry. *Mar. Chem.*, **6**, 243–252.

Elgquist, B., and Wedborg, M. (1979). Stability of calcium sulphate ion pair at the ionic strength of seawater by potentiometry. *Mar. Chem.*, **7**, 273–280.

Eriksson, G. (1979). An algorithm for the computation of aqueous multicomponent, multiphase equilibria. *Anal. Chim. Acta*, **112**, 375–383.

Gamble, D. S., and Schnitzer, M. (1973). The chemistry of fulvic acid and its reactions with metal ions, p. 265–302 in *Trace metals and metal-organic interactions in natural waters* edited by P. C. Singer, Ann Arbor Science Publ., Michigan, USA, 380 pp.

Goldhaber, M. B., and Kaplan, I. R. (1974). The sulphur cycle, p. 569–655 in *The Sea*, Vol. 5: Marine Chemistry edited by E. D. Goldberg, John Wiley & Sons, 895 pp.

Granéli, W. (1977). Measurement of sediment oxygen uptake in the laboratory using undisturbed sediment cores. *Vatten*, **3/77**, 1–15.

Hansson, I. (1973). A new set of acidity constants of carbonic acid and boric acid in seawater. *Deep-Sea Research*, **20**, 461–478.

Hansson, I., and Jagner, D. (1973). Evaluation of the accuracy in Gran plots by means of computer calculations. Application to potentiometric titration of total alkalinity and carbonate content in seawater. *Anal. Chim. Acta*, **65**, 363–373.

Head, P. C. (1976). Organic processes in estuaries, p. 54–91, in *Estuarine Chemistry* edited by J. D. Burton and P. S. Liss, Academic Press, 229 pp.

Ingri, N. W., Kakolowicz, W., Sillén, L. G., and Warnqvist, B. (1967). HALTAFALL, a general program for calculating the composition of equilibrium mixtures. *Talanta*, **14**, 1261–1286. Errata ibid. 15 (xi) (1968).

Irving, H. M. N. H., 1977. *Dithizone*, The Chem. Soc. London, 106 pp.

Jansson, B.-O., and Hagström, Å. (1978). Sea bed respiration: Core samples for oxygen consumption measurements. Proc. XI. Conf. Baltic Oceanographers, Rostock, April 24–27.

Kester, D. R., Byrne Jr., R. H., and Liang, Y.-J. (1975). Redox reactions and solution complexes of iron in marine systems, pp. 56–79 in *Marine chemistry in the coastal environment* edited by T. M. Church, Am. Chem. Soc. Symp. Ser. No. 18.

Liss, P. S. (1976). Conservative and non-conservative behaviour of dissolved constituents during estuarine mixing, pp. 93–130 in *Estuarine Chemistry* edited by J. D. Burton and P. S. Liss, Academic Press, 229 pp.

Lu, J. C. S., and Chen, K. Y. (1977). Migration of trace metals in interfaces of seawater and polluted surficial sediments. *Environmental Sci. Techn.* **11**, 174–182.

Mackenzie, F. T., and Garrels, R. M. (1966). Chemical mass balance between rivers and oceans. *Am. J. Sci.*, **264**, 507–525. Cf. Drever, J. I. (editor) (1977). *Sea water. Cycles of the Major Elements.* Benchmark papers in geology/45. Dowden, Hutchinson & Ross, Inc., Stroudsburg, Pa., USA, 344 pp.

Martell, A. E., and Smith, R. M. (1977). *Critical Stability Constants.* Vol. 3: Other organic ligands. Plenum press, New York and London, 495 pp.

Morris, A. W., Mantoura, R. F. C., Bale, A. J., and Howland, R. J. M. (1978). Very low salinity regions of estuaries: important sites for chemical and biological reactions. *Nature* **274**, 678–680.

Murray, J. W., Grundmanis, V., and Smethie, W. M. (1978). Interstitial water chemistry in the sediments of Saanich Inlet. *Gemochim. Cosmochim. Acta* **42**, 1011–1026

Ottewill, R. H., and Holloway, L. R. (1975). Electrokinetic properties of particles, pp. 599–621 in *The Nature of Seawater* edited by E. D. Goldberg, Dahlem Konferenzen, Berlin, 714 pp.

Parsons, R. (1975). The rôle of oxygen in redox processes in aqueous solutions, pp. 505–522 in *The Nature of Seawater*, edited by E. D. Goldberg, Dahlem Konferenzen, Berlin, 719 pp.

Paulson, A. J. (1978). Potentiometric studies of cupric hydroxide complexation. *Thesis*. Graduate School of Oceanography, University of Rhode Island, USA.

Ringbom, A. (1963). *Complexation in Analytical Chemistry*. Interscience Publ.

Rowe, G. T., Clifford, C. H., Smith Jr., K. L., and Hamilton, P. L., (1975). Benthic nutrient regeneration and its coupling to primary productivity in coastal waters. *Nature*, **255**, 215–217.

Sholkovitz, E. R. (1976). Flocculation of disssolved organic and inorganic matter during the mixing of river water and sea-water. *Geochim. Cosmochim. Acta*, **40**, 831–45.

Sillén, L. G., and Martell, A. E. (1964). *Stability Constants of Metal-ion Complexes*. Spec. publ. No. 17. The Chemical Society, London, 754 pp.

Sillén, L. G., and Martell, A. E. (1971). *Stability Constants of Metal-ion Complexes*. Supplement No. 1. Spec. publ. No. 25. The Chemical Society, London, 865 pp.

Smith, R. M., and Martell, A. E. (1976). *Critical Stability Constants*, Vol. 4: Inorganic complexes. Plenum Press, New York and London, 257 pp.

Stumm, W., and Brauner, P. A. (1975). Chemical Speciation, p. 173–239 in *Chemical Oceanography*, Vol. 1, 2nd ed. by J. P. Riley and G. Skirrow, Academic Press, 606 pp.

Whitfield, M. (1969). *Eh* as an operational parameter in estuarine studies. *Limnol. Oceanogr.*, **14**, 547–558.

Williams, P. J. le B. (1975). Biological and chemical aspects of disssolved organic material in sea water, pp. 301–363 in 2nd ed. Vol. 2 of *Chemical Oceanography* edited by J. P. Riley and G. Skirrow, Academic Press, 647 pp.

Wong, G. T. F., and Brewer, P. G. (1974). The determination of iodate in South Atlantic waters, *J. Mar. Res.*, **32**, 25–36.

Wong, G. T. F., Brewer, P. G., and Spencer, D. W. (1976). The distribution of particulate iodine in the Atlantic Ocean. *Earth Planetary Sci. Letters*, **32**, 441–450.

Wong, G. T. F., and Brewer (1977). The marine chemistry of iodine in anoxic basins. *Geochim. Cosmochim. Acta*, **41**, 151–159.

Zajic, J. E. (1969). *Microbial Biogeochemistry*, Academic Press, 345 pp.

Zirino, A., and Yamamoto, S. (1972). A pH-dependent model for the chemical speciation of copper, zinc, cadmium, and lead in seawater. *Limnol. Oceanogr.*, **17**, 661–671.

Chemistry and Biogeochemistry of Estuaries
Edited by E. Olausson and I. Cato
Copyright © by John Wiley & Sons Ltd.

J. C. DUINKER
Netherlands Institute for Sea Research,
Den Burg, Texel

4

Suspended Matter in Estuaries: Adsorption and Desorption Processes

1 INTRODUCTION

Rivers transport trace metals to the ocean in dissolved, colloidal and particulate forms. In estuaries, where river water and coastal or oceanic waters of widely different compositions are mixed, strong gradients in chemical and physico–chemical properties occur (Dyer, 1972; Phillips, 1972; Burton, 1976). As a result, the relative contributions of the various chemical

                                    J. C. Duinker

species of any element to these forms can be modified considerably. Dissolved and particulate suspended components have different transport mechanisms within the estuarine and coastal regions (Postma, this volume). It is important therefore to understand the effects of estuarine processes on trace metal behaviour for predicting the geochemical behaviour of each individual element and its possible effects on organisms, as well as the role of estuaries in the mass balance between rivers and oceans (Mackenzie and Garrels, 1966; Burton, 1976; Bewers and Yeats, 1977).

The adsorption experiments of Kharkar *et al.* (1968) and the interpretation of estuarine bottom sediment data (de Groot, 1966, 1973) have been quoted frequently to illustrate significant release of trace metals from river-borne suspended matter once in contact with sea water. Liss (1976), in a recent review of studies on the fate of dissolved river-borne trace metals within the estuary, concludes that 'except in the case of iron, present evidence indicates that processes are not of sufficient magnitude to give rise to major deviations from linearity on metal-salinity plots'. However, recent data suggest that large scale removal of dissolved species occurs in a number of estuaries for several more trace elements than for iron alone.

2 MAJOR PROBLEMS IN THE STUDY OF DISSOLVED AND PARTICULATE TRACE METALS IN ESTUARIES

The interpretation of the trace metal composition of estuarine water and particles in terms of relevant processes is fraught with problems. For our purpose, the major ones can be classified as follows:

(i) *The complexity of estuarine processes*

An estuary is a dynamical system in which the distribution patterns of dissolved and particulate components result from the physical mixing characteristics of natural waters of widely different composition and a variety of chemical and biogenic reactions taking place independently and simultaneously. Additionally, variations in meteorological conditions, flow characteristics and biological activity are responsible for variations in the concentrations of trace metals in estuaries, with different time scales (Turekian, 1971). In order to understand the role of estuaries in the biogeochemical cycling of trace elements it is essential, though by no means trivial, to distinguish the effects of biological and chemical from hydrodynamic processes.

(ii) *Non-linearity*

Several properties do not vary linearly with the parameter that characterizes the mixing series between river water and sea water, such as chlorinity or salinity. This has been observed for several basic properties as suspended load, pH and dissolved oxygen concentration.

(iii) *Analytical problems*

The behaviour of a trace element during estuarine mixing depends on its chemical state, whether in solution or particulate. Although techniques are available to determine total concentrations at the levels such as occur in river and sea water solution, no general method is available to distinguish different chemical species in solution, except for a limited number of trace metals (Beneš and Steinnes, 1975; Batley and Florence, 1976; Florence and Batley, 1977; Duinker and Kramer, 1977).

The analysis of dissolved trace metals in a series of estuarine water samples is usually more difficult than for a series of river or sea water samples due to the large range in the amounts of sea salt and dissolved organic matter in the matrix. Similar problems as for dissolved species are met in the analysis of sediment samples when distinguishing different binding sites. However, general chemical methods have been proposed recently (Gibbs, 1973; Gupta and Chen, 1975; Förstner, 1977, this volume). A problem that remains is the distinction of river-borne particles, resuspended bottom material and flocculates derived from river-borne dissolved and colloidal material. Progress has been made by using natural carbon and oxygen isotope ratios to distinguish fluvial and marine sources (Salomons, 1975; Schultz and Calder, 1976).

3 SOURCES, TRANSPORT, AND DISTRIBUTION OF PARTICULATE MATTER

The behaviour of particulate trace metals in an estuary is essentially related to that of total particulate matter. Laboratory experiments have shown that suspensions which are stable in fresh water environments—by virtue of interparticle repulsive forces between the electrically charged particles—are destabilized by double layer compression in even slightly saline water; the rate of flocculation depends on the extent of particle collisions and thus on particulate matter concentration (Carroll and Starkey, 1960; Whitehouse *et al.*, 1960; Postma, 1967). Flocculation is a reversible process: deflocculation can occur when particles are transported back into fresh water. These observations have been often used to state that fine-grained river-borne particulate matter—of which a considerable fraction may be colloidal (e.g. clay minerals)—is subject to flocculation in the early stages of estuarine mixing. Nevertheless, according to Meade (1972), the evidence that the effects of salt flocculation on the concentration of suspended matter can actually be observed in estuaries is very limited.

Field observations of Postma (1961), Nichols and Poor (1967) and Kranck (1974) show a decrease in grain size of suspended matter towards the head of the estuary (Figure 1). Kranck observed that flocs of any size are made up of grains of all sizes down to and including the finest material in the distribution. Similar observations were reported by Sheldon (1968). Data obtained in this laboratory on suspended matter concentrations in the Rhine estuary (Figure 2) show relatively large variations between cruises: in some cases they are consistent with the removal by salt flocculation; in others they are above the straight line that would result in the case of pure physical mixing without other processes occuring. Edzwald *et al.* (1974) emphasized the importance of chemical parameters for the deposition of particularly cohesive suspended particles in estuaries.

Before sediment particles can be flocculated and precipitated to an appreciable extent by inorganic salts in natural situations, a number of conditions should be met: a large proportion of particles should possess appropriate surface properties, the particles should have a considerable

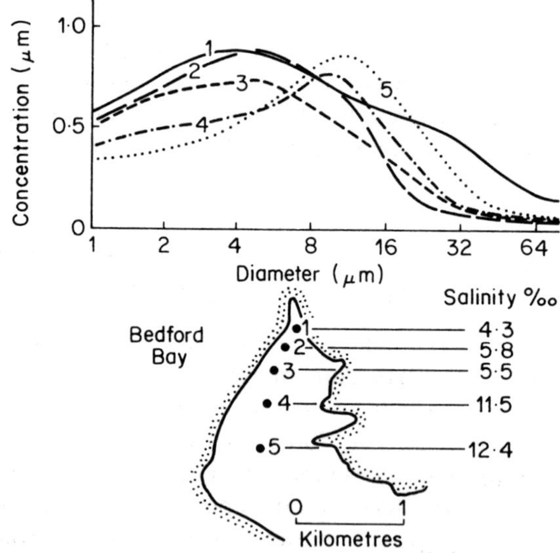

Figure 1. Particle spectra of particulate matter in Bedford Basin, N.S., illustrating change in particle spectra of suspended river sediment on entering a marine environment. The initial decrease in particle concentration is due to settling of large particles out of suspension and the subsequent decrease in small particles and increase in large particles is due to flocculation (Kranck, 1974). (Reproduced by permission of the National Research Council of Canada.)

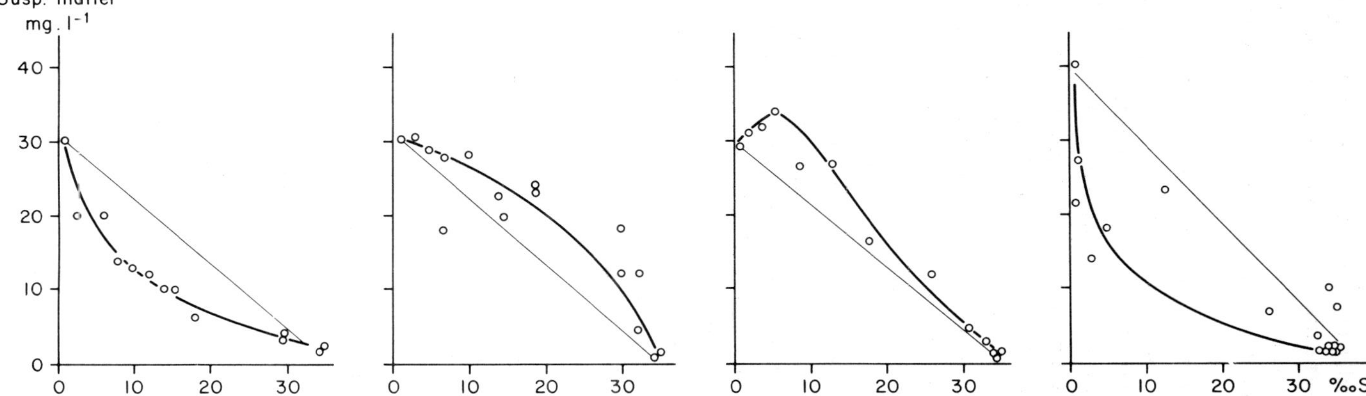

Figure 2. Suspended matter concentrations and salinity in surface layer of the Rhine estuary during four cruises (Duinker and Nolting, 1976). (Reproduced by permission of the Netherlands Institute for Sea Research.)

concentration (in the order of grams per litre) and suitable flow conditions should allow flocs to form and settle out of suspension. These conditions may not be fulfilled during any set of observations, or may be absent permanently in a particular estuary, partly accounting for the different observations (Figure 2).

A contribution to particulate matter in estuaries is derived from originally dissolved components (humic substances) flocculating in the early stages of mixing (Hair and Bassett, 1973), independently from the flocculation of preexisting particles (Eckert and Sholkovitz, 1976; Sholkovitz, 1976). Organic matter can also be precipitated by bacteria (Sheldon *et al.*, 1967).

A common feature of many estuaries is the existence of a turbidity cloud near the limit of sea salt (Postma, 1967). The turbidity maximum can be explained in terms of the dynamic processes that circulate and mix estuarine waters, probably the most important factors in the deposition of sediments in estuaries (Meade, 1972). Sediments in moderately stratified or vertically homogeneous estuaries (Pritchard, 1967) are moved progressively landward along the bottom and they accumulate near the limit of net landward flow. In highly stratified estuaries coarse sediments are trapped near the toe of the salt wedge (Roberts and Pierce, 1976). Suspended matter can be transported landward well beyond the region of salt intrusion even if the transfer of bottom water is seaward (van Veen, 1937; Gorsline, 1967; Postma, 1967). Estuarine bottom sediments are resuspended when bottom currents are sufficiently strong due to river flow or periodically within the tidal cycle. The effects are enhanced by storms and dredging activities (Holmes, 1977). The mixing of resuspended particles into the suspended load of the overlying water, with different composition and history results in a complicated time dependence of trace metal distributions.

Fine sediments being able to escape the estuary into the marine environment with the outflowing water, may carry a higher concentration of trace metals (Cranston and Buckley, 1972; Duinker *et al.*, 1974; Duinker and Nolting, 1976). Filtration and centrifugation of the water samples may then result in different 'particulate' metal concentrations (Duinker *et al.*, 1979).

4 SORPTION PROCESSES OF TRACE METALS

In this subchapter we shall summarize the conflicting evidence that has been presented in the literature to support the transfer of trace metals either from particulates into solution or from solution into particulates under estuarine conditions.

The chemical behaviour of a trace metal during its transport within the estuary is determined to a large extent by its chemical form in which it is transported by the river: (1) in solution as inorganic ion and both inorganic and organic complexes, (2) adsorbed onto surfaces, (3) in solid organic particles, (4) in coatings on detrital particles after coprecipitation with and sorption onto mainly iron and manganese oxides, (5) in lattice positions of detrital crystalline material, and (6) precipitated as pure phases, possibly on detrital particles. This scheme allows the distinction of trace metal fractions that are readily available (dissolved and adsorbed), fractions that become available after chemical changes (organically bound and in oxide coatings) and forms that are practically unavailable for release (in crystal structures of suspended particles).

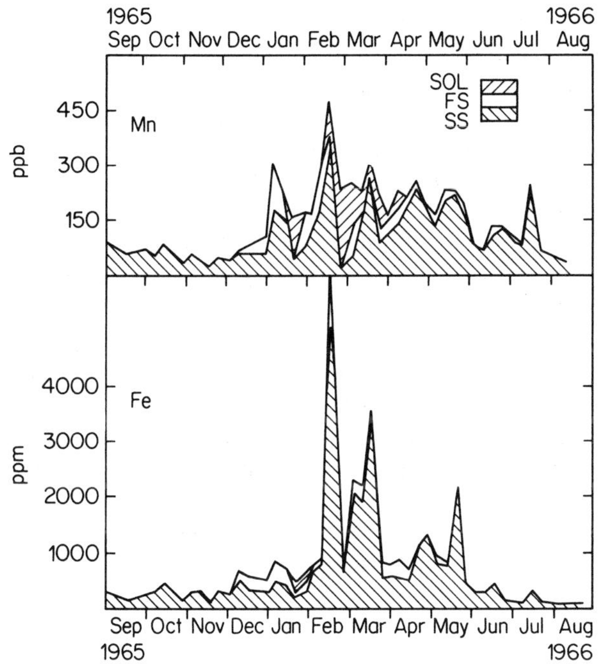

Figure 3. Iron and manganese concentrations in the soluble, filtered-solids, and settled-solids of the samples from Susquehanna river (Reproduced with permission from Carpenter *et al.*, 1975. Copyright by Academic Press Inc. (London) Ltd.)

The contribution of dissolved species to the total amount of each individual element will differ between rivers (Turekian and Scott, 1967; Kharkar *et al.*, 1968; Windom *et al.*, 1971) and may vary drastically over the seasons (Carpenter *et al.*, 1975; Figure 3).

4.1 Hydrous manganese and iron oxides

The solubility of an element depends on its oxidation state, pH, oxygen concentration and the presence of organic and inorganic ligands. The behaviour and properties of Mn(II) and hydrous Mn oxides has been described by Hem (1963), Stumm, and Morgan (1970) and Lee (1975). Under reducing conditions, manganese (IV) oxide is reduced to the soluble species Mn(II). Under oxidizing conditions, both Mn(II) and Mn(IV) can be present: the oxidation in acid-neutral to slightly alkaline environments is slow. The only iron species present in slightly acid to alkaline environments is ferric iron. It precipitates as a hydroxide, usually in an amorphous form (Stumm and Lee, 1960, 1961).

Hydrous metal oxide species of manganese and iron are important for the transport of trace metals in natural waters. Mechanisms through which interaction may occur have been described (Hem, 1977; Posselt *et al.*, 1968; Lockwood and Chen, 1973; Murray, 1975a, 1975b). Jenne (1968) stated that the properties of e.g. Co, Zn, Cu, and Ni in the environment cannot be explained in terms of the contributions from organic matter, clays, carbonates, and precipitation as discrete oxides or hydroxides. The environmental behaviour of Mn and Fe hydrous oxides has to be taken into account as one of the dominant factors. This has been supported by the results of several authors (Posselt *et al.*, 1968; Lockwood and Chen, 1973; Aston and Chester, 1973; Lee, 1975; Berrang and Grill, 1974; Chao and Anderson, 1974; Förstner, 1977). It is important whether the species to be adsorbed is present during formation of the hydrous metal oxides; the amount adsorbed on freshly precipitated oxides is considerably larger than on aged precipitates. The distribution coefficients for heavy metals for freshly precipitated MnO_2 are larger than for alkaline or alkaline earth metals, enabling preferential uptake of heavy metals on hydrous metal oxides in the presence of large amounts of these ions.

4.2 Clay minerals

Clay minerals have a considerable cation exchange capacity. It is unlikely, however, that they play an important role in the transport of heavy metals. Their cation exchange capacity represents only a small part of the sorption capacity of suspended matter for heavy metals. Additionally, trace metals have to compete with the bulk species (Ca, Mg, Na) for exchange sites. Major elements may therefore be expected to occupy most of the sites as the distribution coefficients are of the same order of magnitude (Lee, 1975). The adsorption of heavy metals to clay minerals as determined in laboratory experiments is too small to account for the concentrations in sediments (Hedges, 1977). Their importance might well be in their ability to act as nucleation centres for Fe/Mn oxides in the fresh water region (Förstner, 1977) and during estuarine mixing (Aston and Chester, 1973), as well as centres for the flocculation and precipitation of dissolved and colloidal organic matter during estuarine mixing (Rashid, 1969; Rashid *et al.*, 1972; Sholkovitz, 1976).

4.3 Organic matter

Organic matter in natural waters is able to modify the solubility, the redox potential and precipitation behaviour of metals. It is generally believed that the primary role of natural organics in influencing the transport of heavy metals occurs through complexation mechanisms with sedimentary humic compounds. The concentration of dissolved organic matter in river water exceeds those of dissolved trace metals considerably. Humic substances constituting a large fraction of river water DOM, are complex marcomolecular phenolic carboxylic acids, being transported as true solutions of polyelectrolytes or as hydrophilic colloids (Shapiro, 1957, 1964; Schnitzer and Kahn, 1972). Their anodic character enables them to interact with trace metal cations, forming complex linkages of various kinds by ion-exchange, surface adsorption and chelation. The stability of metal–humic complexes in natural waters is higher than that of corresponding inorganic–metal complexes (Reuter and Perdue, 1977). Complexation can be significant even in excess of major ions (Shapiro, 1964; Rashid, 1971, Rashid and Leonard, 1973, Duinker and Kramer, 1977). Humic materials can also interact with metal oxides, metal hydroxides and clay minerals to form associations (Senesi *et al.*, 1977). The solubility of humic acids is dependent on salinity (Sieburth and Jensen, 1968; Eckert and Sholkovitz, 1976; Gardner and Menzel, 1974). Laboratory and field studies indicate the transition of dissolved humic acids into particulate form during estuarine mixing (Sholkovitz, 1976; Hair and Bassett, 1973; Eckert and Sholkovitz, 1976).

Organically bound metal may be present as insoluble complexes, as peptized colloidal species and in true solution. Iron species can be peptized through interaction with organics and this may influence the interaction between hydrous iron oxide and other species in solution (Shapiro, 1964). Natural organics may be in solution at low iron concentrations and in colloidal form at higher iron concentrations (Hall and Lee, 1974). Organic molecules can be adsorbed by mineral particles by physical adsorption, cation exchange, and chemical reactions. Clay minerals strongly adsorb polymeric molecules, especially cationic polymers by cation exchange: the adsorption of proteins to clays is well documented (Perez Rodriguez *et al.*, 1977). Stable amorphous coatings can be formed by adsorption of humic material on clay surfaces (Jackson, 1969; McKyes *et al.*, 1974). Organic coatings will affect surface adsorption characteristics of particles. Neihof and

Loeb (1972) demonstrated that solid particles, whether they are positively or negatively charged in organic free sea water, are negatively charged in natural sea water.

4.4 Associations

The significance of the Fe/Mn hydrous oxides, organic matter, and clay minerals for the transport of trace metals decreases in the order $MnO_2 >$ humics $> Fe(OH)_3 >$ clay minerals. It is important to realize that their common occurrence in natural waters is in the form of associations. The control of heavy metals by mineral fragments with hydrous oxide coatings may actually be a tertiary or quaternary system, involving colloidal or dissolved organics (Lee, 1975). Gibbs (1973) analysed the distribution of Fe, Ni, Cu, Cr, Co, and Mn over the different forms in which these elements are transported by the Amazon and Yukon rivers. For Cu, Cr and Co, incorporation in the crystalline structures of the sediments is the major transport mechanism; for Fe, Mn, and Ni it is the second most important, precipitated and coprecipitated forms being most important. As the composition of the crystalline material is dictated by the rock composition in the drainage basin of the river, differences are to be expected between different river systems. The effects may be enhanced by different relative contributions of adsorption.

4.5 Electrostatic and specific interactions

Particles suspended in an electrolyte medium carry an electric charge originating from solid structure defects, uncompensated chemical bonds and the substitution of lattice ions with ions of different charge. The ions in the neutralizing electric double layer are exchangeable with ions of the bulk solution. As long as electrostatic adsorption predominates, sorption is a rapid and reversible process (Helfferich, 1962; Parks, 1975).

Electrostatic adsorption in natural waters does occur. It has been suggested that the surface charge of clay minerals and Fe/Mn hydrous oxides can be considered as a reliable guide to their adsorption characteristics (Krauskopf, 1956; Pravdic, 1970). However, electrostatic interaction cannot explain such phenomena as adsorption of ions against electrostatic repulsion and charge reversal in excess concentrations of particular ions. It appears that its contribution to the net free energy of adsorption is usually smaller than that of specific interaction mechanisms such as solvation and hydrophobic, covalent and hydrogen bonding. Specific adsorption may be slow; it is less readily reversible in the sense that equilibrium is not achieved in a reasonable time. Although electrostatic adsorption in natural waters occurs to some extent its contribution is relatively small (Parks, 1975; O'Connor and Kester, 1975; Koppelman and Dillard, 1977).

5 MODELS FOR ESTUARINE BEHAVIOUR OF DISSOLVED AND SUSPENDED TRACE METALS

5.1 Laboratory studies

The strong salinity gradients in estuaries can be expected to result in ion-exchange reactions on particularly fine-grained particles (Wiklander, 1964). Measurements of the electrokinetical potential of sea sediments in sea water of varying salinity showed that most sediments have negative charge in fresh water. Transition to positive charge occurs at salinity values characterizing the early stages of mixing: 1.8–10‰ S (Pravdić, 1970; Martin *et al.*, 1971). A riverborne particle with adsorbed material entering the estuary may undergo recharging in the low salinity range. Cations adsorbed to the negatively charged particles by purely Coulombic forces would then be expected to be at least partially desorbed and the particle would tend to adsorb anions due to its positive charge. This picture may be modified by dissolved organic matter. Murray and Meinke (1976) found that the adsorption of Cd, Co, and Ag to sediments in fresh water is diminished by soluble sewage material, while the adsorption of Zn does not change. We may also expect changes in the trace metal adsorption characteristics with salinity, resulting from changes in adsorption of organics with salinity. Thus, the adsorption of humic acids to clay minerals increases with S (Rashid *et al.*, 1972). The adsorption can be influenced by ionic strength in estuaries (Reimers and Krenkel, 1974).

The pioneering work of Krauskopf (1956) would suggest that a series of lithogenous, biogenous, and hydrogenous materials would adsorb trace metals from sea water. The studies involved the adsorption of Zn, Cu, Pb, Ni, Co, Hg, Ag, Cr, Mo, W, V on hydrated iron oxide, hydrated manganese oxide, apatite, clay, plankton and peatmoss. Iron oxide and manganese oxide were found to be most effective adsorbents; the other ones were relatively inactive except toward Cu, Hg, and Zn. This would be consistent with the strong undersaturation of several heavy metals in sea water with respect to the solubility of relevant mineral phases.

However, experimental work on adsorption and desorption of Co, Ag, Se, Mo, Sb, Cr, Cs, and Rb onto and from montmorillonite, illite, kaolinite, and ferric oxide, manganese oxide, hydrated ferric oxide and peat under conditions that, according to the authors, resembled natural conditions in streams and estuaries, supported the release of some trace elements in sea water (Kharkar *et al.*, 1968). The results indicated the release of a substantial portion of Co, Se, and Ag when the adsorbents, after contact with trace metal solutions for 1 hour, were contacted with sea water for 1 hour. It was concluded that river-borne particles rather than acting as sites for the removal of trace elements in sea water, actually contribute additional soluble amounts of some trace elements to the oceans.

132 *J. C. Duinker*

Van der Weijden *et al.* (1977) studied the desorption of Cr, Mn, Co, Ni, Cu, Zn, Cd, Fe, and Pb from river-borne suspended matter of the Rhine (obtained by centrifugation) in distilled water, in 1:1 diluted artificial sea water, in artificial sea water and 'nitrate sea water'. The degree of mobilization in the diluted and undiluted artificial sea water decreased in the order Cd>Zn>Mn>Ni>Co>Cu>Cr. Iron and lead did not show any tendency for mobilization.

O'Connor and Kester (1975) studied the adsorption of Cu and Co form aqueous solutions on to illite, beach sand, and polyethylene. A model was developed in which trace metal adsorption occurs through a specific pH dependent surface reaction. Thus, knowledge of the pH gradient is essential for understanding the fate of adsorbed trace metals as they are carried across salinity gradients at river mouths. The tendency for desorption appeared to increase with increasing salinity at constant pH; the relative importance of salinity is large in the 0–10‰ range. Electrostatically bound trace metals are desorbed as the medium changes from fresh water to salt water but at environmentally significant pH values they represent a negligible portion of the total adsorbed load.

Radio-tracer experiments (Aston and Chester, 1973) have shown the importance of the concentration of pre-existing particles and increasing salinity for the precipitation rate of iron during estuarine mixing. The

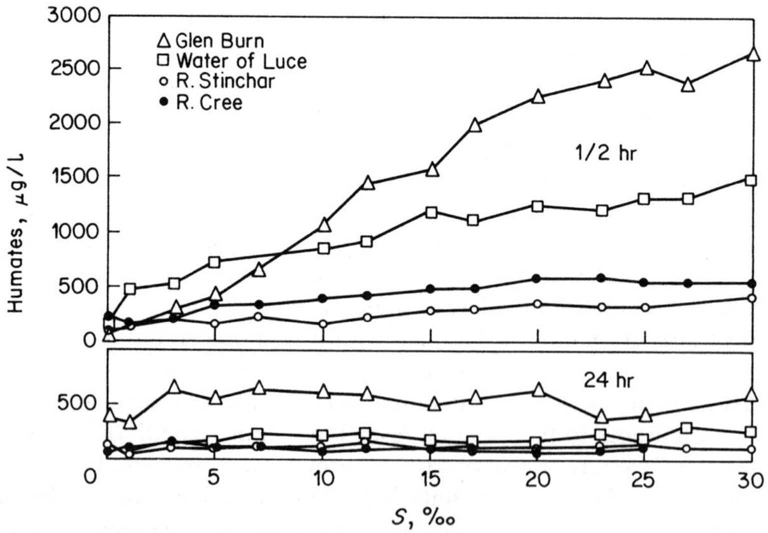

Figure 4. Humate flocculation *vs.* salinity for four river waters. Flocculant concentrations, both ½ hr and 24 hr after mixing with sea water, are given in μg humate/l of river water (Sholkovitz, 1976). (Reproduced by permission of Pergamon Press Ltd.)

importance of the process of coprecipitation of trace metals with iron hydrous oxide may stem from the formation of coatings around sediment particles, shielding its surface from interaction with sea water. Precipitation of riverborne dissolved components may also occur without the presence of pre-existing particles (Eckert and Sholkovitz, 1976; Sholkovitz, 1976). In his study of the products being formed during the mixing of filtered river water and filtered sea water at varying salinities, Sholkovitz (1976) established a rapid flocculation of humates, organic carbon, Fe, Al, Mn, and P. The amounts of flocculates increase as salinity increases from 0‰ to 15–20‰ S, above which little additional flocculation occurs (Figure 4). The extent of flocculation is very salinity dependent indicating the destabilization of river introduced colloidal humic substances during estuarine mixing. The formation of lab-produced flocculates reflects the *in situ* produced estuarine mixing. This process may remove substantial amounts of trace metal by coprecipitation.

5.2 Field studies of estuarine behaviour of dissolved and suspended trace metals

The behaviour of riverborne trace metals during estuarine mixing has been studied by several approaches:

(i) Trace metal concentrations in bottom sediments in the fresh and brackish water region of the rivers Rhine, Scheldt, Ems, and the Wadden Sea, separated 100–200 km from the Rhine river mouth have been interpreted in terms of desorption of trace metals from suspended matter in the estuary (de Groot, 1966, 1973). At each sampling location, a series of samples was taken from the very upper layer of the bottom sediment, the so-called freshly deposited sediment. The trace metal concentrations were corrected for grain size effects. The decrease in these concentrations in going from the fresh water region through the estuary into the coastal area (Wadden Sea) was assumed to be caused by the gradual release of trace metals from riverborne suspended matter once in contact with sea water into solution as metal–organic complexes. Differences between trace metals were attributed to different degrees of mobilization. Manganese was considered to remain fixed to suspended matter and this element was used as a reference for the degree of mobilization of other elements: Fe, Mn, Cu, Zn, Cr, Cd, Pb, Ti, Al.

(ii) The mixing curves of dissolved trace metal concentrations measured *in situ* over the whole estuarine salinity range against a suitable index of mixing (such as salinity) can be used to distinguish conservative from nonconservative behaviour. Criteria for the identification of any nonconservative behaviour have been given by Boyle *et al.* (1974). The main problems in the interpretation of these curves are related to variations in end-member concentrations, the possibility of input of material into the estuary from other sources and the correct choice of end members (Boyle *et al.*, 1974; Liss, 1976). Most major ions of river water and sea water (Na^+, K^+, Ca^{2+}, SO_4^{2-}, Cl^-) are found to behave conservatively; Mg^{2+} and trace metals may show considerable reactivity in the estuary. The principles of the reactant approach and its application to recent work on dissolved trace metals have been recently summarized by Liss (1976).

(iii) The combined analysis of both particulate and dissolved species has resulted in more insight in the dominant processes in estuaries (Bradford, 1972; Carpenter *et al.*, 1975; Duinker and Nolting, 1976, 1977, 1978; Wollast *et al.*, 1979; Duinker *et al.*, 1979).

(iv) A new approach to estuarine studies—called the *product* approach studying the solid particles resulting from mixing sediment free river water and sea water—has been introduced by Sholkovitz (1976). His method may establish (a) the extent and salinity dependence of nonconservative behaviour, (b) the composition and chemical form of removal products and (c) biological removal products (Figure 4). The advantages of this method compared with the *reactant* approach, i.e. the *in situ* measurement of dissolved species, are (a) problems with varying end member concentrations are absent, (b) the chemical form and composition of the products can be obtained by analysis without problems associated with estuarine field studies to distinguish recent flocculation products from riverborne and resuspended particles and (c) the influence of physicochemical conditions on the formation of floccules can be analysed.

The removal of dissolved iron in the early stages of mixing of river water and sea water has been established in quite a number of estuaries (Liss, 1976); additionally (see Table 1) it has also been established in the Rhine estuary (Duinker and Nolting, 1976, 1977, 1978; Figure 5) and in four Scottish rivers (Sholkovitz, 1976). Freshly formed iron containing flocculates may be the dominant factor (Sholkovitz, 1976; Duinker and Nolting, 1976) for the observed removal of dissolved trace elements during estuarine mixing (Duinker and Nolting, 1977, 1978). Manganese, through its oxide and hydroxide species, may be expected to show similar properties as iron; although it may respond differently to changes in environmental parameters such as pH, pE and dissolved oxygen concentrations (Hem, 1963; Wollast *et al.*, 1979). Actual data of dissolved manganese in the Rhine and Scheldt estuaries (Duinker *et al.*, 1979) show a complex dependence on salinity (Figure 6). The positive deviation from the ideal dilution line, in some cases resulting in a clear maximum at low salinities, was interpreted by *in situ* production from either resuspended bottom sediment once in contact with low pH surface water, or from a contribution of interstitially dissolved Mn mixed into the overlying water. The relative contributions of these sources is not yet fully understood. In the lower part of the estuaries, above 15–20‰ S, Mn is removed from solution. Evans *et al.* (1977) observed similar

Table 1. Field observations of dissolved and suspended trace metal behaviour in estuaries.

River estuary	Metal	Process	Authors
Alaskan fjord	Zn	Removal	Burrell, 1973
Satilla, Ogeechee,	Fe	Removal	Windom *et al.*, 1971
Altamata	Mn	Removal to some extent	
Back river	Mn, Fe, Ni, Cu, Zn, Pb Cd	Removal of waste water plant discharge	Helz *et al.*, 1975
Corpus Christi Bay	Cd, Mn	Removal, remobilization from bottom in winter	Holmes *et al.*, 1974
Var	Zn	Release at higher salinities	Fukai *et al.*, 1975
Beaulieu	Fe	Removal	Holliday and Liss, 1976
	Mn	Conservative, possibly *in situ* production at low salinities	
	Zn	Conservative	
Fraser	Zn	Removal and release at higher salinities	Grieve and Fletcher, 1977
Clyde	Zn	Removal	Mackay and Leatherland, 1976
St. Lawrence	Fe	Removal	Subramanian and d'Anglejan, 1976
	Mn	Conservative	
Strait of Georgia off Fraser river	Cu, Zn	Release bottom sediment lower estuary in spring and early summer	Thomas and Grill, 1977
Merrimack	Fe	Removal	Boyle *et al.*, 1974
4 Scottish rivers	Mn, Fe, Al	Removal	Sholkovitz, 1976
Columbia	^{54}Mn, ^{65}Zn	Release	Evans and Cutshall, 1973
Rhine	Fe, Cd, Cu, Pb	Removal Conservative	Duinker and Nolting, 1976, 1977
	Zn	Conservative or Removal	
	Mn	Cycling; *in situ* production at low salinities from bottom sediment	Duinker *et al.*, 1979 Wollast *et al.*, 1979
	Se	Conservative or slight removal	Sloot, v.d., *et al.*, 1978
	Sb	Conservative	
	Mo	Conservative	
Newport	Mn	Cycling; *in situ* production at low salinities from bottom sediment	Evans *et al.*, 1977

J. C. Duinker

Table 1 (*continued*)

River estuary	Metal	Process	Authors
Scheldt	Mn	Cycling, *in situ* production at low salinities from bottom sediment	Duinker *et al.*, 1979 Wollast *et al.*, 1979
	Sb	Removal	Sloot, v.d., *et al.*, 1978
	Mo	Conservative	
	Se	Conservative	
Narbadi, Tapti, Godavari	$^{234}U/^{238}U$, ^{238}U	Conservative	Borole *et al.*, 1977
La Have	Hg	Removal	Cranston, 1976
Gironde	^{238}U	Conservative	Martin *et al.*, 1977
Charente	^{238}U	Removal	Martin *et al.*, 1977

dissolution processes of manganese in the Newport river estuary (see also Postma, this volume p. 182). Holliday and Liss (1976) observed conservative behaviour of manganese in the Beaulieu estuary, with a possible slight desorption at low salinities. Observed dissolved Mn concentrations throughout the Scheldt estuary could be reproduced by equilibrium model calculations using results of laboratory experiments on the effects of pH, Eh, Cl⁻

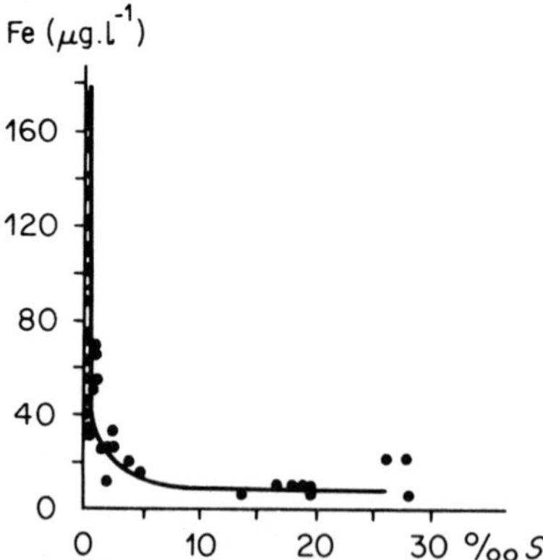

Figure 5. Dissolved iron as a function of salinity in the Rhine estuary in December 1974 (Duinker and Nolting, 1976). (Reproduced by permission of the Netherlands Institute for Sea Research.)

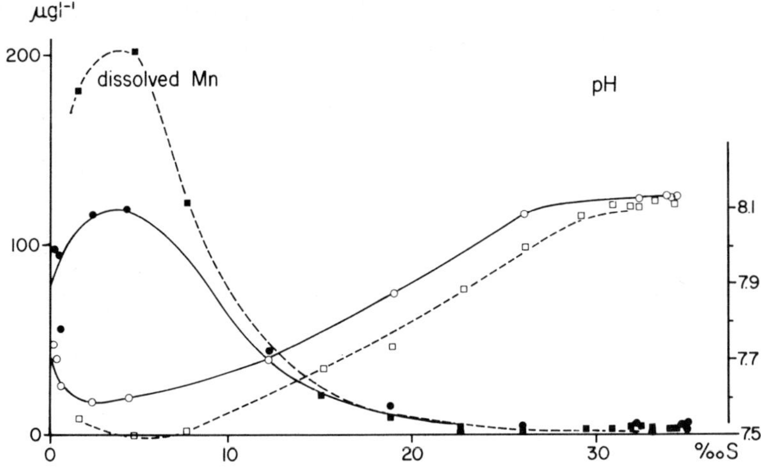

Figure 6. Dissolved manganese concentrations (●, ■) and pH (○, □) in surface water samples of the Rhine (○, ●) and Scheldt (□, ■) estuaries (Duinker *et al.*, 1979.)

and dissolved O_2 concentrations (Wollast *et al.*, 1979). The laboratory results on mobilization of several trace metals from suspended matter of the Rhine into artificial sea water (van der Weyden *et al.*, 1977) are in conflict with their own field observations and those of Duinker and Nolting (1976, 1977, 1978). The former authors suggest that this discrepancy may be interpreted by assuming that original riverborne particulate metals are deposited in bottom sediments at salinity values that are too low to cause significant mobilization. However, a considerable part of the particulate load of the river is transported further out into and through the estuary, so this cannot fully explain the discrepancy. Salomons and Mook (1977) used stable isotope methods to distinguish fluvial and marine sediments in the Rhine estuary. Trace metal concentrations in bottom sediments were found to be higher than on the basis of the mixing ratios of the two end members. This increase could be attributed to precipitation–adsorption processes.

Grieve and Fletcher (1977) studied the behaviour of Zn in the Fraser river estuary. The authors explained the increase in particulate suspended Zn in the mixing zone by adsorption of dissolved species onto hydrous iron-oxide coatings developed as suspended or resuspended sediments encounter new parcels of fresh water. They suggest that the increasing dissolved Zn values at higher salinities would then reflect desorption as suspended sediments move seaward and encounter more saline waters.

Thomas and Grill (1977) found high dissolved concentrations in their study of dissolved copper and zinc off the mouth of the Fraser river in the spring and early summer, the period of maximum water and sediment discharge by the river. Their data of a station in the Strait of Georgia 14 km off the mouth of the main distributory channel (Figure 7) show increasing concentrations at all depths above 150 m but in particular at about 30 m. Values start to decrease in September, remaining low during whole winter. A plot of dissolved metal concentrations of all samples near the river mouth in the freshet period against salinity shows that the highest values occur in the waters that presumably had been formed through the mixing of oceanic with silt laden river water. They occur in the 25–28‰ S range, i.e. in waters with depth roughly equal to that of the outer edge of the banks forming the main platform of the Fraser river delta. This made the authors to suggest that much of the metal was desorbed from sediment which, following discharge, settled onto the delta flats. Such a hypothesis would also account for the layer of high metal water observed in August extending outward in the Strait just below the surface (Figure 8). Trace metals may be solubilized from solid waste material dumped into the sea. Differences in waste treatment techniques may lead to differences in trace metal behaviour. Los Angeles County sewage particulate trace metals—after primary anaerobic treatment mainly present as sulphides and oxides—are not mobilized in the vicinity of the outfall (Morel *et al.*, 1975). 90% of the particulate Cd, Cu, Ni, Pb is mobilized from Los Angeles City sewage upon introduction to sea water (Rohatgi and Chen, 1975, 1976). It has been argued already that riverborne organic matter flocculates during estuarine mixing. Trace elements are removed as well. The extent of removal, relative to river water,

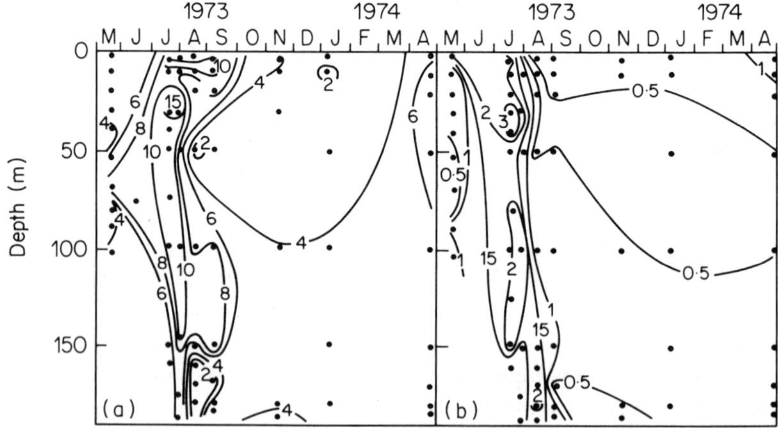

Figure 7. The changes in the (a) dissolved Zn and (b) dissolved Cu distributions observed in the Strait of Georgia at one station between May 1973 and May 1974. The solid dots represent the sampling points. Concentrations in μg/l (Reproduced with permission from Thomas and Grill, 1977. Copyright by Academic Press Inc. (London) Ltd.)

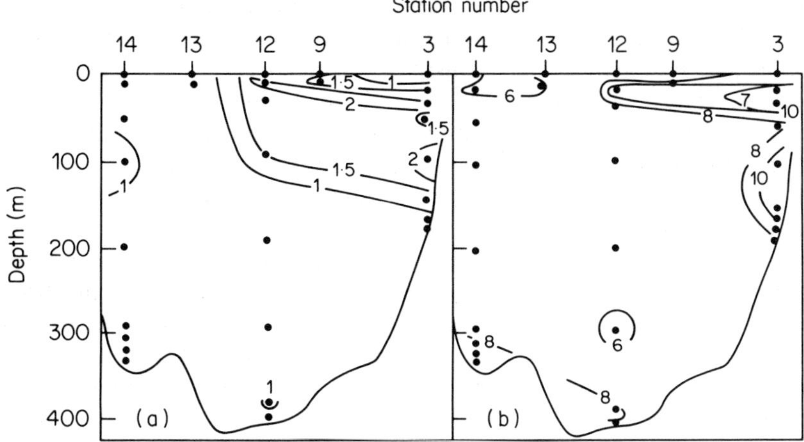

Figure 8. The distribution of (a) dissolved Cu and (b) dissolved Zn observed 1–3 August 1973 in a longitudinal section extending roughly the length of the southern half of the Strait of Georgia. The solid dots represent the sampling points. Concentrations in μg/l (Reproduced with permission from Thomas and Grill, 1977. Copyright by Academic Press Inc. (London) Ltd.)

ranges between 75 and 115% for Fe, 90 and 100% for Mn, 50 and 200% for P, and 10 and 70% for Al (Sholkovitz, 1976). Thus, dissolved organic matter, particularly humic material, plays a major role in controlling the concentrations of inorganic trace elements in river waters and their non-conservative behaviour during estuarine mixing.

The presence of a turbidity cloud within the estuary is likely to enhance the transition of trace metals from solution into particulate form. This would be similar to the laboratory results of Aston and Chester (1973) who found that the rate and extent of flocculate formation is strongly dependent on particle concentration.

It can be concluded from the work of several investigators that trace metal concentrations in solution are the most sensitive parameters for estimating the net effect of removal–mobilization processes in the estuary. Table 1 summarizes the results for several estuaries based on measurements of trace metal concentrations including at least the dissolved species. It can be concluded that mobilization of trace metals from suspended matter into solution is not as common a result as has been indicated earlier.

6 EXCHANGE OF TRACE METALS AT THE ESTUARINE BOTTOM SEDIMENT–WATER INTERFACE

Estuarine bottom sediments are composed of inorganic and organic constituents derived from particulate components of the river and marine environment (Dyer, 1972; Postma, 1967; Gorsline, 1967; Roberts and Pierce, 1976; Salomons, 1975). They act as sinks for riverborne trace metals (Müller and Förstner, 1975; Cranston, 1976; Förstner, 1978). Components which are transported in dissolved form by the river may be included in the bottom sediments as well, by adsorption and coprecipitation (Craig, 1974; Jacobs and Keeney, 1974; Sholkovitz, 1976). Bacteria may play an important role in the deposition of metals (McLerran and Holmes, 1974; Brinckman and Iverson, 1975).

The decomposition of aquatic detritus in bottom sediment results in changes in chemical and physico–chemical properties (pH, concentrations of dissolved O_2, sulphide and organic molecules). This may lead to changes in the chemical form in which trace elements are associated with the sediment particles, possibly shifting them into positions from which they are more readily available to the overlying water, e.g. from exchangeable positions. Enrichment of overlying water in trace metals through desorption from diagenetically altered estuarine sediments has been described by several authors. The large dissolved inorganic zinc concentration just above the sediment-water interface in Chesapeake Bay in May was attributed to desorption from deposited sediment (Bradford, 1972). The sediment had been deposited during discharge of large amounts by the Susquehanna river in March–April. The movement of the salt wedge landward afterwards, caused ion-exchange equilibria to be established resulting in the exchange of Zn for Mg and Ca from the salt wedge. Helz *et al.* (1975) studied the fate of trace metals (Mn, Fe, Cu, Zn, Cd, Pb) from a waste water treatment plant in Backriver. Removal to the sediments is observed. After the initial decrease, concentrations of Mn and Cd rise again, suggesting the remobilization from the bottom. Similar observations were made by Morel *et al.* (1975). Release from deposited sediments was observed for Zn in Chesapeake Bay by Carpenter *et al.* (1975). Thomson *et al.* (1975) studied the accumulation in and release from sediments of Long Island Sound. They found release of U from the bottom sediment and no loss of Cu, Zn, and Pb.

A further result of the diagenetic processes occurring in organic-rich sediments is the increase in the concentrations of dissolved trace metals in the interstitial waters. Considerable evidence is available to demonstrate the existence of a relatively large reservoir of interstitially dissolved trace metals within the upper layers of reducing sediments, including the estuarine environment (Duchart *et al.*, 1973; Elderfield and Hepworth, 1975; Manheim, 1976). The concentrations can be many orders of magnitude larger than the solubility values of the sulphides, even in the presence of hydrogen sulphide within the sediment column. It has been postulated that the formation of stable complexes of trace metals and dissolved organic molecules may account for this observation (Presley *et al.*, 1972). Humic substances are the major components of dissolved organic matter (Krom and Sholkovitz, 1977); their ability to form stable complexes with a range of trace metals may well support this suggestion, as has been shown by several authors (Schnitzer and Kahn, 1972; Rashid and Leonard, 1973; Nissenbaum and Swaine, 1976; Lindberg and Harris, 1974). Laboratory experiments have shown that trace metals like Cu, Mn, Co, Ni, and Zn can be solubilized from their carbonate and sulphide species by the interaction with organic matter (Rashid and Leonard, 1973). Aerobic bacterial activity has also been demonstrated to be effective in solubilizing metals from their precipitated sulphides (Duncan and Trussell, 1964; Fagerström and Jernelöv, 1971). Changes in redox conditions in the sediment induced by large nutrient supply may lead to changes in remobilization of trace metals into interstitial water. Experiments of Lu and Chen (1977) show that Fe and Mn can be released to interfacial sea water under reducing conditions while other elements are removed from sea water as sulphides (Cu, Cd, Ni, Pb, Zn). Competition between inorganic precipitation and organic complexation were also observed by Hallberg (1973). Organic complexation has been observed to decrease with increasing salinity (Lindberg and Harris, 1974).

Enrichment of the overlying water from interstitial water may occur through several mechanisms. The existence of concentration gradients, in particular of dissolved organic matter in the sedimentary column, may be an important pathway through which diffusion into overlying water can take place (Nissenbaum and Swaine, 1976). Remobilization has also been observed as a result of dredging (Bella and McCauley, 1974; Holmes, 1977), biological activity in the bottom (Petr, 1977), compaction (Reinhard and Förstner, 1976) and wave action and erosion (Andersen-Harris, 1973; Wakeman, 1976).

7 DISCUSSION AND CONCLUSIONS

An increasing number of papers has been concerned recently with the fate of trace metals in estuaries. The interpretation of the observed concentrations in estuarine waters is usually a complicated matter. Some important aspects in the interpretation of trace metal data in estuaries can be summarized as follows.

The applicability of laboratory experiments of adsorption and desorption processes to the actual situation in natural waters may be limited. In several cases, chemically pure forms have been selected as adsorbents such as standard clay minerals and iron or manganese oxides. It has been demonstrated that mineral phases in natural waters usually occur in some sort of association. Clay minerals interact with oxide coatings and humic substances (Aston and Chester, 1973; Gibbs, 1973; Johnson, 1974). Humic substances can also interact with metal ions, metal oxides, and hydroxides (Senesi *et al.*, 1977); organic coatings will affect the properties of particles such as the size and settling velocity (Kranck, 1974) and surface adsorption characteristics (Neihof and Loeb, 1972). Thus, the cation-exchange capacity of particles may be modified considerably by organic flocculant coatings (Rashid, 1969). Pillai *et al.* (1971) have demonstrated that marine sediment particles with organic coatings show an increase in the adsorption of Zn, Co, and Mn. It is not unlikely therefore that laboratory experiments on the sorption of trace elements under artificial conditions (Kharkar *et al.*, 1968) cannot be extrapolated to actual estuarine conditions. The use of sophisticated sorption models may prove to be useful (Parks, 1975; O'Connor and Kester, 1976). The complex interplay between physical, chemical, and biological processes in estuaries is the main reason for the problems encountered in designing laboratory experiments that are representative for estuarine situations. Thus, several attempts have been made to study the sorption processes by exposing sediment particles to various solutions and measuring the resultant changes in sediment and solution compositions. In an approach of this kind, involving fluvial sediment and sea water, the effect of important parameters that are known to vary during actual estuarine mixing may not be included, e.g. the initial decrease in pH in the early stages of mixing river water and sea water (Mook and Koene, 1975), and the absence of dissolved organic matter in artificial sea water. Alternatively, in studies involving sediment-free river water and sea water, the effects of pre-existing particles that are known to act as nucleation centres (Aston and Chester, 1973) are not included. Additionally, it is difficult to estimate the contact time of sediment particles with—and their concentrations in—solutions of varying composition (salinity) in the estuary. It is difficult therefore to estimate how well laboratory conditions may represent the actual situation.

In studies on the fate of riverborne trace metals in estuaries, efforts should be made to include both dissolved and particulate forms. In some cases, the evidence given for either release or removal processes would have been supported by additional information, particularly of dissolved species. The value of such integrated studies has been demonstrated in several cases (Carpenter *et al.*, 1975; Troup and Bricker, 1975; Duinker and Nolting, 1976, 1977, 1978; Wollast *et al.*, 1979; Duinker *et al.*, 1979).

Bottom sediment in estuaries is a significant source for trace metals in the overlying water. This has been established in several cases for diagenetically modified anoxic sediments. On the other hand, it has not been taken into consideration in several cases where it might have contributed to an explanation of the data on dissolved and suspended trace metal concentrations. It is essential that a clear distinction is made between the desorption of trace metals from riverborne suspended particles upon contact with sea water and the desorption from bottom sediments or the contributions from interstitial water once the sediments have been diagenetically modified.

The decrease in the concentrations of a range of trace metals in bottom sediments in a seaward direction in the Rhine and Ems estuaries have been interpreted in terms of mobilization from suspended matter into solution as metal–organic complexes (de Groot, 1966, 1973). If the decrease in the trace metal concentrations were due to mobilization from suspended matter into solution, more than 50% of the riverborne suspended Fe, Cd, Zn, Pb, Cr would have gone into solution near the head of the estuary. A relatively large and measurable increase in dissolved concentrations would be expected. No indication of any mobilization of Fe, Cu, Pb, Cd, Zn has been found in the dissolved metal–salinity plots. Instead, Fe, Cu, Cd and in some surveys Zn were found to be removed from solution (Duinker and Nolting, 1977, 1978; Fig. 9). Manganese, that was considered to remain fixed to riverborne suspended matter during its transport through the estuary (de Groot, 1966, 1973) was found to be mobilized into solution (Duinker *et al.*, 1978; Wollast *et al.*, 1979). It cannot be used as a reference for other riverborne trace elements in bottom sediments therefore. The bottom sediment data can be conveniently interpreted in terms of mixing of fluvial and marine sediments (Müller and Förstner, 1975; Duinker and Nolting, 1976, 1977; Salomons and Mook, 1977) with additional contributions from originally dissolved components.

Summarizing these results and those in other estuaries (Table 1) we may conclude that evidence for desorption of trace metals from riverborne suspended matter upon contact with sea water is very limited. On the other hand, conservative behaviour and net removal of riverborne dissolved trace metals has been observed for Fe, Al, Mn, Cd, Cu, Zn in several estuaries. The removal process is apparently more important than any release that might take place. It is not impossible that release and removal of some trace metals (e.g. U) may take place simultaneously (Martin *et al.*, 1977).

Trace metals are subject to internal biogeochemical cycles within the estuary (Turekian, 1977). Physical processes were reported to dominate the total annual budgets of Mn, Fe, and Zn in the Newport river estuary, despite the high biological productivity (Wolfe *et al.*, 1973).

It can be concluded that only a very limited part of the river transported load of trace metals may reach the ocean. Only those elements that form very strong anionic complexes may be transported out of the estuary in dissolved form (Turekian, 1977). Particles, through estuarine flocculation and sedimentation processes, play therefore a dominant role in the observed depletion of the ocean in certain trace metals. This may have important consequences for the residence time of trace metals in ocean water (Bewer and Yeats, 1977).

J. C. Duinker

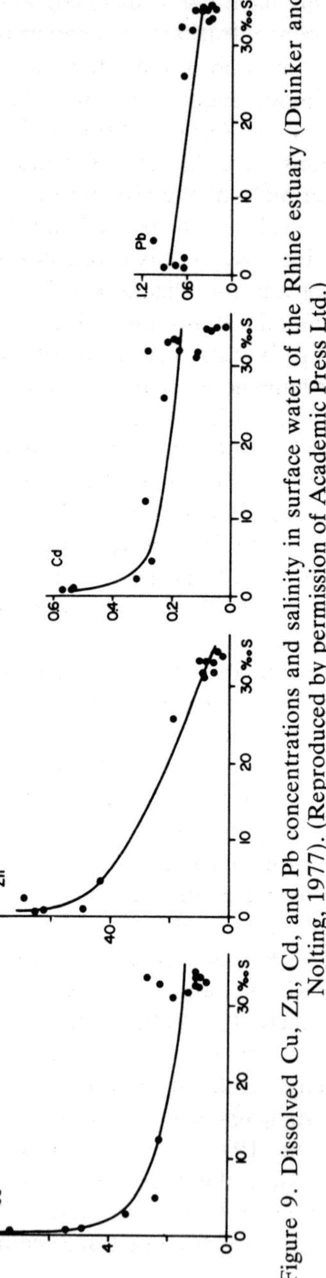

Figure 9. Dissolved Cu, Zn, Cd, and Pb concentrations and salinity in surface water of the Rhine estuary (Duinker and Nolting, 1977). (Reproduced by permission of Academic Press Ltd.)

REFERENCES

Aston, S. R., and Chester, R. (1973). The influence of suspended particles on the precipitation of iron in natural waters. *Est. Coast. Mar. Sci.* **1,** 225–231.

Batley, G. E., and Florence, T. M. (1976). A novel scheme for the classification of heavy metal species in natural water. *Analyt. Lett.,* **9,** 379–388.

Bella, D. A., and McCauley, J. E. (1972). Environmental considerations for estuarine dredging operations. *Proc. World Dredging Conf.,* p. 457–482.

Beneš, P., and Steinnes, E. (1975). Migration forms of trace elements in natural fresh waters and the effect of the water storage. *Water. Res.,* **9,** 741–749.

Berrang, P. G., and Grill, E. V. (1974). The effect of manganese oxide scavenging on Mo in Saanich Inlet, B. C. *Mar. Chem.,* **2,** 125–148.

Bewers, J. M., and Yeats, P. A. (1977). Oceanic residence times of trace metals. *Nature,* **268,** 595–598.

Borole, D. V., Krishnaswani, S., and Somayajuki, B. L. K. (1977). Investigations on dissolved uranium, silicon and on particulate trace elements in estuaries. *Est. Coast. Mar. Sci.,* **5,** 743–753.

Boyle, E. A., Collier, R., Dengler, A. T., Edmond, J. M., Ng, A. C., and Stallard, R. F. (1974). On the chemical mass-balance in estuaries. *Geochim. Cosmochim. Acta,* **38,** 1719–1728.

Bradford, W. L., (1972). A study on the chemical behaviour of zinc in Chesapeake Bay water using anodic stripping voltammetry. *Techn. Rep. Chesapeake Bay Inst.,* no. **76,** 103 pp.

Brinckman, F. E., and Iverson, W. P. (1975). Chemical and bacterial cycling of heavy metals in the estuarine system. In: Church, T. M. (ed)., *Marine chemistry in the coastal environment,* ACS symposium Ser., **18,** 319–342, *Am. Chem. Soc.,* Washington, D. C.

Burrell, D. C. (1973). Distribution patterns for some particulate and dissolved trace metals within an active glacial fjord. In: *Radioactive Contamination of the Marine Environment,* IAEA, Vienna, 89–102.

Burton, J. D. (1976). Basic properties and processes in estuarine chemistry. In: Burton, J. D. and Liss, P. S. (eds.), *Estuarine Chemistry,* Ac. Press London, 1–36.

Carpenter, J. H., Bradford, W. L., and Grant, W. (1975). Processes affecting the composition of estuarine waters (HCO_3, Fe, Mn, Zn, Cu, Ni, Cr, Co and Cd). In: Cronin, L. E. (ed.) *Estuarine Res.,* vol. 1, Ac. Press, 188–214.

Carroll, D., and Starkey, H. C. (1960). Effect of seawater on clay minerals. In: Ingersen, E. (ed.) *Clays and Clay Minerals, Monograph No. 5, Earth Sci Ser.,* Perg. Press, 80–101.

Chao, T. T., and Anderson, B. J. (1974). The scavenging of Ag by Mn and Fe oxides in stream sediments collected from two drainage areas of Colorado. *Chem. Geol.,* **14,** 159–166.

O'Connor, T. P., and Kester, D. R. (1975). Adsorption of copper and cobalt from fresh and marine systems. *Geochim. Cosmochim. Acta,* **39,** 1531–1543.

Craig, H. (1974). A scavenging model for trace elements in the deep sea. *Earth and Planetary Sci. Letters,* **23,** 149–159.

Cranston, R. E. (1976). Accumulation and distribution of total mercury in estuarine sediments. *Est. Coast. Mar. Sci.,* **4,** 695–700.

Cranston, R. E., and Buckley, D. E. (1972). The application and performance of microfilters in analyses of suspended particulate matter. *Unpubl. manuscr. BI-R-72-7,* Bedford Inst. Oceanogr., 14 pp.

Duchart, P., Calvert, S. E., and Price, N. B. (1973). Distribution of trace metals in the pore waters of shallow water marine sediments. *Limnol. and Oceanogr.*, **18**, 605–610.

Duinker, J. C., van Eck, G. T. M., and Nolting, R. F. (1974). On the behaviour of copper, zinc, iron, and manganese in the Dutch Wadden Sea; evidence for mobilization processes. *Neth. J. Sea Res.*, **8**, 214–239.

Duinker, J. C., and Kramer, C. J. M. (1977). An experimental study on the speciation of dissolved zinc, cadmium, lead, and copper in river Rhine and North. Sea water, by differential pulsed anodic stripping voltammetry. *Mar. Chem.*, **5**, 207–228.

Duinker, J. C., and Nolting, R. F. (1976). Distribution model for particulate trace metals in the Rhine estuary, Southern Bight and Dutch Wadden Sea. *Neth. J. Sea Res.*, **10**, 71–102.

Duinker, J. C. and Nolting, R. F. (1977). Dissolved and particulate trace metals in the Rhine estuary and the Southern Bight. *Mar. Pollut. Bull.*, **8**, 65–71.

Duinker, J. C., and Nolting, R. F. (1978). Mixing, removal and mobilization of trace metals in the Rhine estuary. *Net. J. Sea Res.*, **12**, 205–223.

Duinker, J. C., Nolting, R. F., and van der Sloot, H. A. (1979). The determination of suspended metals in coastal waters by different sampling and processing techniques (filtration, centrifugation). *Neth. J. Sea Res.* **13**, 282–297.

Duinker, J. C., Wollast, R., and Billen, G. (1979). Manganese in the Rhine and Scheldt estuaries, Part 2: Geochemical cycling. *Est. Coast. Mar. Sci.* **9**, 727–738.

Duncan, D. W., and Trussell, P. C. (1964). Advances in the microbiological leaching of sulphide ores. *Can. Metall. Qtrly*, **3**, 43–55.

Dyer, K. R. (1972). Sedimentation in estuaries. In; Barnes, R. S. K. and Green, J. (eds.), *The Estuarine Environment*, Appl. Sci. Publ. London, 11–32.

Eckert, J. M., and Sholkovitz, E. R. (1976). The flocculation of iron, aluminium, and humates from river water by electrolytes. *Geochim. Cosmochim. Acta*, **40**, 847–848.

Edzwald, J. K., Upchurch, J. B., and O'Melia, C. R. (1974). Coagulation in estuaries. *Env. Sci. Technol.*, **8**, 58–63.

Elderfield, H., and Hepworth, A. (1975). Diagenesis, metals and pollution in estuaries. *Mar. Pollut. Bull.*, **6**, 85–87.

Evans, D. W., and Cutshall, N. H. (1973). Effects of ocean water on the soluble-suspended distribution of Columbia River radionuclides. In: *Radioactive Contamination of the Marine Environment*, IAEA, Vienna, 125–140.

Evans, D. W., Cutshall, N. H., Cross, F. A., and Wolfe, D. A. (1977). Manganese cycling in the Newport River estuary, North Carolina. *Est. Coast. Mar. Sci.*, **5**, 71–80.

Fagerström, T., and Jernelöv, A. (1971). Formation of methylmercury from pure mercuric sulphide in aerobic organic sediment. *Water Res.*, **5**, 121–122.

Florence, T. M., and Batley, G. E. (1977). Determination of chemical forms of trace metals in natural waters with special reference to Cu, Pb, Cd and Zn. *Talanta*, **24**, 151–158.

Förstner, U. (1977). Sources and sediment associations of heavy metals in polluted coastal regions. *Sec. Symp. Origin and Distr. of the elements*, Unesco Paris, 10–13 May.

Förstner, U. (1980). Inorganic pollutants, particularly heavy metals in estuaries. In: Olausson, E. and Cato, I. (eds.), *Chemistry and Biogeochemistry of Estuaries*. Wiley, Chichester, 307–348.

Fukai, R., Murray, C. N., and Huynh-Ngoc, L., (1975). Variations of soluble zinc in the Var river and its estuary. *Est. Coast. Mar. Sci.*, **3**, 165–176.

Gardner, W. S., and Menzel, D. W. (1974). Phenolic aldehydes as indicators of terrestrially derived organic matter in the sea. *Geochim. Cosmochim. Acta,* **38,** 813–822.

Gibbs, R. J. (1973). Mechanisms of trace metal transport in rivers. *Science,* **180,** 71–73.

Gorsline, D. S. (1967). Contrasts in coastal bay sediments on the Gulf and Pacific coasts. In: Lauff, G. H. (ed.), *Estuaries, Am. Ass. Adv. Sci. Publ.,* **83,** 219–225.

Grieve, D., and Fletcher, K. (1977). Interactions between zinc and suspended sediments in the Fraser River estuary, British Columbia. *Est. Coast. Mar. Sci.,* **5,** 415–419.

de Groot, A. J. (1966). Mobility of trace elements in deltas. Transactions, Commissions II and IV, *Int. Soc. Soil Sci.,* Aberdeen, 267–279.

de Groot, A. J. (1973). Occurrence and behaviour of heavy metals in river deltas with special reference to the Rhine and Ems rivers. In: Goldberg, E. D. (ed.), *North Sea Science,* MIT Press Cambridge, Mass. and London, U.K., 308–325.

Gupta, S. K., and Chen, K. Y. (1975). Partitioning of trace metals in selective chemical fractions of near-shore sediments. *Env. Letters,* **10,** 129–158.

Hair, M. E., and Bassett, C. R. (1973). Dissolved and particulate humic acids in an east coast estuary. *Est. Mar. Coast. Sci.,* **1,** 107–111.

Hall, K., and Lee, G. F. (1974). Molecular size and spectral characterization of organic matter in a meromictic lake. *Wat. Res.,* **8,** 239–251.

Hallberg, R. O. (1973). The microbiological C-N-S cycles in sediments and their effects on the ecology of the sediment-water interface. *OIKOS Suppl.,* **15,** 51–62.

Hedges, J. I. (1977). The association of organic molecules with clay minerals in aqueous solutions. *Geochim Cosmochim. Acta,* **41,** 1119–1123.

Helfferich, F. (1962). *Ion Exchange,* McGraw Hill, New York.

Helz, G. R., Huggett, R. J., and Hill, J. M. (1975). Behaviour of Mn, Fe, Cu, Zn, Cd and Pb discharged from a waste water treatment plant into an estuarine environment. *Wat. Res.,* **9,** 631–636.

Hem, J. D. (1977). Reactions of metal ions at surfaces of hydrous iron oxide. *Geochim. Cosmochim. Acta,* **41,** 527–539.

Hem, J. D. (1963). Chemical equilibria and rates of manganese oxidation. *U.S. Geol. Surv. Water Supply Papers,* **1667A,** A1–A64.

Holliday, L. M., and Liss, P. S. (1976). The behaviour of dissolved iron, manganese and zinc in the Beaulieu estuary, S. England. *Est. Coast. Mar. Sci.,* **4,** 349–353.

Holmes, C. W., Slade, E. A., and McLerran, C. J. (1974). Migration and redistribution of Zn and Cd in marine estuarine systems. *Env. Sci. Techn.,* **8,** 255–259.

Holmes, C. W. (1977). Effect of dredged channels on trace metal migration in an estuary. *J. Res. U.S. Geol. Surv.,* **5,** 243–251.

Jackson, M. L. (1969). *Soil Chemical Analysis.* Prentice-Hall, Englewood Cliffs.

Jacobs, L. W., and Keeney, D. R. (1974). Methyl mercury formation in mercury treated sediments during *in situ* equilibration. *J. Env. Qual.,* **3,** 121–126.

Jenne, E. A. (1968). Controls of Mn, Fe, Co, Ni, Cu and Zn concentrations in soils and water: the significant role of hydrous Mn and Fe oxides. *Adv. Chem. Ser., Vol.* 73: *Trace Inorganics in Water,* 337–387.

Johnson, R. G. (1974). Particulate matter at the sediment-water interface in coastal environments. *J. Mar. Res.,* **32,** 313–330.

Kharkar, D. P., Turekian, K. K., and Bertine, K. K. (1968). Stream supply of dissolved Ag, Mo, Sb, Se, Cr, Co, Rb and Cs to the oceans. *Geochim. Cosmochim. Acta,* **32,** 285–298.

Koppelman, M. H., and Dillard, J. G. (1977). A study of the adsorption of Ni(II) and Cu(II) by clay minerals. *Clays and Clay Min.*, **25**, 457–462.

Kranck, K. (1974). The role of flocculation in the transport of particulate pollutants in the marine environment. In: *Proc. Int. Conf. of Persistent Chemicals in Aquatic Ecosystems*, Ottawa, Nat. Res. Council, section 1, p. 41–46.

Krauskopf, K. B. (1956). Factors controlling the concentrations of thirteen rare metals in sea water. *Geochim. Cosmochim. Acta*, **9**, 1–32B.

Krom, M. D., and Sholkovitz, E. R. (1977). Nature and reactions of dissolved organic matter in the interstitial waters of marine sediments. *Geochim. Cosmochim. Acta*, **41**, 1565–1573.

McKyes, E., Sethi, A., and Yong, R. N. (1974). Amorphous coatings on particles of sensitive clay soils. *Clays and Clay Min.*, **22**, 427–433.

Lee, G. F. (1975). Role of hydrous metal oxides in the transport of heavy metals in the environment. In: Krenkel, P. A. (ed.), *Heavy Metals in the Aquatic Environment*, Proc. Int. Conf. Nashville.

McLerran, C. J., and Holmes, C. W. (1975). Deposition of Zn and Cd by marine bacteria in estuarine sediments. *Limnol. Ocean.*, **19**, 998–1001.

Lindberg, S. E., and Harriss, R. C. (1974). Mercury-organic matter associations in estuarine sediments and interstitial water. *Env. Sci. Technol.*, **8**, 459–462.

Liss, P. S. (1976). Conservative and non-conservative behaviour of dissolved constituents during estuarine mixing. In: Burton, J. D. and Liss, P. S. (eds.), *Estuarine Chemistry*, Ac. Press, 93–130.

Lockwood, R. A., and Chen, K. Y. (1973). Adsorption of Hg(II) by hydrous manganese oxides. *Env. Sci. Technol.*, **7**, 1028–1034.

Lu, J. C. S., and Chen, K. Y. (1977). Migration of trace metals in interfaces of seawater and polluted surficial sediments. *Env. Sci. Technol.*, **11**, 174–181.

Mackay, D. W., and Leatherland, T. M. (1976). Chemical processes in an estuary receiving major inputs of industrial and domestic wastes. In: Burton, J. D., and Liss, P. S. (eds.), *Estuarine Chemistry*, Ac. Press, 185–218.

Mackenzie, F. T., and Garrels, R. M. (1966). Chemical mass balance between rivers and oceans. *Am. J. Sci.*, **264**, 507–525.

Manheim, F. T. (1976). Interstitial waters of marine sediments. In: Riley, J. P., and Chester, R. (eds.), *Chemical Oceanography*, 2nd ed., vol. 6, 115–186.

Martin, J. M., Jednačak, J., and Pravdić, V. (1971). The physico-chemical aspects of trace elements behaviour in estuarine environments. *Thalassia Jugosl.*, **7**, 619–637.

Martin, J. M., Nijampurkar, V., and Salvadori, F. (1978). Uranium and Thorium isotopes behaviour in estuarine systems. In: *Biogeochemistry of Estuarine Sediments*, Unesco.

Meade, R. H. (1972). Transport and deposition of sediments in estuaries. *Geol. Soc. Am. Mem.*, **133**, 91–120.

Mook W. G., and Koene, B. K. S. (1975). Chemistry of dissolved inorganic carbon in estuarine and coastal brackish waters. *Est. Coast. Mar. Sci.*, **3**, 325–336.

Morel, F. M. M., Westall, J. C., O'Melia, C. R., and Morgan, J. J. (1975). Fate of trace metals in Los Angeles county wastewater discharge. *Env. Sci. Technol.*, **9**, 756–761.

Müller, G., and Förstner, U. (1975). Heavy metals in sediments of the Rhine and Elbe estuaries: mobilization or mixing effect? *Env. Geol.*, **1**, 33–39.

Murray, C. N., and Meinke, S. (1976). Influence of soluble sewage material on adsorption and desorption behaviour of cadmium, cobalt, silver, and zinc in sediment—freshwater, sediment–seawater systems. *IAEA Radioact. in the sea* no. 51, 5 pp.

Murray, J. W. (1975a). The interaction of cobalt with hydrous manganese dioxide. *Geochim. Cosmochim. Acta*, **39**, 635–647.

Murray, J. W. (1975b). The interaction of metal ions at the manganese dioxide-solution interface. *Geochim. Cosmochim. Acta*, **39**, 505–519.

Neihof, R. H., and Loeb, G. I. (1972). The surface charge of particulate matter in seawater. *Limnol. Oceanogr.*, **17**, 7–16.

Nichols, M. M., and Poor, G. (1967). Sediment transport in a coastal plain estuary. *Proc. Am. Soc. Civil Eng.*, **93WW4**, 83–95.

Nissenbaum, A., and Swaine, D. J. (1976). Organic matter-metal interactions in recent sediments: the role of humic substances. *Geochim. Cosmochim. Acta*, **40**, 809.

Parks, G. A. (1975). Adsorption in the marine environment. In: Riley, J. P., and Skirrow, G. (eds.), *Chemical Oceanography*, 2nd ed., vol. 1, 241–308.

Perez Rodriguez, J. L., Weiss, A., and Lagaly, G. (1977). A natural clay organic complex from Andalusian Black earth. *Clays and Clay Min.*, **25**, 243–251.

Petr, T. (1977). Bioturbation and exchange of chemicals at the mud-water interface. In: Golterman, H. L. (ed.), *Interactions between Sediments and Freshwater.* Dr. W. Junk, B. V. Publishers, the Hague and Centre for Agricultural Publishing and Documentation, Wageningen, 473 pp.

Phillips, J. (1972). Chemical processes in estuaries. In: Barnes, R. S. K., and Green, J. (eds.), *The Estuarine Environment, Appl. Sci. Publ.* Ltd. Lond.

Pillai, T. N. V., Desai, M. V. M., Mathew, E., Ganapathy, S., and Ganguly, A. K. (1971). Organic materials in the marine environment and the associated metallic elements. *Current Sci.*, **40**, 75–81.

Posselt, H. S., Anderson, F. J., and Weber, W. J. (1968). Cation sorption on colloidal hydrous manganese-dioxide. *Env. Sci. Technol.*, **2**, 1087–1093.

Postma, H. (1961). Transport and accumulation of suspended matter in the Dutch Wadden Sea. *Neth. J. Sea Res.*, **1**, 148–190.

Postma, H. (1967). Marine pollution and sedimentology. In: Olson, T. A., and Burgess, F. J. (eds.), *Pollution and Marine Ecology*, 225–234.

Postma, H. (1980). Sediment transport and sedimentation. This volume p. 153–186.

Pravdić, V. (1970). Surface charge characterization of sea sediments. *Limnol. Oceanogr.*, **15**, 230–233.

Presley, B. J., Kolodny, Y., Nissenbaum, A., and Kaplan, I. R. (1972). Early diagenesis in a reducing fjord, Saanich Inlet, Br. Columbia, II. Trace element distribution in interstitial water and sediment. *Geochim. Cosmochim. Acta*, **36**, 1073–1090.

Pritchard, D. W. (1967). Observations of circulation in coastal plain estuaries. In: Lauff, G. H. (ed.), *Estuaries. Am. Assoc. Adv. Sci. Publ.*, **83**, 757 pp. 37–44.

Rashid, M. A. (1971). Role of humic acids of marine origin and their different molecular weight fractions in complexing di- and trivalent metals. *Soil Sci.*, **111**, 298–306.

Rashid, M. A., Buckley, D. E., and Robertson, K. R. (1972). Interactions of a marine humic acid with clay minerals and a natural sediment. *Geoderma*, **8**, 11.

Rashid, M. A. (1969). Contribution of humic substances to the cation exchange capacity of different marine sediments. *Maritime Sediments*, **5**, 44–50.

Rashid, M. A., and Leonard, J. D. (1973). Modifications in the solubility and precipitation behaviour of various metals as a result of their interaction with sedimentary humic acid. *Chem. Geol.*, **11**, 89–97.

Reimers, R. S., and Krenkel, P. A. (1974). Kinetics of mercury adsorption and desorption in sediments. *J. Wat. Poll. Control. Fed.*, **46**, 353–365.

Reinhard, D., and Förstner, U. (1976). Metallanreicherungen in Sedimentkernen aus Stauhaltungen des mittleren Neckars. *N. Jb. Geol. Mh.*, **H1**, 301–320.

Reuter, J. H., and Perdue, E. M. (1977). Importance of heavy metal–organic matter interactions in natural waters. *Geochim. Cosmochim. Acta*, **41**, 325–334.

Roberts, W. P., and Pierce, J. W. (1976). Deposition in upper Patuxent estuary, Maryland, 1968–1969. *Est. Coast. Mar. Sci.*, **4**, 267–280.

Rohatgi, N. K., and Chen, K. Y. (1976). Fate of metals in wastewater discharge to ocean. *J. Env. Eng. Div. ASCE*, **102 (EE3)**, 675–685.

Rohatgi, N., and Chen, K. Y. (1975). Transport of trace metals by suspended particulate on mixing with sea water. *J. Water. Poll. Contr. Fed.*, **47**, 2298–2316.

Salomons, W. (1975). Chemical and isotopic composition of carbonates in recent sediments and soils from western Europe. *J. Sed. Petrol.*, **45**, 440–449.

Salomons, W., and Mook, W. G. (1977). Trace metal concentrations in estuarine sediments: mobilization, mixing or precipitation. *Neth. J. Sea Res.*, **11**, 199–209.

Schnitzer, M., and Kahn, S. U. (1972). Humic substances in the environment, Dekker.

Schultz, D. J., and Calder, J. A. (1976). Organic carbon $^{13}C/^{12}C$ variations in estuarine sediments. *Geochim. Cosmochim. Acta*, **40**, 381–385.

Senesi, N., Griffith, S. M., Schnitzer, M., and Townsend, M. G. (1977). Binding of Fe^{3+} by humic materials. *Geochim. Cosmochim. Acta*, **41**, 969–976.

Shapiro, J. (1957). Chemical and biological studies on the yellow organic acids of lake water. *Limnol. Oceanogr.*, **11**, 161–179.

Shapiro, J. (1964). The effect of yellow organic acids on iron and other metals in water. *J. Am. Water WKS. Ass.*, **56**, 1062–1082.

Sheldon, R. W., Evelyn, T. P. T., and Parsons, T. R. (1967). On the occurrence and formation of small particles in sea water. *Limnol. Oceanogr.*, **12**, 367–375.

Sheldon, R. W. (1968). Sedimentation in the estuary of the river Crouch, Essex, England. *Limnol. Oceanogr.*, **13**, 72–83.

Sholkovitz, E. R. (1976). Flocculation of dissolved and inorganic matter during the mixing of river water and sea water. *Geochim. Cosmochim. Acta*, **40**, 831–845.

Sieburth, J. McN., and Jensen, A. (1968). Studies on algal substances in the sea. I. Gelbstoff (humic material) in terrestrial and marine waters. *J. Exp. Mar. Biol. Ecol.*, **2**, 174–189.

Sloot, H. A. van der, Massee, R., and Wals, G. D. (1978). Methods for the determination of Se, As and Sb species in natural waters and their application in an environmental study. *Third Int. Conf. on Nuclear Methods in Environmental and Energy Research*, 10–13 Oct. 1977, Columbia (Miss.).

Stumm, W., and Morgan, J. J. (1970). *Aquatic Chemistry*, Wiley, New York, 583 pp.

Stumm, W., and Lee, G. F. (1960). The chemistry of aqueous iron. Schweiz. *Zeitschr. für Hydrology*, **XXI**, 295–319.

Stumm, W., and Lee, G. F. (1961). Oxygenation of ferrous iron. *Ind. Eng. Chem.*, **53**, 143–146.

Subramanian, V., and d'Anglejan, B. (1976). Water chemistry of the St. Lawrence estuary. *J. Hydrol.*, **29**, 341–354.

Thomas, D. J., and Grill, E. V. (1977). The effect of exchange reactions between Fraser River sediment and sea water on dissolved Cu and Zn concentrations in the Strait of Georgia. *Est. Coast. Mar. Sci.*, **5**, 421–427.

Thomson, J., Turekian, K. K., and McCaffrey, R. J. (1975). The accumulation of metals in and release from sediments of Long Island Sound. In: Cronin, L. E. (ed.), *Estuarine Research*, vol. 1, Ac. Press, 28–44.

Troup, N. B., and Bricker, O. P. (1975). Processes affecting the transport of materials from continents to oceans. In: Church, T. M. (ed.), Marine Chemistry in the coastal Environment, *ACS Symp. Ser.*, **18,** Am. Chem. Soc. Washington, D.C.

Turekian, K. K., and Scott, M. R. (1967). Concentrations of Cr, Ag, Mo, Ni, Co and Mn in suspended material in streams. *Env. Sci. Technol.*, **1,** 940–942.

Turekian, K. K. (1971). Rivers, tributaries and estuaries. In: Hood, D. W. (ed.), *Impingement of Man on the Oceans*, Wiley-Interscience, p. 9.

Turekian, K. K. (1977). The fate of metals in the oceans. *Geochim. Cosmochim. Acta*, **41,** 1139–1144.

Veen, J. van (1937). Korte beschrijving der uitkomsten van onderzoekingen in de Hoofden en langs de Nederlandse kust. Tijdschr. *K. Ned. Aardrijksk. Gen.*, **54,** 155–195.

Wakeman, T. (1976). The biological ramifications of dredging and disposal activities. In; *Proc. 7th World Conf. on Dredging: Env. Effects and Technol.*, pp. 53–68, San Francisco.

Weyden, C. H. van der, Arnoldus, M. J. H. L., and Meurs, C. J. (1977). Desorption of metals from suspended material in the Rhine estuary. *Neth. J. Sea Res.*, **11,** 210–225.

Whitehouse, U. G., Jefferey, L. M., and Debbrecht, J. D. (1960). Differential settling tendencies of clay minerals in saline waters. *Proc. 7th Conf. Clays and Clay Mins.*, 1–79.

Wiklander, L. (1964). Cation and anion exchange phenomena. In: Bear, F. E. (ed.), Chemistry of the Soil. *Am. Chem. Soc. Monograph Ser.*, **126,** Reinhold N. Y. 375 pp., 107–149.

Windom, H. L., Beck, K. C., and Smith, R. (1971). Transport of trace elements to the Atlantic Ocean by three southeastern rivers. *Southeast. Geol.*, **12,** 169–181.

Wolfe, D. A., Cross, F. A., and Jennings, C. D. (1973). The flux of Mn, Fe and Zn in an estuarine ecosystem. *Radioactive Contamination of the Marine Env.* (Symp. Seattle, July 1972), IAEA, 159–176.

Wollast, R., Billen, G., and Duinker, J. C. (1979). Manganese in the Rhine and Scheldt estuaries. Part 1: Physico-chemical behaviour. *Est. Coast. Mar. Sci.* **9,** 161–169.

Chemistry and Biogeochemistry of Estuaries
Edited by E. Olausson and I. Cato
Copyright © 1980 by John Wiley & Sons Ltd.

H. POSTMA
Netherlands Institute for Sea Research,
Den Burg, Texel

5

Sediment Transport and Sedimentation

1 ESTUARINE DEVELOPMENT AND RIVER INPUT

In the geological time scale estuaries are only very temporary structures, of which the age is measured in thousands instead of millions of years. A dominating factor is the rise of sea level since the last ice age (see further Fairbridge, this volume). Sea level became comparatively stable only about 3,000 years ago so that most modern estuaries are not older than about twice that age. This fact must be kept in mind when considering estuarine development.

The history of an estuarine system, after it has been established as a geomorphological entity, is largely determined by its sediment supply. The prime source of sediments is obviously the land. Globally, about 12 km^3 or 18×10^9 tons of chiefly fine-grained sediment are carried from the land to the sea every year, mainly by rivers (Kuenen, 1950; Holeman, 1968). A small quantity is transported through the atmosphere. Another source is coastal erosion. To estimate the influence of this transport on estuarine development, one must know, first of all, how much of the total amount is retained in nearshore areas.

There are three possible main regions of deposition: the estuaries, the continental shelf, and the deep sea. The continental shelves are, generally

 H. Postma

speaking, covered with relatively coarse 'relict' sands deposited during lower levels of the sea. Drowned barriers and tidal channels can locally still be discerned (Emery, 1968). The rise of sea level obviously did not create wholly new depositional environment, although the older deposits are partly reworked (Curray, 1969; Swift, 1976) and some fine-grained material is deposited in shelf depressions (Figure 1). The causes of the non-depositional character and partial reworking are strong residual or tidal currents, occasional severe storms and hurricanes. Conditions for deposition are somewhat better on the continental slope but this area is relatively small (McCave, 1972).

As regards deposition in the deep sea, an important clue is found in the distribution of lithogenous material in deep sea sediments. It appears that for clay species and quartz wind transport is of about equal importance as

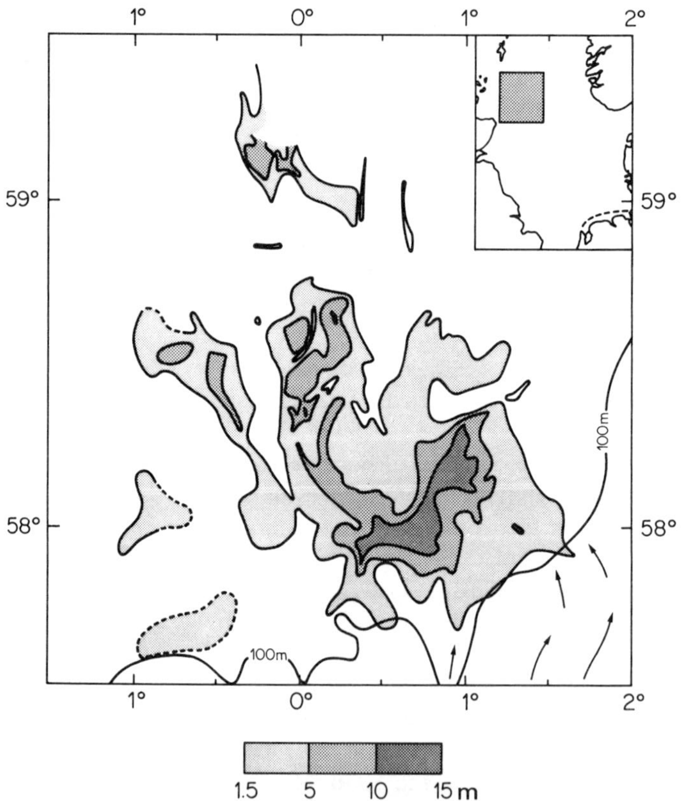

Figure 1. Distribution and thickness of the Witch Deposits in the northern North Sea. Sedimentation is assumed to have started about 18,000 BP when sea level approximately followed the 100 m depth contour (After Jansen, 1976).

continental run-off (Windom, 1976). This indicates that at present relatively little continental material reaches the deep sea. This conclusion is supported by deep sea sedimentation rates. An average rate of lithogenic sedimentation is 0.4 g per cm^2 per thousand years. With a deep ocean area of 270×10^6 km^2 this gives an annual deposition of 1.1×10^9 tons or only 6% of total sediment supply.

The deep sea and the continental shelf thus accomodate only a small part of the present sedimentary input, at best some 10%. This seems to leave estuaries as the main areas of sedimentation; a look at the actual supply points of terrestrial material gives a somewhat differentiated picture.

Of the total sediment input of 18×10^9 tons almost one half (8×10^9 tons) is supplied by about 30 large rivers and within this group an even smaller number of rivers, who have their sources in the highlands of Tibet and in the Himalayas, take the biggest share (5.5×10^9 tons). These are the Indus, Ganges and Brahmaputra, the Irrawaddy and Mekong, the Yellow River and the Yangtze. In addition to big deltas, they have built large submarine fans who often reach into the deep sea and may accommodate most of the sediment they have supplied. Of the rivers outside the big southeast Asian systems only two, the Amazon and the Mississippi, are in the same class with regard to sediment supply, each of them having an annual output of about 0.5×10^9 tons. In the case of the Amazon, large quantities of mud are deposited in a long coastal zone stretching from the mouth of the river up to the Orinoco delta. The mud banks move slowly from east to west with a speed of about 1.5 km per year, moving about 0.05×10^9 tons of sediment (10% of the total Amazon supply; Diephuis, 1966). Another part is deposited on the continental slope and rise and in the abyssal Caribbean. A comparable distribution pattern applies to the Mississippi (Curray, 1960; Figure 2). However, even by taking such areas of sediment accumulations into account, the volume to be stored in estuaries may still come to about half of the total supply.

Most of this amount will be supplied in relatively modest portions by a very large number of relatively small rivers which, added up together, still provide two-thirds of total continental supply. The fact that estuaries still offer so much storage space for new sediments from the land obviously means that in many cases sedimentation has not kept pace with the inundation through the rise of sea level after the last low sea level period. Only part will have reached a new equilibrium.

An interesting illustration of this state of affairs is provided by the series of estuaries along the east coast of the United States between Cape Cod and Cape Canaveral (Meade, 1969; Figure 3). The estuaries in the section of the coast north of Cape Lookout receive much river water but relatively little sediment and the river valleys are still fully visible. About 90% of the sediment supplied is retained in the estuaries (Meade *et al.*, 1975). The

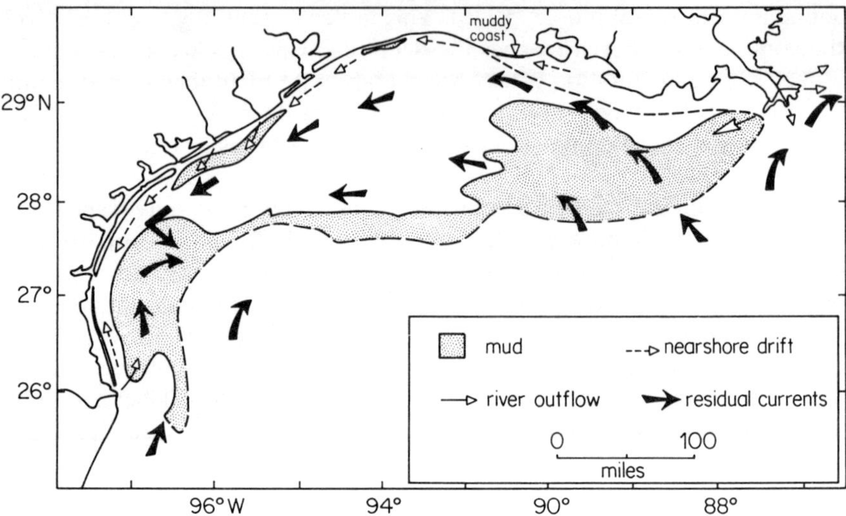

Figure 2. Mud deposits and currents in the northern Gulf of Mexico (After Curray (1960) redrawn by McCave (1972)).

southern section receives less river water but a big load of sediment and here the valleys, although admittedly smaller than in the north, have almost completely filled up and vanished under a mud blanket.

On a small scale, the figure illustrates a phenomenon, that is essentially world-wide. The northern section receives water from drainage areas, scoured by glaciers in the past, which provide little sediment, whereas the southern section has ample sediment supply from less denuded areas which have not been glaciated. The filling of northern estuaries and especially of fjords, who provide ample storage space, proceeds very slowly.

An example of estuarine filling is given for the river Loire (Barbaroux *et al.*, 1974) in Figure 4. These clearly show the magnitude of the deposits following the rise of sea level starting from a wide and relatively deep basement. A well-studied example of estuarine deposition since the last glaciation is Long Island Sound (Bokuniewicz *et al.*, 1976, Figure 5). In shallower estuaries the addition of sediment has not proceeded in such an orderly fashion since the basin, after inundation, is reshaped by currents and waves including the repeated formation of new channels. In such cases much older sediments have been eroded and redistributed and the over-all impression is one rather of erosion than of deposition (Colquhoun *et al.*, 1974).

It should be stressed that, although the overall value for total suspended load carried by rivers to the sea used above may be not far from the truth, accurate data for individual estuaries are often not available. A long series of measurements closely spaced in time is necessary to arrive at a realistic

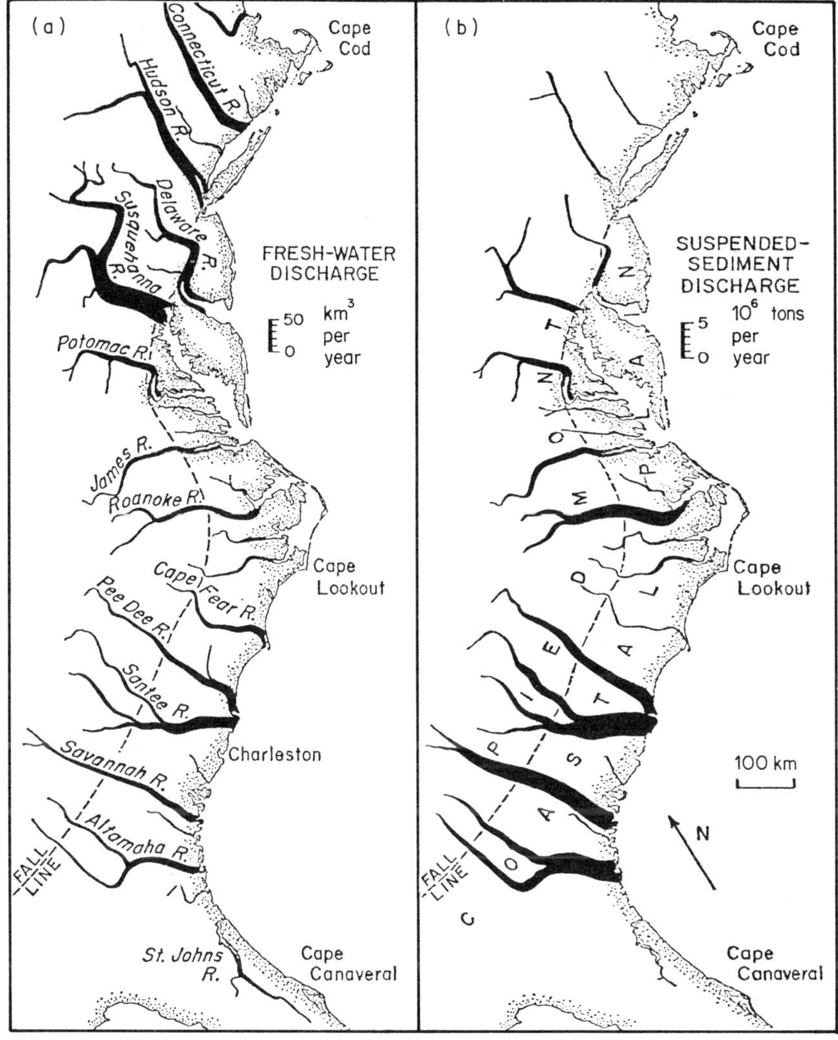

Figure 3. (a) Water and (b) sediment discharged by rivers between Cape Cod and
Cape Canaveral (After Meade, 1969).

input. Abnormally high river run-off may transport more material into an
estuary in a few days than normal run-off in years. A dramatic example is
given by Schubel (1974) for upper Chesapeake Bay. The passage of hurricane 'Agnes' was accompanied by heavy rains which increased the flow of
the Susquehannah River to a peak of 25 times above normal. In one week
more sediment was supplied than in the two or three preceding decades.

H. *Postma*

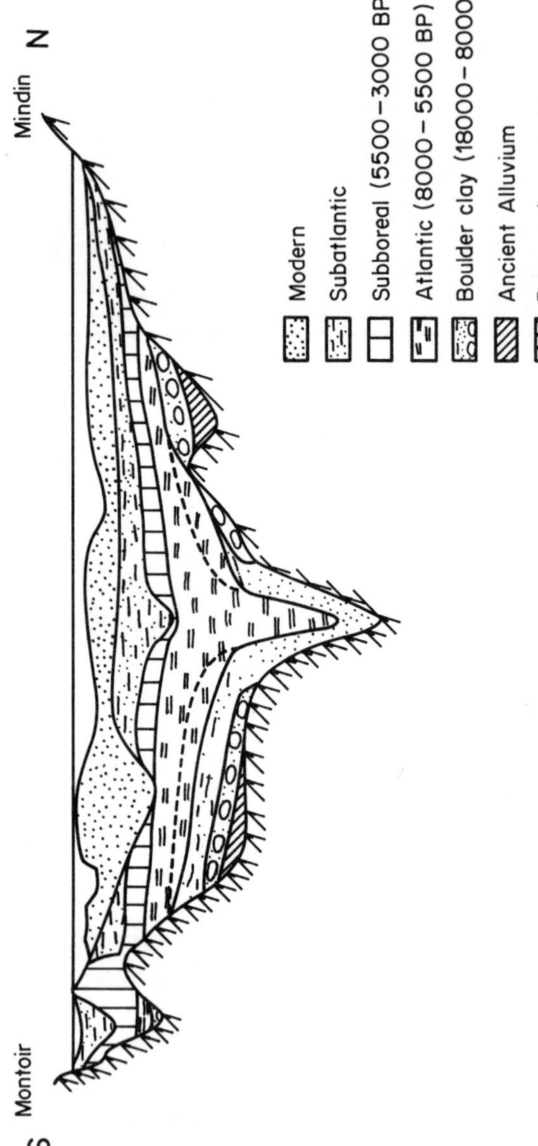

Figure 4. Evolution of the estuary of the Loire during the Quaternary (After Barbaroux *et al.*, 1974).

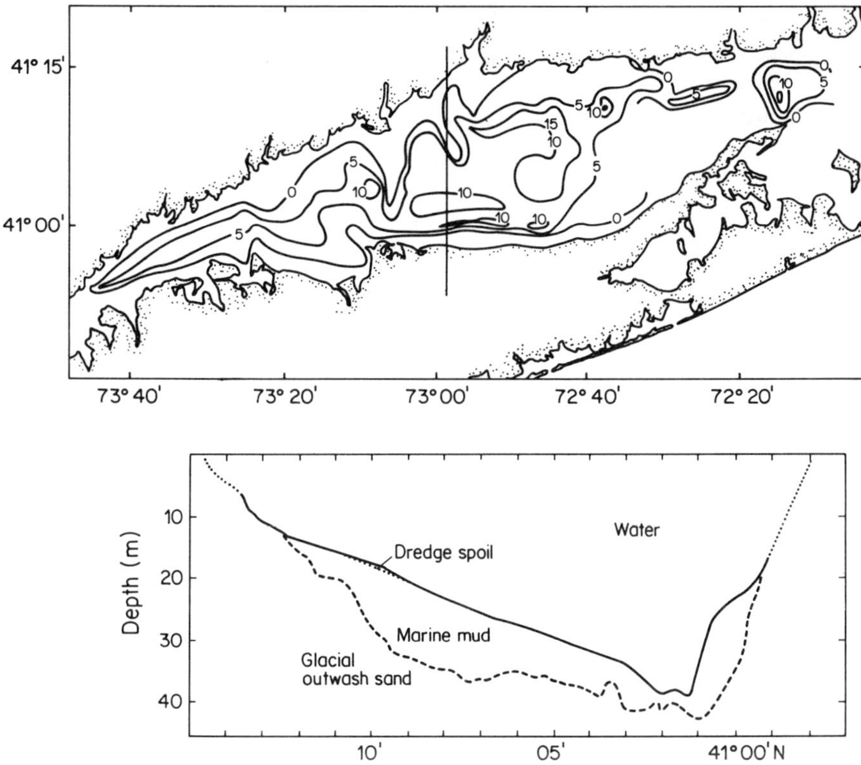

Figure 5. Estuarine filling of Long Island Sound. Upper figure: thickness of sediments above glacial basement. Lower figure: NS cross section (After Bokuniewicz *et al.*, 1976).

Storms can also introduce marine sediment into outer estuarine channels and even by 'washover' of sand bars. Long-term changes in sediment supply can be caused by deforestation or reforestation of the river drainage area, by soil erosion through urban development plans (Roberts and Pierce, 1976) and by the building upstream reservoirs for hydroelectric power and water storage.

2 WATER AND SEDIMENT CIRCULATION PATTERNS

It follows from the foregoing discussion that conditions for sediment deposition in estuaries are much more favourable than on the open shelf although the rise of sea level has influenced both environments in a similar way. An obvious reason is that estuaries offer an effective shelter against strong waves and currents, so that also fine-grained material can settle, whereas shelves are too exposed. This is undoubtedly an important factor, but in addition a number of water and sediment transport mechanisms exist which actively prevent the escape of sediment. Some of these mechanisms extend outward over the inner shelf.

A very old concept is that of (i) *landward transport by waves* touching bottom in shallow water (Figure 6). Such waves have sharp crests with a high speed orbital velocity and long flat troughs with a low velocity. This causes a net onshore movement of especially sand-size material. Equilibrium is attained where this movement is balanced by the offshore pull of gravity on a slope in seaward direction. Generally speaking, wave surge transport does not seem to work effectively in depths greater than 10–15 m, but it is able to push sand and even pebbles to the shore over many kilometres. A climax is reached in the surf zone at the beach. This type of transport is also responsible for the creation of offshore sand bars and, since waves seldom approach a coast at a right angle, for sand transport along the shore. The nearshore barriers erected in this manner have grown above the high tide level along many coasts and over large distances, thus forming the outer defenses of tidal lagoon systems such as those along the south-east coast of the United States and on the continental side of the North Sea. In some cases several sand bars have been built parallel to each other (Curray and Moore, 1964).

These systems are not necessarily connected with a river mouth and the sediments they contain may be of marine instead of fluviatile origin. Many barrier-built coastal embayments do not receive fresh water at all.

A second mechanism that can cause a net transport in one direction is (ii)

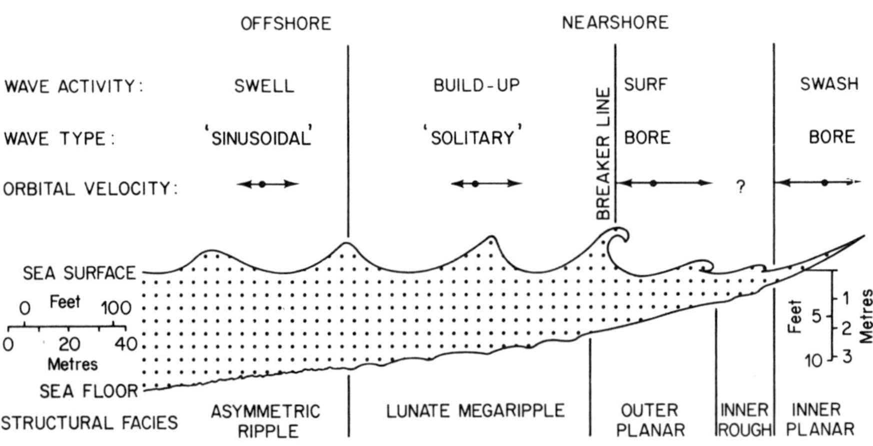

Figure 6. Schematic representation of wave and sea floor characteristics near the shore. (After Clifton *et al.*, (1971) in Swift, 1976).

tidal movement. There are several possibilities which all start from the observation that there is a time lag between the tidal current rhythm and the changes in suspended matter concentration (Figures 7 and 8). This time lag is caused by the fact that an increasing current needs time to stir up sediment from the bottom and carry it into a higher water level, whereas in a decreasing current the suspended sediment needs time to settle. The lag effect is only effective in creating unidirectional transport if combined with a certain tidal asymmetry. A common asymmetry in estuaries with extensive intertidal flats is a slower turn of the tide at high water than at low water (Postma, 1961). An example is given in Figure 10. A numerical calculation is given in Figure 9 (Groen, 1967). The time lag used in this example is 3 hours. Obviously this time lag depends on depth and particle size, but to be effective particles should in any case settle rather slowly. This excludes coarse and median-grained sand above about 100 microns. Probably the greatest effect is obtained for particles between 10 and 50 microns, i.e. for silt and very fine sand, but time lags have also been demonstrated for fine sand (Gry, 1942; Thorn, 1975).

It should be noted that the lag effect described here is simply based on vertical turbulent exchange and settling velocity and that it does not require a certain cohesion of sediment. This property may considerably strengthen the lag effect by increasing the time needed to stir up sediment from the bottom, but it is not a prerequisite.

The influence of cohesion on sediment transport is demonstrated by Figure 10, representing an experiment with Wadden Sea mud in a circular tank (a so-called 'caroussel'; Creutzberg and Postma, 1979) in which a current was generated by means of rotating paddles. The mud was left at rest on the bottom for 16 hours until it formed a layer of 3.5 cm thickness. In the increasing current erosion started at 40–50 cm/sec, whereas in the decreasing current deposition took place at 10–20 cm/sec. The erosion velocity is smaller after shorter consolidation times, but may already be significantly higher than deposition velocity after a few hours.

A second remark to be made here is that the assumed tidal asymmetry, although quite common, is not universal. In large tidal channels the tide is mostly symmetrical and in small gullies the asymmetry may be in the opposite sense, the turn of the tide at HW (high water) being fast and LW (low water) being slow (Boon, 1974; Figure 11).

In the case of a symmetrical tide accumulation cannot be due to the tide itself. Another type of accumulation mechanism however, may take over: in an intertidal flat system the water at HW is spread over the shallow flats, whereas at LW it is retracted in the deep gullies. The difference in depth causes a greater mud deposition at HW than at LW. As a result, a lag occurs only in the first ebb phase and not in the corresponding flood phase.

In the case of an 'opposite' tidal asymmetry there must be a loss of mud in

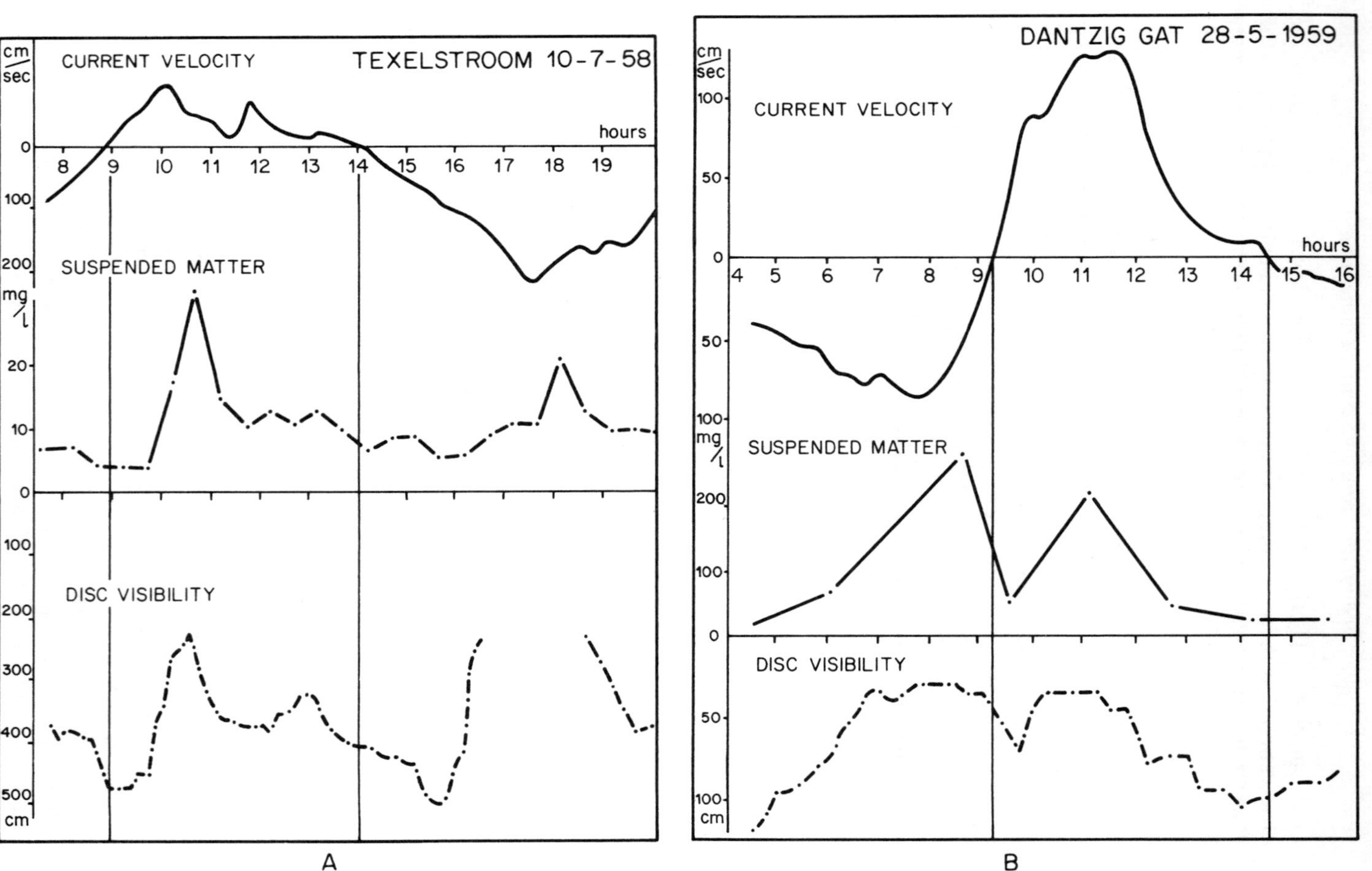
TEXELSTROOM 10-7-58
CURRENT VELOCITY
SUSPENDED MATTER
DISC VISIBILITY
hours
cm/sec
mg/l
cm
A
DANTZIG GAT 28-5-1959
CURRENT VELOCITY
SUSPENDED MATTER
DISC VISIBILITY
hours
cm/sec
mg/l
cm
B

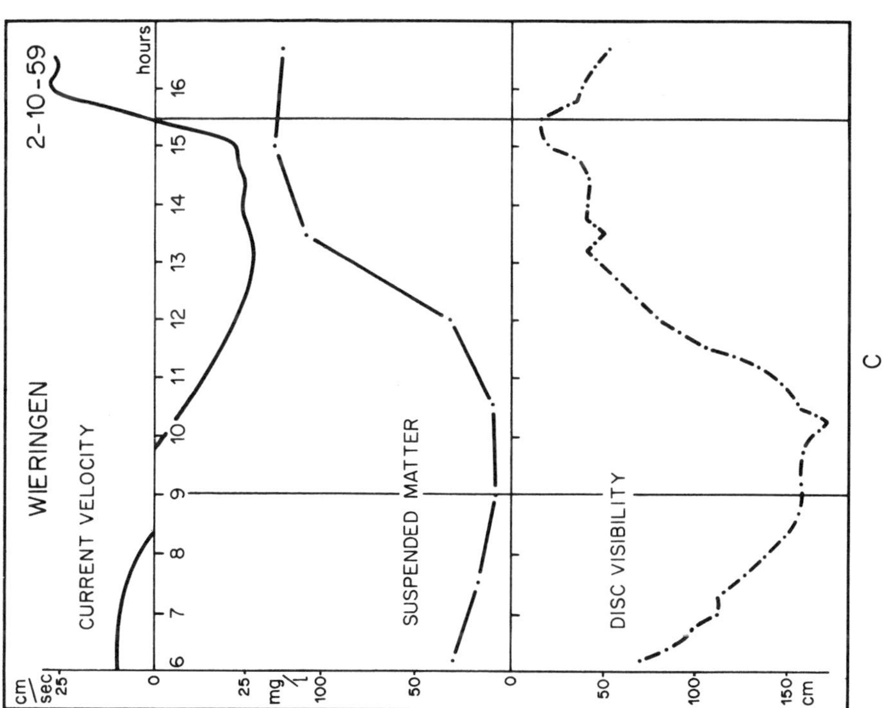

Figure 7. Current velocity, suspended matter concentration (surface) and Secchi disc visibility at three Wadden Sea stations. Note time lag between slack water and sediment concentration and tidal asymmetry at stations B and C (After Postma, 1961).

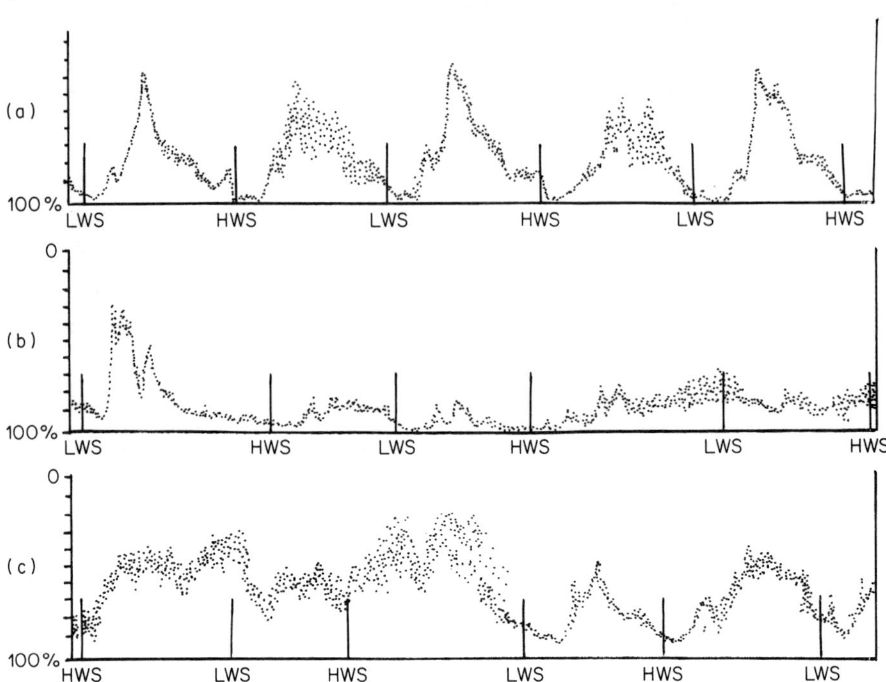

Figure 8. Tidal rhythm of turbidity in the Wadden Sea measured with a transparency meter near a tidal inlet in february 1965. High water slack (HWS) and low water slack (LWS) are indicated. (a) The upper series shows the normal variations. (b) In the middle series a storm from the west pushes clear North Sea water into the inlet (long flood period in first interval). (c) In the lower series the storm has subsided and turbid water from the interior, churned up in shallow water, arrives in the inlet (two long ebb periods). Observations of the writer.

stead of a gain. This situation seems to occur in creeks cut into tidal marshes which normally are hardly flooded at HW. Under these conditions marshes will lose sediment through the creeks. The processes building up such marshes and compensating for these losses are occasional inundations during storms, when the incoming tidal surge is heavily loaded with silt that is subsequently deposited in the marsh.

The tidal accumulating mechanism, like the wave surge transport, is independent of the presence of a river. It is, in fact, most clearly observed in systems behind offshore sand bars dissected by tidal inlets but without fresh water supply.

A third mechanism causing net inward transport of suspended matter is (iii) *the estuarine circulation.* (see further Bowden, this volume) This circulation is essentially a two-layer flow affected by the tide and caused by differences in specific weight between river and sea water; in some cases it is

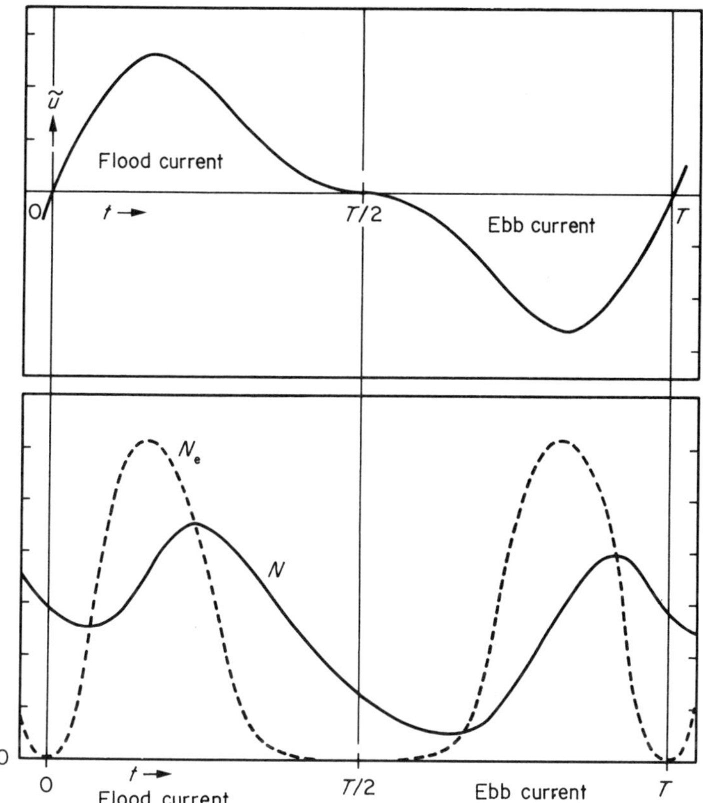

Figure 9. Upper figure: asymmetric tide. Lower figure: sediment concentration without time lag (N_e) and with a time lag of three hours (N). Multiplication of N with current velocity gives sediment movement with the result that flood sediment transport is 38% larger than ebb transport calculations in Groen, 1967.

caused by offshore winds (upwelling; Otto, 1975; Chase, 1975) or temperature differences (Postma, 1965). Fjords are a special case of salt wedge estuaries but with a very deep lower layer that is flushed occasionally over the sill. The different water movements in the three types of estuaries (salt wedge, partially mixed, and well mixed estuaries) result in a different behaviour of the sediment load (Figure 12, 13). In the case of a salt wedge estuary most or all suspended material is fluviatile. Since water movements in the bottom layers are slow the wedge tends to be filled up with sediment settling through the halocline and if sufficient material is supplied the river has from time to time to take a different course. If both the tidal and wave regime in the adjacent sea are mild a delta may be formed. Well studied

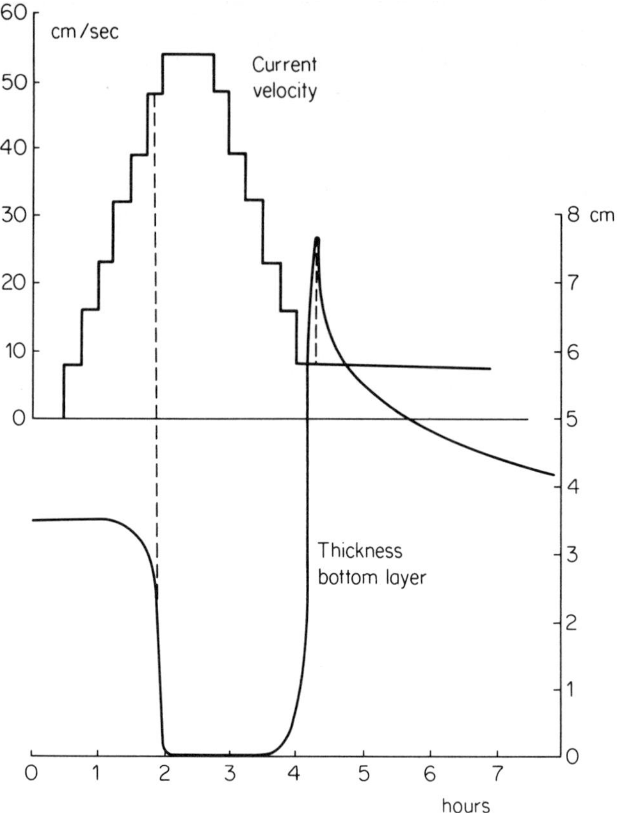

Figure 10. Experiment to measure depositon and erosion velocity of Wadden Sea mud. For explanation see text, p. 161.

examples of such delta's besides that of the Mississipi, are those of the Rhône (van Straaten, 1959), Niger (Allen, 1964), and Orinoco (van Andel, 1967). Fjords are obviously estuaries of the salt wedge type with insufficient sediment supply to build up a substantial deposit.

In partially mixed estuaries the landward bottom flow is sufficiently strong to move suspended sediment up the estuary until the head of the salt intrusion. This sediment may be fluviatile material that has settled from the upper into the lower layer, but it may also be of marine origin, depending on the concentrations of suspended matter in the river or in the sea. Around the head of the salt intrusion concentrations of suspended matter become much higher than in either the landward or the seaward source waters and a so-called 'turbidity maximum' is formed. The estuarine circulation thus acts as a sediment trap, preventing or at least retarding the escape of sediment to

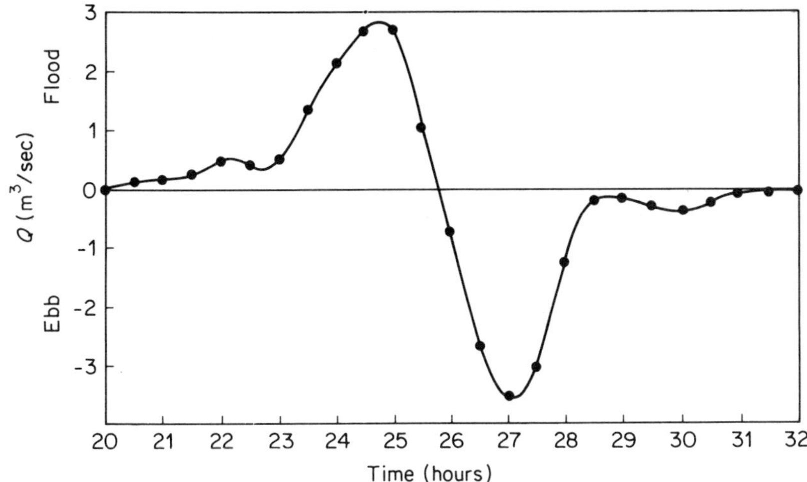

Figure 11. 'Opposite' asymmetry in a tidal channel of a salt marsh drainage
system in Virginia (After Boon, 1974).

the open sea. Classical examples of turbidity maxima are found in the French
Atlantic estuaries such as the Gironde (Glangeaud, 1938; Allen *et al.*, 1974)
and the estuary of the Loire (Berthois, 1964; Gallenne. 1974). Since, such
maxima have been observed in numerous other estuaries (examples in
Figure 14).

The grain-sizes trapped by the estuarine circulation are generally smaller
than those accumulated by the tidal movement since the sediment moves in
a cycle with a slower period and thus can settle more slowly. The turbidity
maximum of the Sacramento river in San Francisco Bay, for example,
accumulates grains of 8μm; that of the Loire seems to prefer 6μm
(calculated from Gallenne, 1974). In the Gulf of San Miguel, Panama, Swift,
and Pirie (1970) find a median equivalent diameter of about 4μm. The
present writer found values of $5–10 \mu$m (Postma, 1967). The average
particle size in the Saint Lawrence estuary is 5μm (d'Anglejan and Smith,
1973). A very interesting steady state numerical model developed by Festa
and Hansen (1978) demonstrates that the magnitude and location of a
turbidity maximum depends on particle size, the amount of suspended
matter available and the strength of the estuarine circulation. A stronger
circulation can handle coarser grains with greater settling velocities. Figure
14 summarizes some of their results.

The location of the zone of high turbidity can be very variable, depending
on the state of the tide and fresh water supply. A strong river flow can push
the turbidity maximum out of the estuary and at very low river flow the
maximum disintegrates. In a wide and deep estuary the turbidity pattern can

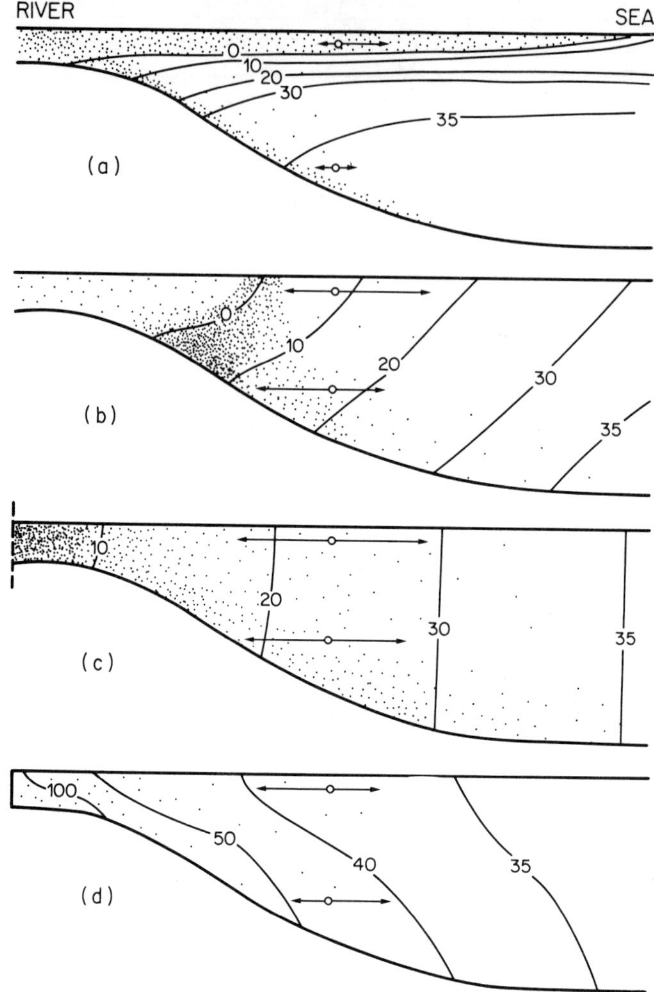

Figure 12. Schematic representation of types of estuaries. The dots indicate sediment concentration and the arrows net water movements over ebb and flood.

(a) Salt wedge estuary. It is assumed that most river sediment is carried to the sea with the river water in the upper layer. An example is the Mississippi.

(b) Partially mixed estuary. The suspended matter is concentrated in a turbidity maximum.

(c) Fully mixed estuary. The suspended matter is concentrated nearshore as in the case of a tidal mechanism without appreciable river flow.

(d) Negative estuary with excess evaporation and high salinities at the head. Net inward flow at the surface and outward flow near the bottom.

be quite complicated. An example is the Saint Lawrence estuary. Downstream from the actual turbidity maximum secondary maxima develop, often disconnected from the channel floor (d'Anglejan and Ingram, 1976; Figure 15). These may derive from the main turbid zone by horizontal advection or by cross-channel transport from mud banks alongside the river. Other examples are in fjords, where bottom sediments are at rest most of the time, but may suddenly be resuspended by momentous inflow of heavier sea water across the sill (Price and Skei, 1975).

In summary, three main mechanisms appear to be responsible for nearshore sediment accumulation: (1) wave transport which acts mainly on coarse material, such as sand, shell remains and pebbles, (2) tidal transport which acts on finegrained sand and silt and (3), estuarine circulation which accumulates silt and mud. Combined, they act on the whole grainsize spectrum, except on grains of micron sizes which practically behave like dissolved material.

It should be added, however, that there are also modes of water circulation which act in the opposite direction. Attention has already been drawn to the possibility that forms of tidal action exist which are neutral or even cause offshore sediment transport. Another specific case is that of *negative estuarine circulation* (Figure 12(d)). In this case the net movement of bottom water is seaward and that of surface water landward. The driving force is excess evaporation causing high salinities—and corresponding high densities—in the inner section of an estuary. In such a situation silt will be exported by the bottom water. The sandy character of negative estuaries is enhanced by their location in arid regions. A negative estuarine circulation does not exclude trapping of mud by the tidal action mechanism, but the amount seems to be small. In an example studied by the writer (Laguna Guerrero Negro in Baja California; Postma, 1965) finegrained suspended matter concentrations were always below 1 mg/l and mostly below 0.1 mg/l, although suspended sand concentrations reached maxima of 200 mg/l.

Negative estuaries, however, offer an excellent environment for the study of tidal action on sand movement. The paper mentioned above is used to demonstrate the relation between current velocity and the amount of sand in suspension (Figures 16 and 17). It shows, firstly, that for sand (median diameter 125 μm) there is only a small time lag between tidal rhythm and concentration and, secondly, that the amount of sand in suspension increases logarithmically with current velocity. This causes suspended sand concentrations in estuaries to be extremely variable and makes budget calculations on the basis of sand concentrations and tidal flow much more difficult than for finegrained material. As a result, careful and repeated mapping combined with grainsize analyses is mostly the only way to determine whether an estuary is gaining or losing sand. An example of an estuary mainly filled with marine sand is Bolinas Lagoon in California (Ritter, 1972).

 H. Postma

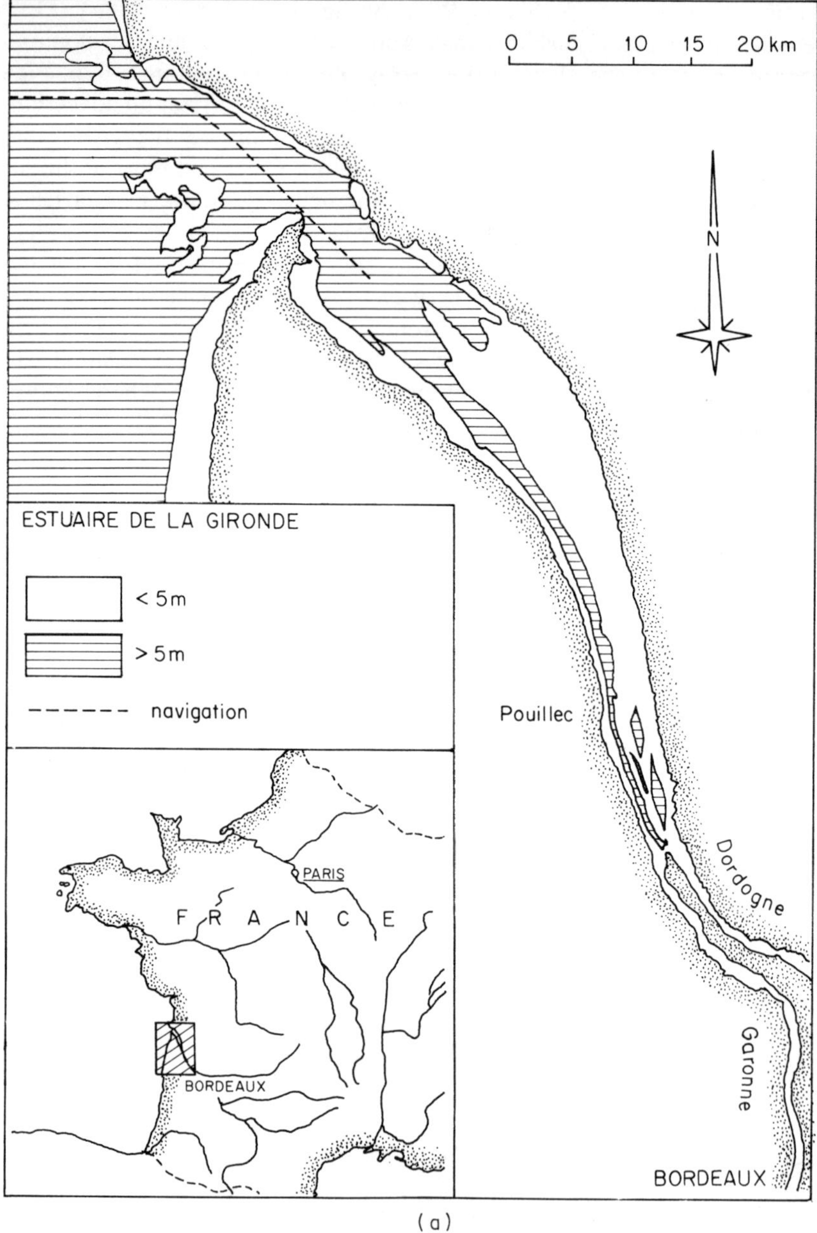

Figure 13. Location of the Gironde estuary (a) and position of the turbidity maximum at HW and LW (b) Average tide and low river flow: 15 September 1961 (After Allen *et al.*, 1974).

RIVER
BORDEAUX
INLET
-30 -20 -10 ← km — 0 — km → 10 20 30 40 50 60 70 80 90
0
1
5
10
20
0.5
HW
10 m
10 km
0
1
5
10
0.5
LW
Turbidity in g/l
< 0.1
0.1–1.0
1.0–10
> 10
Salinity in ‰
···· 0.5 ····
(b)

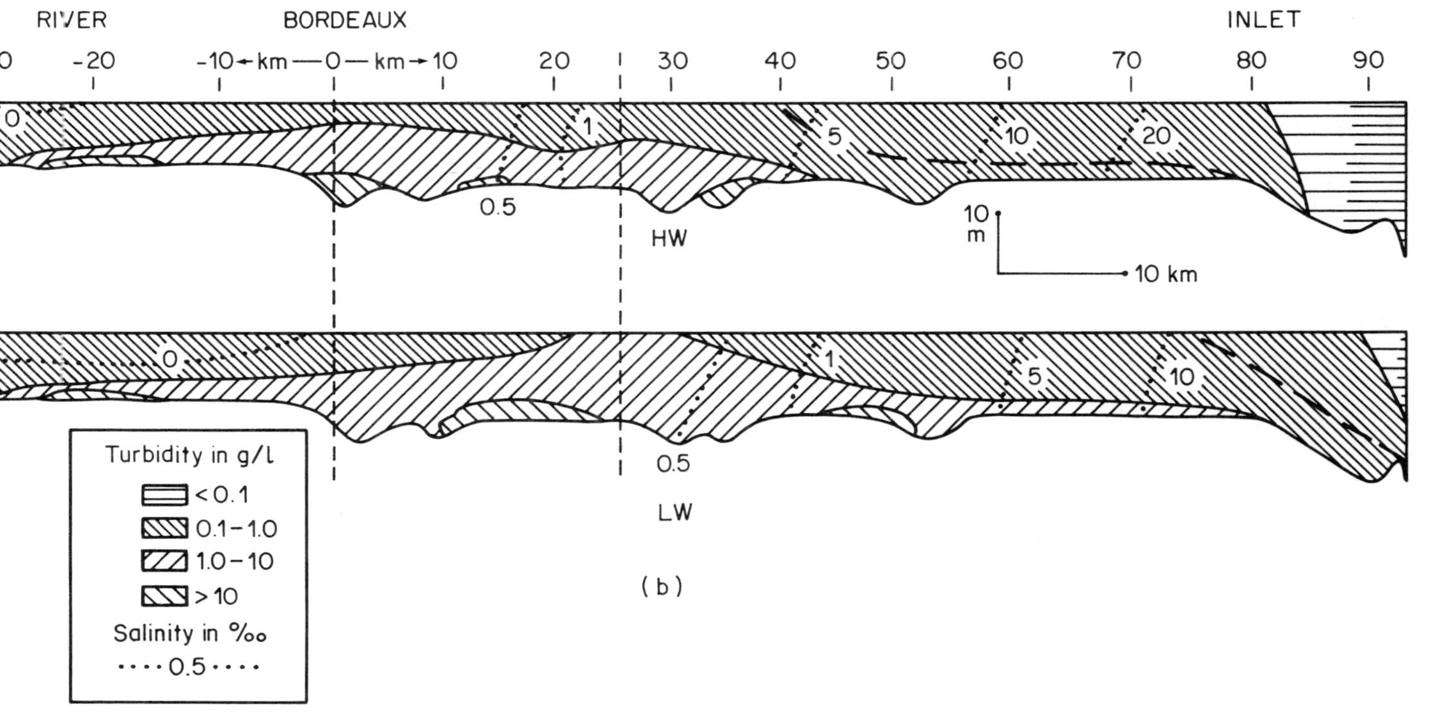

Figure 13(b).

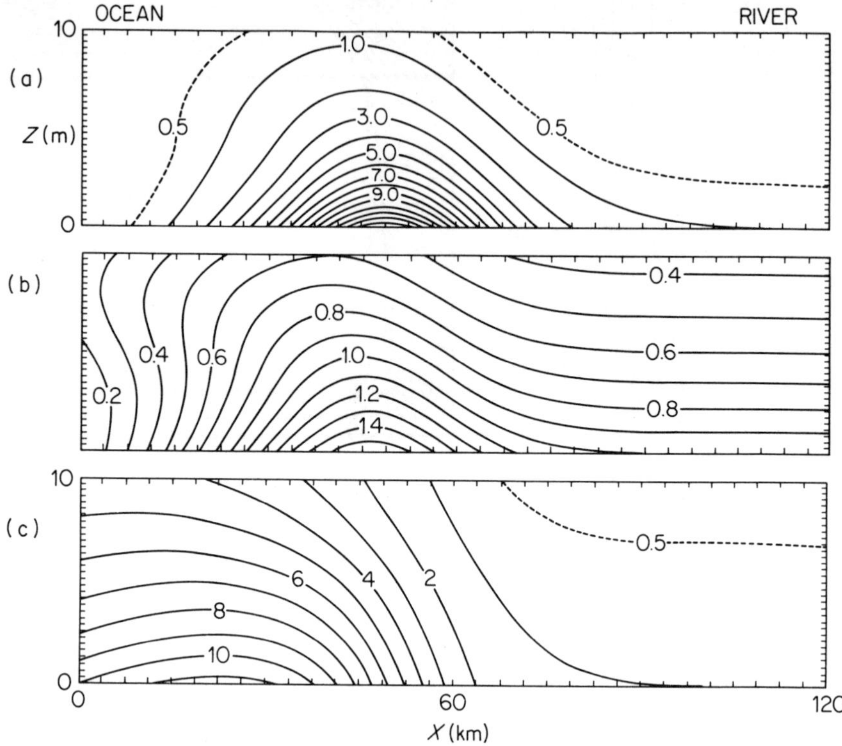

Figure 14. Steady state numerical model of turbidity maximum in a partially mixed estuary. (a) River to ocean source ratio 10; particle settling velocity 3×10^{-3} cm/sec. (b) River to ocean source ratio 10; particle settling velocity 1×10^{-3} cm/sec. (c) River to ocean source ratio 0.1; particle settling velocity 1×10^{-3} cm/sec. Hydrodynamic equations and boundary conditions in Festa and Hansen, 1978.

The accumulation processes assure a greater availability of sediment in estuaries, and they assist in the formation of new estuarine environments, for example by the formation of offshore bars. They are not, of course, the only prerequisite for deposition but they increase the rate of estuarine development. Following this development from an early stage, there is at first ample space for deposition. Most of this space will gradually be filled up through particle by particle accumulation of continental material as happens today in fjords and deep river valleys. Such an environment is therefore well suited for the study of the sedimentary record over the last decades or hundreds or thousands of years, including the history of pollution, providing this is not disturbed by bioturbation (examples in Goldberg, 1978, and this volume). If an estuary approaches equilibrium, however, sedimentary patterns become much more complicated since deposits are more readily

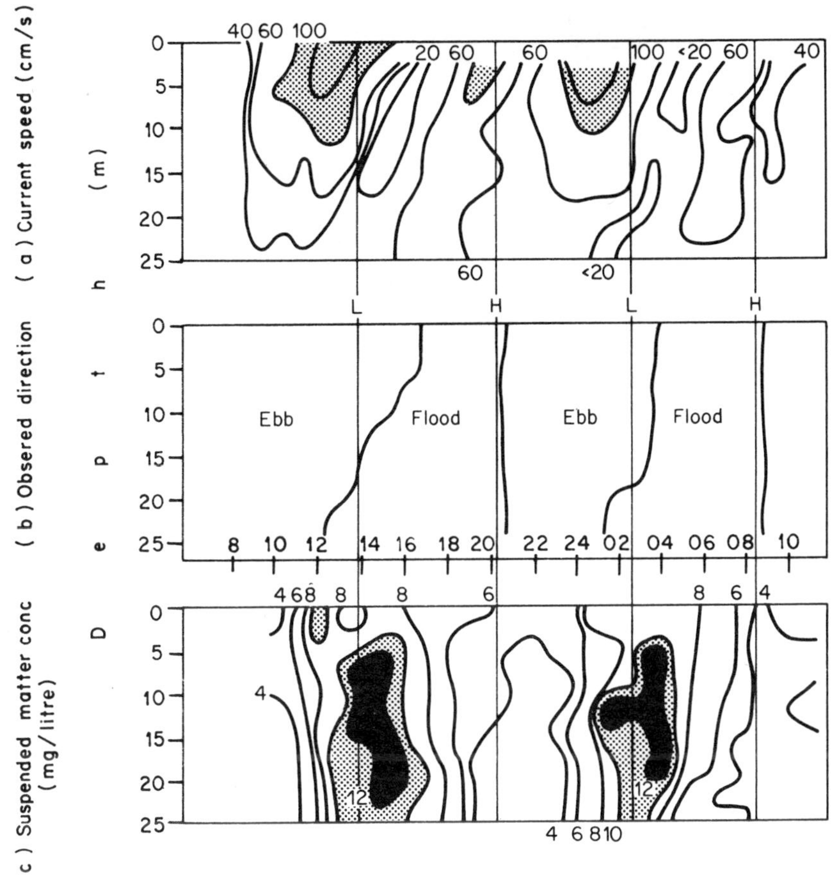

Figure 15. Current velocity and directions and suspended matter over two tidal cycles at a station in the St Lawrence estuary, 26–27 June 1974, 48° N, 70°30′ W (After d'Anglejan and Ingram, 1976).

attacked by waves and currents. This attack causes frequent erosion and redeposition leading to a redistribution according to grain-size and a much better sorting. Moreover, more material is deposited on the shelf in front of the estuary or along the coast which may partly return in a later stage of development as 'marine' sediment and mix with more recent 'fluviatile' material. Particle by particle deposition becomes an exception instead of a rule and it is very difficult to indicate places where undisturbed sedimentation took place over a prolonged time.

In its final stage, an estuary will approach equilibrium with modern sea level. In some cases this will mean that the estuarine basin is converted back

                    H. Postma

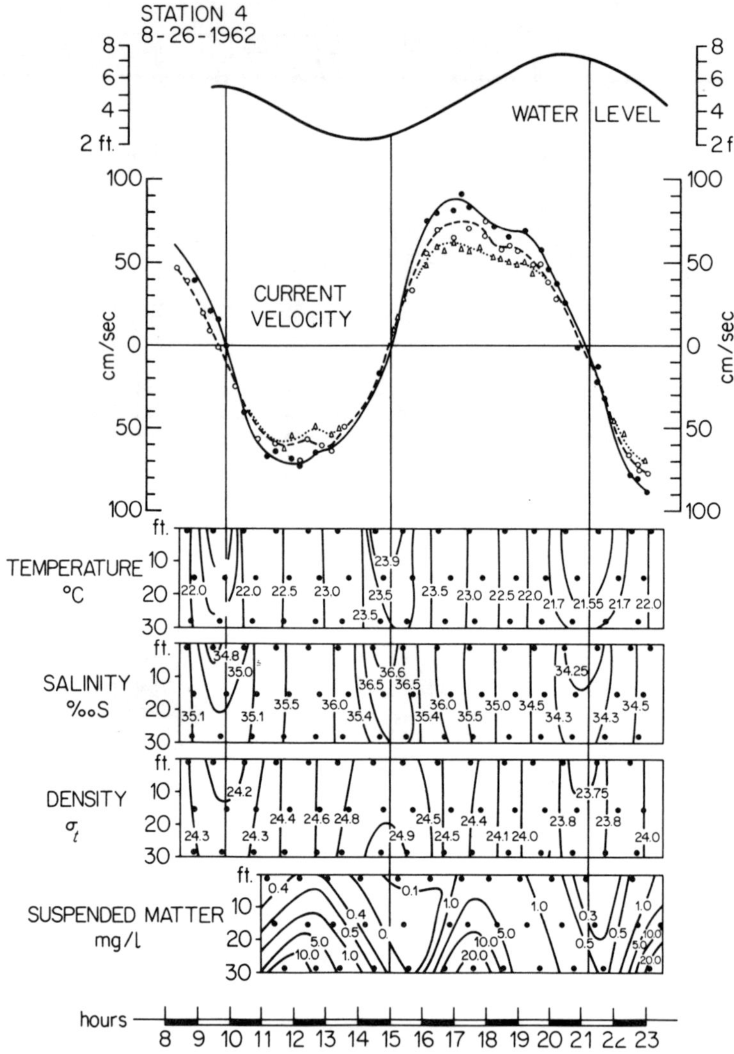

Figure 16. Variations of properties in a hypersaline ('negative') estuary, Laguna Guerrero Negro, Mexico (After Postma, 1965).

into a fluviatile environment (Schubel and Meade, 1977). This happens where hydrodynamic forces other than river flow inside the estuary are not able to halt sedimentation. Generally, the pressure to fill comes both from the land and the sea. The final stage is not attained, however, in cases where the attack of waves or currents becomes stronger with increasing shallowness. In wind exposed areas waves keep intertidal flat levels well below the

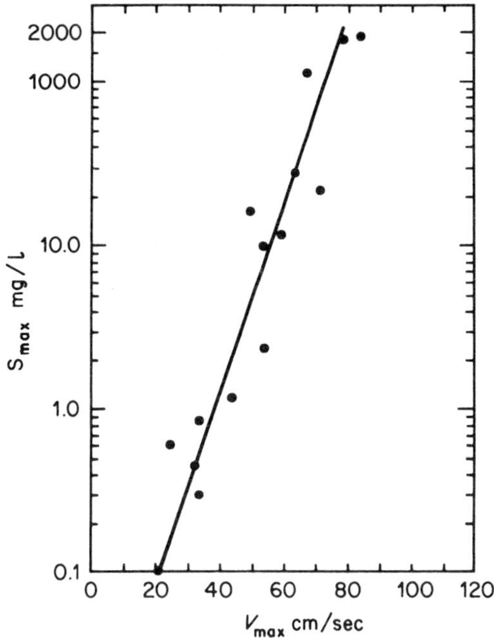

Figure 17. Relation between maximum current velocity (V_{max}) and maximum matter concentration of suspended matter (S_{max}) during flood and ebb in Laguna Guerrero Negro, Mexico. (After Postma, 1965).

high tide mark by preventing deposition (Postma, 1957). This in turn may preserve sufficient intertidal volume in the estuary for the maintenance of a strong tidal regime.

Typical examples of this stage of equilibrium are barrier-built estuaries, the morphology of which is chiefly molded by the action of currents and waves on sand deposits (for a review see Swift, 1976). An essential characteristic of the equilibrium situation is the circulation of sand in closed systems in which wind, waves and currents alternatingly dominate.

3 MODIFICATION OF SEDIMENT PROPERTIES

The sizes of particles introduced in an estuary may vary considerably. Particles in suspension, however, are mostly finer than 100 microns and the bulk of suspended matter has sizes from 1 to 10 microns. Examples have already been given on preceding pages. Most of the coarser material is fine sand, sometimes mixed with shell remains. The finer material is composed of

Table 1 Chemical composition of river inorganic suspended load, in mg/g of dry weight at 600 °C, and average composition of surficial rocks (After Martin and Meybeck, 1978).

	Al	Ca	Fe	K	Mg	Mn	Na	P	Si	Ti
Tropical rivers	111.6	7.6	54.6	19.7	9.4	0.8	5.5	1.6	276	7.4
Temperate and cold rivers	67.5	40.9	38.2	20.5	11.6	1.0	8.8	1.1	309	4.3
Average rivers	89	22.5	46.5	20.3	11.4	0.7	8.2	1.2	293	5.3
Average rocks	69.3	45.0	35.9	24.4	16.4	0.7	14.2	0.6	275	3.8

clays, hydroxides and organic matter. The general chemical composition of river suspended load is given in Table 1.

Since particles are trapped in estuaries their residence time is generally much longer than that of the estuarine water mass, even if they remain in suspension. During their stay they are subject to important changes in size as well as composition. the two most important processes are flocculation and biological grain-size changes.

Clay particles and hydroxides are flocculated by the inorganic salts of sea water, mostly already at very low salinities. In quiet water the floccules can slowly grow to sizes of several hundred microns. If the water is agitated growth can be faster because of more frequent particle collisions, but at the same time this agitation limits the final size of the floccules. Flocculation is also promoted by a greater particle concentration (Dyer, 1972; Kranck, 1973). Floccules disintegrate when they return to fresh water (Nelson, 1960; Postma, 1967).

Floccules have a specific weight of about 1.5 which is smaller than that of their inorganic components (s.w. about 2.6), but they still settle much faster than individual grains. Therefore flocculation promotes estuarine trapping.

An example of floccule behaviour over the seabed is given by Kajehare *et al.* (1974) for a bay near Hokkaiso. Suspended matter forms large floccules which sink down rapidly in the water column. When touching a rough volcanic pebble bottom they disintegrate and the individual lighter particles diffuse upward. The process repeats itself.

An interesting end product of accumulation is so-called 'fluid mud'. This is a highly concentrated mud suspension containing from 20 up to several hundred grams of sediment per litre. It has been found in several rivers and, in cases of very high loads of suspended matter, also outside estuaries on the shelf. An example of the latter is the coast of the Guyana's where banks of Amazon mud are displaced from east to west mainly as fluid mud (Diephuis, 1966).

A well-studied example is the fluid mud in the estuaries of the Gironde and the Loire (Allen *et al.*, 1974; Gallenne, 1974). The body of fluid mud has the form of a lens with a length of 1–10 km and a height of 0.5–2.5 m (Figure 18). It moves back and forth with the tides over only a few hundred metres. It is built up mainly by settling of material from the overlying turbidity maximum at slack tide; enrichment is more important at neap tides and erosion at spring tides. Seasonal shifting of the fluid mud over larger distances takes place as a function of seasonal changes in flow of the Loire. In periods of very high runoff both the turbidity maximum and the fluid mud may be completely expelled from the estuary into the open sea.

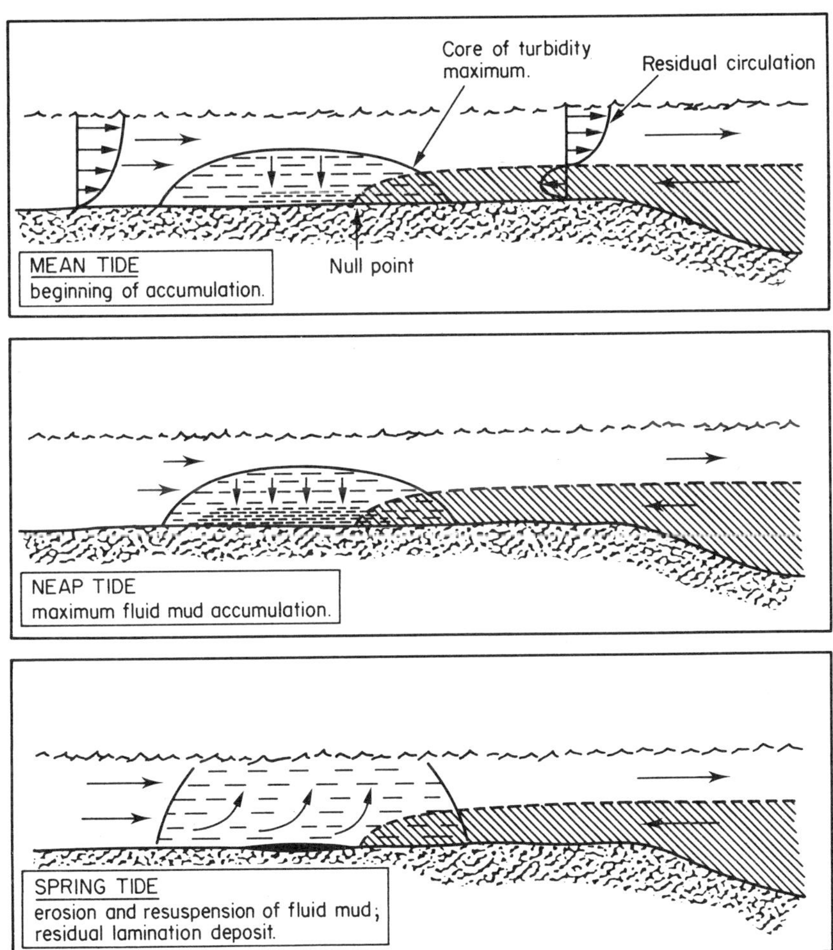

Figure 18. Movement of fluid mud in a partially mixed estuary (After Allen *et al.*, 1974).

How fluid mud can maintain a separate identity between the state of suspension on the one hand and of deposition on the other has been explained by the writer and by Kuenen (1968). In a suspension particles fall unhindered through a fluid which is more or less at rest. However, if particles are at short distances from each other, the fluid has to move upward; this upward flow retards the fall of other particles. This process is enhanced if grains combine in floccules and larger units, forming a spectrum of particle clouds. As the heavier clouds sink, the lighter ones are forced to flow upward carrying their particles along to the top of the suspension. This 'convective stirring' maintains a homogeneous composition from top to bottom in the fluid mud. In experiments the distance between top and bottom of the fluid mud decreases at a constant rate by loss of water through the upper interface and the formation of a deposit at the lower interface (Figure 19; from Kuenen, 1968). No fluid mud formation takes place in fresh water, if the unflocculated particles settle so slowly that they do not stir the fluid.

Besides flocculation a prerequisite for the formation of a distinct body of fluid mud is a very uniform composition with regard to particle size and an identical behaviour of the clouds of floccules, which should easily be formed and easily be destroyed. This uniformity is promoted if the fluid mud is derived from a turbidity maximum, which itself has already made a pre-selection of particles. In the case of the Loire estuary, the percentage of montmorillonite is at least doubled (to 40%) at the expense of other clay

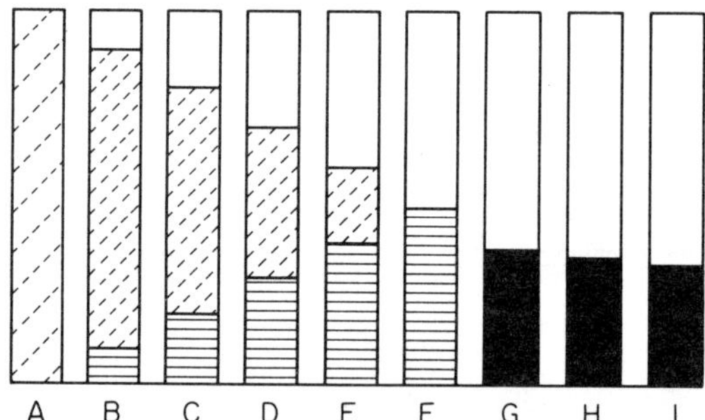

Figure 19. Formation and consolidation of fluid mud; experiment of Kuenen, 1968. (a) homogeneous suspension; (b) start of fluid mud formation: water is expelled from the top, forming a clean upper layer, and from the base, forming sediment. This process continues in (c), (d), and (e), until the upper and lower layer meet (f). (g), (h), and (i) further consolidation of the sediment by expulsion of water. Total duration several hours.

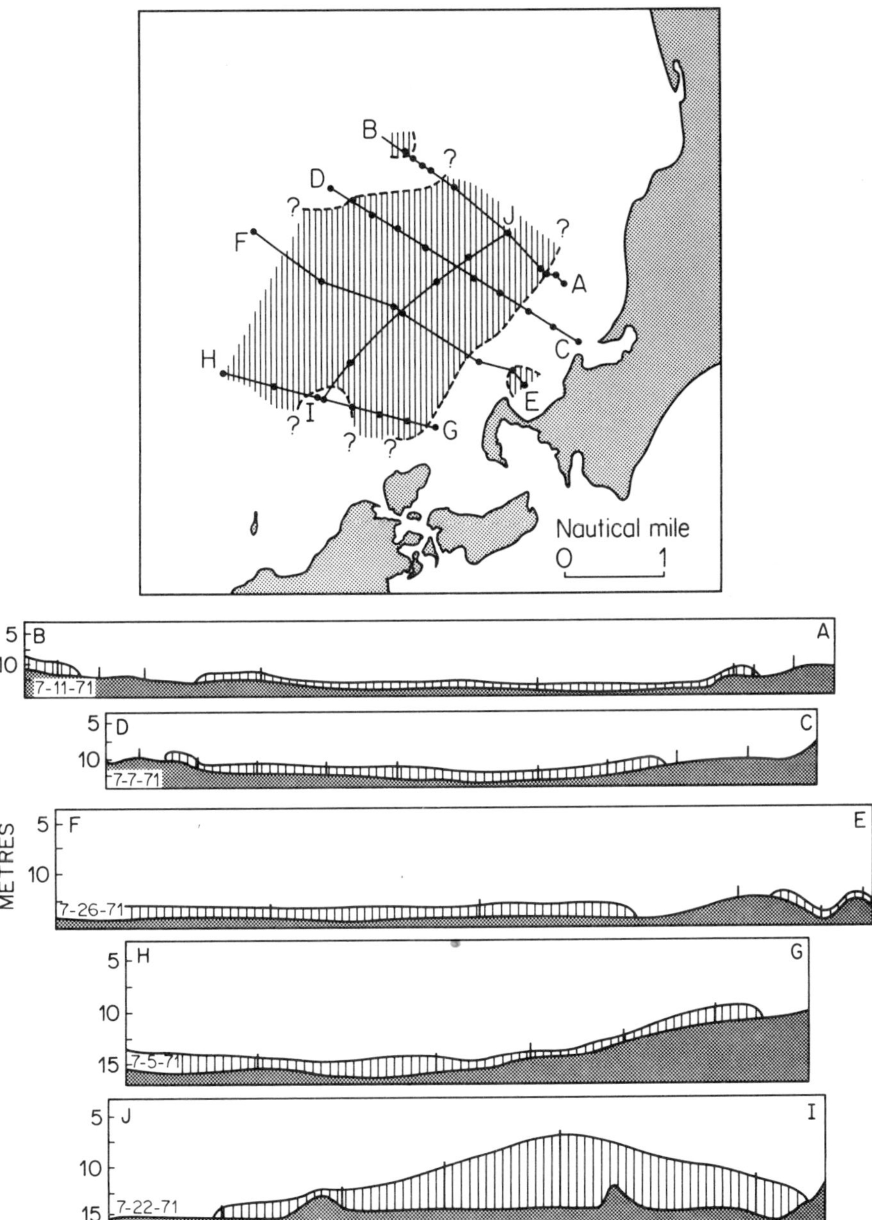

Figure 20. Upper figure: distribution of deposit-feeding benthos community in Buzzard's Bay, Massachusetts, in a silty clay deposit. Lower series: dimensions of cloud of turbidity above the area (After Rhoads, 1973).

minerals (illite, kaolinite, and chlorite). The increase of montmorillonite is also reflected in the composition of the deposits on the mud banks of the estuary. Gallenne (1974) explains the favouring of this mineral above the others by a lower settling velocity. The equivalent grainsize can be calculated to be of the order of 6 microns.

Changes in grain-size by biological activity can take place in several ways. New particles are added to the sediment assemblage by the growth of bacteria and phytoplankton. This can cause an addition of small particles, but also of larger ones such as of several diatom species. Since the addition can amount to several percent of the total quantity of suspended matter the change in the grain-size spectrum can be considerable, especially in summer (Figure 20). Diatom frustules form a more permanent component in many estuarine sediments (Brockmann, 1908).

The macrobenthos in estuaries, specifically suspension and deposit feeders, influence grain-size in two opposite directions. First, small particles are combined to larger ones in faecal pellets. This causes an increase in grain-size comparable with flocculation, but since faecal pellets can be very stable the reverse process proceeds much slower than in the case of chemical flocculation. The ultimate effect of faecal pellet formation is increased stability and cohesion of the estuarine deposits.

Activity of deposit feeders living in the upper bottom layers annihilates this effect to a certain degree in fine-grained sediment and mud (Rhoads and Young, 1970). These animals rework the deposit, decreasing compaction and increasing water content and giving a rough texture to the sediment-water interface. As a result the bottom is easily resuspended even by weak tidal currents. This is shown by experiments with sediment traps and by the existence of a close relationship between this type of deposit and the turbidity of the overlying water (Rhoads, 1973; Figure 20).

Besides by biological activity, new particles can be added to the suspended matter assemblage by flocculation of dissolved matter. This process has long been assumed to be the main cause of the development of turbid zones in an estuary. In fact, very little material precipitates from solution in the mixing zone of fresh and salt water. Some examples for dissolved inorganic chemical species will be given in the next chapter. Regarding organic substances, there is some evidence of precipitation of humic matter by sea water (Jerlov, 1955; Höpner and Orliczek, 1978), but this process does not seem to be of quantitative importance.

4 INFLUENCE ON ESTUARINE CHEMISTRY

A large body of research is carried out on the behaviour of chemical species in the estuarine environment. Attention was first focused on elements participating in the biological cycle, such as nutrients. A main conclusion of this research was that, based on sediment accumulation processes, estuaries act as nutrient traps and that this behaviour is of great importance in increasing estuarine and nearshore fertility (Postma, 1969).

In recent years the interest has shifted to a much wider range of chemical compounds and the role of estuaries in the global geochemical cycle is widely discussed. In addition, marine pollution has stimulated the study of man-made chemical substances. Since these subjects are treated separately in other chapters of this volume only a few general remarks on the relation between estuarine chemistry and sediments are made here.

Chemicals are carried into the estuary partly in solution and partly incorporated in particles. Depending on the type of element and the conditions in the estuary a redistribution between the dissolved and particulate phase takes place. What happens in solution can be derived from the relation between concentration and salinity, which is represented by a straight line if no concentration changes occur, and which deviates from a straight line if a net amount of the element goes into solution or precipitates. A common complication is that there are more than two end members, for example two rivers flowing into the same estuary. In that case one has to find a conservative dissolved substance with a different concentration in the two rivers to distinguish between these two sources.

What happens in suspension can in principle be estimated by plotting concentrations per unit weight of suspended matter against percentages of river *versus* marine material. Again, if no changes occur, the relation should be a straight line. There are, however, several complications to construct such a relation. It is mostly very difficult, if not impossible, to determine exactly what part of particulate matter enters the estuary from the land or from the sea. Different types of suspended matter may be trapped in different sections of an estuary. The amounts and fractions deposited or eroded are mostly not known with sufficient accuracy, and may be quite variable.

An example of the behaviour of a chemical species that can be transferred from the dissolved to the particulate phase and back is given in Figure 21 (modified after Evans *et al.*, 1977). It shows the concentrations of dissolved and particulate manganese in the Newport river estuary (North Carolina) in relation to salinity (see also Duinker, this volume). Concentrations of dissolved manganese are above the mixing line, but approach this line asymptotically at the ocean side. This means that little Mn in addition to Mn from the river is exported.

The particulate Mn (per unit water volume and per unit weight) also shows concentrations above the mixing line, but the maximum lies more seaward than the maximum of dissolved Mn, especially if expressed per unit weight. Total suspended matter increases into the estuary.

　　　　　　　　　　　H. Postma

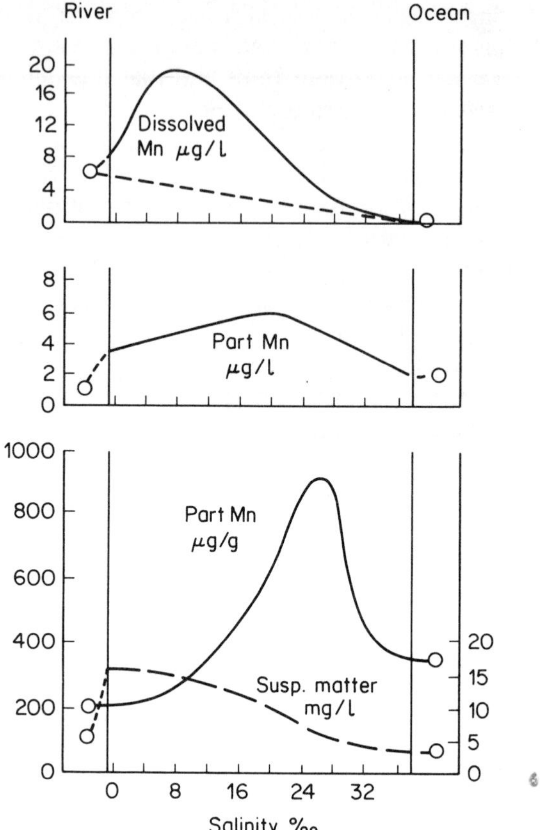

Figure 21. Suspended matter and manganese species plotted against salinity for the fully mixed Newport river estuary, North Carolina (After Evans *et al.*, 1977; slightly modified).

Data on the deposits of the estuary indicate a net deposition of 1–4 mm per year, a concentration of Mn of about 150 μg/g, which is lower than in suspended matter, and a concentration of dissolved Mn of 25 μg/l, which is higher than in the overlying water.

The manganese cycle derived from this information is that Mn goes into solution in the inner part of the estuary, probably by release from the bottom, which below the top 2 or 3 cm is reducing (see also Duinker, this volume). The dissolved Mn^{2+} precipitates in the outer part of the estuary, perhaps by oxidation to MnO_2 or as a coprecipitate with Fe. The particulate Mn thus formed is carried upstream and redeposited, which closes the cycle.

Some losses to the ocean must occur since the bottom sediments are considerably poorer in Mn than in the imported suspended sediments, but in principle the system could run without any loss although concentrations of both dissolved and suspended Mn are much higher than in the source waters!

As the example shows budget calculations obviously require a complete study of the fate of an element in the estuarine cycle, especially in estuaries which are in equilibrium, so that input and output are about equal. The situation is more simple in estuaries with considerable net sedimentation. Ultimately the amount trapped of a chemical compound should be estimated from the concentration in the deposit and by determining the magnitude of this deposit.

The matter of the estuarine budget becomes of special significance in relation to the oceanic budget of chemical species, the oceanic residence time of several metals being of the same order as the age of modern estuaries. However, as already stated in the beginning of this chapter, most estuaries have a short lifespan and, therefore, on a geological time-scale storage of chemical species is mostly only temporarily.

REFERENCES

Allen, G. P., Castaing, P., and Klingebiel, A. (1974). Suspended sediment transport and deposition in the Gironde estuary and adjacent shelf. *Mem. Inst. Géol. Bassin Aquitaine*, **7**, 27–36.

Allen, J. R. L. (1964). Sedimentation in the modern delta of the river Niger, West Africa. In *Deltaic and Shallow Marine Deposits*, ed. L. M. J. U., van Straaten, 26–34, Elsevier.

D'Anglejan, B., and Ingram, R. G. (1976). Time-depth variations in tidal flux of suspended matter in the St. Lawrence estuary. *Est. and Coast. Mar. Sc.*, **4**, 401–416.

D'Anglejan, B., and Smith, E. C. (1973). Distribution, transport and composition of suspended matter in the St. Lawrence estuary. *Can. J. Earth Sc.*, **10**, 1380–1396.

Barbaroux, L., Gallenne, B., Ottman, F., and Margarel, J. P. (1974). Evolution de l'estuaire de la Loire au Quaternaire. *Mem. Inst. Geol. Bassin Aquitaine*, **7**, 267–275.

Berthois, L. (1964). Etudes de dynamique estuarienne dans le fleuve 'la Loire'. *Rapp. 4e Congrès International Portuaire*, Anvers, 12 pp.

Bokuniewicz, H. J., Gebert, J., and Gordon, R. B., (1976). Sediment Mass balance of a large estuary, Long Island Sound. *Est. and Coast. Mar. Sc.*, **7**, 523–536.

Boon, J. D. (1974). Optimized measurements of discharge and suspended sediment transport in a salt drainage system. *Mem. Inst. Geol. Bassin Aquitaine*, **7**, 67–73.

Brockmann, C. (1908). Ueber das Verhalten der Planktondiatomeen des Meeres bei Herabsetzung der Konzentration des Meerwassers und über das Vorkommen von Nordseediatomeen im Brackwasser der Wesermündung. *Wiss. Meeresunters. Helgoland*, **8**, 1–18.

Chase, J. (1975). Wind-driven circulation in a Spanish estuary. *Est. and Coast. Mar. Sc.*, **3**, 303–310.

Clifton, H. E., Hunter, R. E., and Phillips, R. L., (1971). Depositional structures and processes in the non-barred, high energy nearshore. *J. Sed. Petrol,* **41,** 651–670.

Colquhoun, D. J., Comer, C. D., and Pierce, J. W. (1974). Native of estuarine deposits. Atlantic coast of North America. *Mem. Inst. Geol. Bassin Aquitaine,* **7,** 247–252.

Conomos, T. J., and Peterson, D. H., (1977). Suspended-particle transport and circulation in San Francisco Bay. In *Estuarine Processes, Vol. II,* 82–92, Academic Press.

Creutzberg, F., and Postma, H. (1979). An experimental approach to the distribution of mud in the southern North Sea. *Neth. J. Sea Res.,* **13,** 99–116.

Curray, J. R., (1960). Sediments and history of Holocene transgression, continental shelf, northwest Gulf of Mexico. In *Recent sediments, northwest Gulf of Mexico,* Shepard, Phleger and Van Andel, ed., 221–266. Am. Ass. Petr. Geol.

Curray, J. R. (1969). History of continental shelves. In *The New Concepts of continental Margin Sedimentation.* Am. Geol. Inst., Wash., JC6, 1–18.

Curray, J. R., and Moore, D. (1964). Pleistocene deltaic progradation of continental terrace, Costa de Nayarit, Mexico. In *Marine Geol. of the Gulf of California, a Symposium.* Am. Assoc. Petr. Geol., Mem. no. 3, 193–215.

Diephuis, J. G. H. R. (1966). The Guiana Coast. *Tijdschr. Kon. Ned. Aardr. Genootsch.,* **83,** 145–152.

Dyer, K. R. (1972). Sedimentation in estuaries. In *The estuarine environment,* Barnes and Green, ed., 10–31.

Emery, K. O. (1968). Relict sediments on continental shelves of the world. *Bull. Am. Assoc. Pet. Geol.,* **52,** 445–464.

Evans, D. W., Cutshall, N. H., Cross, F. A., and Wolfe, D. A. (1977). Manganese cycling in the Newport river estuary, North Carolina. *Est. and Coast. Mar. Sc.,* **5,** 71–80.

Festa, J. F., and Hansen, D. V. (1978). Turbidity maxima in partially mixed estuaries. *Est. and Coastal Mar. Sc.,* **7,** 42–53.

Glangeaud, L. (1938). Transport et sédimentation dans l'estuaire et à l'embouchure de la Gironde. *Bull. Soc. Geol. France,* **8,** 599–630.

Gallenne, B. (1974). Study of fine material in suspension in the estuary of the Loire and its dynamic grading. *Est. and coast. Mar. Sc.,* **2,** 261–272.

Goldberg, E. D., editor (1978). *Biogeochemistry of Estuarine Sediments.* UNESCO.

Groen, P. (1967). On the residual transport of suspended matter by an alternating tidal current. *Neth. J. Sea Res.,* **3,** 564–574.

Holeman, J. N. (1968). The sediment yield of major rivers of the world. *Water Resourc. Res.,* **4,** 737–747.

Gry, H. (1942). Das Waltenmeer bei Skallingen, No. 1. Quantitative Untersuchungen über den Sinkstofftransport durch Gezeitenströmungen. *Folia Geogr. Danica,* **11,** 1–138.

Höpner, T., and Orliczek, C. (1978). Humic matter as sediment component in estuaries. In *Biogeochemistry of Estuarine Sediments,* ed. E. D. Goldberg, UN-ESCO.

Jansen, J. H. F. (1976). Late Pleistocene and Holocene history of the northern North Sea, based on acoustic reflection records. *Neth. J. Sea Res.,* **10,** 1–43.

Jerlov (1955). Factors influencing the transparancy of Baltic waters. *Meddl. Oceanogr. Inst. Göteborg,* nr. 25, 19 pp.

Kajehara, M., Matsunaga K. and Maita, Y. (1974). Anomalous distribution of suspended matter and some chemical compositions in seawater near the sea bed: transport processes. *J. Oc. Soc. Japan,* **30,** 232–240.

Kraft, J. C., Sheridan, R. E., Moose, R. D., Strom, R. N., and Weil, C. D. (1974).

Middle-late Holocene evolution of the morphology of a drowned estuary system—the Delaware Bay. *Mem. Inst. Géol Bassin Aquitaine*, **7**, 297–305.

Kranck, K. (1973). Flocculation of suspended sediment in the sea. Nature, **246**, 348–350.

Kuenen, Ph. M. (1950). *Marine Geology*, John Wiley and Sons, 551 pp.

Kuenen, Ph. M. (1968). Settling convection and grainsize analysis. *J. Sed. Petrol.*, **38**, 817–831.

McCave, I. N. (1972). Transport and escape of fine-grained sediment from shelf areas. In *Shelf Sediment Transport*; Swift, Duane, and Pilkey ed., 225–247.

Martin, J. M., and Meybeck, M. (1978). Content of major elements in river dissolved and particulate load. In *Biogeochemistry of Estuarine Sediments*; E. D. Goldberg, ed., UNESCO.

Meade, R. H. (1969). Landward transport of bottom sediments in estuaries of the Atlantic coastal plain. *J. Sed. Petrol.*, **39**, 222–234.

Meade, R. H., Sachs, P. L., Manheim, F. T., Hathaway, J. C. and Spencer, D. W. (1975). Sources of suspended matter in waters of the middle Atlantic Bight. *J. Sed. Petrol.*, **45**, 171–188.

Nelson, B. W. (1960). Clay mineralogy of the bottom sediments, Rappahannock River, Virginia. *Proc. Seventy Nat. Conf. on Clays and Day Minerals*, 135–147.

Otto, L. (1975). Oceanography of the Ria de Arosa, (NW Spain). *Thesis*, Utrecht, pp. 211.

Postma, H. (1957). Size and frequency distribution of sands in the Dutch Wadden Sea. *Arch. Néerl. Zool.*, **12**, 319–349.

Postma, H. (1961). Suspended matter and Secchi disc visibility in coastal waters. *Neth. J. Sea Res.*, **1**, 359–390.

Postma, H. (1965). Water circulation and suspended matter in Baja California Lagoons, *Neth. J. Sea Res.*, **2**, 566–604.

Postma, H. (1967). Sediment transport and sedimentation in the marine environment. In *Estuaries*, G. H. Lauff ed., publ. no. 83 A. A. A. S., 158–179.

Postma, H. (1969). Suspended matter in the marine environment. In *Morning Review Lectures of the Second International Oceanographic Congress*, Moscow 1966, 213–219. Unesco, Paris.

Postma, H. (1967). Chemistry of coastal lagoons. In *Lagunas Costeras, un Symposio*, UNAM–UNESCO, Mexico, 421–430.

Price, N. B., and Skei, J. M. (1975). Areal and seasonal variations in the chemistry of suspended particulate matter in a deep water fjord. *Est. and Coast. Mar. Sc.*, **3**, 349–369.

Rhoads, D. C. (1973). The influence of deposit-feeding benthos on water turbidity and nutrient recycling. *Am. J. Science*, **273**, 1–22.

Rhoads, D. C., and Young, D. K. (1970). The influence of deposit-feeding organisms on sediment stability and community trophic structure. J. Mar. Res., 28, 150–178.

Ritter, J. R. (1972). Sediment transport in a tidal inlet. *Proc. 13th Coastal Engineering conference Am. Soc. Civil Eng.*, Vancouver, 823–842.

Roberts, W. P., and Pierce, J. W. (1976). Deposition in upper Patuxent estuary, Maryland, 1968–1969. *Est. and Coast. Mar. Sc.*, **4**, 267–280.

Schubel, J. R. (1974). Effects of tropical storm Agnes on the suspended solids of the northern Chesapeak Bay. *Mar. Sc.*, **4**, 113–132.

Schubel, J. R., and Meade, R. H. (1977). Man's impact on estuarine sedimentation. *Proc. Conf. Est. Poll. Control and Assess*, vol. I, US env. Prot. Agency, Wash., 193–209.

Swift, D. J. P. (1976). Continental shelf sedimentation. In *Marine Sediment Transport and Environmental Management*, Stanley and Swift ed., John Wiley, 311–351.

Swift, D. J. P., Stanley, D. J., and Curray, J. R. (1971). Recent sediments on continental shelves: a reconsideration. *J. Geol.*, **79,** 322–346.
Swift, D. J. P., and Pirie, R. G. (1970). Fine-sediment dispersal in the Gulf of San Miguel, Western Gulf of Panama: a reconnaissance. *J. Mar. Res.*, **28,** 69–95.
Thorn, M. F. C. (1975). Deep tidal flow over a fine sand bed. *Proc. XVIth Congress Int. Ass. Hydraul. Res.*, Sao Paulo.
Van Andel, Tj. H. (1967). The Orinoco Delta. *J. Sed. Petrol.*, **37,** 297–310.
Van Straaten, L. M. J. U. (1959). Littoral and submarine morphology of the Rhone Delta. *Proc. 2nd Coast. Geomorph. Conf.*, 233–264.
Windom, H. L. (1976). Lithogenous material in marine sediments. In *Chemical Oceanography*, Riley and Chester, ed., Acad. Press, 103–135.

Chemistry and Biogeochemistry of Estuaries
Edited by E. Olausson and I. Cato
Copyright © 1980 by John Wiley & Sons Ltd.

B. J. PRESLEY and J. H. TREFRY
Department of Oceanography,
College of Geosciences,
Texas A and M University, Texas

6

Sediment—water interactions and the geochemistry of interstitial waters

1 INTRODUCTION AND SIGNIFICANCE OF INTERSTITIAL WATER STUDIES

Most of the sedimentary rocks exposed at the earth's surface, and those encountered in oil well drilling, were originally deposited in shallow marine basins, including estuaries. Furthermore, much of the water found with petroleum in reservoir rocks is thought to have been trapped at the time of deposition, and thus to be ancient seawater. Yet ancient sediments and oil well waters differ from recent marine muds and modern seawater in a number of ways. Modern unconsolidated marine muds, compared to ancient sediments, contain an abundance of igneous rock fragments, amorphous materials, recognizable organic compounds, remains of organisms and aragonite and high Mg calcite. Oil well brines are usually more saline than modern seawater and are enriched in Ca, Sr, Ba, Li, Br, Mn, Si, and HCO_3^-, but depleted in SO_4^{2-}, Mg, and K. Thus, chemical reactions (diagenesis) affect both solid and fluid phases of marine sediments after deposition.

Reactions affecting sediments have more than theoretical interest because they can control the formation of economic mineral deposits, the acoustic and engineering properties of the sediment, exchanges between the sediment and the overlying water and other important aspects of sediment formation. Many diagenetic reactions are thought to begin soon after sediment deposits, but identifying incipient reactions in the solid phase is difficult. Chemical changes are easier to detect in the fluid phase of a water saturated sediment because the relative changes in concentration are much greater in this more chemically dilute phase.

Interstitial water (or pore water) refers to water trapped in sediment pores during the accumulation of sediment particles, and is thus initially bottom water from the depositional site. Marine scientists have since the work by Murray and Irvine (1895) been interested in the chemistry of pore water. Recent reviews of marine pore water studies have been published by Manheim and Sayles (1974) and by Manheim (1976) and should be consulted for more details on the history of pore water studies, especially with regard to the deep sea and work by Russian investigators. This review shows how pore water studies can aid in understanding estuarine chemistry and geochemistry, both for theoretical interest and for such practical purposes as mineral exploration, pollution control, and planning of harbour and offshore construction projects.

2 SAMPLING SEDIMENT AND INTERSTITIAL WATER

In many nearshore environments, sediment may be sampled by directly shoveling or scooping at low tide, or by simply pushing an open metal or plastic tube into the sediment by hand. The latter operation can be performed from a small boat or by SCUBA divers (Aller and Cochran, 1976) thus allowing considerable flexibility in use. Nevertheless, many, perhaps most, sampling programs cannot be conveniently carried out with hand-held equipment but rely on samplers lowered to the bottom on cables or allowed to free-fall with flotation devices for automatic return to the surface. Without expensive camera or television monitors the investigator cannot see a remote sampler at work and this leads to questions as to the representativeness of the recovered sample. Data to be presented in this chapter will show that properly designed and operated equipment can take representative and essentially undisturbed samples from the sea floor in any depth of water, but nevertheless care must be exercised in any sampling operation.

A number of devices have been designed to 'core' an undisturbed small section of sea bottom thereby providing samples from progressively greater depths and more ancient times (see Bouma, 1969; Duxbury, 1971; McQuillin and Ardus, 1977). Where rates of sediment accumulation are low or where chemical reactions are occurring in the top centimetres of sediment, many investigators have expressed concerns about the loss of surface layers or other disturbance during sampling. As a result, box coring devices such as the ones described by Jonasson and Olausson (1966), Bouma (1969), Addy and Ewing (1974), and others are usually preferred. The typical box corer pushes a stainless steel box, 10 or more cm on a side and 50 cm high, into the sediment and closes it with a spade or jaws during withdrawal. The relatively large size of the box core provides enough material for many different analytical determinations, even when the core is sliced into thin sections to give good time definition to the process under study. An obvious shortcoming of the box corer, in addition to its unwieldiness, is the limited depth to which it samples. Conventional gravity corers, consisting of little more than a weighted pipe, can easily sample three or four times deeper than the box corer, are much easier to use, and have been found to be efficient at recovering the sediment surface (Flanagen, 1970). Schink (un-published data) in a detailed study of radon, silicon, and other parameters in gravity cores finds no evidence for loss or disturbance during coring. In fact, much of the data to be presented in this chapter is from gravity cores and the concentration profiles normally seen clearly show that no large scale loss of material or disturbance occurred during sampling. Nevertheless to ensure recovery of the upper-most mm of sediment and to allow collection of uncontaminated sub-samples, a box core is still ideal.

To obtain older (deeper) samples from the sediment column a piston corer of the type developed by Kullenburg in the 1940s or a corer that uses springs, hydralic pressure or impact to aid penetration (see Bouma, 1969, for a description of various coring devices) may be used. A typical piston corer can sample 10 to 20 metres of the sediment column, depending on circumstances and sediment type. Sampling deeper than 20 metres or so into the sea floor can generally only be accomplished by a drilling rig similar to those used to drill oil or water wells. Such a device might prove useful in estuarine areas where sediment is accumulating at rates >1 cm/yr. Other-wise it may not be possible to sample sediment more than a few tens of years old. Piston corers, and especially drill rigs, unless carefully designed can grossly disturb and contaminate sediment samples, and each such device should be evaluated for possible adverse effects before it is used.

That uncontaminated samples can be recovered by drilling is attested to by results from the experimental Mohole drilling off Mexico (Rittenberg *et al.*, 1963) and similar drilling off Florida in 1965. In the latter project, fresh water (60 ppm Cl) was found in sediments 120 km from shore during drilling using seawater in the drill mud. No seawater contamination of samples recovered by wire line coring was detected (Manheim, 1967). Manheim's success with the Florida drilling programme lead him and others to plan an interstitial water program for the Deep Sea Drilling Project (DSDP) which started in 1968. Hundreds of samples from the DSDP have been analyzed by Manheim and Sayles and by Presley and Kaplan (see Manheim and Sayles, 1974 for a review) culminating with a 'Geochemical Leg' in the Caribbean Sea involving many other investigators (see individual reports in Heezen and MacGregor, 1973). The DSDP interstitial water program very clearly demonstrated that samples suitable for a host of geochemical measurements can be recovered by more or less conventional wire line coring during a drilling operation. Drilling has not been extensively used in estuarine and nearshore geochemical studies of unconsolidated sediment due to the costs involved and the fear of blowouts from high pressure gas; but there is no alternative if sediment more than a few tens of years old is to be recovered from area of rapid sedimentation.

Unfortunately, even when all the sampling problems just discussed have been overcome, other problems unique to pore water studies can complicate interpretation of resulting data. For example, the method used to remove pore water from sediment can bias the data as can the seemingly innocuous factor of the temperature during pore water extraction. The large number of literature descriptions of pore water extractors ('squeezers') is one indicator of the appreciation of sampling problems by practitioners in the field. In fact, it seems as though almost everyone who has engaged in pore water studies has designed his own squeezer due to dissatisfaction with previous designs!

The study of pore water chemistry expanded rapidly in the 1950s and 1960s, and so did the approaches to pore water removal. Commercially available gas or piston presses were modified for use (Luscyznski, 1961; Hartmann, 1965; Manheim, 1966) and similar devices were machined from stainless steel (Siever, 1962) or plastic (Presley *et al.*, 1967). Each of the several designs offers certain advantages, but the present authors feel that the one described by Reeburg (1967) gives the best combination of ease of use, flexibility, low cost and freedom from contamination. The original Reeburg (1967) design is shown in Figure 1, from which the essentials of operation can be grasped. Recent modifications by various investigators include a semi-permanently attached diaphragm built into a screw-on top and a self-seating plastic sealing ring which makes the squeezer very easy to use. The Reeburg squeezer can only be used at low pressures (maximum of 1 megapascal ($= 1\,MPa$) or so), whereas the Manheim (1966) design can safely operate at more than $140\,MPa$ and would therefore be preferred for semi-consolidated sediment. Squeezing pressures of more than $0.7\,MPa$ give only limited increases in pore water extraction speed in most unconsolidated sediment, due to the self-filtering action of the sediment which slows flow regardless of the pressure applied. As would be expected, squeezing speed is strongly dependent on lithology and grain size, but about 50 per cent of the total original pore water can be removed from a typical sample with a Reeburgh squeezer in 10 to 20 minutes. Removal of more than 50 per cent

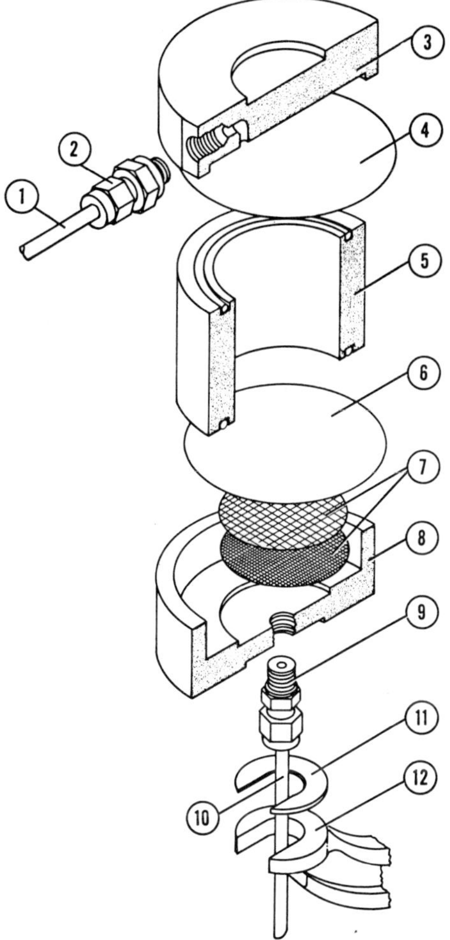

Figure 1. Exploded isometric drawing of squeezer. (1) nylon gas inlet tube, (2) O-ring seal male plug (Swagelok), (3) Delrin cap, (4) dental dam rubber diaphragm, (5) nylon sample retainer with O-rings, (6) filter, (7) nylon screens, (8) Delrin base, (9) nylon male plug (Swagelok), (10) nylon sample drain tube, (11) rubber or cork pad, and (12) modified C-clamp. (From Reeburg, 1967.)

of the original water from clay-rich sediment is a very slow process regardless of the pressures applied and is not worth the effort in most studies. It is very important to emphasize, however, that long squeezing times and pressures in excess of 140 MPa have little affect on the chemistry of the expelled pore water, despite concerns that have been expressed as to the 'semi-permable membrane effect' of clay minerals under pressure. Manheim (1966) quotes extensive Russian literature and his own work to show the validity of data from squeezed pore water, and other investigators (Presley, 1969; Sayles, 1970) have confirmed these results. This is not to say that no artifacts are introduced by the squeezing pressures. Effects on pressure sensitive solubility relationships, especially for gases may be a problem, but these have not been adequately documented. Pressure effects of the carbonate system should be suspected with high pressures and long times (e.g. Hammond, 1973), but the data of Presley and Kaplan (1968) give no indication of problems with shallow water sediments squeezed at low pressures.

Despite the popularity of squeezers for removing pore water from sediments, other techniques have been used. Powers (1957) and Rittenberg (1963), among others, have used centrifugation, a method judged by most to be less convenient and efficient than squeezing. An old, and controversal technique is to wash the pore water out with a large volume of distilled water, followed by filtering or centrifugation. The washing (or leaching) technique was applied by early workers in soil solution chemistry (e.g. Burgess, 1922), was used by Emery and Rittenberg (1952) and advocated as being more reliable than squeezing by Murthy and Ferrell (1972). Manheim (1974) rebutted the Murthy and Ferrell paper, pointing out that the cationic composition of leachates of sediments is influenced by the type of sediment, dilution, temperature, length of leaching, and other factors, whereas pore water squeezed from sediments gives reproducible data representative of the actual chemistry of the 'free' water in the sediment. Manheim's observations are consistent with those of Sayles and Mangelsdorf (1977) who show that rapid cation exchange occurs when clays are immersed in a solution of different composition than the pore water.

As discussed above, workers in the 1950s and 1960s gave considerable attention to possible artifacts due to squeezing pressures used in extracting pore water. On the other hand, temperature effects were largely ignored. Thus, the discovery of Mangelsdorf *et al.* (1969) that temperature changes could induce large changes in pore water chemistry was particularly enlightening. Details of the temperature effects and their implications will be discussed in a later section but brief mention of techniques and instruments for overcoming them will be made here. The first and most obvious approach is to simply maintain or return the sediment to *in situ* conditions during squeezing, a method that has been followed by most investigators since 1969 (e.g. Kahil and Goldhaber, 1973; Horowitz *et al.*, 1973). Regardless of the method used to return a sediment to *in situ* conditions, nothing will be accomplished unless the temperature effects are reversible and almost no work has been done on this point. Presley (1969) and Fanning and Pilson (1971) have shown evidence of reversibility in temperature effects, but more work is needed. An additional point to be stressed is that many constituents are affected very little by temperature changes and, of course, in shallow water, *in situ* sediment temperature can be approximately room temperature, thus avoiding the problem completely.

A novel way to avoid temperature of squeezing effects and at the same time pressure effects, storage effects and other problems, is to separate the pore water from sediment *in situ.* Early workers who removed a plug of sediment from above the water line and allowed pore water to seep in were sampling *in situ*, but a device to accomplish this feat remotely in any depth of water must be fairly complicated. The first widely used *in situ* sampler (Sayles *et al.*, 1973) is a harpoon device which sucks pore water into evacuated chambers after it has penetrated the sediment. Hesslein (1976) and Moyer (1976) both built devices which would expose dialysis bags to the sediment at various depths in the sediment column. The *in situ* samplers can be rather simple in design if they are to be emplaced by divers, thus seem to be ideally suited for estuarine work. They are not as convenient to use as conventional squeezers, and may introduce unknown artifacts when left to equilibrate with the sediment for days or weeks as they often are. At this stage of development, therefore, they have great promise, but have not been fully tested.

3 COMPONENTS OF ESTUARINE INTERSTITIAL WATER

Sediments in estuaries can be exposed to overlying water of widely varying salinity due to tides and changes in fresh water input. In other respects the estuarine environment is similar to any other nearshore marine environment and thus the following discussion will not always be restricted to strictly estuarine sediments. Some of the earliest interstitial water studies considered estuarine situations wherein pore water chemistry varied in response to changes in salinity of the overlying water (e.g. Reid, 1930). It was found that changes in the sediment pore water were less drastic than those in the overlying water, thereby producing less stress on benthic organisms.

Pore water chemistry responds not only to physical factors such as advection of new water and molecular diffusion, but to chemical interactions between the water and the surrounding solid phases of the sediment. Over geologic time these reactions produce consolidated sediments and formation waters. At the same time, the concept of a steady state ocean requires removal of dissolved constituents as they are added (Clarke, 1924; Conway, 1943; Rubey, 1951; Sillen, 1961), and presumably most removal involves interaction with sediments. Mineralogists have looked for newly formed or altered minerals in sea floor sediment since the 1940s (e.g. Grim, 1968) and biologists and microbiologists have been concerned with organic decomposition in sediments for at least as long (e.g. Emery and Rittenberg, 1952). Over the past 15 years or so a number of investigations have studied concentration changes in sediment pore water as an indicator of the extent and nature of solid–aqueous reaction. Pore water concentration changes are very sensitive, compared to changes in the solid phase. For example, precipitation of all the calcium in the pore water of a marine sediment with 50 per cent porosity would add only 0.1 per cent $CaCO_3$ to the solid phase, an amount almost impossible to detect. Similar examples could be given for almost any component of the solid phase, thus pore water studies are a powerful addition to many geochemical investigations.

3.1 The major cations—sodium, potassium, calcium, and magnesium

The four ions, Na^+, K^+, Ca^{++}, and Mg^{++}, account for more than 99 per cent of the positive charge of sea salts and are major constituents of river water salts, the earth's crust and body fluids of organisms. Understanding their behaviour is thus fundamental to understanding the overall chemistry of the sea.

Manheim (1976) quoting extensively from the original Russian literature notes that Russian geochemists were interested in pore water chemistry in the 1930s; however, the earliest references to data on major cations he gives are from the 1950s. In the U.S., Powers (1957) reported that pore waters in the James River estuary were depleted in Mg and K due to uptake by clays. Powers' work deserves special recognition here because it was the forerunner of many later attempts to recognize early diagentic changes by pore water analyses, and because it was an estuarine study. Perhaps the first major paper published in English which deals extensively with the major components of marine pore water is by Siever *et al.* (1965). They found Na concentrations similar to those in the overlying water, Mg usually depleted by about 100 ppm, Ca depleted by variable amounts and K enriched by variable amounts. Siever *et al.* (1965) interpreted their data in terms of uptake and release from silicate and carbonate minerals, and in the case of Ca, possible artifacts produced by storage.

The Russian workers in the 1960s, and Shishkina in particular (see Manheim, 1976) generally reported smaller variations in major cation concentrations than those noted by Siever *et al.* (1965), but found some of the same trends. Similarly, Presley (1969) and his co-workers (Brooks *et al.*, 1968; Presley and Kaplan, 1968) found only slight changes in major cations, except for the case of Ca in reducing sediments.

Table 1. Changes in composition of pore water resulting from raising the temperature from 4 °C to 22 °C during extraction. Expressed as percentage increase (+) or decrease (−) at 22 °C relative to 4 °C (From Sayles *et al.*, 1973).

Element	Site 147 Clays and Marls		Site 148 Clays and Marls		Site 149 Calcareous		Site 149 Siliceous Biogenic	
	% Change	Δ meq.	% Change	Δ meq.	% Change	Δ meq.	% Change	Δ meq.
K	+18	+1.4	+24	+1.8	+16	+1.2	+12	+0.6
Na	+0.9	+4.4	+1.3	+5.9	+1.0	+4.3	+0.5	+1.0
Ca	−	−0.4	−6.5	−1.1	−3.1	−1.3	−1.0	−1.1
Mg	−7.3	−5.1	−7.3	−6.2	−3.1	−2.6	−1.4	−0.7
Net Δ meq.		+0.3		+0.4		+1.6		−0.2

The important discovery of Mangelsdorf *et al.* (1969) that temperature could affect pore water chemistry casts doubt on all the earlier pore water data, because in no case was squeezing controlled at the *in situ* temperature. Of course, many of the nearshore sediments were probably inadvertently squeezed at near *in situ* temperatures. Nevertheless any reported concentration difference between pore water and the overlying seawater must be interpreted in terms of a temperature effect that causes a release of Na and K and an uptake of Mg and Ca with increasing temperature.

The magnitude of the temperature effect can be judged from Table 1, based on data from Sayles *et al.* (1973). The table clearly shows a dependence on lithology, thus adding another variable. An even more serious problem, however, is the scarcity of data on the reversibility of the temperature effect, for use in cases where sediment is artificially heated or cooled to return it to *in situ* conditions. Table 2 from Presley (1969) gives results from a simple experiment supporting reversibility for major cations, and Fanning and Pilson (1971) show similar results for silica. Nevertheless, a more systematic investigation of this phenomenon is still needed.

In order to avoid pressure and temperature artifacts Sayles *et al.* (1973) used an *in situ* pore water sampler. With it they show depletion of pore

Table 2. Change in concentration of cations in interstitial water squeezed from mud after exposure to various temperatures for 24 hours (From Presley, 1969).

Temperature of squeezing	K (p.p.m.)	Ca (p.p.m.)	Mg (p.p.m.)	Na (p.p.m.)
5 °C	405	525	1,235	10,300
23 °C	470	510	1,195	10,300
30 °C	515	490	1,170	10,300
37 °C	550	475	1,140	10,300
5[a]	415	485	1,250	10,300

[a] 24 hours at 5 °C after 37 °C treatment.

B. J. Presley and J. H. Trefry

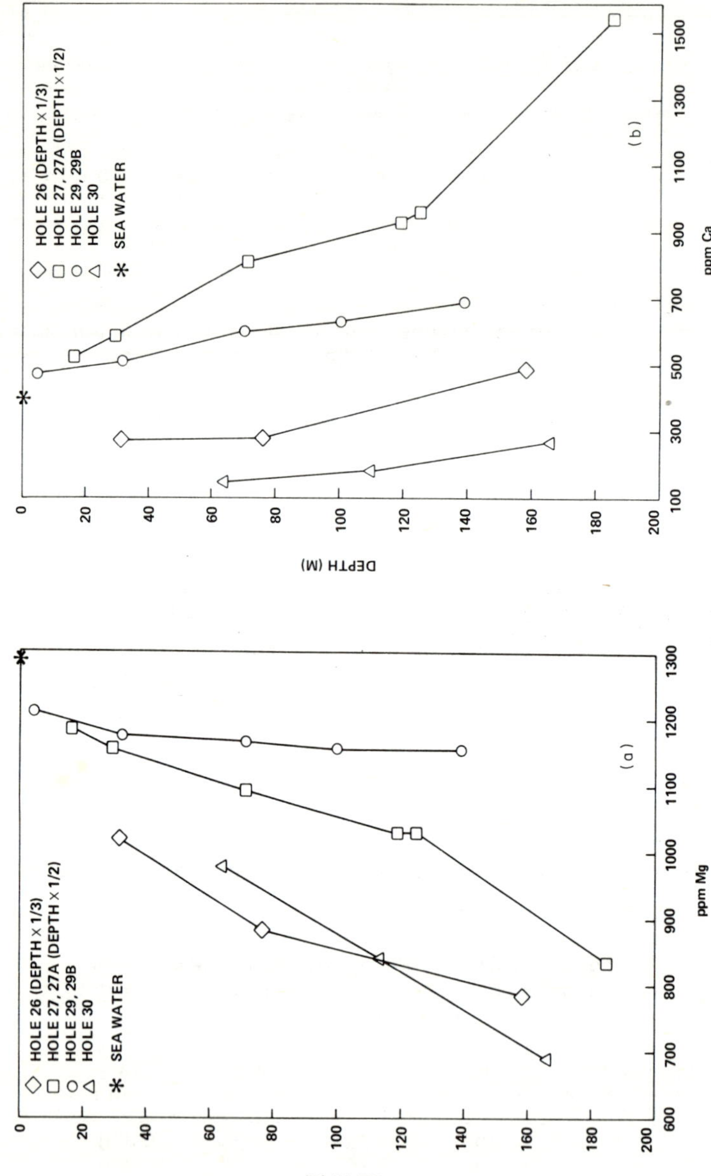

Figure 2. (a, b, and c). Magnesium, calcium, and potassium variations with depth in DSDP drill holes. (From Presley and Kaplan, 1970.)

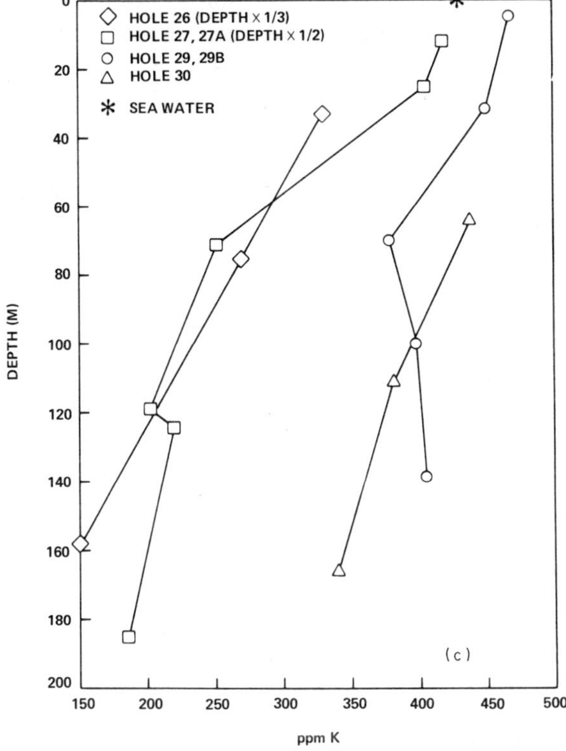

Figure 2(c)

water K and enrichment of Ca in mostly deep-sea environments. Sodium and Mg were not determined. The trends found by Sayles *et al.* (1973) are opposite to those reported by most earlier workers, and unfortunately have not been sought by other workers using *in situ* devices (Barnes, 1973; Mayer, 1976; Hesslein, 1976). Thus, while it seems clear that major cation concentration changes are small in the upper few metres of oxidizing sediments, the exact nature of the changes is still in doubt. Because fluxes into or out of sediments equal to the entire world riverine flux will result from even small concentration gradients (0.1 per cent per cm) these need to be carefully determined.

A clue to the direction of changes of major cation concentrations is given by the analyses of Deep Sea Drilling Project (DSDP) samples (see Manheim and Sayles, 1974 for a review). Analysis of hundreds of DSDP samples has confirmed the trends found early in the project by Presley and Kaplan (1970). In all rapidly accumulating nearshore sediments K and Mg concentrations decrease with depth, whereas Ca increases (Figure 2(a), (b) and (c)).

Sodium normally shows slight decreases that are difficult to detect analytically. The presence of evaporites or volcanic activity can alter the normal trends.

The 'normal' trends found for the DSDP samples bring the sediment pore water closer to formation water in composition and would have been predicted by most geochemists. It seems likely, then, that near surface trends should be similar, but more subtle.

Reducing conditions in the sediments can affect major cation concentration and many other aspects of pore water chemistry. Berner (1964) noted calcium depletion in sulphide-rich Gulf of California sediment pore water and later (Berner, 1966) suggested carbonate precipitation from biogenically produced carbonate as a possible explanation. Presley and Kaplan (1968) were able to more carefully document the extent and nature of $CaCO_3$ precipitation while working with nearshore sediment off southern California. Several more recent studies have found Ca depletions of up to 75 percent in reducing sediments (e.g. Bischoff *et al.*, 1970; Nissembaum *et al.*, 1972), but other investigations have found constant values or slight increases (e.g. Manheim and Chan, 1974). The large carbonate increase resulting from sulphate reduction should precipitate Ca and it is not clear why this does not happen in all environments. Similarly, the pore water Mg loss by substitution for Fe in clays found by Drever (1971) has not been confirmed by most other workers, despite the feasibility of the phenomenon.

3.2 The major anions—chloride and sulphate

The major cations, Na^+, K^+, Ca^{2+}, and Mg^{2+}, account for more than 99 per cent of the positive charge in seawater salts, and Cl^- and SO_4^{2-} account for more than 99 per cent of the negative charge. However, unlike the cations, chlorine and sulphur are not abundant in most igneous and sedimentary rock minerals and were not originally derived from rock weathering (Rubey, 1951). Thus, Cl^- and SO_4^{2-} do not interact strongly with silicate minerals during diagenesis as the major cations do, but rather vary only with salinity, or in the case of SO_4^{2-} with biological activity.

The total salinity of marine pore water has been of interest for several reasons. Biologists interest in salinity variations in pore water compared to overlying estuarine water was cited earlier in this chapter. Geochemists and biologists are also interested in the history of salinity variations in selected basins and have considered pore water chemistry as a possible 'paleo-salinity' indicator since the time of Kullenberg's (1952) classic work on the Baltic and Mediterranean.

In some nearshore areas pore water evidence of past salinities might be difficult or impossible to interpret due to mixing from laterally moving waters, but this is probably not common. On the other hand, diagenetic reactions commonly affect many components of pore water, although chloride appears to be relatively inert based on thousands of analyses of pore water from all over the world (Manheim, 1976 for a summary). Little evidence of Cl⁻ reactions except for dissolution of evaporites in specialized environments has been found, thus making Cl⁻ a possible indicator of 'paleo-salinity'. An example of short-term Cl variations in estuarine pore water is given in Figure 3. Here near suface Cl concentrations are low in the spring and high in the fall, reflecting riverine input, but at depth concentrations are relatively constant.

Chloride concentration profiles like any profile exhibiting a concentration gradient, are affected by molecular diffusion. Kullenberg was well aware of this when he tried to interpret paleo-salinities in 1952. He pointed out that

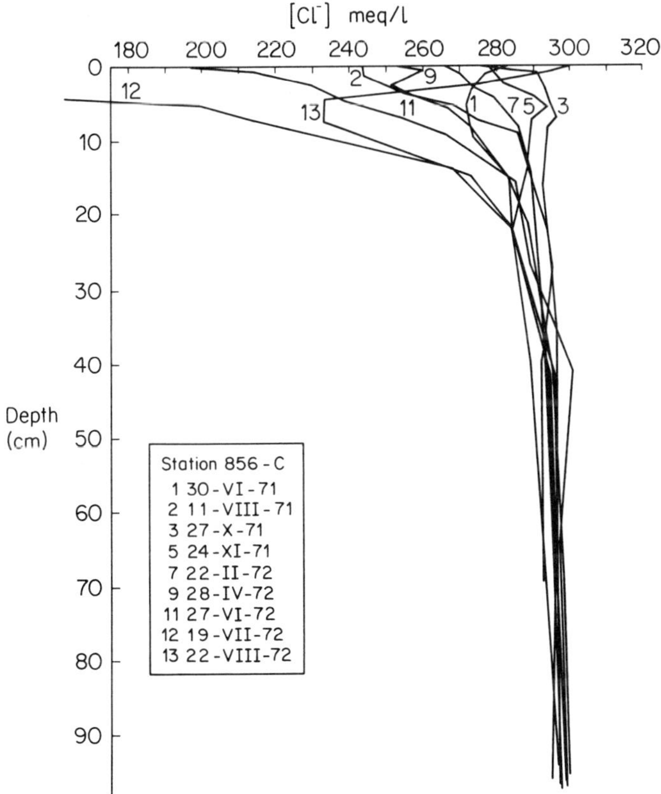

Figure 3. Temporal variation of chloride ion concentration at Station 856-C. Dates are given as day-month (in Roman numerals)-year. Note excellent agreement at depth, but seasonal variation in the upper 20 cm. (From Matisoff *et al.*, 1975). (Reproduced by permission of the American Chemical Society.)

salinity differences in the upper metres of sediment would be obscured by diffusion in a few thousand years. A recent example of the use of Cl^- gradients to calculate paleo-salinity is the paper by Manheim and Chan (1974). They used Lerman and Weiler's (1970) solution to the 'diagenetic equation' (see mathematical treatment section) along with approximated diffusion coefficients, sedimentation rates, and Black Sea chlorinity to generate curves of present-day pore water chlorinity *versus* depth. A comparison of the calculated curves to actual measured concentration profiles show a best fit when 8,000 year old bottom water is assumed to be 3.5 g/kg in Cl.

Calculations similar to the ones used by Manheim and Chan (1974) can be used to solve essentially the inverse problem, that is, in calculating the depth to a hidden object that is feeding an observed concentration gradient. For example, a salt dome or plug at depth should saturate the pore water with respect to NaCl (about 160 g/Kg Cl) and set up a concentration gradient away from the saturated water. Gradients set up by known or inferred buried salt deposits were found by Manheim and Bischoff (1969) in oil company drill holes in the Gulf of Mexico and were encountered in several locations around the world by the DSDP (see Manheim, 1976; for a review). Calculations based on measured concentration gradients must include certain assumptions about the constancy of sedimentation rates and diffusion coefficients, as well as establishment of steady state, but they can provide useful approximations of the time needed to set up a gradient or the depth over which it extends using only easily measured Cl gradients, as was done by Manheim and Sayles (1970).

Based on the above discussion it can be seen that Cl concentration should always be measured in pore water studies. In the absence of perturbing influences such as nearby fresher or more saline water, Cl concentrations should be very close to those in the overlying water, thereby providing a check on analytical procedures, evaporation during storage and the like. Furthermore, measured Cl gradients can be used in diffusion–advection calculations to solve a number of problems in estuarine geochemistry.

Sulphate, like chloride, is rare in common silicate minerals and abundant in evaporites. Unlike Cl^-, however, SO_4^{2-} can be chemically reduced at earth surface conditions. This reduction produces sulphide, provides a separate major reservoir for sulphur, and greatly complicates the geochemisty not only of sulphur, but of a number of other chemical constituents which are involved in the sulphate reduction process.

Depletion of pore water SO_4^{2-} concentration and the accompanying increase in carbonate alkalinity was one of the first observations ever made on pore water chemistry (Murray and Irvine, 1895). The sulphate reduction process has been studied by a host of investigators since Murray and Irvine's classic work, and much of the work has been reviewed and summarized by Goldhaber and Kaplan (1974). It is generally agreed that the reactions leading to and accompanying sulphate reduction are largely biologically mediated, that is, in spite of the state of chemical disequilibrium existing in recently deposited marine sediments, many reactions would be exceedingly slow at low temperature in the absence of enzymes and catalysts produced by organisms (especially bacteria). The bacteria responsible for sulphate reduction require anoxic conditions because they are unable to compete for food with oxygen utilizing organisms in the presence of oxygen. Anoxic conditions result when oxygen is used faster than it can be renewed from the atmosphere, and are rare in marine waters (Richards, 1965). In the sediment column, however, oxygen renewal by advection is greatly restricted and establishment of anoxic conditions is common. Oxygen use comes primarily by organic matter decomposition, thus factors which promote burial of organic matter in sediments lead to greater oxygen consumption. High surface water plankton production provides organic matter than can be buried and shallow water, high total sedimentation rate and restricted water circulation all decrease organic matter decomposition in the water column and therefore promote burial. Because the factors promoting organic matter burial are common in estuaries, sediments there are likely to be organic-rich. Oxygen and nitrate will be quickly depleted in the pore water of organic-rich sediments and sulphate reducing bacteria, which require organic matter as well as anoxic conditions, will begin to multiply. While the supply of sulphate lasts, sulphate reducers appear to be the dominant life form in the sediments.

The process of sulphate reduction can be illustrated in a simplified way by the following reaction:

$$2CH_2O + SO_4^{2-} \rightarrow H_2S + 2HCO_3^- \tag{1}$$

Organic matter is symbolized as CH_2O to avoid showing complicated details of the process, some of which will be discussed below in the sections on nutrients and alkalinity. As is shown by equation (1) a primary result of sulphate reduction is the conversion of sulphate to sulphide on a mole for mole basis, and in many samples H_2S can be readily detected by its characteristic odour. However, in other samples the extent of sulphate reduction cannot be detected, let alone quantified, by the presence of sulphide. Rather, the total reduced sulphur ($H_2S + HS^- + S^{2-}$) concentration is very much less than equivalent to the sulphate reduced. The primary sulphide removal mechanisms is reaction with iron to produce iron sulphides. Iron sulphide formation, and conversion to pyrite, is an extremely complex process which is not completely understood (Sweeney, 1972), but a likely first step is reaction between dissolved sulphide and iron oxide.

$$2FeO \cdot OH + 3H_2S \rightarrow 2FeS + S^\circ + 4H_2O \tag{2}$$

The abundance and type of oxide, or other iron compounds in the sediment, control the kinetics of sulphide precipitation, whereas the kinetics of sulphide production depends primarily on the abundance and nature of organic matter. The dissolved sulphide profile thus represents a balance between production and consumption, and the complicating factors of diffusion and sediment accumulation.

Sediments that contain little reactive iron allow dissolved sulphide to accumulate to near stoichiometric amounts based on sulphate reduced, as was found by Thorstenson and Mackenzie (1971) for $CaCO_3$ sediments from Bermuda. At the other extreme are sediments which keep dissolved sulfide near zero, even as sulphate is being completely removed. This strange phenomenon of no dissolved sulphur, either sulphate of sulphide, was found in some of the sediments from Saanich Inlet, an anoxic fjord (Nissenbaum *et al.*, 1972). A more common situation, however, is for sulphide to build up to a maximum within the top metre or two of the sediment column, then to decrease as sulphate reduction slows or stops. Figure 4 shows a sharp maximum of 50 micromolar concentration at 50 cm depth in Gulf of California sediments. Note the sharp decrease in dissolved sulphide below 50 cm despite continued sulphate reduction as evidenced by continued

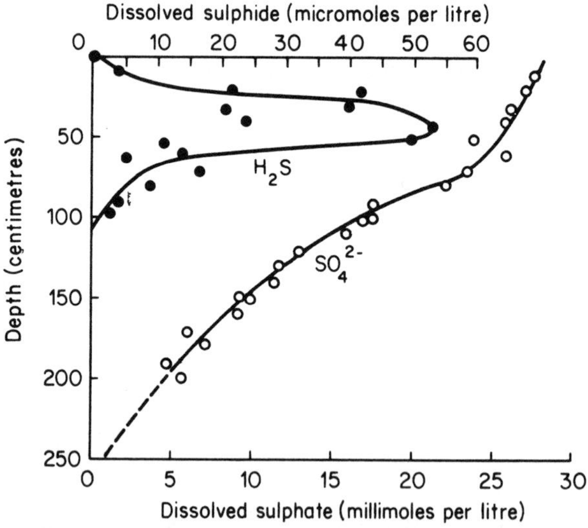

Figure 4. Concentrations of sulphate and sulphide in porewaters of a core from the Gulf of California. Note that the maximum sulphide concentration attained is only 54 micromolar. Below 1 m, sulphide is not present, although sulphate reduction is continuing as evidenced by continuing sulphate depletion. (From Goldhaber and Kaplan 1974). (Reproduced by permission of John Wiley and Sons, Inc.)

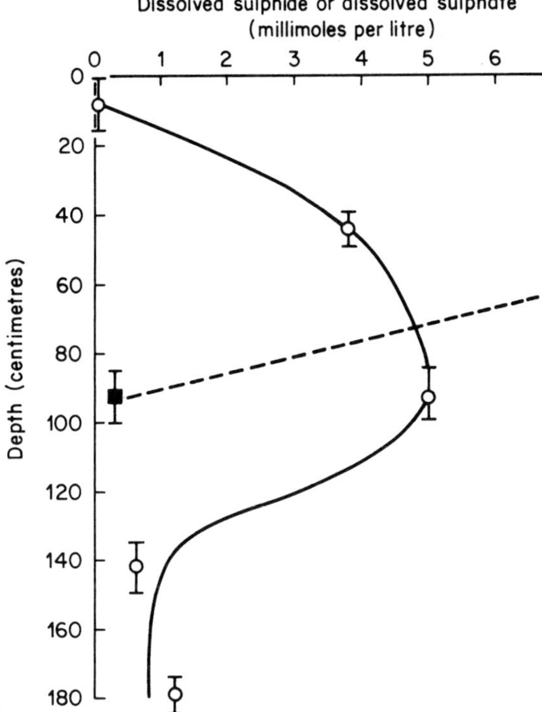

Figure 5. Concentrations of sulphate and sulphide in porewaters of a core from Saanich Inlet, British Columbia. ▪, sulphate. ⌀, sulphide. Dashed line represents extrapolated sulphate. Note that as the sulphate concentration approaches zero, at 90 cm depth, the sulphide concentration exhibits a maximum, and below this depth decreases. Data from Nissenbaum *et al.* (1972). From Goldhaber and Kaplan (1974). (Reproduced by permission of John Wiley and Sons, Inc.)

decrease in sulphate concentration. Only uptake by sediments can adequately explain the sulphide decrease. A somewhat similar situation is shown in Figure 5, except that the sulphide concentration at the maximum is 100 times larger, and the maximum is slightly below the point of complete disappearance of dissolved sulphate. Sulphate reduction is so rapid in the Saanich Inlet sediments shown in Figure 5 that sulphide builds up to high levels send drops only after reduction has essentially stopped.

Another point to note in Figures 4 and 5 is that dissolved sulphide is non-detectable in the upper few cm of the sediment column. In some environments anoxic conditions are established only after some depth of burial, but in others, as here, sulphate reduction occurs near the sediment water interface as is evidenced by other measured parameters. The presence of solid phase reduced sulphur (mainly FeS and FeS_2) is especially informative in identifying regions of sulphate reduction, as these minerals are almost never detrital. FeS and FeS_2 are, however, often abundant in the upper layers of sediment where no dissolved sulphide can be detected. Dissolved sulphide can be lost during sampling, but is also particularly subject to reaction with active iron compounds in the upper few cm of sediment. Temporal variations at a given site can also lead to the seeming paradox of reduced sulphur in oxidizing environments. For example, Goldhaber *et al.* (1977) found very different sulphide and sulphate profiles in Long Island Sound sediment in the winter than in the summer. The differences were best explained as due to greatly increased mixing of the upper six to eight cm of sediment by macroinfauna during the summer which effectively homogenized the upper sediment layers. At the same time, however, rates of sulphate reduction were more rapid in the summer.

In summary, dissolved sulphide profiles are not a reliable indicator of the extent of sulphate reduction, and even dissolved sulphate profiles can be misleading. Solid phase reduced sulphur is perhaps the best measure, although it integrates over a considerable time period, and may be produced in reducing 'micro-environments' such as the interior of radiolarian shells (Emery and Rittenberg, 1952).

3.3 Alkalinity, pH, and Eh

Alkalinity, pH and Eh are interrelated properties of seawater and pore water. Their values are critical to living organisms, and at the same time both living organisms and inorganic reactions affect the distribution of their values. The fact that pore water can have an alkalinity much higher than seawater was noted by Murray and Irvine (1895) who inferred that the increase was due to carbonate. The mechanism for carbonate production can be seen in the simplified reaction for sulfate reduction given earlier (equation (1)].

$$2CH_2O + SO_4^{2-} \rightarrow H_2S + 2HCO_3^- \tag{1}$$

Carbonate production during sulfate reduction can be contrasted with organic matter (carbohydrate) oxidation with molecular oxygen,

$$CH_2O + O_2 \rightarrow CO_2 + H_2O \tag{3}$$

which produces CO_2, thereby affecting pH but not alkalinity. The critical factor in sulphate reduction is the transfer of charge from SO_4^{2-}, a species that does not significantly react with protons in the pH range of interest to species that do (HCO_3^-, CO_3^{2-}, HS^-, etc.).

A number of other reactions can also affect alkalinity or pore water. Liberation of ammonia by hydrolysis of basic nitrogen compounds increases alkalinity by the following reactions:

$$CH_2NH_2COOH + 2(H) \rightarrow NH_3 + CH_4 + CO_2 \tag{4}$$

$$NH_3 + CO_2 + H_2O \rightarrow NH_4^+ + HCO_3^- \tag{5}$$

Organic phosphate release probably does not affect alkalinity, and in any case is quantitatively much smaller than HCO_3^- and NH_3 production. Reactions involving cations can affect alkalinity by adding or removing anions of weak acids, for example:

$$Ca^{2+} + CO_3^{2-} \rightarrow CaCO_3 \tag{6}$$

$$Al_2Si_2O_5(OH)_4 + 5Mg^{2+} + 10HCO_3^- + SiO_2$$
$$\rightarrow Mg_5Al_2Si_3O_{10}(OH)_8 + 10CO_2 + 3H_2O \tag{7}$$

A number of other reactions are possible, depending on the solid phase in the sediment and the Eh and pH of the environment.

Berner *et al.* (1970) introduced the concept of alkalinity modelling when they compared alkalinities measured in pore water to those calculated from an expression similar to the following:

$$Alk = Alk_0 + \Delta C_{NH_4} + 2\Delta C_{Ca^{2+}} + 2\Delta C_{Mg^{2+}} - 2\Delta C_{SO_4^{2-}} \tag{8}$$

Where Alk_0 is the alkalinity expected for seawater of a given chlorinity and ΔC is the difference between the observed concentration of a given ion and the concentration expected for seawater of the observed chlorinity. Berner *et al.* (1970) found that alkalinities calculated by equation (8) agreed well with those measured in organic rich nearshore muds from off the northeast coast of the United States (Figure 6).

The high alkalinity values shown in Figure 6 result primarily (80–90 per cent) from a progressive sulphate depletion with depth in the sediment column, although ammonia build-up at depth contributes 10–20 per cent. The pore water Ca and Mg changed little with depth in these sediments and were not important to alkalinity. On the other hand, Sholkovitz (1973), working with sediments from the Santa Barbara Basin off Southern California, found it necessary to include measured depletions in pore water Ca in order to balance measured alkalinities with those calculated by the model (Figure 7). Sholkovitz assumed that pore water Ca depletions he measured were due to $CaCO_3$ precipitation, as shown by equation (6). He also found Mg depletions, but because these seemed not to affect alkalinity he felt they were not due to authigenic silicate formation of the type shown in equation (7).

Mackenzie and Garrels (1966) suggested that authigenic silicate formation was a major sink for cations, silica and bicarbonate, but, as just mentioned,

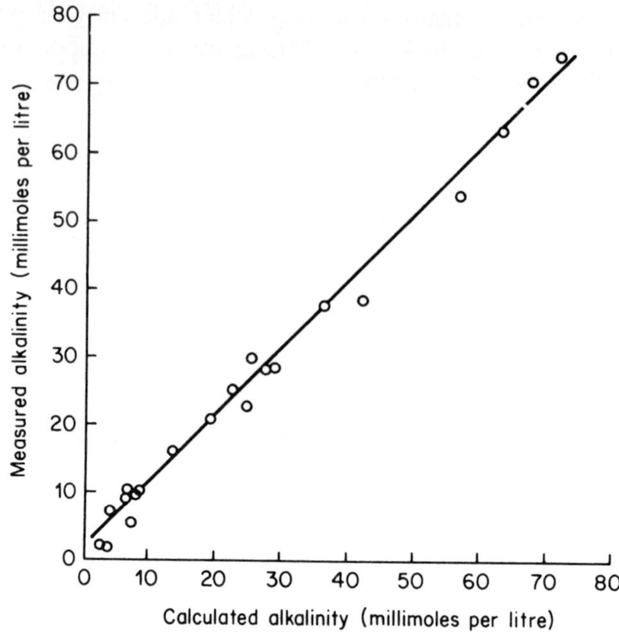

Figure 6. Plot of titration alkalinity versus alkalinity predicted by taking into account sulphate reduction and ammonia generation. (After Berner *et al.*, 1970.)

Berner *et al.* (1970) and Sholkovitz (1973) could find no evidence for such reactions in their data. Sholkovitz's data does, however, suggest the needed reaction in that a definite Mg depletion was found in most cases, but the trends are subtle and he uses another reaction to explain the apparent loss of alkalinity. Previously unpublished data obtained by Presley on three cores from the Cariaco Trench (Figure 8) are shown in Table 3. These cores show a much greater discrepancy between calculated and measured alkalinities than was found by Sholkovitz, and the discrepancy cannot be resolved by considering Ca and Mg losses. Measured K and Na concentrations showed no consistent depletion with depth, although the needed Na depletion could have been hidden by the poor precision of the measurement. Support for an actual loss of K and Na accompanying an alkalinity loss comes from the DSDP (Manheim, 1976) where easily measured losses have been recorded over the greater sediment depths involved. At the same time, however, the DSDP results show a variety of relationships between major cations, alkalinity, ammonia, sulfate, and CO_2, thus complicating attempts to infer diagenetic reactions from observed pore water concentrations.

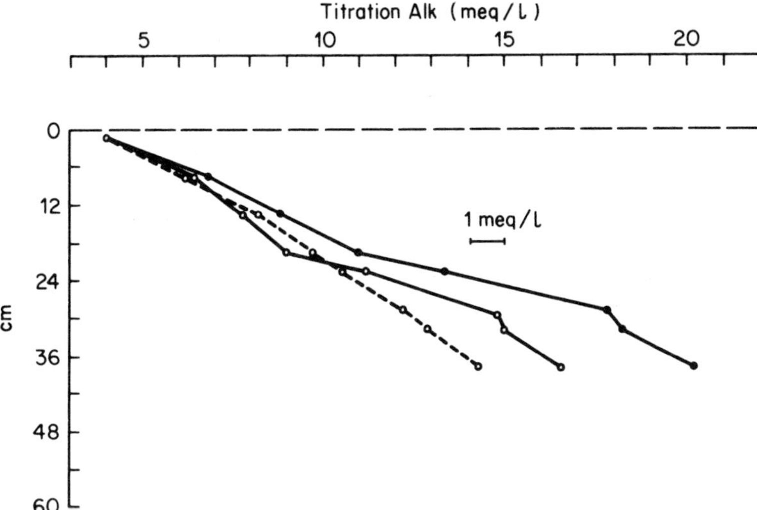

Figure 7. Calculated and observed alkalinities in Santa Barbara Basin sediments. (From Sholkowitz, 1973). (—●—) Alk. produced by SO_4^{2-} reduction and NH_4^+ formation. (—○—) Alk. resulting from SO_4^{2-} reduction and NH_4^+ formation and authigenic $CaCO_3$ precipitation. (---○---) Measured interstitial water alkalinity. (From Sholkovitz, 1973). (Reproduced by permission of Pergamon Press, Ltd.)

Only under unusual circumstances is the pH of seawater outside the range 7.8 to 8.3 (Skirrow, 1965). This constancy is due mainly to the bicarbonate ion, which buffers against pH change. The buffer capacity is, however, much more limited than is commonly realized, only about 0.5 m mol/Kg/pH at normal seawater pH (Ben-Yaakov, 1973). This means that the addition of only 0.5 m mol/Kg of CO_2 will lower the pH of seawater by one pH unit. Few, if any, processes are operative which would add 0.5 m mol/Kg of CO_2 to open ocean water, but by contrast 54 m moles of CO_2 will be produced by equation (1) when the 27 m mol/Kg of sulphate in normal seawater are completely reduced, a common phenomenon in sediment pore water. In addition, 27 m moles of H_2S will be produced, as well as large amounts of ammonia and phosphate. Such large additions to pore water are well demonstrated by the data of Nissenbaum *et al.* (1972) shown in Table 4.

The maximum CO_2, NH_3, and PO_4^{3-} concentrations given in Table 4 are somewhat higher than those found in most studies, but similar values have been noted by Berner *et al.* (1970), Presley and Kaplan (1968), Matisoff *et al.* (1975), Goldhaber *et al.* (1977) and others. The pH values given in Table 4 are also similar to those reported in the references just listed and in many

     B. J. Presley and J. H. Trefry

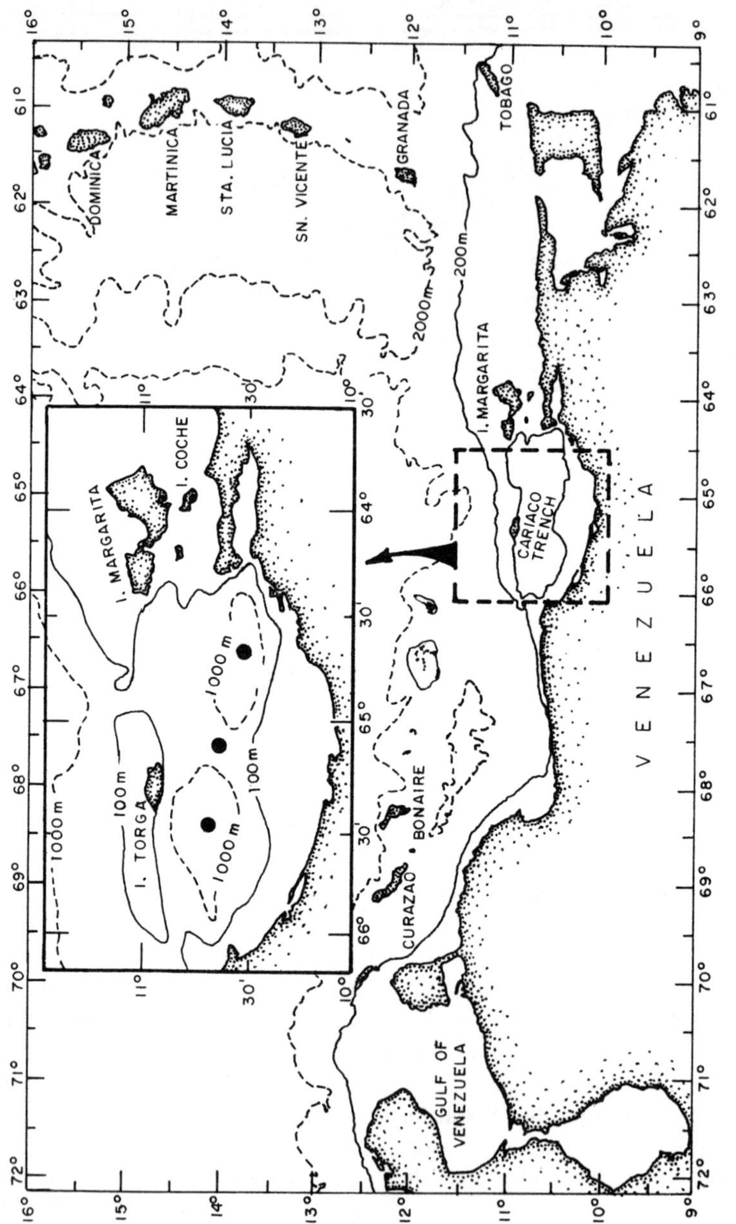

Figure 8. Core locations for data shown in Table 3.

Table 3. Selected components of anoxic pore water from the Cariaco Trench. Δ SO$_4$, Δ Ca, and Δ Mg refer to differences between the concentrations expected based on Cl and that measured. A_0 is the alkalinity calculated by the method of Berner *et al.* (1970), assuming the original value was 2.4, A_0 is the actual observed alkalinity and Δ A is the difference between the two.

Depth (cm)	Δ SO$_4$ mM	NH$_3$ mM	PO$_4$ μM	Δ Ca mM	Δ Mg mM	A_c meq/1	A_0 meq/1	Δ A meq/1
Eastern core								
1–5	3.4	0.56	—	1.50	0	6.8	—	—
5–10	5.6	0.75	36.7	1.37	0	11.6	8.4	3.2
10–20	9.7	1.22	52.1	2.20	0	18.6	12.0	6.6
20–30	13.1	1.38	57.0	3.19	0	23.6	—	—
40–50	17.7	2.28	76.0	5.42	0	29.2	23.6	5.6
70–80	29.2	3.30	97.5	5.83	0	52.4	25.4	27.0
100–110	29.2	3.80	114.5	6.50	0	51.6	29.8	21.8
130–140	29.2	4.25	128.0	6.50	0	52.1	29.8	22.3
150–160	29.2	4.90	141.0	6.50	0	52.7	29.3	23.4
Saddle core								
0–10	4.7	0.74	23.5	0.42	0.82	10.1	5.8	4.3
20–30	7.3	1.09	31.2	1.31	0.82	13.8	9.0	4.8
50–60	12.5	1.53	39.4	2.69	1.64	20.2	10.2	10.0
80–90	13.9	1.65	43.3	3.66	3.70	17.1	—	—
100–110	18.1	2.00	45.6	4.36	2.88	26.1	16.8	9.3
120–130	21.1	2.18	45.4	4.69	2.88	31.6	20.0	11.6
140–150	24.7	2.65	53.0	5.46	4.11	35.3	23.0	12.3
Western core								
0–6	1.9	0.33	18.7	0.67	0.50	4.2	4.6	−0.4
6–10	5.2	0.69	29.7	1.54	0.52	9.4	8.1	1.3
10–20	7.2	1.00	38.2	3.12	0.54	10.5	11.4	−0.9
30–40	10.9	1.51	41.7	3.50	1.00	16.7	13.4	3.4
50–60	14.0	1.81	51.7	4.05	2.50	19.1	16.4	2.7
80–90	18.5	2.28	60.0	4.69	2.51	27.3	15.0	12.3
100–110	22.4	2.63	66.0	5.52	2.65	33.5	16.8	16.7
120–130	25.3	3.02	72.5	6.12	3.75	36.3	21.4	14.9
140–150	26.3	3.46	78.5	6.94	3.75	37.1	23.0	14.1
150–160	27.5	3.62	82.0	6.76	4.05	39.4	24.2	15.2

other references (e.g. Emery and Rittenburg, 1952; Siever *et al.*, 1965; Sholkovitz, 1972). Almost no pH values outside the range 7.0–8.0 can be found in the hundreds of reported measurements. The constancy of pore water pH in spite of the potentially disturbing additions of CO_2, NH_3, etc. can be partially explained by the calculations of Nissenbaum *et al.* (1972) which show that pH of 7.0 should result when the products of marine organic matter decomposition equilibrate in water. Ben-Yaakov (1973) extended the calculation to include the effect of H_2S loss by iron sulphide formation, as is shown in Figure 9.

Thorstenson and MacKenzie (1971) allowed marine algae to decompose anaerobically in seawater and followed the pH change to a minimum of 6.7,

Table 4. pH, Eh and related variables in the interstitial water of Saanich Inlet sediments (From Nissenbaum *et al.*, 1972.)

Depth (cm)	pH	Eh (mV)	ΣCO_2 (mM/1)	ΣS^{2-} (mM/1)	SO_1^{2-} (mM/1)	NH_3 (mM/1)	PO_4 (μM/1)	SiO_2 (mM/1)
Core 1								
0–15	7.6	+330	2.7	tr.	22.9	0.15	42	—
40–50	8.0	−120	13.6	3.8	11.5	0.53	35	—
85–100	7.9	−140	23.7	5.0	0.3	2.05	155	—
135–150	8.0	—	29.7	0.6	0.7	3.50	73	—
175–185	8.0	−120	36.2	1.2	0.3	4.26	122	—
Core 2								
0–15	7.8	−60	40.6	tr.	1.3	4.05	120	—
75–85	7.8	+260	49.2	0	tr.	8.58	115	—
150–165	7.7	+250	—	0	tr.	10.0	91	—
225–235	7.7	+260	51.2	0	tr.	14.0	136	—
Core 3								
0–10	8.0	−100	30.2	5.3	1.4	3.71	195	—
50–60	7.6	−140	40.6	4.4	0.3	—	280	—
100–110	7.6	−130	—	3.1	0.3	5.24	325	—
150–160	7.8	—	36.2	3.1	tr.	6.41	420	—
190–200	7.7	−100	39.1	2.8	tr.	6.71	375	—
Core 3B								
790–820	7.5	+40	49.5	1.3	0	5.0	400	2.72
1710–1740	7.7	+340	66.0	0	0	0	240	2.50
2620–2650	7.8	+380	52.7	0	2.0	0	160	2.04
3445–3475	7.9	+370	44.6	0	tr.	0	200	2.04
Core 4								
0–15	7.7	−120	32.4	4.1	2.2	4.26	383	—
50–65	7.7	−110	34.6	4.1	1.9	7.92	358	—
100–110	7.8	−120	36.8	3.8	1.2	10.4	333	—
150–160	7.7	−135	39.7	3.8	tr.	10.6	320	—
200–210	7.9	−110	42.4	3.4	0.2	9.8	363	—
240–250	7.7	−110	40.1	3.1	0.2	9.1	291	—

giving laboratory support for the Ben-Yaakov (1973) model. As has been mentioned, reported field measurements lend qualitative support to the model, but a rigorous field test is almost impossible due to difficulties in measuring some of the constituents and because so many possible reactions involving protons could be occurring simultaneously in the sediment-water mixture.

Theoretical difficulties in measuring redox potential (Eh) are even greater than those of measuring pH, for some of the same reasons. An almost unlimited number of reactions can involve uptake and release of electrons, as is true of protons, making it very difficult to determine which reactions actually control the system. Morris and Stumm (1967, page 279) state

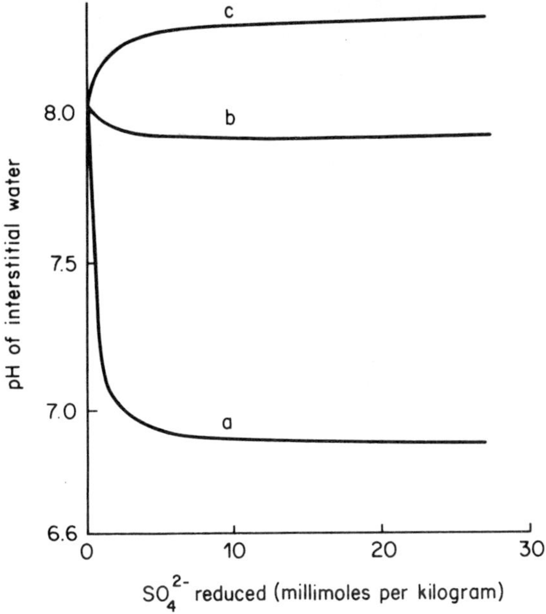

Figure 9. Predicted pH of porewater of anoxic marine sediments as a function of the amount of sulphate reduced, for different sulphide concentrations: $\sum H_2S = -\Delta SO_4^{2-}$ (curve a), $\sum H_2S = -0.2 \Delta SO_4^{2-}$ (curve b), $\sum H_2S = 0$ (curve c). (After Ben-Yaakov, 1973.)

additional considerations in the following way:

'Since natural waters are generally in a dynamic rather than equilibrium condition, even the concept of a single oxidation–reduction potential characteristic of the aqueous environment cannot be maintained. At best, measurements can reveal an Eh value applicable to a particular system or systems in partial chemical equilibrium and then only if the systems are electrochemically reversible at the electrode surface at a rate that is rapid compared with the electron drain or supply by way of the measuring electrode.'

A large number of Eh measurements have been made in estuarine and other nearshore environments despite the cautions about data interpretation expressed above. Interestingly enough, ZoBell (1946), perhaps the first user of Eh in marine sediments, anticipated problems in quantifying the measurement when he stated 'while the Eh values obtained for sediment samples are

                    B. J. Presley and J. H. Trefry

more descriptive than physicochemically exact, such values may prove to be a useful means of characterizing sediments'. Whitfield (1969) gives a good, concise discussion of the theoretical problems with Eh measurements, then goes on to show their operational usefulness. He made several dozen Eh and pH measurements in an Australian estuary and found, as had Baas Becking *et al.* (1960) earlier, that Eh varied much more widely than pH and thus could be used to describe the sediments. Furthermore, because Eh could be measured quickly and easily it was especially useful in a pilot study to identify potentially stagnant areas of the estuary.

As is illustrated in Figure 10, a decrease in Eh with depth in the sediment column is common in nearshore sediments, but the sharpness of the decrease and the minimum value reached can vary widely (Bass Becking *et al.*, 1960; Berner, 1963; Nissenbaum *et al.*, 1972; Matisoff *et al.*, 1975 and many others).

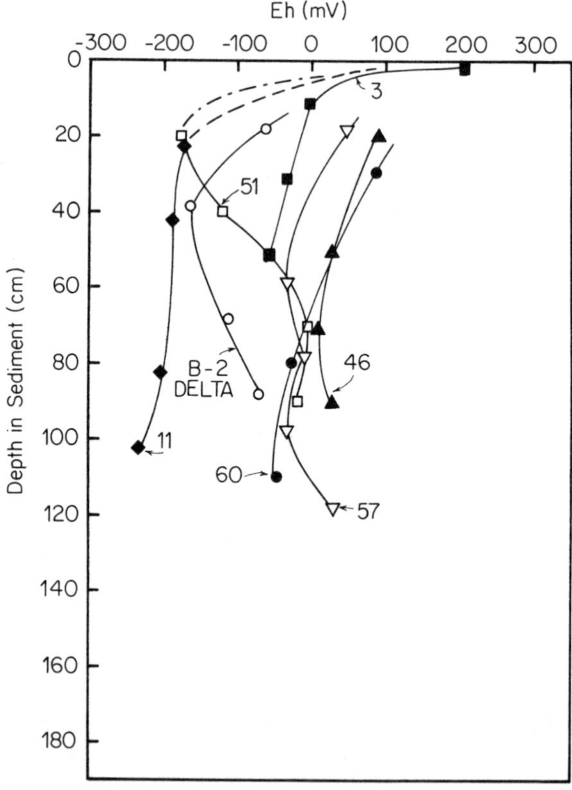

Figure 10. Eh in near-shore sediments from the Gulf
of Mexico (from McKee, 1977.)

It is generally agreed that reduction of the various oxidized species in a sediment–water mixture will proceed in a stepwise manner, roughly in the sequence expected from thermodynamic considerations (e.g. Latimer, 1952). Thus, dissolved oxygen will be depleted first as Eh is lowered, followed by nitrate, manganese oxides, iron oxides, and finally sulphate. Attempts to define critical Eh values at which one redox system gives way to another have been only partially successful, as would be expected from the discussion on Eh interpretation given earlier. Nevertheless, work by Takai and Kamura (1966), Turner and Patrick (1968), Ponnamperund (1972) and others shows that oxygen disappears at Eh values below about $^+250\,\text{mV}$, nitrate and manganese oxides begin to reduce at values slightly below this ($\sim$250–300 mV), iron oxides can only be reduced below $^+100\,\text{mV}$ and sulphate only below $^-150\,\text{mV}$. Some overlap occurs, for example sulphate reduction certainly starts before all iron oxide is reduced, but neither iron oxide or sulphate is reduced in the presence of oxygen or nitrate.

Berner (1963) found a direct relationship between Eh and the sulphide concentration such that:

$$\text{Eh} = -0.485\,\text{V} + 0.0295\,\text{pS}^{2-} \tag{9}$$

We have failed to find this relationship in most of the anoxic environments we have studied, but we agree with an earlier observation by the senior author (Nissenbaum *et al.*, 1972) that Eh is always negative in the presence of sulphide and positive in its absence. We have found sulphide absent in several locations where it has obviously been present, as evidenced by greatly reduced sulphate concentration. Pore water from such environments gives positive Eh values, no doubt due to artifacts introduced during sampling and handling of what is a very poorly poised system. When sulphide is present the pore water is much better poised, and gives stable and reproducible negative Eh readings.

3.4 Nutrients and organic carbon

Nutrient and organic carbon concentrations in pore water are closely related to some of the parameters discussed above, especially alkalinity, Eh and sulphate. Most of the silicon, nitrogen, and phosphorus incorporated into phytoplankton in the euphotic zone of the water column is subsequently released (regenerated) during sinking of organic detritus (Riley, 1951; Redfield *et al.*, 1963; Menzel, 1974). In nearshore areas, however, enough organic detritus can reach the bottom to make regeneration on and in the sediments a significant source of nutrients. Hale (1974) found that nutrient release from Narragansett Bay sediments could supply 80 per cent of the nitrogen and 200 per cent of the phosphorus needed by bay phytoplankton. Rowe *et al.* (1975) calculated that the nitrogen flux from New York Bight sediments was 200 per cent of that needed for photosynthesis. Other investigators have found nearshore sediments a much less important source of nutrients to the overlying water column (Rittenberg *et al.*, 1955; Hartwig, 1968), but nevertheless sediment pore waters are almost always greatly enriched in nutrients compared to the overlying water (Rittenberg *et al.*, 1955; Brooks *et al.*, 1968; Sholkovitz, 1973; Matisoff *et al.*, 1975; Goldhaber *et al.*, 1977 and many others). The phosphate and ammonia values given in Tables 3 and 4 are typical of those in many nearshore areas and demonstrate the high enrichments that can be found.

In the presence of molecular oxygen, organic decomposition is dominated by organisms using oxygen as an electron acceptor (e.g. Goldhaber and Kaplan, 1974). Richards (1965) represented organic decomposition, oxygen use and nutrient regeneration by the following equation:

$$(CH_2O)_{106}(NH_3)_{16}(H_2PO_4) + 138O_2$$
$$\rightarrow 106CO_2 + 122H_2O + 16HNO_3 + H_3PO_4 \quad (10)$$

The $C:N:P$ ratios are $106:16:1$ as was found in living plankton by Flemming (1940).

When molecular oxygen has been used up, organic matter decomposition continues, mediated by organisms which use progressively less efficient electron acceptors (nitrate, manganese oxides, iron oxides, sulphate, and carbon dioxide). Sulphate is especially important in marine systems, due to its high concentration in normal seawater. Furthermore, sulphate reduction produces H_2S which is toxic to most forms of life. Richards (1965) represented sulphate reduction in the following way:

$$(CH_2O)_{106}(NH_3)_{16}(H_3PO_4) + 53SO_4^{2-} \rightarrow 106CO_2 +$$
$$53S^{2-} + 16NH_3 + 106H_2O + H_3PO_4 \quad (11)$$

The products shown in equation (11) will, of course, react with each other to produce an equilibrium assemblage of the various possible species, as discussed by Nissenbaum *et al.* (1972), but such reactions do not affect the overall stoichiometry. Richards and Vaccaro (1956) found PO_4^{3-} concentrations in the anoxic waters of the Cariaco Trench close to those predicted for carbon oxidation by equation (11), ammonia, however, was only three to four times greater than phosphate, rather than 16. It was suggested that nitrogen gas formation might account for the missing ammonia, thus pointing out one of many processes than can complicate the simple picture given by equation (11).

A number of papers (e.g. Redfield, 1934; Cooper, 1937; Redfield *et al.*, 1963) have considered ratios of nutrients released to the water column during organic decomposition, but the first attempts to model pore water nutrient concentration from the stoichiometry of decomposition seem to be those of Sholkovitz (1973) and Hartmann *et al.* (1973). Figure 11 shows

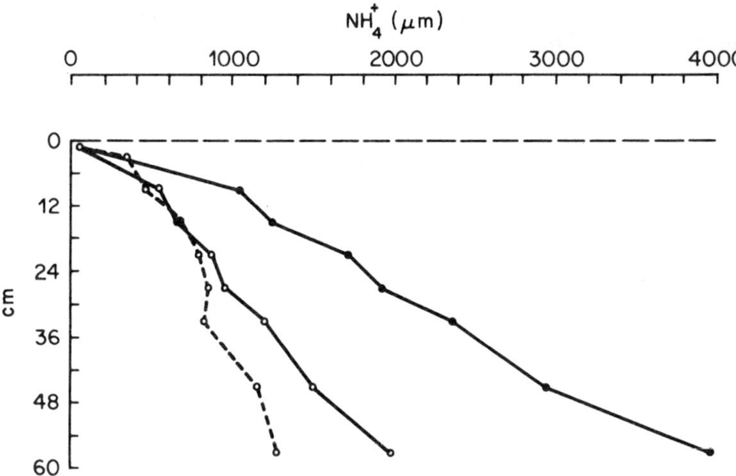

Figure 11. (—O—) NH$_4^+$ produced assuming a C:N = 6.6. (—·—) NH$_4^+$ produced assuming a C:N = 13.2 (---O---) Measured interstitial water NH$_4^+$. Calculated and measured ammonia in a core from Santa Barbara Basin (from Sholkovitz, 1973). (Reproduced by permission of Pergamon Press, Ltd.)

Sholkovitz's (1973) measured ammonia and that calculated from the measured sulphate depletion using two different C:N:P ratios. The 212:16:1 values provide a much better fit to the data, and at the same time are closer to the ratios measured in sedimentary organic matter from the study area.

Hartmann *et al.* (1973) used different, but equivalent, data treatment on pore water from the West African continental margin. They simply plotted all CO_2, NH_3, and PO_4^{3-} data, corrected for initial concentrations, against sulphate lost (reduced). The resulting regression line for NH_3 *vs.* SO_4^{2-} gave a near 0 intercept and a slope indicating 8 moles of NH_3 produced for each 53 moles of sulphate reduced, identical to Sholkovitz (1973) result. However, Hartmann *et al.* (1973) show by second order regression analysis that the pore water SO_4^{2-}:NH_3 ratio changes from about 14 near the sediment surface to about 4 at depth, thus implying a change in the solid phase organic matter with depth. Pore water PO_4^{3-} and CO_2 concentrations in the West African sediments also show changing ratios with depth, and in fact the same thing is indicated by the increasing divergence between Sholkovitz's measured and calculated values (Figure 11).

In order to use the stoichiometry shown by equation (11) to interpret pore water nutrient data it may be necessary to consider processes in addition to those discussed above. Ammonia can be taken up by clay minerals (McBeth, 1917), phosphate can precipitate as Ca or Fe salts (Bray *et al.*, 1973) and

                    B. J. Presley and J. H. Trefry

both PO_4^{3-} and NH_3 are released by fermentative reactions other than sulphate reduction. This latter process can be seen in the data Table 3 and 4 where PO_4^{3-} and NH_3 continue to increase below the depth where sulfate completely disappears. At the same time several processes can affect pore water CO_2 concentrations (carbonate equilibria, methane formation, etc.) making C:N and C:P ratios ambiguous.

The CO_2 shown in equation (11) would be mostly in the form of HCO_3^- at typical pore water pH values, with only minor amounts of CO_2 and CO_3^{2-}. Total CO_2 ($CO_2 + HCO_3^- + CO_3^{2-}$) can be calculated from measurements of alkalinity and pH, but Presley and Kaplan (1968) used a more direct method on Southern California continental borderland sediments. They simply acidified filtered pore water and measured the CO_2 evolved. The CO_2 was then analyzed for its carbon isotope composition with the results shown in Figure 12. The total CO_2 concentration at the bottom of the cores is 10–20 times greater than that of the overlying seawater, with carbon isotope ratios matching those of organic carbon from the area.

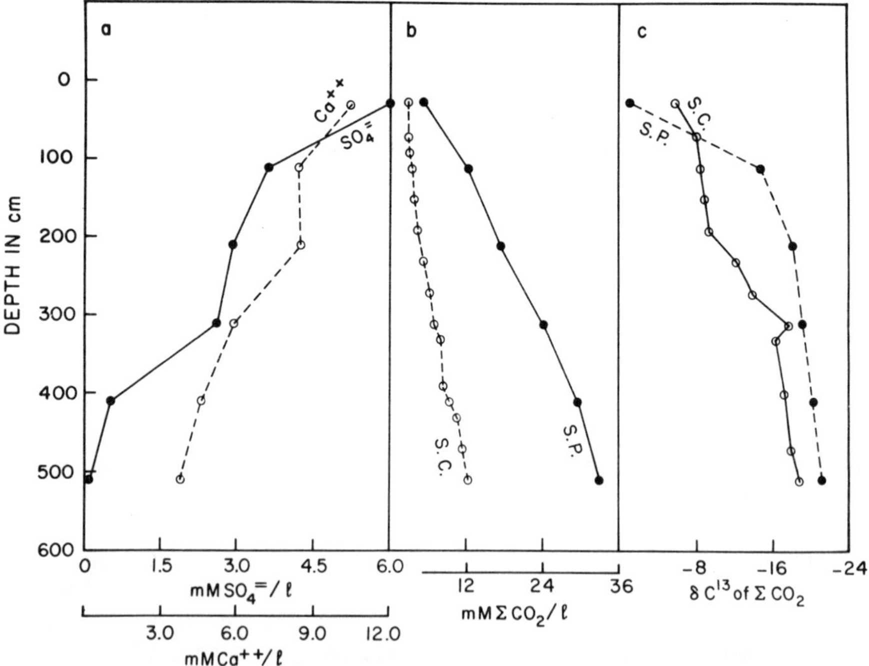

Figure 12. Concentrations of Ca, SO_4^{2-}, total CO_2 and $\delta\ C^{13}$ in San Pedro Basin (SP) and Santa Catalina Basin (SC) sediment pore water. Plot a shows San Pedro only. $\delta\ C^{13}$ relative to PDB. From Presley and Kaplan (1968). (Reproduced by permission of Pergamon Press, Ltd.)

It is clear from the above that organic matter is decomposed as it is buried in the sediment column, producing dissolved CO_2 and nutrients. Dissolved organic matter (DOM) is also produced, according to the few measurements of this parameter that have been made. In seawater DOM values are relatively constant at 2–3 mg C/1, but Nissenbaum *et al.* (1971) found 50–150 mg C/1 in pore waters from Saanich Inlet. Jeffrey (unpublished data) found about 30 mg C/1 in Cariaco Trench sediments and only 5–10 mg C/1 in most Gulf of Mexico sediments, but she did note increases in concentration with depth, confirming that observation of Nissenbaum *et al.* (1971).

Almost all published reports show that dissolved silica concentrations are much higher in pore water than in the bottom water overyling sediments (Siever *et al.*, 1965; Hurd, 1973; Schink *et al.*, 1974, 1975). Often a sharp increase is seen even in the top cm of the sediment column (Schink *et al.*, 1975) thus implying a considerable flux of silica from sediments to the overlying water. Some of this observed increase across the sediment-water

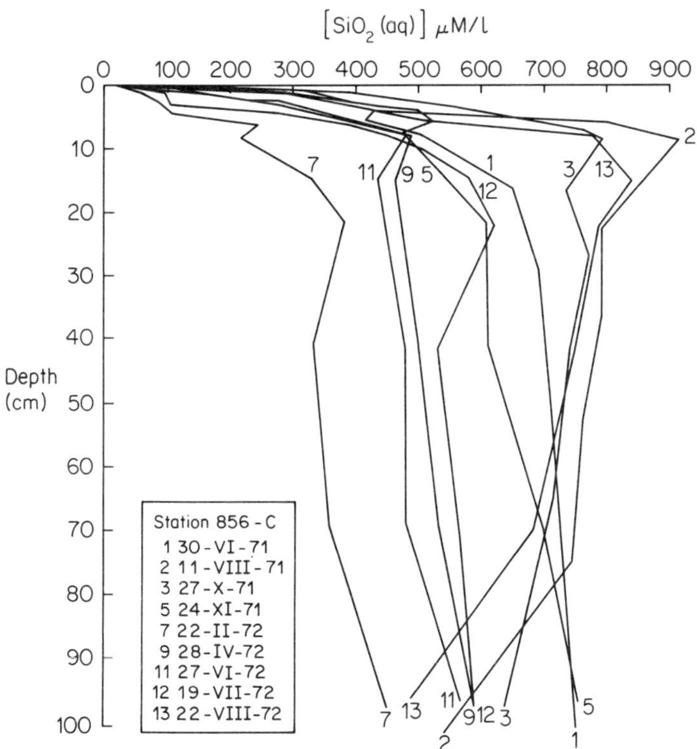

Figure 13. Temporal variation of dissolved silica at Station 856-C. Note strong seasonal variation (from Matisoff *et al.*, 1975). (Reproduced by permission of the American Chemical Society.)

interface is an artifact of warming the sediment before pore water extraction (Fanning and Pilson, 1971), but strong enrichments remain after correction for the temperature artifact. Matisoff *et al.* (1975) report little spatial variability in dissolved silica in several cores from one location in Chesapeake Bay, but extreme temporal variability over a one year period (Figure 13). Summer concentrations were as much as 450 per cent higher than those in winter and, unlike winter profiles, decrease with depth, a behaviour which cannot be explained by squeezing artifacts. Over larger geographic areas silica concentrations vary widely from place to place, as was noted by Siever *et al.* (1965) and emphasized by Schink *et al.* (1974).

Dissolution of siliceous organisms can release silica to the pore water, whereas clays and other silicates can either release or take up silica depending on circumstances (e.g. Helgeson and MacKenzie, 1970). Both dissolution of organism test and silicate reactions are complicated functions of pore water composition, pH, temperature, sediment composition and grain size, thus the explanation for any given observed pore water profile is extremely complex. Nevertheless, the statement of Siever *et al.* (1965) that 'it appears that the dominant control on dissolved silica is the abundance and rate of dissolution of diatoms in the sediment' gives at least a qualitative explanation. Schink *et al.* (1975) give a more quantitative treatment of pore water silica profiles by presenting a detailed mathematical model which includes effects of (1) advection of fluids, (2) advection of solids, (3) diffusion of fluids, (4) diffusion of solids, (5) adsorption–desorption, and (6) dissolution–precipitation. A further discussion of the Schink model is given in the mathematical Treatment Section.

3.5 Dissolved gases

The distribution and behavior of CO_2 and H_2S in pore water has been discussed above. Other gases also occur in sediments, and attempts to identify and quantify these by analysing escaping bubbles have been made since the work of Shaw (1914). Workers at Nagoya University (Koyama, 1953) did pioneering work on lake and shallow-water marine sediments, whereas the first work on deep-water marine sediments appears to be that of Emery and Hoggan (1958). These early workers slurried sediment with gas-free water, removed gases by vacuum or carrier gas stripping and analyzed by manometry, mass spectrometry, or gas chromatography. The next advance in sampling technique was made by Reeburg (1967) who designed a device for conveniently squeezing pore water from sediment, measuring its volume, and removing gases by carrier gas stripping. Analysis was by gas chromatography. A number of workers who were active in the 1960s and early 1970s described their techniques and data in a book edited by Kaplan (1974).

Gases that have been identified in marine sediments include Ar, H_2, He, N_2, O_2, CO_2, H_2S, NH_3, Rn, CH_4 and other hydrocarbons. These gases can dissolve in seawater from the atmosphere and ultimately be trapped in pore water of despositing sediments. Some of the gases can also be generated by thermocatalytic and volcanic processes deep in the earth and migrate up to near-surface layers. Finally, biological processes which are most active in the near-surface layers of sediment, can produce or consume various gases. The biological processes have been of special interest due to the large volumes of gases that can be generated or consumed and because they are of practical significance in soil and plant science, sewage, and wastewater treatment, engineering properties of sediments and other fields.

With time (depth of burial in the sediment column) a series of reactions which involve gases occurs. At the sediment surface, oxygen is consumed by respiration producing CO_2 which can then react in a number of ways. The small amount of oxygen (usually $<0.2\,mM$) dissolved in pore water is quickly used up in most estuarine sediments and organisms then convert dissolved NO_3^- to N_2. This reaction is quantitatively unimportant due to the low concentration of NO_3^- available for reduction, but its completion allows for further lowering of the Eh and additional reactions. The reactions following nitrate reduction are initiated by fermentative degradation of organic matter which releases CO_2 and H_2. The hydrogen released is apparently immediately used by micro-organisms in reducing sulphate to H_2S, which in turn reacts with iron oxides to form iron sulphides in most sediments. The CO_2 is meanwhile accumulating as HCO_3^- or being precipitated as $CaCO_3$. Finally, when all sulphate has been reduced, hydrogen is used by another group of micro-organisms to reduce CO_2 to CH_4. The details of all of there reactions are extremely complicated and the subject of much controversy. There is a little doubt, however, that CH_4 and N_2 become the dominant gases in estuarine pore water at a few cm depth in the sediment column.

Measurements of CH_4 concentrations have been made by Reeburg (1968, 1974) who found up to $150\,ml/1$ in pore water from Chesapeake Bay sediments, but most values were much lower than the maximum amount. Bernard *et al.* (1978) measured values of $50–400\,\mu l/1$ CH_4 in 12 cores from the Gulf of Mexico shelf and slope. The latter authors also report values for ethene, ethane, propene, and propane, all of which are in much smaller concentrations than that of CH_4, but which seem to have a similar biological origin. Martens and Berner (1977) show that CH_4 builds up to bubble saturation in Long Island Sound sediments, as was found for Chesapeake Bay by Reeburg (1974). The resulting bubbles can strip other gases from the pore water further complicating observed distributions.

In summary, the processes which determine the distribution of gases in the pore water of estuarine sediments are complicated and not completely understood. At the same time these largely biological processes can strongly disturb the sediment both chemically and physically.

3.6 Trace metals

Trace metal concentrations in estuarine interstitial water are often reported to be several orders of magnitude higher than in the overlying water (Brooks *et al.*, 1968; Presley *et al.*, 1972; Duchart *et al.*, 1973; Elderfield and Hepworth, 1975). This enrichment is most striking for Mn and Fe which are very susceptible to dissolution under the reducing conditions generally found in estuarine sediments. Steep concentration gradients may thus be established and significant remobilization of reduced species and fluxes of metals from pore water to overlying waters may be observed.

Figure 14 shows, for example, some typical Mn profiles. Case I would be

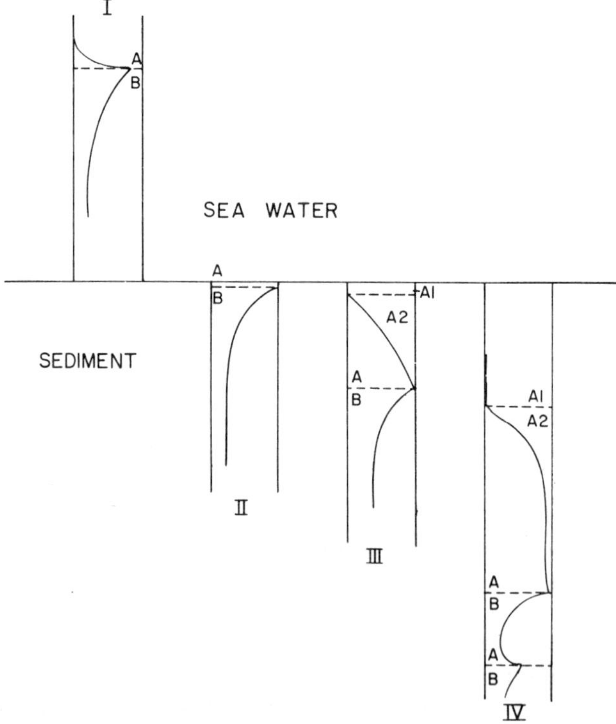

Figure 14. Typical interstitial Mn profiles (from Elderfield, 1976), showing variations in the anoxic–oxic (A–B) boundary (A1-A2 in cases III and IV) and the regions of maximum Mn remobilization which vary as a function of sediment (or water column) redox conditions (from Elderfield, 1976.)

observed where a stratified water column with anoxic bottom water is found (e.g. the Black Sea). Here, Fe and Mn and perhaps other metals may be reduced and solubilized in the water column. When oxic waters overlie anoxic sediment (Case II), maximum metal remobilization occurs at the sediment–water interface or in the top millimetres of sediment. Any net diffusive flux from the sediment is a function of the depth of the oxidized layer. In Case III, mildly reducing conditions allow a well defined oxic zone (A-1) to develop wherein remobilized metals may be trapped and greatly concentrated over natural levels. Robbins and Callendar (1975) showed that the negative interstitial Mn gradient observed below the A–B boundary (Figure 14, Cases II and III) may be controlled by precipitation of rhodochrosite ($MnCO_3$). Finally, Case IV exemplifies a feasible pattern for slowly accumulating sediments where a thick oxidized zone occurs with possible discrete anoxic segments.

Cases II and III (Figure 14) are typical for estuarine environments (Li *et al.*, 1969; Calvert and Price, 1972; Robbins and Callender, 1975; Holdren *et al.*, 1975). Data from Trefry (1977) in Figure 15 compare interstitial and sediment Mn distribution in rapidly accumulating Mississippi Delta sediment with that in deeper water Mississippi Fan sediment. Anoxic conditions are found in both cases, however, the fan sediment has a thin oxic zone (A-1) whereas the delta core is reducing even at the sediment–seawater interface. In the Mississippi Delta example, a calculated flux of Mn on the order of $1,000$ μg cm^{-2} y^{-1} is passing into the overlying water and consequently the bottom sediment has a total Mn concentration ($\sim$660 μg/g) which is about one-half that in the particulate flux ($\sim$1,300 μg/g) to the sediments. A comparable flux of Fe is also passing from these nearshore sediments and similar fluxes of Mn have been reported for Narragansett Bay (Graham *et al.*, 1976) and the Tees and Conway estuary (Elderfield and Hepworth, 1975). Furthermore, Elderfield and Hepworth (1975) also suggest that significant fluxes of Cu, Zn, Co, Ni, Pb, and Fe pass from TEEs and Conway estuary pore water. This is somewhat contradictory to the work of Lu and Chen (1977) who found from laboratory experiments that Fe and Mn were released from reducing sediments, that Cd, Cu, Ni, Pb, and Zn were released from oxidizing seiments and that Cr and Hg showed no significant change.

Identifying the specific chemical reactions which control interstitial metal concentrations and subsequent metal mobility remains a somewhat elusive goal. The presence or absence of free oxygen is certainly a primary factor in the behaviour of Mn and Fe and perhaps other metals. Where interstitial O_2 and nitrate are depleted, concentrations of dissolved Mn and Fe clearly increase and *vice versa*. Yet, the role of carbonate, sulphide and phosphate ions, organic ligands and adsorption in controlling these metals is only partly understood.

            B. J. Presley and J. H. Trefry

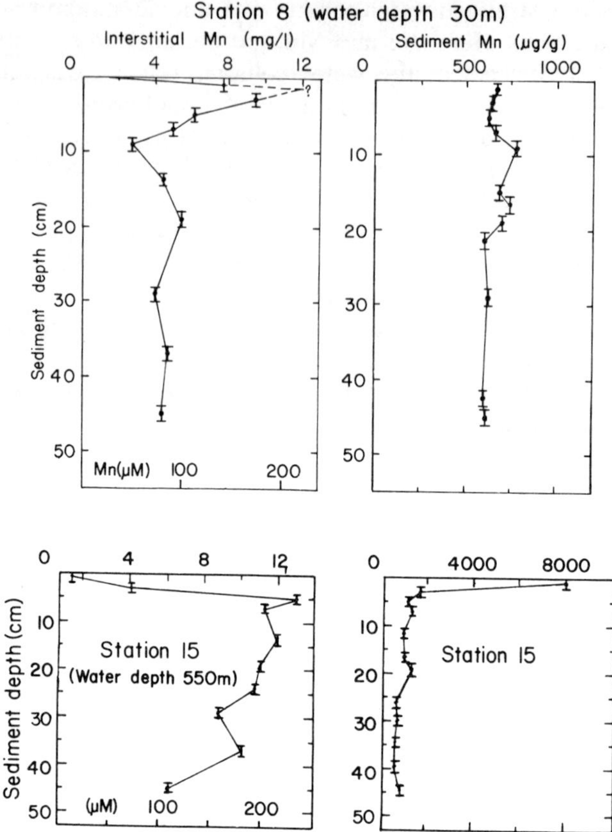

Figure 15. Interstitial water and sediment Mn concentrations for samples from the Mississippi River Delta (water depth 30 m) and Mississippi Fan (water depth 550 m.)

Metal sulphide solubility is one important constraint on the activity of several metals in sulphate-reducing sediments. Copper, Pb, Cd, and Zn sulphides, for example, have a very low solubility and might be expected to remain relatively immobile in reducing estuarine sediments. Despite this prediction, Presley *et al.* (1972) and Elderfield and Hepworth (1975) contend that measured interstitial Co, Cu, Fe, Ni, Pb, and Zn concentrations exceed calculated solubility levels by several orders of magnitude. These investigators believe that metal complexation with natural organic ligands can maintain metals at observed levels. Bender and Klinkhammer (1970), however, found very low concentrations of interstitial Cd (<0.1 ppb), Cu (<1 ppb) and Ni (<2 ppb) in sulphate-reducing sediments from Narragansett Bay, Rhode Island. Comparison data (Bender and Klinkhammer, 1978) from non-sulphate reducing sediments support, in principle, the importance of sulphides as one controlling factor of the interstitial chemistry of these metals. Furthermore, Bender and Klinkhammer's (1978) data for oxygen and nitrate-reducing sediments from the eastern tropical Atlantic show high interstitial Cu (6–8 μg/1) which they believe reflects copper input from degraded organic matter.

Detailed interstitial metal profiles with supporting data are scarce in the literature and subject to the many problems of trace metal analysis. At present, our understanding of interstitial Fe and Mn behaviour is broadly outlined, however, any comprehensive picture for most of the other metals awaits further research efforts.

4. MATHEMATICAL TREATMENT OF INTERSTITIAL WATER DATA

Descriptive analysis of the vertical distributions of interstitial water ions and compounds and the reactions in which they participate may be augmented by the use of theoretical models. Such quantitative treatment may consider, for example, simple diffusion of chloride ions into estuarine pore water from an overlying brackish water mass (Lerman and Weiler, 1970) or incorporate the effects of input fluxes, dissolution rates, bioturbation, adsorption, and diffusion into modelling interstitial silica (Schink *et al.*, 1975). These models, in turn, may be used to gain a more detailed representation of the chemical exchanges which occur across the sediment-water interface and how they affect the overall chemical balance of an estuary.

Idealized kinetic models for the behaviour of a dissolved species during early diagenesis have followed from the initial work of Berner (1964) and may be expressed mathematically by a general 'diagenetic equation':

$$\frac{\partial C}{\partial t} = \frac{1}{\phi} \frac{\partial\left(\phi D_s \frac{\partial C}{\partial x}\right)}{\partial x} - \omega \frac{\partial C}{\partial x} + R \tag{12}$$

where C = concentration of dissolved species (moles cm^{-3}).

t = time (sec).

ϕ = porosity (interstitial water volume fraction; dimensionless).

D_S = diffusion coefficient in pore water (cm^2 sec^{-1}) corrected for temperature and the ratio of the diffusion path length through a porous medium to the straight path distance (tortuosity).

x = distance below the sediment–water interface (cm).

ω = apparent velocity of sediment and pore water as viewed from the advancing sediment-water interface, the sedimentation rate (cm sec^{-1}).

R = applicable reaction terms (mole cm^{-3} sec^{-1}) which affect the concentration of the species considered, including oxidation reduction, adsorption-ion exchange, mineral dissolution, etc.

The terms of equation (12) (from left to right) account for pore water diffusion, advective supply of dissolved species by continual sediment accumulation, and likely chemical reactions.

Appropriate diffusion coefficients may be calculated from interstitial profiles (e.g. Berner, 1974; Holdren *et al.*, 1975) or derived experimentally (Li and Gregory, 1974; Goldhaber *et al.*, 1977). Berner (1974) calculates a diffusion coefficient for sulphate of 2×10^{-6} cm^2 sec^{-1} by using data from the Santa Barbara Basin and combined solutions for the steady-state diagenetic equation, (incorporating a term for sulfate reduction)

$$D\frac{\partial^2 C}{\partial x^2} - \omega\frac{\partial c}{\partial x} + \frac{dc}{dt_{\text{biol}}} = 0 \tag{13}$$

where

$$\frac{dc}{dt_{\text{biol}}} = \frac{-K_s G_s}{2}$$

K_s = sulphate reduction rate (sec^{-1}).

G_s = organic carbon available for sulfate reduction (moles cm^{-3}).

and an expression for the steady state input of metabolizable organic water (where G_s is nondiffusible)

$$-\omega\frac{\partial G_s}{\partial x} - K_s G_s = 0 \tag{14}$$

Using experimental concentration and depths, choosing values for ω and $G_{s,0}$ and fitting the solution equations to the empirical data the resultant coefficient may be obtained.

Li and Gregory (1974), using radioactive tracers in a cell filled with red clay, related diffusion coefficients of ions in sediment pore waters (D_s) to those in seawater (D_m) by:

$$D_s = \phi D_m / \theta^2 \tag{15}$$

where ϕ = porosity (dimensionless).

       θ = tortuosity (dimensionless).

For sulphate, $D_m = 6.2 \times 10^{-6}$ cm^2 sec^{-1} at 6.3 °C, the temperature of bottom water in the Santa Barbara Basin (Rittenberg *et al.*, 1955). With $\phi = 0.8$ (Rittenberg *et al.*, 1955) and $\theta = 1.15$ (Van Brakel and Heertjes, 1974) $D_s = 3.8 \times 10^{-6}$ cm^2 sec^{-1} for the Santa Barbara Basin pore waters, in reasonable agreement with that obtained by Berner (1974).

From laboratory diffusion experiments for sulphate in Long Island Sound sediments, Goldhaber *et al.* (1977) found $D_s = 4 \times 10^{-6}$ cm^2 sec^{-1} at 20 °C and $D_s = 2.4 \times 10^{-6}$ cm^2 sec^{-1} at 6 °C, again in good agreement with the above results. The Long Island Sound values also point out the distinct increase in diffusion rates which occur at higher temperatures according to the Stokes–Einstein relation:

$$\left(\frac{D_m\eta}{T}\right)_{T_1} = \left(\frac{D_m\eta}{T}\right)_{T_2} \tag{16}$$

where η = the viscosity of seawater (centipoises) at a given temperature, $T(°K)$.

Such increased diffusion rates would be characteristic of an estuarine system relative to the deep sea since the difference between 25 °C and 2 °C temperatures would result in a doubling of estuarine diffusion coefficients. Reliable values for D_s of varied chemical species in estuarine sediments are needed. Permeability, constrictivity and other controlling factors still must be addressed *via* real systems. Indeed, the higher sulphate diffusion coefficient calculated at 6 °C by the Li and Gregory (1974) technique relative to the empirical work of Berner (1974) and Goldhaber *et al.* (1977) may be linked to such constraints.

The advective term (for sediment–water deposition) in the diagenetic equation $\left(\omega\dfrac{\partial c}{\partial x}\right)$ has generally been ignored in deep-sea sediment work because of very slow sediment accumulation rates (e.g. Berner, 1964; Schink *et al.*, 1975). In estuaries, however, sedimentation rates in excess of 0.1 cm y^{-1} may increase the significance of the advective term for a given species and therefore should be considered in the overall model.

In addition to the rates of diffusion and sediment advection, the rates of production and removal of a particular species are important components of a diagenetic model. The previously shown reaction expression for sulphate reduction (equation 13) is one example of a biogenic mechanism wherein the process is mediated by bacteria or other microorganisms. Biogenic production of methane, ammonia and phosphate can be similarly treated (Berner, 1974; Martens and Berner, 1977; Goldhaber *et al.*, 1978). Abiogenic production/removal mechanisms modelled include the dissolution of opaline silica (Schink *et al.*, 1975) or calcium carbonate (Berner, 1966), adsorption (Berner, 1976), and oxidation and reduction of manganese (Holdren *et al.*, 1975; Aller, 1977). Even though these last reactions are abiogenic, they may have been brought about by biogenic processes in the sediments.

Observed burrows, reworking and x-radiographic absence of discrete laminae provide physical evidence for one more complication in interstitial modelling, organism mixing of sediments. Schink and Guinasso (1975, 1977) and Aller (1977) have modelled the effects of vertical sediment mixing to establish a bioturbation rate (D_B) for inclusion in the overall diagenetic equation.

Production of ammonia, phosphate, manganese and other species in estuarine sediments may establish large concentrations gradients between the interstitial water and the overlying sediment. This leads to diffusive fluxes of disolved species across the sediment–seawater interface. These may be quantified by:

$$F = -D_s \frac{dc}{dx} \tag{17}$$

Where D_s = diffusion coefficient, corrected for porosity and tortuosity $(cm^2 \, sec^{-1})$

C = concentration of dissolved species $(mol \, cm^{-3})$

x = thickness of sediment from surface to concentration maximum at depth (cm)

Interstitial fluxes have also been measured directly by implementation of a benthic chamber. Graham *et al.* (1976), for example, placed open-bottom PVC chambers (about $25l$ volume) on the sediment for 3 hours. Benthic fluxes of Mn found for Narragansett Bay sediments ($\sim 2 \pm 1 \, \mu g \, cm^{-2} \, d^{-1}$) are quite comparable with previously given, calculated values for Mn in Mississippi Delta sediments ($\sim 1000 \, \mu g \, cm^{-2} \, y^{-1}$). Benthic fluxes from estuarine sediments can also show large seasonal variations as a function of temperature and diagenetic activity (Goldhaber *et al.*, 1977; Aller, 1977).

Interstitial water studies remain an important key to understanding the complex chemical activity occurring in sediments and biogeochemical cycling of elements in an estuary. We have merely highlighted some of the work completed and problems yet unsolved. The reader is enthusiastically referred to the many references given below for more detail on any given area.

5 REFERENCES

Addy, S. K., and Ewing, M. (1974). A new box corer design for the investigation of manganese-nodule distribution in a sediment column. *Marine Geology*, **17**, M17–M25.

Alexander, W. B., Southgate, B. A., and Bassindale, R. (1932). The salinity of the water retained in the muddy foreshore of an estuary. *J. Mar. Bio. Assoc., U.K.*, **18**, 297–298.

Aller, R. C. (1977). The influence of macrobenthos on chemical diagenesis of marine sediments. *Ph.D. Dissertation*, Yale Univ., New Haven, 600 pp.

Aller, R. C., and Cochran, J. K. (1976). $^{234}Th/^{238}U$ Disequilibrium in nearshore sediment: Particle reworking and diagenetic time scales. *Earth and PLanetary Science Letters*, **29**, 37–50.

Baas Becking, L. G. M., Kaplan, I. R., and Moore, V. (1960). Limits of the natural environment in terms of pH and oxidation-reduction potential. *Jour. of Geol.*, **68**, 243–284.

Barnes, R. D. (1973). An in situ interstitial water sampler for use in unconsolidated sediments. *Deep Sea Res.*, **20**, 1125–1128.

Bender, M. L., and Klinkhammer, G. P. (1978). Effect on the diagenetic status of sediments on the concentration of trace metals in pore water. In: *First American–Soviet symposium on chemical pollution of the marine environment*, K. K. Turekian and A. I. Simonov (eds). Environmental Protection Agency, Gulf Breeze Florida publication 600/9-78-038.

Ben-Yaakov, S. (1973). pH buffering of pore water of recent anoxic marine sediments. *Limnol. Oceanog.*, **18 (1)**, 86–94.

Bernard, B. B., Brooks, J. M., and Sackett, W. M. (1978), Light hydrocarbons in recent Texas continental shelf and slope sediments. *Jour. of Geophy. Res.* **83**, 4053–4061.

Berner, R. A. (1963). Electrode studies of hydrogen sulfide in marine sediments. *Geochim. et Cosmochim. Acta*, **27**, 563–575.

Berner, R. A. (1963). An idealized model of dissolved sulfate distribution in recent sediments. *Geochim. et Cosmochim. Acta*, **28**, 1497–1503.

Berner, R. A. (1964). Distribution and diagenesis of sulfur in some sediments from the Gulf of California. *Mar. Geol.*, **1**, 117–140.

Berner, R. A. (1966). Chemical diagenesis of some modern carbonate sediments. *Am. J. Sci.*, **264**, 1–36.

Berner, R. A. (1974). Kinetic models for the early diagenesis of nitrogen, sulfur, phosphorus and silicon in anoxic marine sediments. In, *The Sea*, Vol. 5, E. D. Goldberg (ed.), John Wiley–Interscience, 427–450.

Berner, R. A. (1976). Inclusion of adsorption in the modeling of early diagenesis. *Earth Planet. Sci. Letters*, **29**, 333–340.

Berner, R. A., Scott, M. R., and Thomlinson, C. (1970). Carbonate alkalinity in the pore waters of anoxic marine sediments. *Limnol. Oceanog.*, **15**, 544–549.

Bischoff, J. L., Greer, R. E., and Luistro, A. O. (1970). Composition of interstitial water of marine sediments: Temperature of squeezing effect. *Science*, **167**, 1245–1246.

Bouma, A. H. (1969), *Methods for the Study of Sedimentary Structures*, Wiley-Interscience, New York, 458 pp.

Bray, J. T., Bricker, O. P., and Troup, B. N. (1973). Phosphate in interstitial water of anoxic sediments: oxidation effects during sampling procedure. *Science*, **180**, 1362–1364.

Brooks, R. R., Presley, B. J., and Kaplan, I. R. (1968). Trace elements in the interstitial water of marine sediments. *Geochim. et Cosmochim. Acta*, **32**, 397–414.

Bruce, J. R. (1928). Physical Factors on the Sandy Beach. Part II Chemical changes, carbon dioxide concentration and sulphides. *Jour. Mar. Biol. Assoc.*, *U.K.*, **15**, 553–565.

Burgess, P. S. (1922). The solid solution extracted by Lipman's direct pressure method compared to 1–5 water extracts. *Soil Sci.*, **14**, 191–216.

Calvert, S. E., and Price, N. B. (1972). Diffusion and reaction profiles of dissolved manganese in the pore water of marine sediments. *Earth Planet. Sci. Letters*, **16**, 245–249.

Clarke, F. W. (1924). The data of geochemistry, 5th edition. *U.S. Geol. Surv. Bull.*, **770**, 841 p.

Conway, E. J. (1943). The chemical evolution of the ocean. *Proc. Roy. Irish Acad. Sect. B*, **48**, 161–212.

Cooper, L. H. N. (1937). On the ratio of nitrogen to phosphorus in the sea. *J. Mar. Biol. Assoc.*, *U.K.*, **22**, 177–182.

Drever, J. I. (1971). Magnesium iron replacement in clay minerals in anoxic marine sediments. *Science*, **172**, 1334–1336.

Duchart, P., Calvert, S. E. and Price, N. B. (1973). Distribution of trace metals in the pore waters of shallow marine sediments. *Limnol. and Oceanogr.*, **18**, 605–610.

Duxbury, A. C. (1971). *The Earth and Its Ocean.* Addison-Wesley Publishing Co., Reading, Mass. 318 pp.

Elderfield, H. (1976). Manganese fluxes to the oceans. *Mar. Chem.*, **4**, 103–132.

Elderfield, H., and Hepworth, A. (1975). Diagenesis, metals and pollution in estuaries. *Mar. Pollut. Bull.*, **6**, 85–87.

Emery, K. O., and Rittenberg, S. C. (1952). Early diagenesis of California basin sediments in relation to origin of oil. *Bull. Amer. Assoc. Petrol. Geol.*, **35**, 735–806.

Emery, K. O., and Hoggan, D. (1958). Gases in marine sediments. *Bull. Amer. Assoc. Petrol. Geol.*, **42**, 2174–2188.

Fanning, K. A., and Pilson, M. E. Q. (1971). Interstitial silica and pH in marine sediment: Some effects of sampling procedures. *Science*, **173**, 1228–1231.

Flannagan, J. F. (1970). Efficiencies of various grabs and corers in sampling freshwater benthos. *J. Fish. Res. Board, Canada*, **27**, 1961.

Fleming R. H. (1940). The composition of plankton and units for reporting population and production. *Proc. Sixth Pacific Sci. Cong.*, 535–540.

Goldhaber, M. B., and Kaplan, I. R. (1974). The Sulfur Cycle. In, *The Sea, Vol.* 5, E. D. Goldberg (ed.) John Wiley and Sons, New York, 566–655.

Goldhaber, M. B., Aller, R. C., Cochran, J. K., Rosenfeld, J. K., Martens, C. S., and Berner, R. A. (1977). Sulphate reduction, diffusion and bioturbation in Long Island Sound sediments. *Am. Jour. of Sci.*, **277**, 193–237.

Graham, W. F., Bender, M. L., and Klinkhammer, G. P. (1976). Manganese in Narragansett Bay. *Limnol. and Oceanogr.*, **21**, 655–673.

Grim, R. E. (1968). *Clay Minerology*, McGraw-Hill Book Co., New York, 596 pp.

Hale, S. S. (1974). The role of benthic communities in the nutrient cycles of Narragansett Bay. *M.S. Thesis*, Univ. of Rhode Island, Kingston. 123 pp.

Hammond, D. E. (1973). Interstitial Water Studies, Leg 15—a comparison of the major element and carbonate chemistry data from sites 147, 148 and 149. In, Heezen, B. C., and I. D. MacGregor, (eds.) *Initial Reports of the Deep Sea Drilling Project, Vol.* 20, Washington, D.C. (U.S. Government Printing Office) pp. 831–850.

Hartmann, M. (1965). An apparatus for the recovery of interstitial water from recent sediments. *Deep-Sea Res.*, 12, 225–226.

Hartmann, M., Muller, P., Suess, E., and Van der Weijden, C. (1973). Oxidation of organic matter in recent marine sediments. '*Meteor*' *Forschungsergeb.*, *Reihe C*, **12**, 74–86.

Hartwig, E. O. (1976). The impact of nitrogen and phosphorus release from a siliceous sediment on the overlying water. *In*: *Estuarine Processes*, vol. 1, edited by M. Wiley, Academic Press, New York, pp. 103–117.

Heezen, B. C., and MacGregor, I. D. (1973). *Initial Reports of the Deep Sea Drilling Project, Vol.* 20, Washington (U.S. Government Printing Office) pp, 751–904.

Helgeson, H. C., and Mackenzie, F. T. (1970). Silicate-sea water equilibria in the ocean system. *Deep-Sea Res.*, **17**, 877–892.

Hesslein, R. H. (1976). An in situ sampler for close interval pore water studies. *Limnol. and Oceanogr.*, **21**, 912–914.

Holdren, Jr., G. R., Bricker III, O. P., and Matisoff, G. (1975). A model for the control of dissolved manganese in the interstitial waters of Chesapeake Bay. In,

Marine Chemistry in the Coastal Environment, T. M. Church (ed.), Am. Chem. Soc., Washington, p. 364–381.

Horowitz, R. M., Waterman, L. S., Broecker, W. S., and Bopp, R. (1973). Interstitial water studies, Leg 15, new procedures and equipment. In, Heezen, B.C., and MacGregor, I. D. (eds.), *Initial Reports of the Deep Sea Drilling Project, Vol.* 20, Washington, D. C. (U.S. Government Printing Office) pp. 757–763.

Hurd, D. C. (1973). Factors affecting solution rate of biogenic opal on seawater. *Earth Planet. Sci. Letters,* **15,** 411–417.

Jonasson, A., and Olausson, E. (1966). New devices for Sediment sampling. *Marine Geology,* **4,** 365–372.

Kahil, E. K., and Goldhaber, M. (1973). A sediment squeezer for removal of pore waters without air contact. *Jour. Sed. Petrol.,* **43,** 553–557.

Kaplan, I. R. (1974). *Natural Gases in Marine Sediments.* Plenum Press, New York, 324 pp.

Koyama, T. (1953). Measurement and analysis of gases in sediments. *J. Earth Sci.,* Nagoya Univ., **1,** 107–118.

Kullenberg, B. (1952). On the salinity of the water contained in marine sediments. *Meddelanden fran Oceanografisk Institut Göteborg,* No. **21,** 38 pp.

Latimer, W. M. (1952). *Oxidation Potentials* (2nd ed.), Prentice-Hall, New York.

Lerman, A., and R. R. Weiler, (1970). Diffusion and accumulation of chloride and sodium in Lake Ontario. *Earth and Planetary Sci. Letters,* **10,** 150–156.

Li, Y., Bischoff, J., and Mathieu, G. (1969). The migration of manganese in the Arctic Basin sediment. *Earth Planet. Sci. Letters,* **7,** 265–270.

Li, Y., and Gregory S. (1974). Diffusion of ions in sea water and in deep-sea sediments. *Geochim. Cosmochim. Acta,* **38,** 703–714.

Lu, J. C. S., and Chen, K. Y. (1977). Migration of trace metals in interfaces of seawater and polluted surficial sediments. *Environ. Sci. Technol.,* **11,** 174–182.

Lusczynski, N. J. (1961). Filter-press method of extracting water samples for chloride analysis. *U.S. Geol. Surv. Water Supply Paper No.* 1544-A 8 pp.

Mackenzie, F. T., and Garrels, R. M. (1966). Chemical mass balance between rivers and ocean. *Amer. J. Sci.,* **264,** 507–525.

Mangelsdorf, P. C., Wilson, T. R. S., and Dantell, E. (1969). Potassium enrichments in interstitial waters of recent marine sediments. *Science,* **165,** 171–173.

Manheim, F. T. (1966). A hydraulic squeezer for obtaining interstitial water from consolidated and unconsolidated sediments. *U.S. Geol. Survey Prof. Paper* 550-C, 256–261.

Manheim, F. T. (1967). Evidence for submarine discharge of water on the Atlantic continental coast of the southwestern United States and suggestions for further search. *Trans. New York Acad. Science ser. II,* **29 (7),** 839–853.

Manheim, F. T. (1974). Comparative studies on extraction of sediment interstitial waters. Discussion and comment on the current state of interstitial water studies. *Clays and Clay Minerals,* **22,** 337–343.

Manheim, F. T. (1976). Interstitial waters of marine sediments. In, *Chemical Oceanography,* Riley, J. P., and Chester, R. (ed.), *Academic Press,* New York, 115–181.

Manheim, F. T., and Sayles, F. L. (1970). Brines and interstitial brackish water in drill cores from the deep Gulf of Mexico. *Science,* **170,** 57–61.

Manheim, F. T., and Sayles, F. L. (1974). Composition and origin of interstitial waters of marine sediments, based on Deep Sea Drill Cores. In, *The Sea, Vol.* 5, E.d. Goldberg (ed.) Wiley–Interscience, New York, 527–568.

Manheim, F. T., and Bischoff, J. L. (1969). Geochemistry of pore waters from Shell

Oil Company drill holes on the continental slope of the northern Gulf of Mexico. *Chemical Geology*, **4**, 63–82.

Manheim, F. T., and Chan, K. M. (1974). Interstitial water of Black Sea sediments: New data and review. In, E. T. Degens and D. A. Ross (eds.) The Black Sea—geology, chemistry, and biology. *Am Assoc. Petrol. Geol., Tulsa, Okla.*, 155–180.

Martens, C. S., and Berner, R. A. (1977). Interstitial Water Chemistry of Anoxic Long Island Sound Sediments. 1. Dissolved gases. *Limnol. and Oceanogr.*, **22**, 10–25.

Matisoff, G., Bricker III, O. P., Holdren, Jr., G. R., and Kaerk, P. (1975). Spacial and temporal variations in the interstitial Water Chemistry of Chesapeake Bay sediments. In *Marine Chemistry in the Coastal Environment*, T. M. Church (ed.), Am. Chem. Soc., Washington, D. C., 343–363.

Mayer, L. M. (1976). Chemical water sampling in lakes and sediments with dialysis bags. *Limnol. and Oceanogr.*, **21**, 909–911.

McBeth, I. G. (1917) Fixation of Ammonia in soils. *J. Agr. Res.*, **9**, 141–155.

McKee, T. R. (1977). Ferromanganese remobilization in recent sediments of the Gulf of Mexico. Ph.D. dissertation, Texas A & M University, College Station, Texas.

McQuillin, R., and Ardus, D. A. (1977). Exploring the Geology of Shelf Seas. Graham and Trotman Limited, London, 234 pp.

Menzel, D. W. (1974). Primary productivity, dissolved and particulate organic matter, and the sites of oxidation of organic matter. In, The Sea, Vol. **5**, E. D. Goldberg (ed.), 659–678.

Morris, J. C., and Stumm W. (1967). Redox equilibria and measurements of potentials in the aquatic environment. Advan. Chem., Ser 67, *Am. Chem. Soc.*, 270–285.

Murray, J., and Irvine, R. (1895). On the chemical changes which take place in the composition of seawater associated with blue muds on the floor of the ocean. *Trans. Roy. Soc. Edinburgh*, **37**, 481–508.

Murthy, A. S. P., and Ferrell, Jr., R. E. (1972). Comparative chemical composition of sediment interstitial waters. *Clays and Clay Minerals*, **20**, 317–321.

Nissenbaum, A., Presley, B. J., and Kaplan, I. R. (1972). Early diagenesis in a reducing fjord; Saanich Inlet, British Columbia—I. Chemical and isotopic changes in major components of interstitial water. *Geochim. et Cosmochim. Acta*, **36**, 1007–1027.

Nissenbaum, A., Baedecker, M. J., and Kaplan, I. R. (1971). Studies on dissolved organic matter from interstitial water of a reducing marine fjord. *Adv. in Org. Geochem*, Gaerther, L., and Wehner, H. (eds.), Pergamon Press, p. 427–440.

Ponnamperund, F. N. (1972). The chemistry of submerged soils. In, *Advances in Agronomy*, Vol. 24, N. C. Brady (ed.), Academic Press, New York.

Powers, M. C. (1957). Adjustment of land derived clays to the marine environment. *Jour. of Sed. Petrology*, **27**, 355–372.

Presley, B. J. (1969). Chemistry of interstitial water from marine sediments. *Ph.D. dissertation*, University of California, Los Angeles.

Presley, B. J., and Kaplan, I. R. (1968). Changes in dissolved sulfate, calcium and carbonate from interstitial water of nearshore sediments. *Geochim. et Cosmochim. Acta*, **32**, 1037–1048.

Presley, B. J., Brooks, R. R., and Kappel, H. M. (1967). A simple squeezer for removal of interstitial water from ocean sediments. *Jour. Marine Res.*, **25**, 355–357.

Presley, B. J., and Kaplan, I. R. (1970). Interstitial water chemistry: deep sea drilling project, Leg 4. In *Initial Reports of the Deep Sea Drilling Project, Vol. IV*, 415–430.

Presley, B. J., Kolodny, Y., Nissenbaum, A., and Kaplan, I. R. (1972). Early diagenesis in a reducing fjord, Saanich Inlet, British Columbia–II. Trace element distribution in interstitial water and sediment. *Geochim. Cosmochim. Acta,* **36,** 1073–1090.

Reeburg, W. S. (1967). An improved interstitial water sampler. *Limnol. and Oceanog.,* **12,** 163–165.

Reeburg, W. S. (1976). Methane consumption in Cariaco Trench waters and sediments. *Earth Planet. Sci. Lett.,* **28,** 1–9.

Redfield, A. C. (1934). On the proportions of organic derivatives in seawater and their relation to the composition of plankton. *James Johnston Memorial Volume,* Liverpool University Press, 177–192.

Redfield, A. C., Ketchum, B. H., and Richards, F. A. (1963). The influence of organisms on the composition of seawater. *In, The Sea,* Vol. 2, M. N. Hill (ed.), Interscience Pub., New York.

Reid, D. M. (1930). Salinity interchanges between seawater in sand and overflowing freshwater at low tide. Jour. Mar. Biol. Assoc., *U.K.,* **16,** 609.

Richards, F. A. (1965). Anoxic basins and fjords. *In, Chemical Oceanography,* Riley and Skirrow (eds.) Academic Press, 611–645.

Richards, F. A., and Vaccaro, R. F. (1956). The Cariaco Trench, an anoxic basin in the Caribbean Sea. *Deep Sea Res.,* **3,** 214–228.

Riley, G. A. (1951). Oxygen, phosphate and nitrate in the Atlantic Ocean. *Bulletin of Bingham Oceanographic Collection,* **13,** 1–126.

Rittenberg, S. C., Emery, K. O., Hulsemann, J., Degens, E., Fay, R., Reuter, J., Grady, J., Richardson, S., and Bray, E. (1963). Biogeochemistry of sediments in Experimental Mohole.' *J. Sediment Petrol.,* **33,** 140–172.

Rittenberg, S. C., Emery, K. O., and Orr, W. L. (1955). Regeneration of nutrients in sediments of marine basins. *Deep Sea Research,* **3,** 23–45.

Robbins, J. A., and Callender E. (1975). Diagenesis of manganese in Lake Michigan sediments. Am. J. Science **275,** 512–533.

Rowe, G. T., Clifford, C. H., and Smith, K. L. (1975). Benthic nutrient regeneration and its coupling to primary productivity in coastal waters. Nature **255,** 215–217.

Rubey, W. W. (1951). The geologic history of seawater—An attempt to state the problem. Bull. Geol. Soc. Am. **62,** 1111–1148.

Sayles, F. L. (1970). Preliminary geochemistry. In, Bader, R. G. *et al.* (eds.), Initial Reports of the Deep Sea Drilling Project, Vol. **4,** Washington, D. C. (U.S. Government Printing Office), pp. 645–655.

Sayles, F. L., and Mangelsdorf, Jr., P. C. (1977). The equilibrium of clay minerals with seawater: exchange reactions. Geochim. et Cosmochim. Acta **41,** 950–960.

Sayles, F. L., Manheim, F. T., and Waterman, L. S. (1973a). Interstitial Water studies on small core samples, Leg 15. Heezen, B. C. and MacGregor, I.D. (ed.) U.S. Government Printing Office, 783–804.

Sayles, F. L., Wilson, T. R. S., Hume, D. N., and Mangelsdorf, P. C. (1973). In situ sampler for marine sedimentary pore waters: Evidence for potassium depletion and calcium enrichment. Science **181,** 154–156.

Schink, D. R., Fanning, K. A., and Pilson, M. E. Q. (1974). Dissolved silica in the upper pore waters of the Atlantic Ocean floor. *Jour. Geophys. Res.,* **79,** 2243–2250.

Schink, D. R., and Guinasso, Jr., N. L. (1975). Quantitative estimates of biological mixing rates in abyssal sediments. *J. Geophys. Res.,* **80,** 3032–3043.

Schink, D. R., and Guinasso, Jr., N. L. (1977). Effects of bioturbation on sediment-seawater interaction. *Mar. Geol.,* **23,** 133–154.

Schink, D. R., Guinasso, Jr., N. L., and Fanning, K. A. (1975). Processes affecting the concentration of silica at the sediment-water interface of the Atlantic Ocean. *Jour. of Geophys. Res.*, **80**, 3013–3031.

Shaw, E. W. (1914). Gas from mud lamps at the mouth of the Mississippi. *U.S. Geol. Survey Bull.* 591-A, 21–45.

Sholkovitz, E. R. (1972). The chemical and physical oceanography and the interstitial water chemistry of the Santa Barbara Basin, California. *Ph.D. thesis*, University of California at San Diego, Scripps Institute of Oceanography, pp. 110–183.

Sholkovitz, E. (1973). Interstitial water chemistry of the Santa Barbara Basin sediments. *Geochim. Cosmochim. Acta*, **37**, 2043–2073.

Siever, R. (1962). A squeezer for extracting interstitial water. *Jour. Sed. Petrology*, **32**, 329–331.

Siever, R., Beck, K. C., and Berner, R. A. (1965). Composition of interstitial waters of modern sediments. *J. Geol.*, **73**, 39–73.

Sillen, L. G. (1961). The physical chemistry of seawater. In, Sears, Mary (ed.) Oceanography. *Am. Assoc. Adv. Sci. Pub.*, **67**, 547–581.

Skirrow, G. (1965). The dissolved gases—carbon dioxide. In, *Chemical Oceanography*, G. Skirrow, and J. P. Riley (eds.), Academic Press.

Sweeney, R. E. (1972). Pyritization during diagenesis of marine sediments. Ph.D. thesis, Dept. of Geology, University of California, Los Angeles, 184 pp.

Takai, Y., and Kamura, T. (1966). The mechanism of reduction in waterlogged paddy soil. *Folia Microbiol. (Prague)*, **11**, 304–313.

Thorstenson, D. C., and Mackenzie, F. T. (1971). Experimental decomposition of algae in seawater and early diagenesis. *Nature*, **234**, 543–545.

Trefry, J. H. (1977). The transport of heavy metals by the Mississippi River and their fate in the Gulf of Mexico. *Ph.D. dissertation*, Texas A & M University, College Station, 233 pp.

Turner, F. T., and Patrick, Jr., W. H. (1968). Chemical changes in waterlogged soils as a result of oxygen depletion. *Trans. 9th Int. Congr. Soil Sci. (Adelaide, Aust.)*, 53–65.

Van Brakel, J., and Heertjes, P. M. (1974). Analysis of diffusion in macroporous media in terms of porosity, a tortuosity and a constrictivity factor. *Int. J. Heat Mass Transfer*, **17**, 1093–1103.

Whelan, Thomas III (1974). Methane and carbon dioxide in coastal marsh sediments. In, I.R. Kaplan (ed.), *Natural Gases in Marine Sediments*. Plenver Press, New York, 47–61.

Whitfield, M. (1969). Eh as an operational parameter in estuarine studies. *Limnol. and Oceanogr.*, **14**, 547–558.

ZoBell, C. E. (1946). Studies on redox potential of marine sediments. *Bull. Am. Assoc. Petrol. Geologists*, **30**, 447–513.

Chemistry and Biogeochemistry of Estuaries
Edited by E. Olausson and I. Cato
Copyright © 1980 by John Wiley & Sons Ltd.

S. R. ASTON

Department of Environmental Sciences,
University of Lancaster,

7

Nutrients, Dissolved Gases, and General Biogeochemistry in Estuaries

1 INTRODUCTION

The distributions of nutrients and dissolved gases in estuarine waters are controlled, in common with other chemical parameters, by the nature of estuarine circulation, mixing, and other physical processes, together with biological, sedimentological, and chemical effects. Some aspects of these physical processes have been described in Chapter 2, and it is not the purpose of the present discussion to examine these features in any detail. There are, however, some physical, chemical, sedimentological, and biological features of the estuarine environment which must be compiled at the outset, since they are potentially important controls of the behaviour of the nutrients, dissolved gases, and the general biogeochemistry in estuaries.

                                   S. R. Aston

These controlling influences may be summarized as follows—

(a) The tidal mixing of fresh and sea waters on a semi-diurnal or diurnal time scale, with corresponding changes in the volume of water in an estuary, produce temporal changes in the contributions of nutrients and dissolved gases from marine and fresh water sources. For example, estuaries are generally enriched in nutrients with respect to ocean waters due to the local influences of land drainage and often pollution.

(b) The circulation, and especially the stratification of some estuaries, generates the possibility of vertical and horizontal variations of the concentrations of nutrients and dissolved gases within an estuary.

(c) Estuarine topography may give rise to particularly restricted circulations, e.g. in fjords, where the mixing of external sea water with the estuarine waters is greatly reduced, and the restricted mixing leads to unusual chemical environments, e.g. anoxic waters.

(d) The current regime in coastal waters and estuaries leads to the deposition of various types of sedimentary material. The deposition and resuspension of sediments in estuaries may influence the budgets of dissolved constituents in estuarine waters, including nutrients and gases (see also Chapter 5).

(e) Chemical reactions occurring during the mixing of river water with sea water may lead to the removal or addition of the dissolved nutrients. Also, the changes in temperature and salinity during estuarine mixing influence the solubility of dissolved gases and thus influence their removal or addition in an estuary.

(f) Biological production and metabolism have significant influences on the occurrence and distribution of nutrients and some gases, e.g. carbon dioxide and oxygen, in estuarine waters. Estuaries are environments with special problems for biological organisms, so that there is a tendency to a decrease in species diversity in estuaries. This decrease does not, however, imply that productivity is low. The biological productivity, distribution of organisms and their interactions are discussed in Chapter 8.

When considering the influences of physical, topographic, biological, and chemical processes on nutrients and dissolved gases in estuaries, it is important to remember that no two estuaries are alike with respect to these influences (see e.g. Howes, 1939; Nash, 1947; Rochford, 1951; Postma, 1954; Hulbert, 1956; and Perkins *et al.* 1964). It has long been realized that the factors (a) to (d) above are physical and topographic controls which will strongly depend on local oceanographic and geomorphological conditions. While their influence on nutrient and dissolved gas *distributions* may be highly significant for individual estuaries, it is extremely difficult and even undesirable to try to formulate their effects on chemical *processes* in estuaries. In recognizing the individual characters of estuaries, the purpose of the present review is to illustrate the chemical and biological processes which occur in all or at least most estuaries. The reader is reminded that these general principles may then be used to explain the nutrient and dissolved gas behaviour in an individual estuary for which data has been acquired, or is already published. Recent examples of such work in specific estuaries include—Hobbie *et al.* (1975), Mackay and Leatherland (1976), Duedall *et al.* (1977), Ho and Barret (1977), Taft and Taylor (1977), Furnas *et al.* (1977).

2 NUTRIENTS IN THE ESTUARINE ENVIRONMENT

2.1 Silicon biogeochemistry in estuaries

Silica has received more attention than most of the chemical constituents of estuaries including the other nutrients. Silica is an important nutrient to some marine organisms, e.g. diatoms, radiolaria, and sponges, the dissolved silica being removed by such organisms in order to provide skeletal material for extra-cellular structures. Thus biological processes have a control on the dissolved silica budget of estuaries. There is, however, a paucity of information on the biological mechanisms of silica removal (see e.g. Spencer, 1975), and considerable debate as to the role of non-biological removal of dissolved silica by precipitation during estuarine mixing. This latter topic has recently been reviewed by Liss (1976) and Aston (1978). It is pertinent to examine the evidence for the possible removal of silica from solution by inorganic processes in estuaries before proceeding to examine its biological cycle.

2.1.1 *Inorganic removal of silicon during estuarine mixing*

Silicon is present in river water in three main forms—detrital quartz, alumino-silicates (clays), and dissolved silicon derived from the weathering of silicate rocks. For waters of pH <9, which is the usual situation in natural waters, Siever (1971) has demonstrated that the dissolved silicon will be present as silicic acid (H_4SiO_4). Krauskopf (1956) suggested that during the mixing of river water with sea water in estuaries, the increased electrolyte concentration leads to the formation of polymeric, colloidal form of silicic acid which flocculates and removes dissolved silica. This idea of dissolved silicon removal in estuaries was fairly widely adopted by early workers as an explanation of the lower dissolved silicon concentrations found in sea water with respect to the *average* concentration in river waters. Later studies by Burton *et al.* (1970) and Siever (1971) have demonstrated that there is,

                                    S. R. Aston

however, no evidence to support the hypothesis that the silicic acid in either river or sea waters is in a polymeric form. In addition, when artificial polymers of silicic acid are added to river, sea, or estuarine waters there is a rapid depolymerization to monomers of silicic acid. These observations suggest that the estuarine removal of dissolved silica by the processes of polymerization and flocculation of silicic acid are unlikely.

While the mechanism(s) of the possible removal of silica from solution by inorganic processes in estuaries is not understood, several studies have been aimed at determining whether or not removal actually does or does not take place. Early results for several Japanese estuaries (Maéda, 1952, 1953; Makimoto *et al.*, 1955, Maéda and Tsukamoto, 1959; and Maéda and Takesue, 1961) suggested that there is a linear relationship between dissolved silicon concentration and the chlorinity of the waters. These workers interpreted their findings as evidence for a lack of dissolved silicon removal during estuarine mixing, the chlorinity of the water samples being used as an index of conservative mixing. These early Japanese data are, however, of little real use in the interpretation of conservative/non-conservative behaviour in estuaries, due to the very restricted range of chlorinities examined. For a discussion of the problems of defining theoretical indices of estuarine mixing, see Aston (1978).

Studies on the Mississippi by Bien *et al.* (1958) suggested that the removal of dissolved silicon during estuarine mixing is extensive, and in some situations almost complete. The data of Bien *et al.* (1958) have been reinterpreted by Schink (1967) who criticized Bien *et al*'s conclusions and suggested that the Mississippi data show evidence of only $\sim$10–20% removal of dissolved silicon. Later studies of the Mississippi estuary by Fanning and Pilson (1973) have indicated much lower removals of about 7% of the dissolved silicon in river water. Other field studies on silicic acid removal from solution in estuaries have included those of Burton (1970), Burton *et al.* (1970), and Burton and Liss (1973), with typically $\sim$10–20% silicon removal during mixing. Liss (1976) has reviewed these various studies, and after incorporating his recent results (see e.g. Liss and Pointon (1973)), has concluded that there is adequate evidence for the non-conservative behaviour of silicon in estuaries, with a removal range of 0–30%. Figure 1 illustrates a typical relationship of dissolved silica and salinity from the surveys of the Alde estuary by Liss and Pointon (1973).

There have been various attempts to simulate the behaviour of dissolved silicon during estuarine mixing using laboratory experiments. Liss and Spencer (1973) and Bien *et al.* (1958) have demonstrated silicon removal from solution, while in contradiction Fanning and Pilson (1973) have found no evidence for the non-conservative behaviour of silicon. These contrasting results may be attributed to the differences in experimental conditions, especially temperature, adopted by these workers.

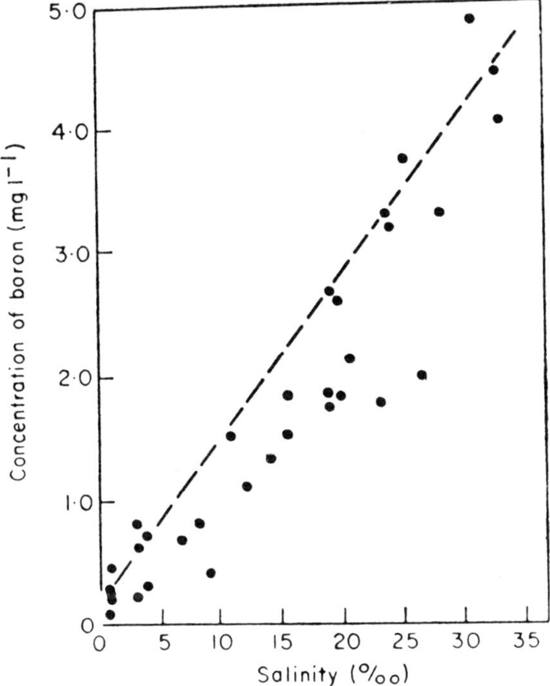

Figure 1. The variations of dissolved silica concentrations with salinity for samples collected in the Alde
estuary, England, during two surveys. The Alde estuary is virtually unpolluted, and the data shown
represents winter conditions for both surveys (After
Liss and Pointon, 1973.) (Reproduced by permission
of Pergamon Press Ltd.)

Recently, Sholkovitz (1976) has attempted to use a *product* approach to
the study of the conservative/non-conservative behaviour of dissolved silicic
acid and other chemical species during estuarine mixing. Sholkovitz's
methodology differs from other experimental and field studies on estuarine
chemistry in so far as that it examines the production of particulate matter
during the mixing of river and sea waters instead of determining the
changing concentrations of the dissolved constituents. Before considering
the results of the *product* approach for silica, it is necessary to consider some
more general aspects of this methodology and its relationships to the
reactant approach adopted by other workers.

Sholkovitz (1976) has criticized the *reactant* approach, *i.e.* that based
upon the deviations of the concentrations of dissolved constituents from
those predicted by a theoretical dilution curve, on four grounds. These may

be briefly summarized as follows:– (1) the composition of the river water 'end-member' of a dilution curve can exhibit large variations and make the line difficult to define; (2) the contribution of more than one source of river water often complicates the mixing assumptions; (3) there is some difficulty in defining the removal products as particulate matter or sediments (see e.g. Coonley *et al.*, 1971, Hair and Bassett, 1973, Bewers *et al.*, 1974); and (4) the *reactant* approach will not give an indication of the mechanisms of removal.

While these may be valid criticisms of the *reactant* approach to estuarine chemistry, it should be pointed out that the *product* approach itself is not devoid of significant problems. First, in common with other laboratory experiments which attempt to model geochemical processes in natural waters there is a basic doubt as to whether the experimental conditions are physically, kinetically and thermodynamically representative of the conditions in the real situation. More specifically, the *reactant* approach depends for its success on the removal and analysis of particulate matter from the reaction system, and on the assumption for either pre-existing particulate matter which has been transported into the estuary has either no effect on the conservative/non-conservative behaviour of the constituents of interest, or that if there is an influence exerted by particulate matter it can be adequately accounted for. The problem of the removal of the *products* from laboratory experiments has been dealt with in a simplistic way by Sholkovitz (1976). 0.45 μm cellulose acetate membrane filters were used to define the 'dissolved' and 'particulate' products. Such filtration techniques are subject to variations in particle size retention and the *product* approach assumes that the particulate products may be defined by an arbitary choice of particle size. The problems of the practical aspects of sea water filtration have been reviewed by Riley (1975).

On the question of the influence of pre-existing particulate matter on estuarine reactions, Sholkovitz (1976) has assumed that this is of no influence and even used pre-filtered waters for the laboratory experiments. For some species it is very reasonable to assume that mineral particles in suspension in estuaries will exert an important influence on flocculation and precipitation processes. Aston and Chester (1973) have shown, for example, that suspended sediment particles act as electronegative nuclei for iron hydrolysis and precipitation in estuaries. By acting as sites for ion adsorption and electrostatic attraction, pre-existing particles probably exert an influence on many chemical species, including nutrients, in the estuarine environment.

Bearing in mind the limitations of the *product* approach, Sholkovitz's data on silica removal in estuaries suggests that the extent of the removal is approximately 3–6%. A typical series of results obtained on the flocculation of silica as a function of salinity are shown in Figure 2.

The overall picture of inorganic silica removal in estuaries is still not clear,

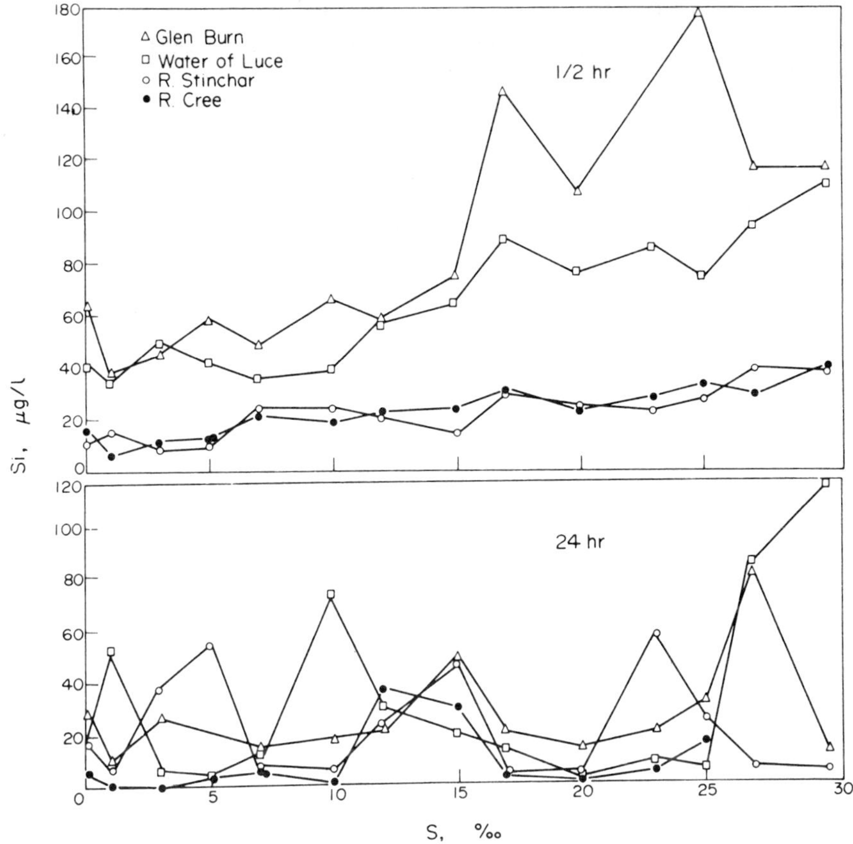

Figure 2. Silica flocculation as a function of salinity for four Scottish river systems. Results for $\frac{1}{2}$ hour and 24 hour experimental runs are illustrated (After Sholkovitz, 1976.) (Reproduced by permission of Pergamon Press Ltd.)

but the field and experimental data obtained to date suggest that if removal occurs it is likely to be limited to a few per cent of the total dissolved silica present, and at a maximum up to 30%.

2.1.2 *Biological removal and cycling of silicon*

Liss and Spencer (1970) have pointed out that dissolved 'reactive' silica is utilized by siliceous diatoms, silico-flagellates and sponges in marine waters, and this will lead to the removal of silicon supplied to the oceans by rivers. The extent to which this biological removal of silicon is important in the estuarine environment depends on four factors: (1) the populations of siliceous organisms present in the estuary; (2) the rate of growth of the siliceous organisms; (3) the influence exerted by physical processes on the growth of the organisms, e.g. temperature and nutrient supply; and (4) the resolution of silica from the skeletal remains of dead organisms.

The large scale biological removal of silica in estuaries has been dismissed by Liss and Spencer (1970) on the assumption that the initial mixing of river water and sea water will usually be too rapid for any biological processes to produce apparent effects in the time available. The biological utilization of reactive silica is, however, a seasonal effect for most estuaries (the exceptions being certain tropical estuaries). This seasonal effect is one of rapid growth and therefore rapid silica removal in spring and summer.

Very little field data is available on the seasonal biological controls of dissolved silica in estuaries, although the recent study by Peterson *et al.* (1975) has provided useful information on the San Francisco Bay estuary, where major variations in the dissolved silica distribution are seasonal and are related to the variations in rates of river supply and silica utilization by phytoplankton. Peterson *et al.* (1975) have shown that when the rate of silica supply by river input is large compared to the rate at which silica is used in the estuary, the decrease in dissolved silica as salinity increases is controlled primarily by the mixing of river and sea waters. When the biological demand for silica increases in spring-summer, the dissolved silica concentration in the estuary is considerably less than that predicted by simple mixing, and is often lower in concentration than dissolved silica in near-surface ocean waters. Figure 3 illustrates the longitudinal distributions of salinity and 'silicate-silicon' during typical winter and summer conditions in San Francisco Bay estuary. Several features are evident: (a) an inverse relationship of salinity and silica during winter (A and C); (b) higher silica concentrations during winter than during summer (C and D); (c) a rapid decrease in silica concentration with distance in the upper portion of the estuary in summer (D); and (d) lower silica concentrations during summer within the estuary than at the seaward boundary (D).

In summary, the extent to which the rapid removal is capable of influencing the dissolved silica budget of estuaries in general is difficult to assess, especially when the environmental factors which control the growth and populations of silica secreting organisms may vary so widely from one estuary to another.

The degree of regeneration of dissolved silicon in estuaries by the decomposition of siliceous organisms is obscure. Although no direct evidence is available on the dissolution of diatoms, etc., in real estuaries, there have been various attempts to measure the regeneration of dissolved silica from dead organisms in laboratory experiments. The early results of Atkins (1945) suggested that diatom frustules were not capable of dissolution, even at the high pH value of sea water. Later experiments with the marine diatom *Ditylum brightwelli* demonstrated silica losses of ~50% in their frustules

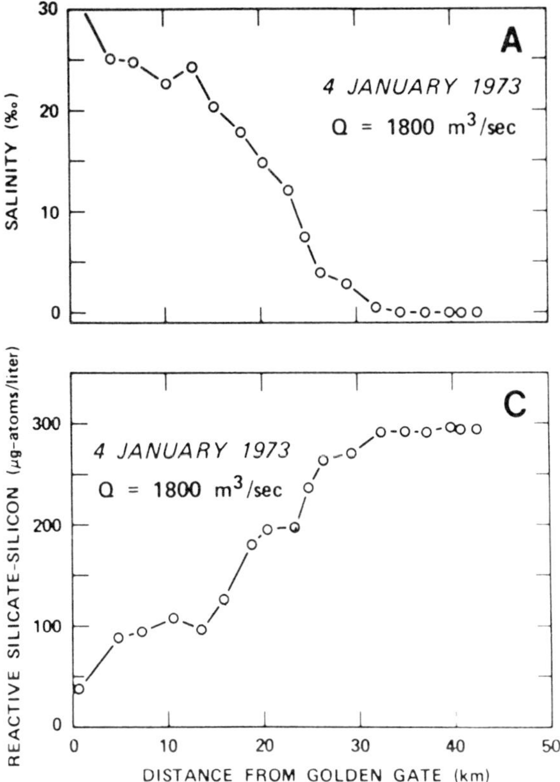

Figure 3. Longitudinal distribution of salinity and silicate–silicon at 2m depth during typical winter and summer conditions in the northern reach of San Francisco Bay. Q indicates combined mean monthly discharge from the Sacramento and San Joaquin rivers (After Peterson *et al.*, 1975.) (Reproduced by permission of Academic Press, Inc.)

after two months in sea water (Harvey, 1955). Similar experiments performed by Jørgensen (1955) with other marine diatoms showed that silica losses from dead organisms were slow and highly variable. Fresh water diatoms tend to dissolve slowly at pH values similar to those formed in brackish-waters and sea water, the process of dissolution apparently being retarded by various cations, e.g. Fe^{3+} and Al^{3+}, and the presence of organic coatings on the amorphous silica frustules (Lewin, 1961). All the available evidence suggests that the contribution of silica to solution in estuaries is unlikely to be significant. The abundant remains of diatoms, radiolaria, and other silica secreting organisms in marine sediments are further evidence

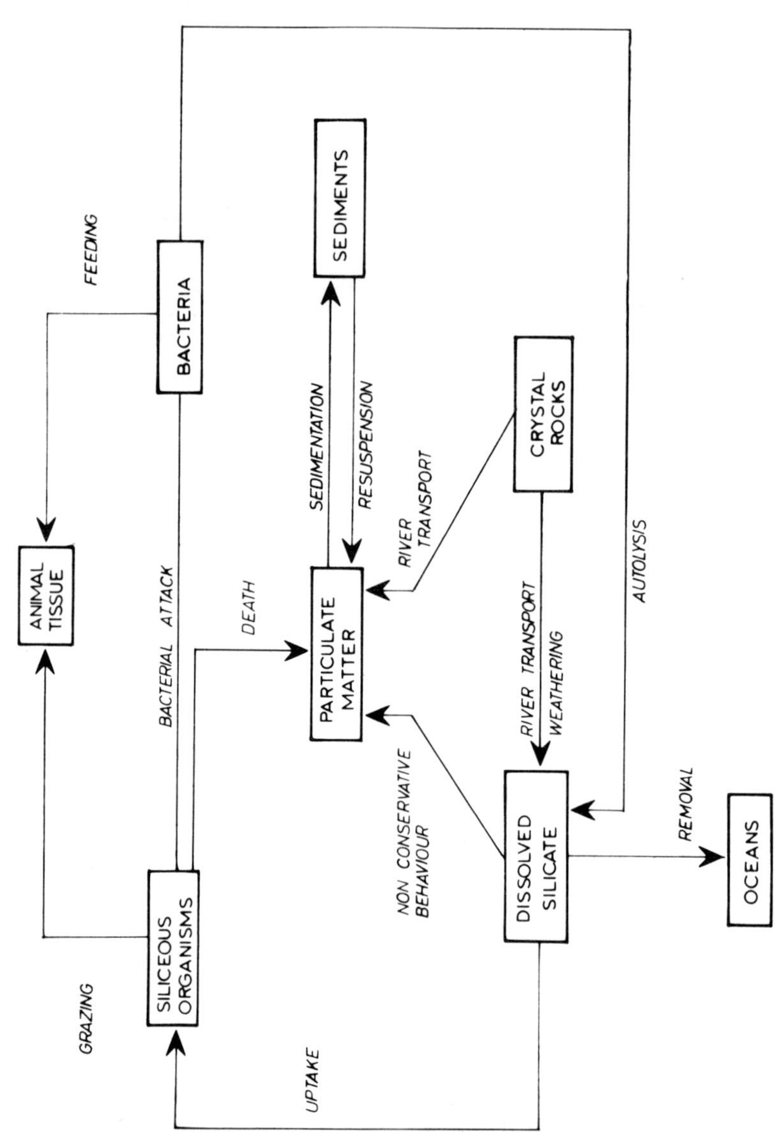

Figure 4. The silicon cycle in the estuarine environment.

that the production of biogenous silica is probably a means of the net removal of dissolved silica in estuaries.

In summary, the estuarine cycling of silica, as depict in Figure 4, is a subject of considerable conflict and poor understanding, and the biogeochemistry of silica in estuaries still remains a most suitable area for research.

2.2 Phosphorus biogeochemistry in estuaries

The weathering of the Earth's crust and surface water transport deliver phosphorus to estuaries as phosphate minerals in suspended detritus and dissolved phosphate in river water. Also, phosphorus is contributed to estuaries from domestic sewage and industrial effluent disposal into rivers and tidal waters. Through biological cycling some of this inorganic phosphorus is converted into dissolved and particulate organic phosphates (see section 2.2.2). Before considering the biogeochemical cycle of phosphorus in estuaries, it is worthwhile paying some attention to non-biological factors which may control the budget of dissolved, suspended particulate, and sedimentary phosphorus in estuaries.

2.2.1 *Inorganic controls on phosphorus*

The almost ubiquitous nature of phosphate pollution from detergents, fertilisers, and domestic and industrial effluents in estuarine and coastal waters has generated recent interest in this important nutrient. The first detailed study of inorganic phosphate in estuaries was probably that of Steffánsson and Richards (1963), who noted that there was a distinct lack of variation of phosphate concentrations in estuarine waters of different salinities taken from the Columbia River Estuary. Steffánsson and Richards (1963) proposed that the observed lack of variation of dissolved phosphate could be attributed to two main causes: (1) the similarity of inorganic phosphate concentrations in the river and sea waters which contribute to the estuary; and (2) the existence of a 'buffering' effect which maintains the average phosphate concentration in the estuary at $\sim 37\,\mu\mathrm{gPl}^{-1}$.

Phosphate 'buffering' by solution-mineral reactions have been cited frequently for fresh water environments (Gessner, 1960; Mortimer, 1971; and Patrick and Khalid, 1974), and recent studies suggest that a similar mechanism can occur in the estuarine environment (Stirling and Wormald, 1977). The removal of dissolved phosphate from sea water has been demonstrated to be adsorption onto solid mineral surfaces (phosphatic and non-phosphatic), and in common with other adsorption processes the extent of removal depends on the phosphate concentration in solution and the mineral surface area available (Carritt and Goodgal, 1954; Burns and Salomon,

1969; and Chen *et al.*, 1973). The laboratory experiments of these workers suggest that the adsorption is not of a simple physical type since this is an exothermic process while phosphate adsorption increases with increasing temperature. The adsorption is at a maximum for the pH range 3–7 (Carritt and Goodgal, 1954; Jitts, 1959; and Burns and Salomon, 1969), so that phosphate removal will be greater in fresh and brackish waters (pH = 3–8) than in sea water (pH = 8.1). The adsorption is also favoured by the presence of phosphate as $H_2PO_4^-$ rather than HPO_4^{2-} ions which dominate the phosphate dissociation systems at higher pH values (Kester and Pytkowicz, 1967).

There is further evidence to suggest that at a fixed pH and temperature, phosphate adsorption is depressed by increasing salinity (Caritt and Goodgall, 1954; and Burns and Salomon, 1969). This may be a reflection of an ion-exchange adsorption mechanism in which there is greater competition for ion-exchange sites by other anions, e.g. Cl^-, SO_4^{2-}, Br^-, at higher salinities. The present evidence on the factors controlling phosphate removal in estuaries suggests that removal will be favoured under low salinity, low pH, and high phosphate concentration conditions, i.e. those conditions experienced towards the head of a typical estuary.

If the observed constancy of phosphate distributions is to be accounted for by a so-called 'buffering' mechanisms, there must be an opportunity for phosphate contribution to estuarine waters by desorption. The desorption of phosphate from solid mineral phases present in sediments or detritus has received little attention. Mortimer (1971) has suggested that redox potential (Eh) is important in controlling the release of phosphate from lake sediments (and presumabley estuarine sediments), reducing conditions favouring phosphate desorption. Also, Bray *et al.* (1973) have found that the oxidation of reducing anoxic sediments leads to the removal of phosphate from pore water solution. This suggests that in anoxic layers of estuarine sediments, phosphate is mobile in the pore waters and may migrate by diffusion and advection to the sediment surface. The removal of phosphate from solution in the pore waters under oxic conditions is probably due to the formation of insoluble $Fe^{3+}PO_4^{3-}$ complexes (Upchurch *et al.*, 1974).

Pomeroy *et al* (1965) have attempted to measure the release of phosphate from suspended sediments to estuarine waters by laboratory experiments, and have found that a 'buffering' effect was operating. The equilibrium concentration of inorganic phosphate in estuarine waters was ~22–28 $\mu g \times Pl^{-1}$. In a similar series of experiments Butler and Tibbitts (1972) reported phosphate 'buffering' in the range ~22–46 $\mu g \times l^{-1}$ for estuarine waters. These 'buffered' phosphate concentrations are similar to the observed concentrations reported for actual estuarine waters, and it appears that phosphate 'buffering' by sediment adsorption/desorption may well be an important control on this nutrient in estuaries.

The phosphate 'buffering' mechanism is of considerable importance when the effects of sewage pollution on estuaries are considered, and the role of sediments in retaining pollutant phosphorus. Aston and Hewitt (1977) have studied the dispersion of dissolved and particulate phosphorus from sewage discharge sites adjacent to a semi-enclosed tidal bay, with particular reference to phosphorus distributions in inter-tidal sediments. The coastal effluent outfalls were found to be important sources of particulate phosphate in the estuary, with the deposition of particulate phosphate under the strong influence of channel geometry. Shallow broad channels concentrating phosphate at their heads, while narrow steep sided and deeper channels concentrated phosphate deposition at their mouths. These features of phosphate deposition have been explained by Aston and Hewitt (1977) in terms of the particle size distributions of suspended matter and its relationship to phosphate concentrations in the suspended particles.

2.2.2 *Biological cycling of phosphorus*

Figure 5 represents the phosphorus cycle for estuaries in a generalized diagrammatic form. The cycle is very similar to the cycle of phosphorus in ocean waters, except for a few significant features. These may be summarized as follows: (a) the contribution of phosphate from sewage is far more important in the estuarine environment compared to the open oceans; (b) the input of river transported phosphorus in solution and particulate forms is very variable in estuaries on a relatively short time sacle, and is a function of catchment processes. In the open ocean, such changes in river input are intergrated; (c) estuarine waters have a greater contact with their underlying sediments compared to ocean waters, and phosphate exchange is probably more effective in the estuarine situation; (d) the overall rate biological cycling of phosphorus in estuaries is dependent on environmental factors which can change dramatically on a short time scale compared to that of the open oceans.

Recently, several studies have been made of phosphorus biological cycles in individual estuaries, and have provided the basis for a discussion of this important nutrient. Some of the more general and salient points which have come out of these studies are summarized here.

Correll *et al.* (1975) have investigated the phosphorus flux and cycling in the Rhode River sub-estuary of Chesapeake Bay, including a study of the role of mud-flat periphyton and plankton communities in the phosphorus cycle. From their field studies of the Rhode River sub-estuary and laboratory experiments with ^{32}P radiotracer on biological phosphorus cycling in the estuarine environment, Correll *et al.* (1975) have concluded that orthophosphate is taken up from solution in brackish waters by bacteria, mainly located on the surfaces of organic detritus and suspended mineral particles,

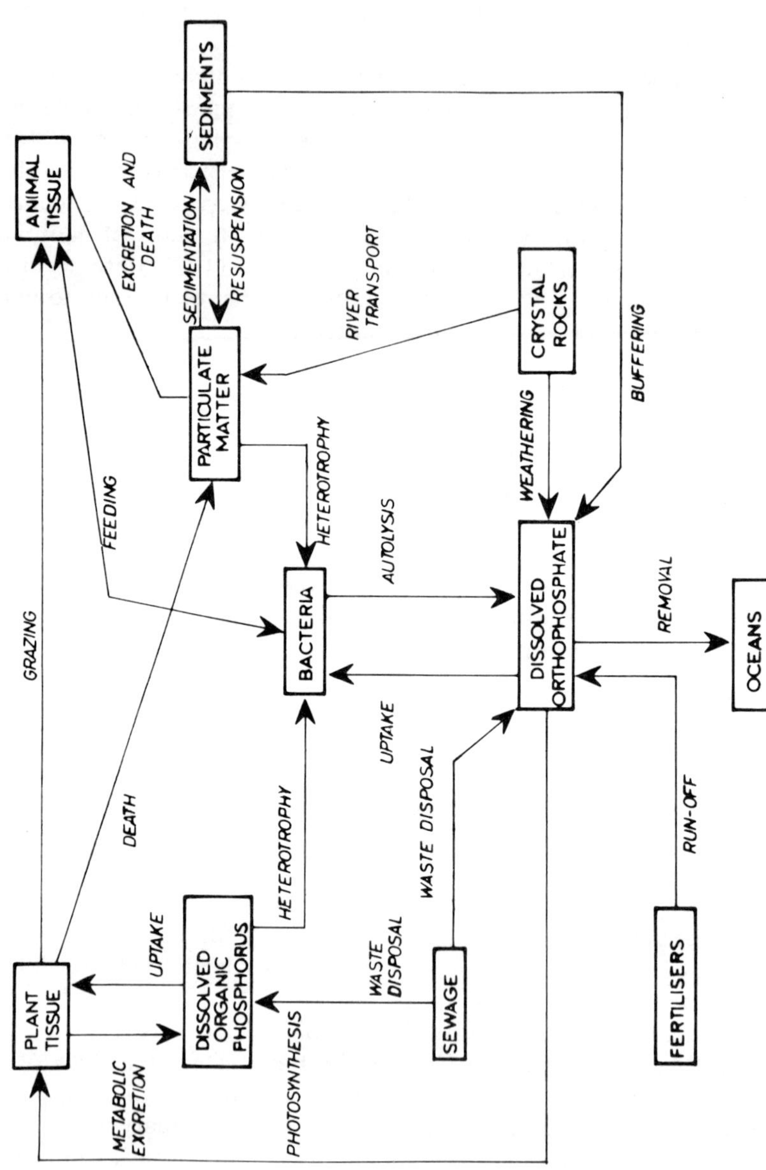

Figure 5. The phosphorus cycle in the estuarine environment.

and by phytoplankton under the appropriate light intensity conditions. Subsequently, the bacteria and phytoplankton are consumed by filter feeding organisms, notably ciliate protazoans, which then release phosphorus back into the estuarine waters as orthophosphate ions and dissolved and particulate organic phosphorus compounds.

In reporting their investigations, Correll *et al.* (1975) have been critical of the earlier experimental studies of Pomeroy and his co-workers (e.g. Pomeroy *et al.*, 1963; 1965; 1966; and 1969) which attempted to illustrate some aspects of the estuarine biological cycling of phosphorus by the use of ^{32}P radiotracer techniques. The main criticisms leveled by Correll *et al.* (1975) are that these earlier studies omitted kinetic analyses of the phosphorus fluxes, and that the rates of phosphorus uptake, turnover times, and transport times reported by Pomeroy and his co-workers are highly variable. Correll *et al.* (1975) have cited the work of Rigler (1956, 1964) and Lean (1973) who have emphasized the importance of very detailed time series if rates of phosphorus uptake and transport are to be quantitively evaluated for aquatic ecosystems.

Hobbie *et al.* (1975) have investigated the sources and fates of phosphorus (and nitrogen, see Section 2.3.2) in the Pamlico River estuary (North Carolina), and have estimated that ~60% of the ortho-phosphate entering the estuary is retained by the estuarine sediments. In this estuary, nitrogen rather than phosphorus is the nutrient which stimulates and controls phytoplankton blooms. In such estuaries as that of the Pamlico River, which has phosphate inputs from sewage, phosphate mining, and industrial sources, it is exceedingly likely that biological activity will play a minor role in the overall phosphorus budget. The mechanism of phosphate 'buffering' between waters and sediments, as discussed above in Section 2.2.1., is probably dominant in the phosphorus cycle of estuaries with high phosphate inputs.

Several coastal plain estuaries of the Eastern United States have been studied by Taft and Taylor (1976) with respect to phosphorus biogeochemistry. These authors have shown that the coastal plain estuaries exhibit an annual phosphorus cycle with maximum concentrations of dissolved phosphate in summer. The data used by Taft and Taylor (1976) included their own and that from various other studies of the estuaries of the Eastern U.S.A., which include the Chesapeake Bay, Delaware Bay, South River (Maryland), Potomac River, and Pamlico river Estuaries (Newcombe and Lang, 1939; Pomeroy *et al.*, 1956; Smayda, 1957; Jeffries, 1962; Patten *et al.*, 1963; Reimold, 1965; Whaley *et al.*, 1966; Copeland and Hobbie, 1972; Hobbie *et al.*, 1972; and Shlopak, 1972). Taft and Taylor (1976) concluded that phosphate can act as a regulator for estuarine phytoplankton biomasses, a conclusion which is supported by phytoplankton nitrogen to phosphorus ratios and by alkaline phosphatase enzyme activity which is an indicator of the state of cell inorganic phosphate nutrition. These authors have, however, pointed out that no conclusions can be reached from their own data concerning phosphorus regulation of primary productivity during spring, phosphorus monoesters being available at all times in low concentrations to support sustained primary productivity.

2.3 Nitrogen biogeochemistry in estuaries

Nitrogen is supplied in elemental (molecular) and chemically combined forms to estuaries. The main form of combined nitrogen is the dissolved nitrate, which is derived from rock weathering, and pollution sources such as the application of nitrogenous fertilizers to agricultural land. There are relatively few data available on the biogeochemistry of nitrogen compounds in estuaries, but some of the studies of the nitrogen cycles in the oceans and fresh waters can be applied to the estuarine situation.

2.3.1 *Inorganic controls on nitrogen*

Very few studies on the conservative/non-conservative behaviour of nitrate in estuaries have been made. In the Columbia River estuary, an almost linear relationship between dissolved nitrate and salinity was reported by Stefánsson and Richards (1963), with some minor deviations from the linear relationship at low salinities which may indicate some non-conservative behaviour. Mackay and Leatherland (1976) have reported that nitrate frequently behaves in a non-conservative manner in the Clyde estuary, Scotland. The removal is however restricted to bottom waters in which denitrification appears to occur at tines of low oxygen content when nitrate reduction can occur.

The influence exerted by sewage effluent and fertilizer run-off on the abundance of nitrate in individual estuaries will obviously vary geographically, but at present there is little evidence to suggest that there are any chemical or sedimentological controls on the nitrogen cycle in estuaries. In conclusion, it is reasonable to assume that the estuarine distribution of nitrate from both natural and pollutant sources will be controlled by the essentially physical processes of circulation, tidal flow, etc. The biological modifications to the distribution and season occurrence of nitrate and other nitrogen species in estuaries are discussed below.

2.3.2 *Biological cycling of nitrogen*

The behaviour of nitrogen compounds as nutrients in estuaries has received attention from several workers in the last few years. The studies have been based on relatively few estuaries, but do provide some ideas of how these nutrients may behave in the general estuarine environment. The chemical forms in which nitrogen is available for biological utilization are manyfold, and this variety of speciation complicates the study of the element's biogeochemistry. The most abundant form of nitrogen in estuaries is as the elemental gas, derived from the solution of atmospheric nitrogen in marine, fresh, and brackish waters. Spencer (1975) has pointed out that elemental nitrogen is certainly removed and produced by biological action in the marine environment. The extent to which this process is important in the oceans and estuaries is not known, but is unlikely that it is very great.

The much more important forms of nitrogen for biogeochemical processes in estuaries are the dissolved inorganic species, e.g. nitrate and nitrite, ammonia, and organic nitrogen compounds (in dissolved and particulate forms). The general nitrogen cycle which describes the transformations of nitrogen in aquatic environments, modified for estuaries is shown in Figure 6. Two notable features of the cycle are: (a) the greater number of transformations involved in the nitrogen cycling of inorganic species, compared to those of the phosphorus and silicon cycles. The latter two nutrient cycles are simplified by the lack of the chemical species involved; and (b) the important input of *pollutant* nitrogen compounds in the estuarine cycle. Many estuaries receive a considerable quantity of anthropogenic nitrogen, usually in the form of nitrate, and derived from agricultural run-off i.e. fertilizer residues, and domestic and industrial sewage disposal. The total combined nitrogen concentration of river waters on a world-side basis is approximately $225\ \mu g\,l^{-1}$ (Turekian, 1971), while surface sea water contains, on average, $500\ \mu g\,l^{-1}$ total combined nitrogen (Riley and Chester, 1971). In highly polluted situations, the concentrations of nitrate entering estuaries in river water may, however, be several times greater than the average world-wide value for river water. Topping (1977) has summarized the estimates on the input of nitrogen and other nutrients to estuaries from river water and pollution. Nitrogen supply to estuaries is important for primary production, e.g. in a recent study of nutrient–phytoplankton relationships in Narragansett Bay, Furnas *et al.* (1975) have shown that the levels of phytoplankton production which occur furing summer require a very frequent, e.g. 12 hour replenishment of nitrogen from external sources, or that the *in situ* inorganic nitrogen + urea concentrations must be replenished every 3 to 4 hours. Furnas *et al.* (1975) have concluded that a high and persistent estuarine phytoplankton production, such as that observed in Narragansett Bay, requires a very rapid remineralization of nitrogen. This conclusion is supported by experimental data on plankton production, and Furnas *et al.* (1975) dismiss the possibility that the phytoplankton may use a reservoir of dissolved organic nitrogen (Guillard, 1963).

Other recent studies on the supply of nitrogen as a nutrient to estuaries have been carried out by Hobbie *et al.* (1975), Haines (1975), and Dunstan

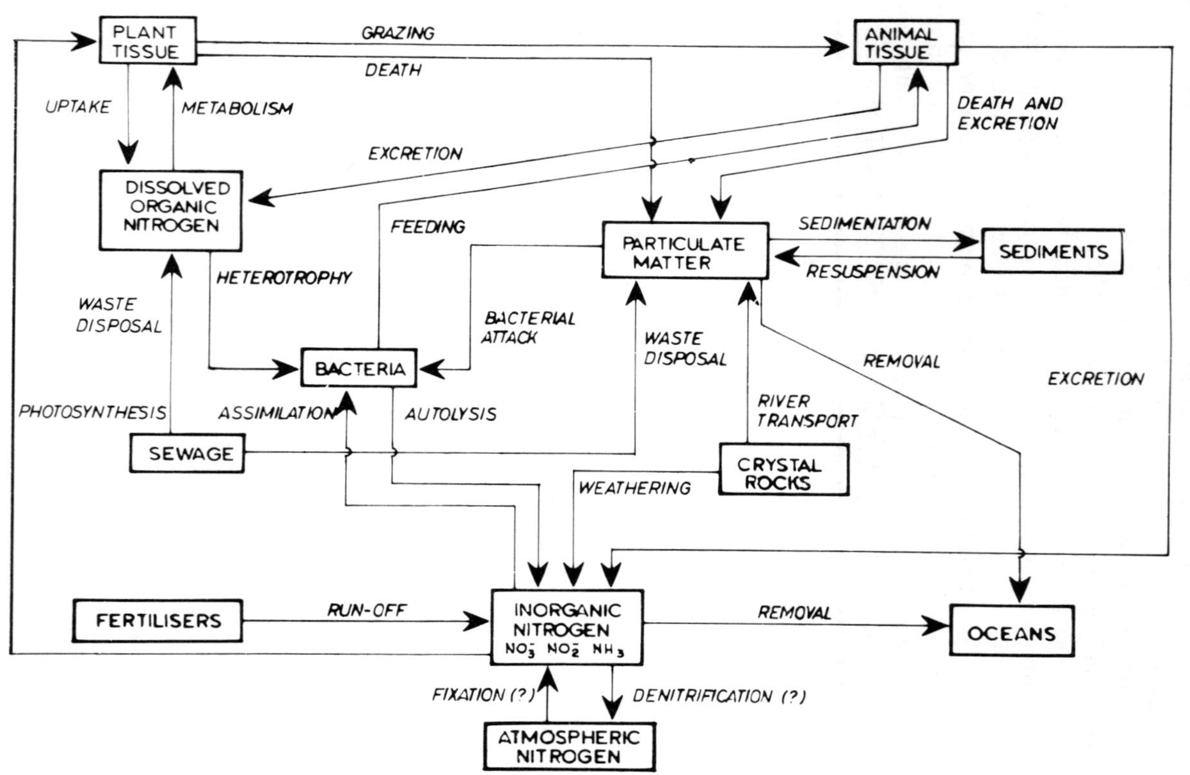

Figure 6. The nitrogen cycle in the estuarine environment.

and Atkinson (1976). In the Pamlico River estuary, the availability of free nitrate was only dominant in winter months, while ammonia was abundant all year (Hobbie *et al.* 1975). The two algae (dinoflagellate) blooms showed good correlations with the appearance of nitrate in the middle reaches of the estuary. The total organic nitrogen (dissolved and particulate) was found to be as great as the inorganic nitrogen, but the organic nitrogen did not appear to be biologically active. This latter observation is consistent with the conclusions of Furnas *et al.* (1975), see above. Haines (1975) has proposed, however, that only biological regeneration is capable of maintaining the high rates of production in coastal waters associated with salt-marsh estuaries. In contrast to the Pamlico River estuary, where all but 5% or less of the nitrogen required for production is accounted for by regeneration, the Georgia Bight has been estimated to receive 40–50% of the nitrogen as nitrate brought in by Gulf Stream intrustions (Dunstan and Atkinson, 1976; Steffansson *et al.*, 1971; and Blanton, 1971). The recent observations on nitrogen supply as an estuarine nutrient which are summarised above, serve to illustrate that the productivity of a given estuary may depend on such extreme conditions as very effective regeneration on a rapid time-scale or the marine and riverine supply of *new* nitrogen.

The biological removal of nitrogen from estuarine waters may be achieved by sedimentation and burial. This process provides a nutrient rich detrital resource for benthic productivity, and stimulates bacterial growth at the water–sediment interface (Rhoads *et al.*, 1975). The release of previously deposited nitrogen compounds from estuarine sediments is a possible source of nutrient for productivity in the overlying waters. Hartwig (1976) has shown that sediment exchange of nitrogen compounds was dominated by ammonia release and nitrate uptake for a coastal siliceous sediment, La Jolla, California. The supply of nutrient nitrogen gives rise, however, to a negligible (0.4%) fraction of that contained in the overlying water body.

3 DISSOLVED GASES IN ESTUARIES

3.1 Gases other than oxygen and carbon dioxide

In contrast to oxygen and carbon dioxide, the other gases dissolved in estuarine waters are not strongly influenced by biological processes. Exceptions to this broad statement are the specialized anoxic environments within which the activity of anaerobic bacteria leads to the production of hydrogen sulphide and methane. Generally, the concentrations and distributions of gases other than oxygen and carbon dioxide will be functions of physico–chemical factors, notably those which influence their solubilities.

The solubilities of gases in natural waters are functions of salinity, temperature and pressure. Three laws define the effects of these factors on the solubility of gases, and may be summarized as follows:

(1) *Salinity*—the presence of dissolved ions in natural waters lowers the solubility of atmospheric gases according to the empirical law of Setchénow, which may be stated as the equation

$$b_2 \log_e (B)_T = b_1 + b_2 S\text{‰} \tag{1}$$

where B is the Bunsen coefficient, b_1 and b_2 are constants for the particular gas in water of salinity S‰ at a given temperature (T)

(2) *Temperature*—the solubility of gases in natural waters decreases with increasing temperature, and as a first approximation, follows the law

$$\log_e B = K . \frac{1}{T} \tag{2}$$

where B is the Bunsen coefficient for the water, T is the absolute temperature, and K is a constant for the particular gas.

(3) *Pressure*—the solubility of a gas is directly proportional to the partial pressure it exerts in the gaseous phase (i.e. in the atmosphere above the water surface), and when equilibrium is achieved the partial pressures in the gaseous and aqueous phase are equal i.e. Henry's law is obeyed.

Since there is little variation in the partial pressures of atmospheric gases overlying estuarine waters it it is more important to consider the effects of salinity and temperature which are highly variable in estuaries. The spatial and temporal (both short term, eg. tidal, and long term, eg. seasonal) changes in an estuary are typically in the approximate ranges of <1 to 35‰ and 0–30 °C for salinity and temperature respectively. The changes in salinity are especially rapid and extreme for some locations within an estuary. It is also pertinent to consider the possibility that the control of atmospheric gas solubilities is a kinetic one, the estuarine system having insufficient time to completely respond to the effects of changing salinity and temperature during a tidal cycle. If this situation arises, the rate controlling step may be diffusion through the air–sea surface interface (see below). First consider the control of gas solubility as influenced by salinity and temperature, and ignoring kinetic effects.

Weiss (1970) has used the data of Douglas (1964, 1965) and Murray and Riley (1970) together with his own observations on the solubilities of nitrogen and argon (and oxygen, see Section 3.2.) in sea water to derive a set of equations to show the dependence of solubility on salinity and temperature. Using the Setchénow equation

$$\log_e (B)_T = b_1 + b_2 S\text{‰} \tag{3}$$

and the Vant Hoff equation to define temperature dependence at constant salinity (S%)

$$\log_e (B)_S = a_1 + \frac{a_2}{T} + a_3 \log_e T \tag{4}$$

where T is the absolute temperature, and a_1, a_2, a_3 are constants for the gas, Weiss (1970) produced a combined equation of the form

$$\log_e B = A_1 + A_2\left(\frac{100}{T}\right) + A_3 \log\left(\frac{T}{100}\right) + S\text{‰}B_1 + B_2\left(\frac{T}{100}\right) + B_3\left(\frac{T}{100}\right)^2 \tag{5}$$

where the temperature is numerically scaled by a factor of 100. Weiss (1970) extended this equation to compensate for the dependence of vapour pressure and the density of sea water on temperature.

$$\log C = A_1 + A_2\left(\frac{100}{T}\right) + A_3 \log_e\left(\frac{T}{100}\right) + A_4\left(\frac{T}{100}\right)$$
$$+ S\text{‰}B_1 + B_2\left(\frac{T}{100}\right) + B_3\left(\frac{T}{100}\right)^2 \tag{6}$$

Equation (6) may be used to express the cocentration of a given gas in sea water in μmol kg^{-1}. Typical results of the application of this equation for the gases nitrogen, oxygen, argon, neon, and helium, are shown in Table 1. The effect of temperature changes which might occur as seasonal extremes in estuaries e.g. from 5° to 30 °C, are capable of producing substantial changes in the solubilities of these gases. For example, nitrogen exhibits solubilities of 549.6 and 383.4 μmol kg^{-1} at 5° and 30 °C respectively for $S = 37$‰. The temperature effect is, however, much less marked for the other gases, e.g. helium. The sea water solubilities of the rarer atmospheric

Table 1. The solubilities of nitrogen, oxygen, argon, neon, and helium in sea water at 35% salinity.

| | μmol kg^{-1} | | | nmol kg^{-1} | |
$t(°C)$	C_{N_2}	C_{O_2}	C_{Ar}	C_{Ne}	C_{He}
0	616.4	349.5	16.98	7.88	1.77
5	549.6	308.1	15.01	7.55	1.73
10	495.6	274.8	13.42	7.26	1.70
15	451.3	247.7	12.11	7.00	1.68
20	414.4	225.2	11.03	6.77	1.66
25	383.4	206.3	10.11	6.56	1.65
30	356.8	190.3	9.33	6.36	1.64

Data from Kester (1975), gas solubilities are calculated from equation (6). (Reprinted with permission from Kester (1975). Copyright by Academic Press Inc. (London) Ltd.).

gases are less well known, some data on krypton and xenon are available (Benson, 1965; Konig, 1963; and Wood and Caputi, 1966).

Gas exchange between estuarine waters and the atmosphere is probably kinetically controlled by the air–sea interface exchange processes. Kester (1975) has adopted the simple model of Whitman (1923) to conceptulise the exchange of gases across the air–sea interface. This model recognizes three regions. A turbulent atmospheric zone, a turbulent liquid zone, and a laminer zone separating the two turbulent zones. The laminar zone has all liquid motion parallel to the air–sea water interface, and uniform partial pressures in the turbulent zones. The rate determining step is the molecular diffusion of gas through the laminar zone. Combining Fick's law and Henry's law, the rate of gas flux through the air-sea interface is given as

$$\frac{\mathrm{d}g}{\mathrm{d}t} = \frac{ADg(p_g - P_g)}{K_g.T} \tag{7}$$

where A is the surface area, D_g is a diffusion coefficient and K_g is the Henry's law constant. Downing and Truesdale (1955) have reported that the experimental gas exchange between air and sea water increases by a factor of two between 5° and 25°, which suggests that estuaring mixing and seasonal changes in estuarine conditions will significantly affect the rate of gas exchange. Another important control on the exchange rate is wind speed, Dowing and Truesdale (1955) have shown at $\mathrm{d}g/\mathrm{d}t$ increases rapidly for wind speeds in the range 3–$15\ \mathrm{m\,s^{-1}}$.

The influence of surface films on gas exchange is not clearly understood, but there is some evidence to suggest that oil pollution and other organic films on estuarine waters are unlikely to reduce gas exchange. A laminar oil zone or film of at least 10^{-4} cm. is required before any depletion of gas exchange is observed, and Downing and Truesdale (1955) have concluded that such films are likely to be very rare.

A further influence on atmospheric gas dissolution in estuarine waters is that of submerged air bubbles. During intensive disturbance of the water surface, e.g. storm conditions, air bubbles are forcibly submerged in estuarine waters. Gas exchange between a submerged bubble and the water is more rapid than at the air-water surface due to two physical factors. The partial pressure of the air in the bubble will be greater than that in the free air, and the laminar zone surrounding the bubble will be reduced in thickness by the turbulent flow past the rising bubble (Kanwisher, 1963). These factors lead to increased solubility for bubble air due to the Henry's law and diffusion effects. There is evidence that these processes may lead to the differential dissolution of gases from submerged bubbles as a result of differences in their diffusion coefficients and Henry's law constants (Wyman *et al.*, 1952).

3.2 Oxygen (and carbon dioxide)

Oxygen and carbon dioxide concentrations in natural waters, including estuaries, are strongly influenced by biological cycles, especially primary productivity. The carbon dioxide system in estuaries is discussed in Chapter 9 as part of the CO_2–$CaCO_3$ system, and will not be dealt with in the present discussion. Before considering the influences of biological cycles on the estuarine chemistry of these gases, it is worthwhile noting that the dissolution of O_2 and CO_2 from the atmosphere will be controlled by the same physical processes as those discussed in Section 3.1.

The data on the solubilities of oxygen and carbon dioxide in marine waters have recently been reviewed by Kester (1975) and Skirrow (1975). Figure 7 shows a comparison of measured oxygen solubilities with those calculated from the Setchénow equation (see Section 3.1) as a function of salinity at 5 °C and 20 °C. The range of salinity covers that which would be encountered in any estuarine situation, while the two temperatures are reasonable approximations to winter and summer temperatures in temperate regions. At 20 °C the measured oxygen solubilities show no common systematic departures from the solubilities predicted by Weiss's (1970) modified Setchénow equation (6). At 5 °C the theoretical predictions differ from the observed values in both sets of data by about $+0.03 \, \mathrm{cm^3 \, 1^{-1}}$ at low salinities and $-0.03 \, \mathrm{cm^3 \, 1^{-1}}$ at 20% salinity, but it is unlikely that such deviations will be of any real significance in estuarine applications. In conclusion, the Setchénow equation (6) in its modified form (Weiss, 1970) is adequate for the prediction of oxygen solubility in estuarine waters at the salinities and temperatures which are commonly encountered.

3.2.1 *Biological and pollutant controls on oxygen in estuaries*

The surface sea waters and river waters mixing together in estuaries will normally be completely or nearly saturated with oxygen dissolved from the atmosphere. However, oxygen is consumed by natural biological consumption and by the oxidation of pollutant organic matter, e.g. sewage, if present. At present many estuaries receive both organic and nutrient pollutants from domestic and industrial sources, and the combination of these two contaminants can lead to major changes in the biological productivity and its associated oxygen demand. The high productivity in estuaries may, under certain circumstances of restricted circulation and flushing, lead to eutrophication (see e.g. Topping, 1977).

The quantitative determination of dissolved oxygen consumption by organic matter in natural waters has been discussed by Richards (1965). Using the average atomic C:N:P ratios for marine plankton as 106:16:1 (Fleming, 1940), and assuming the oxidation state of carbon to be that in

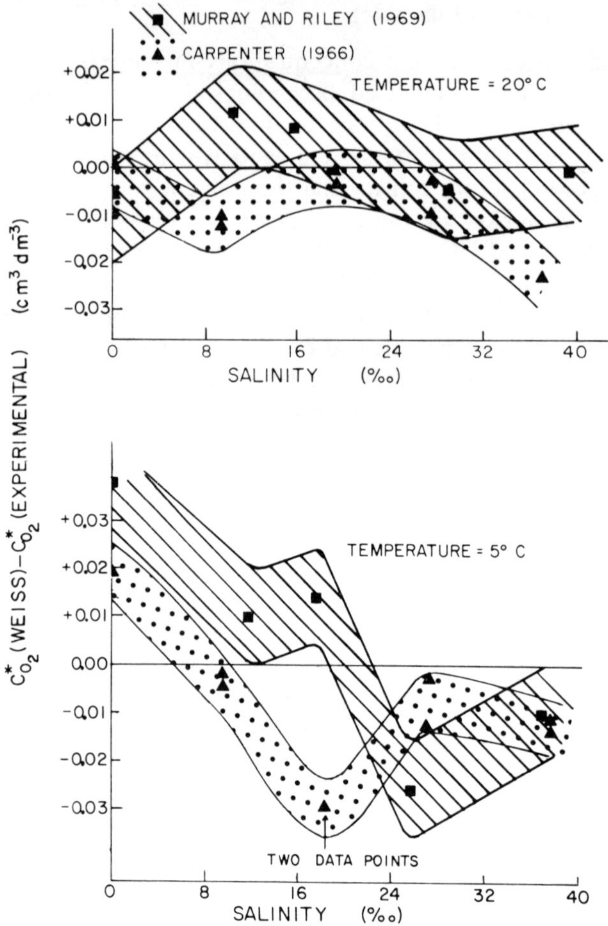

Figure 7. A comparison of measured oxygen, solubilities with those calculated from Weiss's (1970) modification of the Setchénow equation as a function of salinity at 5 °C and 20 °C. The width of the textured bands represents the estimated accuracies of the experimental data (Reproduced with permission from Kester, 1975. Copyright by Academic Press Inc. (London) Ltd.)

carbohydrate ie.–CH_2O, Richards (1965) has calculated the amount of oxygen required for the oxidation of the hypothetical composition of plankton $(CH_2O)_{106}(NH_3)_{16}H_3PO_4$. In waters containing dissolved free oxygen the microbial decomposition of organic debris is then represented by an overall equation

$$(CH_2O)_{106}(NH_3)_{16}H_3PO_4 + 138O_2 \rightarrow 106CO_2 + 122H_2O + 16HNO_3 + H_3PO_4$$

The oxygen demand in this reaction is effectively $2.67\,gO_2\,g^{-1}$ carbon and $4.57gO_2\,g^{-1}$ nitrogen. Head (1976) has pointed out that while the organic matter content of open ocean surface waters is rarely greater than $2.5\,mg\,l^{-1}$, estuarine waters often contain up to $10\,mg\,l^{-1}$ even in the absence of any obvious pollution. The typical oxygen content of natural waters $(\sim 11\,mg\,l^{-1})$ is sufficient to oxidize about $4\,mg$ carbon or $2\,mg$ nitrogen, so that in estuaries, especially those which receive a significant input of pollution, the dissolved oxygen can in theory be easily removed by the oxygen demand of organic debris. Very few instances of de-oxygenation by sewage have been reported (Johnston, 1977), and Head (1976) has suggested that in practice the oxidation does not proceed quickly enough to exceed the supply of oxygen from surrounding water masses and the atmosphere. Under conditions of poor water exchange in estuaries, i.e. fjords and other basins connected to the open sea by shallow sills, the complete removal of all free dissolved oxygen can occur (Richards, 1965; Richards *et al.*, 1965; Richards *et al.*, 1971). The water quality standards outlined by Batelle (1971) recommend that dissolved oxygen concentrations should not fall below $4\,mg\,l^{-1}$ in estuaries and tidal tributaries for the protection of biological resources.

The extent of estuarine pollution by organic matter from sewage and other sources is highly variable, and it is not feasible to discuss the effects of such pollution or even natural organic matter on the oxygen distributions in individual estuaries. An example of a well documented case history is that of the Thames Estuary, England (Department of Scientific and Industrial Research, 1964), the results of this extensive study have been summarized by Head (1976).

A further demand on the oxygen dissolved in estuarine waters is that due to the botton sediments, or rather the bacterial populations within the surface of the sediments. Breuer *et al.* (1977) have demonstrated that the removal of dissolved oxygen to fresh water sediments is a function of the standing crop of aerobic and heterotrophic bacteria at the sediment surface. There are no data available on the oxygen demand of estuarine sediments, but is reasonable to assume that the demand approximates to that found in fresh water conditions.

REFERENCES

Aston, S. R. (1978). In *Chemical Oceanography*. Volume 7, sec. edn., J. P. Riley and R. Chester (editors). Academic Press, London.

Aston, S. R., and Chester, R. (1973). The influence of suspended particles on the precipitation of iron in natural waters. *Estuarine Coastal Mar. Sci.*, **1**, 225–231.

Aston, S. R., and Hewitt, C. N. (1977). Phosphorus and organic carbon distributions in a polluted coastal marine environment. *Estuarine Coastal Mar. Sci.*, **5**, 243–254.

Atkins, W. R. G. (1945). Autotrophic flagellates as the major constituent of the oceanic phytoplankton. *Nature, Lond.,* **156,** 446–447.

Batelle Columbus Laboratories (1971). *Water Quality Criteria Data Book,* Volume 3. Environmental Protection Agency, Project No. 18050 G.M.V. Contract No. 68–1–0007.

Benson, B. B. (1965). In *Occasional Publications, Univ. Rhode Island, No.* 3. D. R. Schink and J. T. Corless (editors), Univ. Rhode Island, Kingston. 91–107.

Bewers, J. M., Macawley, I. D., and Sundby, B. (1974). Trace metals in the waters of the Gulf of St. Lawrence. *Can. J. Earth Sci.,* **11,** 939–950.

Bien, G. S., Contois, D. E., and Thomas, W. H. (1958). The removal of soluble silica from fresh water entering the sea. *Geochim. Cosmochim. Acta,* **14,** 35–54.

Blanton, J. (1971). Exchange of Gulf Stream water with North Carolina shelf water in Onslow Bay during stratified conditions. *Deep-Sea Res.,* **18,** 167–178.

Bowman, M. J. (1977). Nutrient distributions and transport in Long Island sound. *Estuarine Coastal Mar. Sci.,* **5,** 531–548.

Bray, J. T., Bricker, O. P., and Troup, B. N. (1973). Phosphate in interstitial waters of anoxic sediments: Oxidation effects during sampling procedure. *Science, N.Y.,* **180,** 1362–1364.

Breuer, W. S., Abernathy, A. R., and Paynter, M. J. B. (1977). Oxygen consumption by freshwater sediments. *Water Res.,* **11,** 471–473.

Burns, P. A., and Salomon, (1969). Phosphate adsorption by kaolin in saline environments. *Proc. Nat. Shellfish. Ass.,* **59,** 121–125.

Burton, J. D. (1970). The behaviour of dissolved silicon during estuarine mixing. II. Preliminary investigations in the Vellar Estuary, Southern India. *J. Cons. Perm. Int. Explor. Mer.,* **33,** 141–148.

Burton, J. D., Leatherland, T. M., and Liss, P. S. (1970). The reactivity of dissolved silicon in some natural waters. *Limnol Oceanogr.,* **15,** 473–476.

Burton, J. D., and Liss, P. S. (1973). Processes of supply and removal of dissolved silicon in the oceans. *Geochim. Cosmochim. Acta,* **37,** 1761–1773.

Butler, E. I., and Tibbitts, S. (1972). Chemical Survey of the Tamar Estuary I. Properties of the waters. *J. Mar. Biol. Ass. UK.,* **52,** 681–699.

Carritt, D. E., and Goodgal, S. (1954). Sorption reactions and some ecological implications. *Deep-Sea Res.,* **1,** 224–243.

Chen, Y. S. R., Butler, J. N., and Stumm, W. (1973). Kinetic study of phosphate reaction with aluminium oxide and kaolinite. *Environ. Sci. Technol.,* **7,** 327–332.

Coonley, L. S., Jr., Baker, E. B., and Holland, H. D. (1971). Iron in the Mullica River and in Great Bay, New Jersey. *Chem. Geol.,* **7,** 51–63.

Copeland, B. J., and Hobbie, J. E. (1972). *Phosphorus and entrophication in the Pamlico River estuary, N. C., 1966–1969—A summary.* Water Resources Research Institute, Univ. North Carolina, Rept. No. 65.

Correll, D. L., Faust, M. A., and Severn, D. J. (1975). *In Estuarine Research.* Volume 1. L. E. Cronin (editor). Academic Press, New York. 738 pp.

Department of Scientific and Industrial Research (1964). Effects of polluting discharges on the Thames estuary. *Water Pollution Research Technical Paper No. 11.,* H. M. S. O., London. 45 pp.

Douglas, E. (1964). Solubilities of oxygen, argon, and nitrogen in distilled water. *J. Phys. Chem.,* **68,** 169–174.

Douglas, E. (1965). Solubilities of argon and nitrogen in sea water. *J. Phys. Chem.,* **69,** 2608–2615.

Downing, A. L., and Truesdale, G. A. (1955): Some factors affecting the rate of solution of oxygen in water. *J. Appl. Chem.,* **5,** 570–581.

Duedall, I. E., O'Connor, H. B., Parker, S. H., Wilson, R. E., and Robbins, A. S.

(1977): The abundances, distribution and flux of nutrients and chlorophyll a in the New York Bight Apex. *Estuarine Coastal Mar. Sci.*, **5**, 81–105.

Dunstan, W. M., and Atkinson, L. P. (1976). In *Estuarine Processes.* Volume 1. M. Wiley (editor). Academic Press, New York. 541 pp.

Fanning, K. A., and Pilson, M. E. Q. (1973). The lack of inorganic removal of dissolved silica during Mier-Ocean mixing. *Geochim. Cosmochim. Acta*, **37**, 2405–2415.

Fleming, R. H. (1940). The composition of plankton and units for reporting population and production. *Proc. 6th Pacific Sci. Cong. Calif.*, 1939, **3**, 535–540.

Furnas, M. J., Hitchcock, G. L., and Smayda, T. J. (1976). In *Estuarine Processes.* Volume 1. M. Wiley (editor). Academic Press, New York. 541 pp.

Gessner, F. (1960). Untersuchungen Über Den Phosphathaushalt des Amazonas. *Int. Revue Gesamten Hydrobiol. Hydrogr.*, **45**, 339–345.

Guillard, R. R. L. (1963). In *Symposium on Marine Microbiology.* C. H. Oppenheimer (editor). C. C. Thomas, Springfield. 104 pp.

Haines, E. B. (1975). In *Estuarine Research.* Volume 1. L. E. Cronin (editor). Academic Press, New York. 738 pp.

Hair, M. E., and Bassett, C. R. (1973). Dissolved and particulate humic acids in an East Coast estuary. *Estuarine Coastal Mar. Sci.*, **1**, 107–111.

Hartwig, E. O. (1976). In *Estuarine Processes.* Volume 1. M. Wiley (editor). Academic Press, New York. 541 pp.

Harvey, H. W. (1955). *The Chemistry and Fertility of Sea Water*, Cambridge University Press, London. 224 pp.

Head, P. C. (1976). In *Estuarine Chemistry.* J. D. Burton and P. S. Liss (editors). Academic Press, London. 229 pp.

Ho, C. L., and Barrett, B. B. (1977). Distributions of nutrients in Louisiana's coastal waters influenced by the Mississippi River. *Estuarine Coastal Mar. Sci.*, **5**, 173–195.

Hobbie, J. E., Copeland, B. J., and Harrison, W. G. (1972). *Nutrients in the Pamlico River Estuary, N. C.*, 1969–1971. Water Resources Research Institute, Univ. North Carolina, Rept. No. 76.

Hobbie, J. E., Copeland, B. J., and Harrison, W. G. (1975). In *Estuarine Research.* Volume 1. L. E. Cronin (editor). Academic Press, New York. 738 pp.

Howes, N. H. (1939). The ecology of a saline lagoon in south-east Essex. *J. Linn. Soc. (Zool.)*, **40**, 383–445.

Hulbert, E. M. (1956). Distribution of phosphorus in Great Pond, Massachusetts. *J. Mar. Res.*, **15**, 181–192.

Jeffries, H. P. (1962). Environmental characteristics of Raritan Bay, a polluted estuary. *Limnol, Oceanogr.*, **7**, 21–31.

Jitts, H. R. (1959). The adsorption of phosphate by estuarine bottom deposits. *Aust. J. Mar. Fresh water Res.*, **10**, 7–21.

Johnston, R. (1977). In *Marine Pollution.* R. Johnston (editor). Academic Press, London. 729 pp.

Jørgensen, E. G. (1955). Solubility of the silica in diatoms. *Physiol. Plant.*, **8**, 846–851.

Kanwisher, J. (1963). On the exchange of gases between the atmosphere and the sea. *Deep-Sea Res.*, **10**, 195–207.

Kester, D. R. (1975). In *Chemical Oceanography.* Volume 2. Sec. ed. J. P. Riley and G. Skirrow (editors). Academic Press, London. 606 pp.

Kester, D. R., and Pytkowicz, R. M. (1967). Determination of the apparent dissociation constants of phosphoric acid in sea water. *Limnol. Oceanogr.*, **12**, 243–252.

Ketchum, B. H. (1954). Relation between circulation and planktonic populations in estuaries. *Ecology*, **35**, 191–200.

Konig, H. (1963). Ueber die Löslichkeit der Edelgase in Meerwasser. *Z. Natur.*, **18a**, 363–367.

Krauskopf, K. B. (1956). Dissolution and precipitation of silica at low temperatures. *Geochim. Cosmochim. Acta*, **10**, 1–26.

Lean, D. R. S. (1973). Phosphorus dynamics in lake water. *Science, N. Y.*, **179**, 678–680.

Lewin, J. C. (1961). The dissolution of silica from diatom walls. *Geochim. Cosmochim. Acta*, **21**, 182–198.

Liss, P. S. (1976). In *Estuarine Chemistry*. J. D. Burton and P. S. Liss (editors). Academic Press, London. 229 pp.

Liss, P. S., and Pointon, M. J. (1973). Removal of dissolved boron and silicon during estuarine mixing of sea and river waters. *Geochim. Cosmochim. Acta*, **37**, 1493–1498.

Liss, P. S., and Spencer, C. P. (1970). Abiological processes in the removal of silicate from sea water. *Geochim. Cosmochim. Acta*, **34**, 1073–1088.

Mackay, D. W., and Leatherland, T. M. (1976). In *Estuarine Chemistry*. J. D. Burton and P. S. Liss (editors). Academic Press, London, 229 pp.

Maéda, H. (1952). The relation between chlorinity and silicate concentration of waters observed in some estuaries. *Publ. Seto Mar. Biol. Lab.*, **2**, 249–255.

Maéda, H. (1953). The relation between chlorinity and silicate concentration of waters observed in some estuaries. II. *J. Shimonoseki Coll. Fish.*, **3**, 167–180.

Maéda, H., Tsukamoto, M. (1959). The relation between chlorinity and silicate concentration of waters observed in some estuaries. IV. *J. Shimonoseki Col. Fish.*, **8**, 121–134.

Maéda, H., and Takesue, K. (1961). The relation between chlorinity and silicate concentration of waters observed in some estuaries. V. *Rec. Oceanogr. Works Jap.*, **6**, 112–119.

Makimoto, H., Maéda, H., and Era, S. (1955). The relation between chlorinity and silicate concentration of waters observed in some estuaries. III. *Rec. Oceanogr. Works Jap.*, **2**, 106–111.

Mortimer, C. H. (1971). Chemical exchanges between sediments and water in the Great Lakes—speculations on probable regulatory mechanisms. *Limnol. Oceanogr.*, **16**, 387–404.

Murray, C. N., and Riley, J. P. (1970). The solubility of gases in distilled water and sea water—III. Argon. *Deep-Sea Res.*, **17**, 203–209.

Nash, C. B. (1947). Environmental characteristics of a river estuary. *J. Mar. Res.*, **6**, 147–174.

Newcombe, C. L., and Lang, A. G. (1939). The distribution of phosphates in the Cheseapeake Bay. *Proc. Am. Phil. Soc.*, **81**, 393–420.

Patrick, W. H., and Khalid, R. A. (1974). Phosphate release and sorption by soils and sediments: Effect of aerobic and anaerobic conditions. *Science, N. Y.*, **186**, 53–55.

Patten, B. C., Mulford, R. S., and Warinner, J. E. (1963). An annual phytoplankton cycle in the lower Chesapeake Bay. *Chesapeake Sci.*, **4**, 1–20.

Perkins, E. J., Bailey, M., and Williams, B. R. H. (1964). *H. M. S. O., U. K. A. E. A., P. G. Rept.*, **604 (cc)**, 58 pp.

Peterson, D. H., Conomos, T. J., Broenkow, W. W., and Scrivani, E. P. (1975). In *Estuarine Research*. Volume 1. L. E. Cronin (editor). Academic Press, New York. 738 pp.

Pomeroy, L. R., Haskin, H. H., and Ragotzkie, R. A. (1956). Observations on dinoflagellate blooms. *Limnol. Oceanogr.*, **1**, 54–60.

Pomeroy, L. R., Mathews, H. M., and Min, H. S. (1963). Excretion of phosphate and soluble organic phosphorus compounds by zooplankton. *Limnol. Oceanogr.*, **8**, 50–55.

Pomeroy, L. R., Smith, E. E., and Grant, C. M. (1965). The exchange of phosphate between estuarine water and sediments. *Limnol. Oceanogr.*, **10**, 167–172.

Pomeroy, L. R., Odum, E. P., Johannes, R. E., and Roffman, B. (1966). Flux of ^{32}P and ^{65}Zn through a saltmarsh ecosystem. In *Disposal of Radioactive Wastes into Seas, Oceans, and Surface Waters.* I. A. E. A., Vienna, 898 pp.

Pomeroy, L. R., Johannes, R. E., Odum, E. P., and Roffman, B. (1969). In *Symposium on Radioecology, Proc. 2nd Natl. Symp.*, D. J. Nelson and F. C. Evans (editors). Ann Arbor, Michigan. 774 pp.

Postma, H. (1954). Hydrology of the Dutch Wadden Sea. *Archs. Neerl. Zool.*, **10**, 405–511.

Reimold, R. J. (1965). An evaluation of inorganic phosphate concentrations of Canary Creek Marsh. *M. S. Thesis.* Univ. Of Delaware. 61 pp.

Rhoads, D. C., Tenore, K., and Browne, M. (1975). In *Estuarine Research.* Volume 1. L. E. Cronin (editor). Academic Press, New York. 738 pp.

Richards, F. A. (1965). In *Chemical Oceanography* Volume 2. 1st Edition. J. P. Riley and G. Skirrow (editors). Academic Press, London. 508 pp.

Richards, F. A., Cline, J. D., Broenkow, W. W., and Atkinson, L. P. (1965). Some Consequences of the decomposition of organic matter in Lake Nitinat, an anoxic fjord. *Limnol. Oceanogr.*, **10, Suppl.,** R181–R205.

Richards, F. A., Anderson, J. J., and Cline, J. D. (1971). Chemical and Physical observations in Golfo Dulce, an anoxic basin on the Pacific coast of Costa Rica. *Limnol. Oceanogr.*, **16**, 43–50.

Rigler, F. H. (1956). A tracer study of the phosphorus cycle in lake water. *Ecology*, **37**, 550–562.

Rigler, F. H. (1964). The phosphorus fractions and the turnover time of inorganic phosphorus in different types of lakes. *Limnol. Oceanogr.*, **9**, 511–518.

Riley, J. P. (1975). In *Chemical Oceanography*, Volume 3. Sec. ed. J. P. Riley and G. Skirrow (editors). Academic Press, London. 564 pp.

Riley, J. P., and Chester, R. (1971). *Introduction to Marine Chemistry.* Academic Press, London. 465 pp.

Rochford, D. J. (1951). Studies in Australian estuarine hydrology. I. Introductory and comparative features. *Aust. J. Mar. Freshwat. Res.*, **2**, 1–116.

Schink, D. R. (1967). Budget for dissolved silica in the Mediterranean Sea. *Geochim. Cosmochim. Acta.*, **31**, 987–999.

Shlopak, G. P. (1972). An evaluation of the total phosphorus concentration in the waters of two southern Delaware salt marshes. *M. S. Thesis.* Univ. of Delaware. 114 pp.

Sholkovitz, E. R. (1976). Flocculation of dissolved organic and inorganic matter during the mixing of river water and sea water. *Geochim. Cosmochim. Acta*, **40**, 831–845.

Siever, R. (1971). In *Handbook of Geochemistry.* K. H. Wedepohl (editor). Springer-Verlag, Berlin. 442 pp.

Smayda, T. J. (1957). Phytoplankton Studies in lower Narragansett Bay. *Limnol. Oceanogr.*, **2**, 342–359.

Spencer, C. P., (1975). In *Chemical Oceanography.* Volume 2. Sec. ed. J. P. Riley and G. Skirrow (editors). Academic Press, London. 647 pp.

Stéfansson, U., and Richards, F. A. (1963). Processes contributing to the nutrient distributions off the Columbia River and Strait of Juan de Fuca. *Limnol. Oceanogr.*, **8**, 394–410.

Stéfansson, V., Atkinson, L. P., and Bumpus, D. F. (1971). Seasonal studies of hydrographic properties and circulation of the North Carolina shelf and slope waters. *Deep-Sea Res.*, **18**, 383–420.

Stirling, H. P., and Wormald, A. P. (1977). Phosphate/sediment interaction in Tolo and Long Harbours, Hong Kong and its role in estuarine phosphorus availability. *Estuarine Coastal Mar. Sci.*, **5**, 631–642.

Taft, J. L., and Taylor, W. R. (1976). In *Estuarine Processes.* Volume 1. M. Wiley (editor). Academic Press, New York. 541 pp.

Topping, G. (1977). In *Marine Pollution.* R. Johnston (editor). Academic Press, London. 729 pp.

Turekian, K. K. (1971). In *Impingement of Man on the Oceans.* D. W. Hood (editor). Wiley-Interscience, New York. 738 pp.

Upchurch, J. B., Edzwald, J. K., and O'Melia, C. R. (1974). Phosphates in sediments of Pamlico Estuary. *Environ. Sci. Technol.*, **8**, 56–58.

Weiss, R. F. (1970). The solubility of nitrogen, oxygen and argon in water and sea water. *Deep-Sea Res.*, **17**, 721–735.

Whaley, R. C., Carpenter, J. H., and Baker, R. L., (1966). *Nutrient data summary 1964, 1965, 1966: Upper Chesapeake Bay.* Chesapeake Bay Institute Special Rept. No. 12., John Hopkins Univ.

Whitman, W. G. (1923). Preliminary experimental confirmation of the two-film theory of gas absorption. *Chem. Metall. Eng.*, **29**, 146–148.

Wood, D., and Caputi, R. (1966). Technical Rept. No. 988, U.S. Naval Radiological Defence Laboratory, San Franccsco, 14 pp.

Wyman, J., Scholander, P. F., Edwards, G. A., and Irving, L., (1952). On the stability of gas bubbles in Sea Water. *J. Mar. Res.*, **11**, 47–62.

Chemistry and Biogeochemistry of Estuaries
Edited by E. Olausson and I. Cato
Copyright © 1980 by John Wiley & Sons Ltd.

W. J. WOLFF
Research Institute for Nature Management,
Den Burg, Texel

8

Biotic Aspects of the Chemistry of Estuaries

                         W. J. Wolff

1 INTRODUCTION

In general estuaries contain more living organisms than any other part of the seas and oceans. For that reason organisms influence the chemistry of estuaries mostly stronger than is the case in other parts of the sea.

The biotic aspects of the chemistry of estuaries may be grouped into two main categories: biotic transports and biotic transformations. Biotic transports include all processes in which living organisms transport or are the cause of the transport of any material from one place in the estuary to another place or from the estuary to an area outside the estuary and *vice versa*. Biotic transformations comprise those processes in which organisms transform any material from one state into another state.

Since the author is more familiar with European estuaries than with those in other parts of the world, many examples to illustrate the phenomena mentioned in the text are drawn from research done in Europe. It is believed, however, that these examples will illustrate general principles.

Because the field to be covered is so wide, this chapter is restricted to illustrate the main lines of the biotic aspects of the chemistry of estuaries. Many details have been sacrificed in order to keep this chapter within reasonable limits.

2 BIOTIC TRANSPORTS

2.1 Transports in and out of estuaries

2.1.1 *Animals moving between estuaries and the sea*

Many species of animals move actively between the sea and its adjacent estuaries. Such movements have been reported to occur in seals, whales, fishes, crustaceans, molluscs, polychaetes, and other organisms. In many cases these movements may be considered as a kind of overflow during a population peak in either the sea or the estuary to the area where the population normally does not occur (Witte and Zijlstra, 1978). Such movements may be called dispersion. In other cases the movements between the sea and the estuary are integrated into the life-cycle of the organism and in these cases these movements take more the form of migrations. Among these migrating organisms two categories may be distinguished. The first group comprises animals spawning in estuaries, but living in the open sea as adults. The second group spawns in the open sea, lives in the estuaries as juveniles, and retreats to the open sea as adults.

Estuarine spawners are, for instance, anchovy (*Engraulus engrasicolus*) and coastal-spawning herring (*Clupea harengus*) around the North Sea. The herring race spawning before 1932 in the former Zuiderzee, The Netherlands, supplied the estuarine fisheries in that area on average with about 10,000 tons per year (Postuma and Rauck, 1970). This implies that every year the biomass in the Zuiderzee estuary temporarily showed an increase of the order of magnitude of 10 g wet organic matter per m^2. Such spawning migrations therefore may have a measurable influence on the material balance of an estuarine area.

The second group, which uses the estuary as a nursery, reaches the estuarine areas as larvae or juveniles. These live for some months or a few years within the estuary and then leave the area for the open sea. Examples of such species are menhaden (*Brevoortia tyrannus*), croaker (*Micropogon undulatus*), and white shrimp (*Penaeus fluviatilis*) in American estuaries and plaice (*Pleuronectes platessa*), sole (*Solea solea*), and brown shrimp (*Crangon crangon*) in European estuaries. Mullets (e.g. *Mugil cephalus*) occur on both sides of the Atlantic and elsewhere.

Creutzberg *et al.* (1978) report that in 1974, 1975, and 1977 respectively 53, 256, and 1033 million plaice larvae entered the estuarine Wadden Sea, The Netherlands, through one of the tidal inlets. The numbers of flounder (*Platichthys flesus*) larvae were of the same order of magnitude. This implies that as an order of magnitude this estuary is supplied with 1 larva per m^2. However, since these larvae concentrate on the intertidal flats the density on these flats should be of the order of magnitude of 3–4 larvae per m^2. Probably due to an extremely high mortality during the first days of their life Kuipers (1977) found only about 0.2 0-group plaice per m^2 on the Balgzand tidal flats of this estuary. Assuming an influx of 1 larva per m^2 and a wet weight of 0.1 g per larva it appears that the influx of larvae from the open sea to estuaries probably is not very important for the material budgets of estuaries.

In their third year the plaice leave the Wadden Sea. Before departure their density is of the order of magnitude of 0.01–0.1 fish per m^2 and their weight about 50 g (Kuipers, 1977). This means a loss of 0.5–5 g wet organic matter per m^2 for the tidal flats considered. In conclusion, the larval influx hardly influences the material budgets of estuaries, but the departure of the full-grown juveniles has a slight effect. This conclusion may be compared to the results of Nixon and Oviatt (1973) who found that juvenile fish were responsible for an import of about 1 g dry organic matter m^{-2} year^{-1} to a salt marsh cove in Massachusetts, and for an export of the same magnitude.

2.1.2 *Animals migrating between estuaries and fresh water*

Several species of fish spawn in fresh water and live as adults in estuaries or the sea. These species are known as anadromous fish. Their counterpart are the katadromous species which spawn in the sea and live as adults in fresh water. For both groups the estuary is a passage area, but it may also be a spawning area for the katadromous species or an area of residence for the anadromous organisms.

Well-known anadromous fish are salmon (*Salmo salar*), sturgeon (*Acipenser sturio*) and sea lamprey (*Petromyzon marinus*). Katadromous species are eel (*Anguilla anguilla, A. rostrata*) and Chinese mitten crab (*Eriocheir sinensis*).

Anadromous as well as katadromous species displace organic matter from fresh water to estuaries and the sea and *vice versa*. No data are available to estimate the effect on estuaries, but such effects are likely to be very small. The decrease or even extermination of anadromous fish species in several rivers (Lelek, 1976; Wolff, 1978) are an additional factor making this biotic transport of little importance.

2.1.3 *Birds*

Large numbers of birds are a characteristic of many estuaries. Ducks, geese, waders, gulls, and terns are often the most important species, but many other species may also occur. All species may feed on organisms living within the estuary: ducks and geese mainly on plants, the other species on animals. Birds may transport material in two ways. In the first place they may, by migration, transport organic matter in the form of their own body from an estuary to elsewhere or *vice versa*. Secondly, birds may transport the undigested parts of their food from the estuary to their high-tide roosts outside the estuary or from outside areas to the estuary when the latter functions as a resting area.

The biomass of birds in estuaries has been determined by Hulscher (1975), Swennen (1976), and Wolff *et al.* (1976). The values are of the order of magnitude of 0.1 g wet weight per m^2. Even if all birds left the estuary this biotic transport of organic matter would not be very important.

Van Haperen (1974) gives preliminary data on the transport by means of faeces. For a four-month period the salt-water lake Grevelingen he studied, received 0.03 g organic carbon per m^2 (shells excluded) from outside the area. Herring gulls (*Larus argentatus*) transported in the same period 0.06 g calcium carbonate in the form of shells from this estuarine lake to areas outside. Transports of all other compounds were estimated to be much smaller.

2.2 Transport within estuaries

2.2.1 *Horizontal transports*

The horizontal biotic transports within estuaries are connected to either the tidal cycle or to the cycle of the seasons.

Several species of fish actively migrate with the tides up and down the tidal flats. This is especially well-known for the plaice (*Pleuronectes platessa*) which was extensively studied by Kuipers (1975, 1977). The plaice larvae settle at the flats during high tides in spring. During their first tidal cycles they do not migrate, but stay back in shallow pools and wet sediments. Apparently very many young plaice die in these first days. After a few days, however, the young plaice develop a horizontal migration over the tidal flats to those parts of the tidal channels still containing water at low tide. Here the young plaice stay until the next flood tide permits them to visit the intertidal area again (Figure 1). Kuipers estimates that these animals removed annually 1.8–5.0 g dry organic matter m^{-2} from these flats. Similar behaviour has been demonstrated for flounder (*Platichthys flesus*).

Also other species of fish visit the intertidal area, but for these species it is not known whether they do so each tidal cycle and whether a large part of the population takes part in these movements. Regularly encountered on the flats are mullets (*Mugil ramada*), eel (*Anguilla anguilla*), herring (*Clupea harengus*) and whiting (*Merlangius merlangus*) (Wolff, unpublished data).

Not all species of estuarine fish show this behaviour. For instance, sole (*Solea solea*), although abundant in estuaries, has never been observed in intertidal areas.

Also invertebrates take part in daily tidal migrations in estuaries. This is for example the case in brown shrimp (*Crangon crangon*) (Hartsuyker, 1967) and common shore crab (*Carcinus maenas*) (Broekhuijsen, 1936; Breteler, 1975).

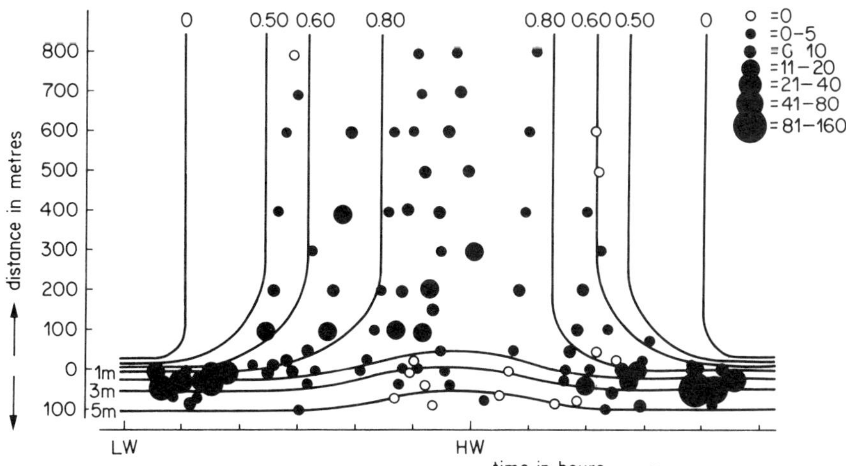

Figure 1. Distribution of I-group plaice during the tidal cycle on the tidal flats of Balgzand. The circles, which indicate one sample each, are plotted against the stage of the tide (abscissa) and the distance from the channel's edge (ordinate): curved lines are isopleths.

The net transport of material undoubtedly is from the flats to deeper water, since all species mentioned apparently come to the flats to feed.

Another horizontal transport occurs on an annual time-scale. Several invertebrates settle as larvae on the highest parts of the tidal flats, where they grow for some time until they migrate to lower tidal levels. Beukema (1973) demonstrated this convincingly for the small bivalve *Macoma balthica*. This species settles in spring on the flats just below high water mark. It grows there all summer. When the water temperatures drop, the animals crawl out of the sediment and are transported by the ebb currents to lower tidal levels as well as subtidal areas where they resettle and live for the rest of their life. In this way large areas of estuarine habitat are dependent on rather small high-level areas of flats for their supply with young *Macoma*. Similar behaviour has been demonstrated for the lugworm *Arenicola marina*.

2.2.2 *Vertical transport in the water*

Many organisms move actively up and down the water masses. Well-known are the diurnal movements of zooplankton and fishes connected with the day-night cycle. In estuaries these movements do not seem to have importance.

Bousfield (1955) postulated for barnacle larvae a behaviour in which the younger stages would live in the upper water layers whereas the older larvae tended to occur in the bottom layers. The estuarine residual circulation pattern then would tend to keep the larvae within the estuary. De Wolf (1973) however, demonstrated that the retention of barnacle larvae in estuaries can be explained by physical factors alone and that the behaviour described by Bousfield has no effect in strong and turbulent tidal currents. Grindley (1972) found that estuarine phytoplankton displays a day-night vertical migration, which in some cases stopped at surface water layers of lower salinity.

Active vertical movements of organisms do not seem to constitute an important form of biotic transport in estuaries.

2.2.3 *Transport from the water to the bottom*

Biotic transport from the water to the bottom occurs when free-swimming larvae of benthic organisms settle and start their bottom life. Although this may concern large quantities of larvae, the transport does not seem to be important on a weight basis.

More important is the action of filter-feeders. Phytoplankton and detritus suspended in estuarine waters move up and down the water column under the influence of the turbulence generated by the wind and the tides. This movement does not result in a net transport up or downwards. However, when filter-feeders on the bottom filter the suspended material out of the passing water mass when it is temporarily close to the bottom, a net downward transport is created (Wolff *et al.*, 1976). This downward transport under the influence of active filter-feeders should not be confused with the net downward transport through sinking under the influence of gravity. It was tentatively estimated by Wolff *et al.* (1976) that about $\frac{3}{4}$ of the food uptake of bottom-living filter-feeders in the salt-water lake Grevelingen, The Netherlands, could be attributed to the former process whereas the rest should be provided by the latter process.

2.2.4 *Transport from the bottom to the water*

Several mechanisms exist for biotic transport from the bottom to the water.

Firstly, eggs, sperm, or larvae of benthic animals or even adult individuals released into the water constitute a net transport from the bottom to the overlying water. De Wilde *et al.* (1978) estimated for the bivalve *Macoma balthica* that about 25% of the total biomass was released in the form of sexual products.

Similarly benthic organisms may release dissolved organic matter or other dissolved products. McRoy *et al.* (1972) estimated for the seagrass *Zostera marina* that this plant moved $0.066\,g$ phosphorus $m^{-2}\,day^{-1}$ from the bottom to the overlying water. A similar process has been described for the salt marsh cord grass *Spartina alterniflora* (Reimold, 1972).

Animals maintaining a flow of water through their burrows may cause chemical changes in the sediment. In a thin layer around the burrows the sediment becomes oxidized, whereas it may be assumed that this process also influences the exchange of dissolved compounds across the water sediment interface.

Rhoads and Young (1972) describe how benthic deposit-feeders may loosen the coherence of muddy sediments after which tidal currents may bring this sediment in suspension. Indeed, disappearance of the benthic fauna after an oil-spill resulted in stabilization of the sediment and clearer water (Rhoads, 1976). Such actions of animals may have a far-reaching effect on the transports between water and bottom.

2.2.5 *Transport within the bottom*

Many organisms move within the bottom. Well-known are the benthic diatoms moving to the surface layer of the sediment of tidal flats during low tide and moving towards slightly deeper sediment layers during high tide.

Also benthic animals move up and down in their burrows in relation to the tides (Vader, 1964) and in relation to seasonal temperature changes.

Bottom animals move also horizontally within the sediment, but no instances are known of any net horizontal biotic transport.

2.3 Concentration within estuaries

Organisms may actively concentrate organic matter in certain places. This is very clear for filter-feeding molluscs, such as oysters, mussels, and cockles. Verwey (1952) estimated that the filter-feeders of the Dutch western Wadden Sea were able to strain the entire volume of this estuary every two weeks. The suspended organic matter thus retained from the water is partly mineralized, partly rejected as pseudofaeces and partly changed into molluscan biomass. In this way the primary production of a large estuarine area and even of the adjacent open sea is concentrated and accumulated in the form of shellfish beds in much smaller areas. The oyster reefs in North American estuaries are the result of the same process.

3 BIOTIC TRANSFORMATIONS

3.1 Photosynthesis

3.1.1 *Introduction*

Photosynthesis is defined as the process in which plants or bacteria transform inorganic compounds into organic material with light as a source of energy. Together with chemosynthesis (Section 3.2) this process is also known as the primary production of organic matter. Secondary production is the transformation of plant organic material into animal organic material.

In photosynthesis CO_2 is used as an electron acceptor and a compound of the form H_2A is an electron donor. For most plants (vascular plants, mosses, algae) the electron donor is H_2O. Green and purple sulphur bacteria (*Chlorobiaceae* and *Chromatiaceae*), however, use H_2S as an electron donor. Other photosynthetic bacteria (*Athiorhodaceae*) use an organic compound in the same way.

The inorganic materials used in the process of photosynthesis by green plants are CO_2, H_2O, nitrogen, phosphorus, and a range of micronutrients. Some plants require special elements; diatoms for example need silica to build their frustules.

When H_2O functions as an electron donor, O_2 is liberated, whereas H_2S as a donor leads to the formation of sulphur.

As in other habitats green plants are the major phososynthesizers in estuaries and there exist no fundamental differences between the photosynthesis process in estuaries and that elsewhere. There are differences, however, in the share of the different groups of plants, ecologically as well as taxonomically. The major ecological groups of photosynthesizing primary produces in estuaries are suspended algae, bottom-living algae and vascular plants. The suspended algae are called phytoplankton, all plants living on the bottom phytobenthos. The latter category comprises unicellular and multicellular algae as well as seagrasses which are vascular plants. Other vascular plants occur in the salt marshes and mangrove forests which are partly terrestrial in nature.

The primary production process is measured in different ways, according to the group of plants studied. Hall and Moll (1975) described the various methods in use. In plankton studies as well as in work on benthic algae two methods are still widely used, viz. the oxygen method and the ^{14}C method. The first method basically employs three bottles containing water and plankton from the same sample. The first bottle is used to determine the oxygen content at the start of the measurement. The two other bottles are resuspended at the place where the sample came from and incubated for some time. One bottle is kept dark, the other is left clear. After incubation photosynthesis is determined from the oxygen changes. The difference between the initial bottle and the light one gives the net photosynthesis; initial bottle minus dark one gives respiration and light bottle minus dark bottle shows gross photosynthesis. Respiration is the amount of oxygen used by the plant in metabolic processes other than photosynthesis. In fact, part of results of gross photosynthesis are used for maintenance of the plant itself.

The ^{14}C method is based on measuring the amount of ^{14}C incorporated into the plant material some time after it has been added to the sample as radioactive bicarbonate. The results of the ^{14}C method are neither gross production nor net production of the O_2 method but somewhere in between. Currently it is believed that the ^{14}C method comes close to net production of the O_2 method.

3.1.2 *Phytoplankton*

Phytoplankton is a collective name for mainly unicellular algae suspended in the water. Usually it is subject to passive transport by all water movements. For that reason phytoplankton in estuaries with a short residence time of the water usually is a mixture of phytoplankton from the adjacent sea and that from the rivers discharging into that estuary. When the residence time of estuarine water is longer than a few weeks an autochthonous estuarine plankton may develop. In the first situation the estuary is an area where marine and freshwater plankton die off and where low primary

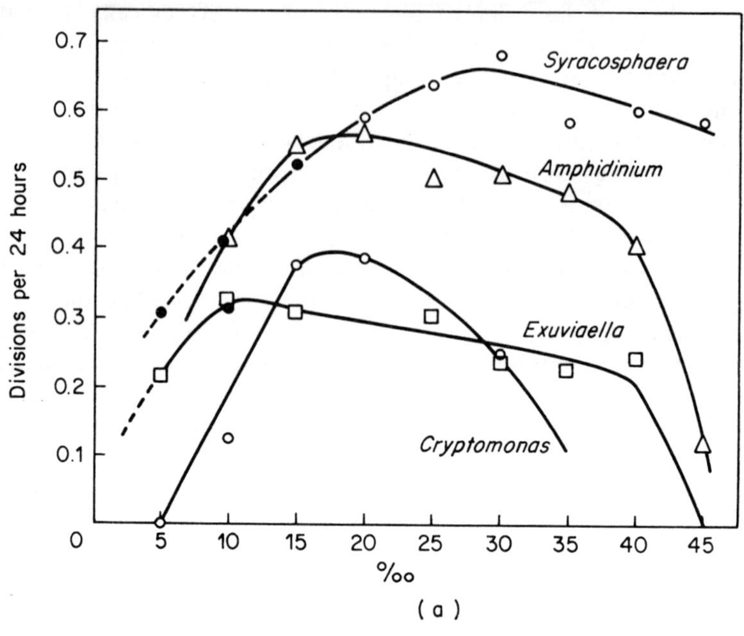
0.7
0.6
0.5
0.4
0.3
0.2
0.1
Divisions per 24 hours
Syracosphaera
Amphidinium
Exuviaella
Cryptomonas
0
5
10
15
20
25
30
35
40
45
‰
(a)

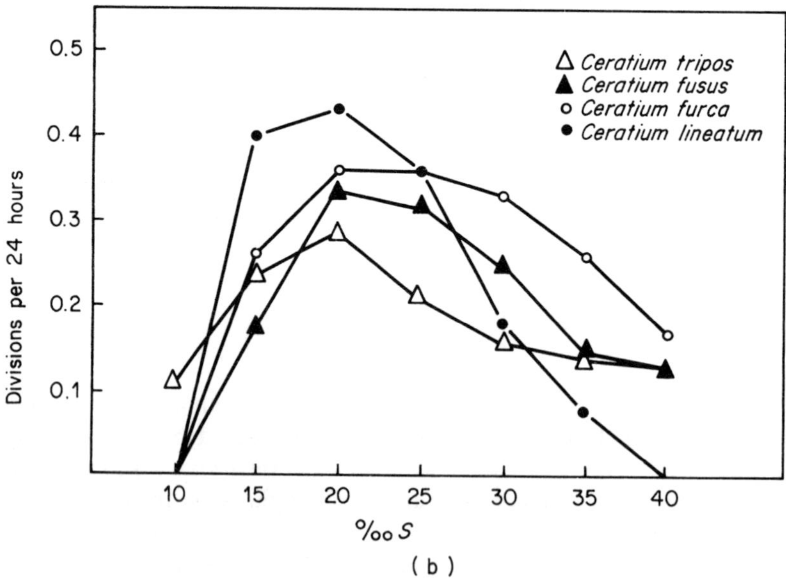
0.5
0.4
0.3
0.2
0.1
Divisions per 24 hours
Ceratium tripos
Ceratium fusus
Ceratium furca
Ceratium lineatum
10
15
20
25
30
35
40
‰ S
(b)

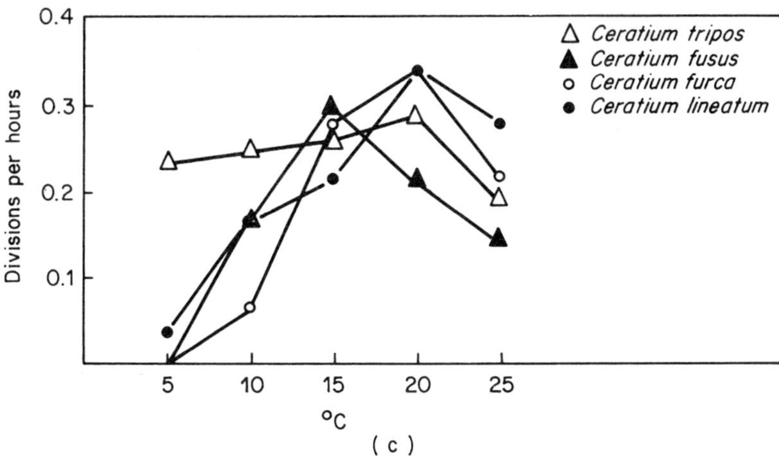

Figure 2. (a) Salinity-curves for the planktonic marine algae species *Syracosphaera carterae, Cryptomonas* sp., *Amphidinium* sp., and *Exuviaella baltica* (after Braarud, 1973). The division rates of *Ceratium* spp. (b) at various salinities, (c) at various temperatures (After Nordli, 1970). (Reproduced from E. J. Perkins, The Biology of Estuaries and Coastal Waters, Academic Press, 1976. Copyright by Academic Press Inc. (London) Ltd.)

production values occur. In the second situation all parts of the estuary may contain normally developing plankton and production values may be high all over the area.

These differences are based on the characteristics of the species constituting the phytoplankton. Most species show fairly wide limits of salinity between which growth and reproduction is possible (Braarud, 1951; Nordli, 1957; Vosjan and Siezen, 1968), although an optimum usually occurs at the salinity the species normally lives in (Figures 2, 3).

In estuaries with short residence times an area of intensive die-off may be expected to occur at intermediate salinities. Low primary production values may be expected to occur there as well. Vegter (1968) presents some measurements from the estuarine area in the SW-Netherlands confirming this expectation (Table 1), but his low production area coincides also with the high turbidity part of the estuary. Indeed, Cadée and Hegeman (1974) as well as Van der Hoek *et al.* (1979) explain the low phytoplankton production in the Ems estuary, The Netherlands, by the high turbidity in its brackish part.

Phytoplankton development in estuaries differs from that in coastal waters in that estuarine phytoplankton show a continuous high productivity between spring and autumn, whereas in coastal waters a spring peak usually is followed by a second peak in late summmr (Riley, 1967; Cadée and Hegeman, 1974). These patterns are caused by the interaction of temperature,

 W. J. Wolff

1.0
0.9
0.8
0.7
0.6
0.5
0.4
0.3
0.2
0.1
0
Relative production rate
Sea water dilutions
Na Cl media
0 2 4 6 8 10 12 14 16 18 20 22 24 26
‰ Salinity
(a)
1.0
0.9
0.8
0.7
0.6
0.5
0.4
0.3
0.2
0.1
Relative production rate
O₂
C - 14
0 10 20 30 40 50 60 70 80 90 100
‰ Salinity
(b)
1.0
0.9
0.8
0.7
0.6
0.5
0.4
0.3
0.2
0.1
Relative production rate
Sea water dilutions
Na Cl media
2 4 6 8 10 12 14 16 18 20 22 24 26
‰ Salinity
(c)

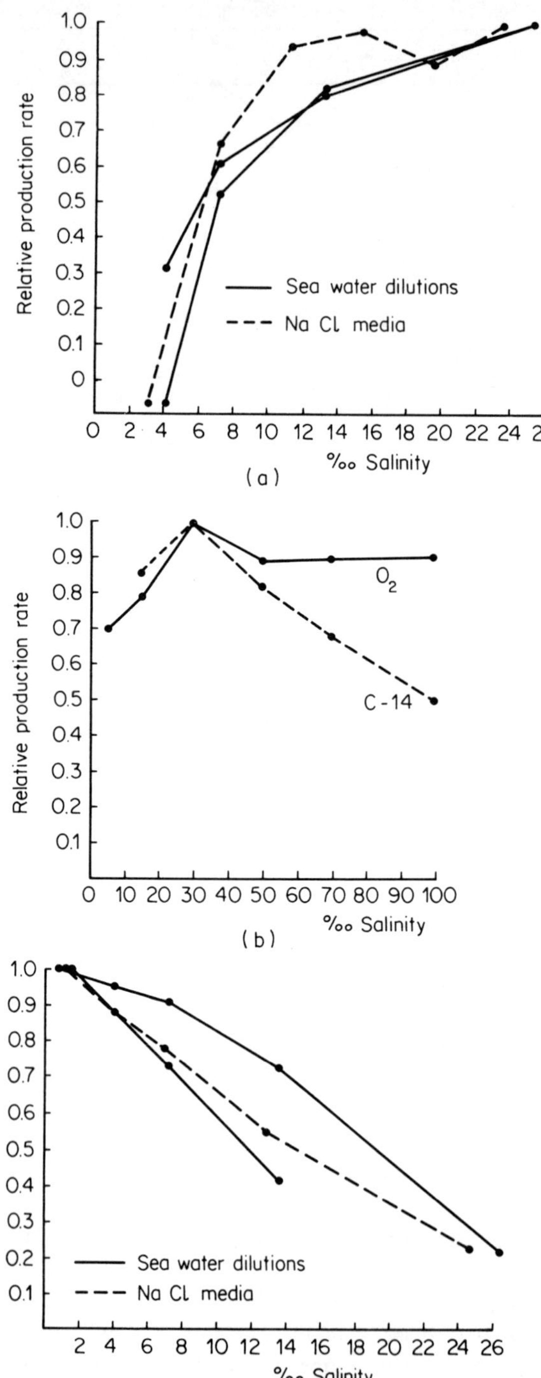

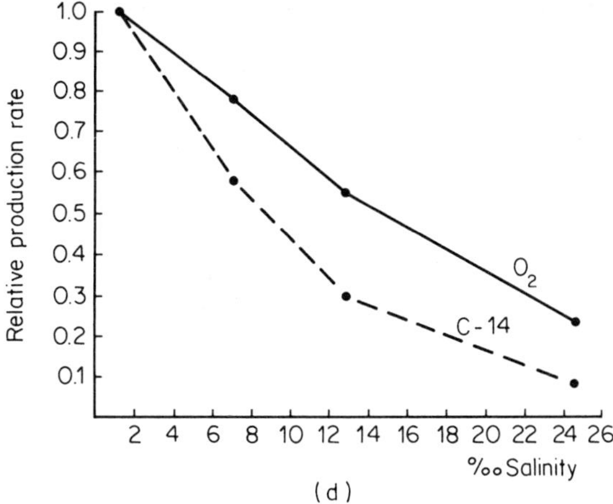

Figure 3. (a) Relative production rate of *Chlamy-domonas* at different salinities; measured by the oxygen method.
(b) Relative production rate of *Chlamydomonas* at different salinities; measured by the oxygen and ^{14}C methods.
(c) Relative production rate of *Scenedesmus* at different salinities; measured by the oxygen method.
(d) Relative production rate of *Scenedesmus* at different salinities; measured by the oxygen and ^{14}C methods. (After Vosjan and Siezen, 1968.)

light, nutrients, grazing, and mixing processes, all of which may cause limitation of phytoplankton growth. Temperature limitation does not seem to be very important in temperate estuaries since phytoplankton development may start in the middle of winter (Riley, 1967). Light limitation is very common in many estuaries and it may occur long before nutrients become limiting (Cadée and Hegeman, 1974). When nutrients become limiting, most often nitrogen compounds seem to be the chief limiting factor in coastal waters and estuaries (Ryther and Dunstan, 1971; Conover, 1975), but phosphate and silicate may be limiting as well, the latter for diatoms only

Table 1. Primary production of phytoplankton in the estuarine area of the S.W. Netherlands. After Vegter, 1968.

Salinity	16‰	3‰	Tidal fresh water	
Potential production	80	20	75	mg C.m^{-3} 2 hr^{-1}
In situ production	175	90	310	mg C.m^{-2} 2 hr^{-1}

Table 2. Phytoplankton primary production in some estuarine areas. All values expressed as grams carbon m^{-2} year^{-1}.

Area	Production	Reference	Method
Long Island Sound, USA	380	Riley, 1956	oxygen
St. Margaret's Bay, Canada	190	Platt, 1971	^{14}C
Cochin Backwater, India	124	Qasim, 1973	^{14}C
Loch Etive, Scotland	70	Wood *et al.*, 1973	^{14}C
Wadden Sea, Netherlands	100–200	Cadée and Hegeman, 1974	^{14}C
Ems estuary, Netherlands	13–55	Cadée and Hegeman, 1974	^{14}C
Grevelingen estuary, Netherlands	146–200	Vegter, 1977	^{14}C

(Van Bennekom *et al.*, 1974). It has to be stressed that different phytoplankton species require different amounts and ratios of nutrients. A nutrient supply limiting the growth of one species, may be plentiful for another. Grazing of phytoplankton by zooplankton as well as zoobenthos (Cadée and Hegeman, 1974; Wolff *et al.*, 1976) also may be limiting to phytoplankton primary production rates. Mixing processes may dilute estuarine phytoplankton blooms by transport to open sea.

Phytoplankton growth is one of the main biological processes affecting the chemistry of estuaries. Photosynthesis results in decrease of the concentrations of carbon dioxide, phosphate, ammonia, nitrate, and sometimes silicate as well as of a range of micronutrients. At the same time oxygen is increasing, often to supersaturation. During the night these processes are reversed. The decrease of the nutrients may go down to below the limit of detection, but general rules applying to all estuaries are difficult to formulate. The production of dissolved organic products will be discussed in Chapter 3.1.6.

To give some impression of the role of phytoplankton production in estuaries some data are given above. All data are expressed as grams carbon produced per square metre per year (Table 2).

3.1.3 *Benthic algae*

Benthic algae form a very diverse group ranging from unicellular diatoms and other species to multicellular plants of several meters length. Unicellular species are especially prominent on soft sediments, whereas the larger species usually need a solid substratum to keep the plant from being washed away. These solid substrates may be anything such as shells, pebbles, rocks, and wooden poles.

On soft sediments the unicellular algae often are visible as a brown or green surface layer of diatoms, euglenophytes, or blue-green algae. These tiny plants are able to migrate back and forth between the surface of the sediment and deeper layers. This migration is tied to the tidal cycle in such a way that the algae emerge when the tide is out.

The primary production of benthic micro algae has been measured on several occasions (Table 3). The values found do not appear to be very high, which rises the question what factors limit the production. Shading by higher plants and flooding during high tide certainly have effects (Pomeroy, 1959). This author also claims light inhibition during low tide, but this could not be confirmed by Cadée and Hegeman (1974a). Colijn and Van Buurt (1975) found no light inhibition in field populations, but a strong inhibition for an unialgal diatom culture. Admiraal (1977) demonstrated temperature to be limiting, whereas the same author (Admiraal, 1977a) showed that salinity only becomes limiting at very large departures from normal values. Admiraal (1977b) also investigated the possibility of limiting nutrient concentrations. He found phosphate concentrations below 0.3 μg at P-PO$_4$ per litre to be limiting for the diatom *Navicula arenaria*. Although phosphate concentrations in estuarine water and sediments usually are higher, phosphate might be limiting in dense growths of diatoms where molecular diffusion of phosphate may be too slow for replenishment. Such depletion of essential growth requirements by intertidal diatoms has been observed to occur for carbon dioxide by Pomeroy (1959) and Colijn (in Admiraal, 1977b). Such carbon limitation will be terminated by the next incoming flood, however. It seems that as a general rule sediment stability has much more influence as a limiting factor than has any other factor.

Table 3. Primary production of benthic microalgae in some estuaries. All values expressed as grams carbon m^{-2} year^{-1}.

Area	Production	Reference	Method
Sapelo Island, Georgia, USA	180	Pomeroy, 1959	O$_2$; net production
Danish fjords	116	Grøntved, 1960	^{14}C
Danish Wadden Sea	115–178	Grøntved, 1962 (compare Pamatmat, 1968)	^{14}C
False Bay, Washington, USA	143–226	Pamatmat, 1968	O$_2$
Ythan estuary, Scotland	31	Leach, 1970	^{14}C
Southern New England, USA	81	Marshall *et al.*, 1971	^{14}C
Dutch Wadden Sea	101±39	Cadée and Hegeman, 1974a	^{14}C
Delaware, USA	160	Gallagher and Daiber, 1974	^{14}C

The larger multicellular algae in estuaries are not yet well-investigated. Mann (1973) and Bunt (1975) review this topic and record some examples of very high production values. However, such algal vegetations usually are absent in estuaries and it seems that in most estuaries the primary production by benthic macroalgae is unimportant relative to other ecological groups.

3.1.4 *Seagrasses*

Seagrasses are vascular plants living submerged, at least temporarily, in coastal waters and estuaries. Species of the genera *Zostera*, *Thalassia*, and *Cymodocea* occur in higher salinities, whereas *Ruppia* and *Zannichellia* are characteristic for brackish water.

Seagrasses are rooted in the sediment and are able to take up chemical compounds from the bottom. Through leakage by the leaves such compounds, e.g. phosphate, are transported from the bottom to the overlying water (McRoy *et al.*, 1972; Section 2.2.4).

The biomass of seagrasses can be very high with maximum values of about 500 g C m^{-2}. Mann (1972) gives a review of the productivity of seagrasses and gives production values ranging from 58 to $1,500 \text{ g C m}^{-2} \text{ year}^{-1}$. Many of the values quoted by Mann (1972) are based on the assumption that annual production is twice the maximum biomass. Sand–Jensen (1975) mentions a ratio of 2.5, however. The highest production estimates originate mainly from tropical and subtropical waters. However, McRoy (1970) using ^{14}C methods and Barsdate *et al.* (1974) arrive at production values of about $1500 \text{ g C m}^{-2} \text{ year}^{-1}$ for the arctic Izembek Lagoon at the Behring Sea, using a daily turnover of 2 per cent of the biomass during 165 days per year. A production of twice the average biomass results in only about $900 \text{ g C m}^{-2} \text{ year}^{-1}$, however.

These high production values make seagrasses in several areas the top producers. For example, Barsdate *et al.* (1974) present data from which may be derived that over 90% of the primary production in Izembek Lagoon, Alaska, results from the growth of *Zostera*.

However, this high production is only for a small part consumed directly. Wood *et al.* (1969) mentions sea urchins, gastropod molluscs, fishes and sea turtles as grazers. Other grazers are ducks, geese, and sirenians. Nienhuis and Van Ierland (1978) quantified the consumption of *Zostera marina* by animals in Lake Grevelingen, The Netherlands, and estimated that about 4% of the total production was consumed directly. Birds took about 1% and the isopod *Idotea chelipes* about 3%. This means that the great majority of the seagrass will enter in detritus food chains. This is confirmed in several studies (Wood *et al.*, 1969; Mann, 1972; Barsdate *et al.*, 1974).

3.1.5 *Salt marshes and mangrove forests*

Estuaries are usually bordered by salt marshes or mangrove forests. The latter only grow in areas without frost, so they are restricted to tropical and subtropical climates. Salt marshes are mainly found north and south of the area of distribution of mangroves and range into arctic and antarctic environments.

Salt marshes and mangrove forests exist of plants of terrestrial origin, which have adapted to life in saline environments. Their photosynthesis process proceeds in the absence of above-ground water and largely stops when the plants are covered by a flooding tide.

Salt marshes and mangroves belong to the most productive environments in estuaries. Primary production values for American salt marshes range from 300 to 4,000 g dry organic matter m^{-2} year^{-1} (Keefe, 1972; Turner, 1976). For European marshes values between 11 and 1,100 g m^{-2} year^{-1} have been found (Wolff *et al.*, in press).

The primary production by mangroves probably is much higher, although detailed figures are lacking. Most of this production is in the form of wood. Leave biomass seems to average 1–2% of the total biomass. The annual leaf fall in three different studies was 470–730 g dry organic matter m^{-2} year^{-1} (Odum and Heald, 1975).

As for seagrasses relatively few of the biomass of salt marshes and mangroves is consumed directly by animals. The main consumers are insects, birds and terrestrial mammals. For a *Spartina-alterniflora* marsh in Georgia, USA, Teal (1962) estimated that 4.6% of the net production of *Spartina* was eaten by insects, whereas the remainder entered the detritus food chain. In a review Joenje and Wolff (1979) present some data on the marshes along the European Wadden Sea. Geese may take up to 80% of the standing crop. Rabbits and hares may consume up to 45–55% of primary production, but for some insects only 4.4% of the primary production was found.

The major part of the primary production of salt marshes and mangroves apparently enters in most areas the detritus food chains. Until a few years ago it was believed that this detritus was transported from the marshes and mangroves to coastal waters, but recent work (Haines, 1977; Woodwell *et al.*, 1977; Wolff *et al.*, in press) has demonstrated that at least in some salt marshes the net transport of detritus is from the estuary to the marsh.

3.1.6 *Extracellular production*

All kinds of aquatic plants may release dissolved organic products. Unfortunately the measurement of the amounts released is faced with several pitfalls, as several artifacts may influence the results (Steemann Nielsen, 1975).

The amounts released seem to vary widely. Values reported range from nearly zero to about 40% of the amount of carbon assimilated. Healthy, rapidly photosynthesizing natural phytoplankton excretes about 5–10% of the total production, but older phytoplankton populations may release much higher amounts, constituting 22–38% of the assimilated carbon (Hellebust, 1967). Thomas (1971) mentions an excretion of 7% for estuarine phytoplankton in Georgia estuaries. Cadée and Hegeman (1974a) found that estuarine benthic microalgae excreted 0.1–9.5% of the daily primary production. Since the high excretion rates coincided with low phytoplankton production the annual excretion amounted to approximately 1% of annual production. For a macroalga (*Fucus vesiculosus*) occurring in estuaries Sieburth and Jensen (1969) report an excretion of 40% of the primary production. Gallagher *et al.* (1976) report that *Spartina alterniflora* vegetations in Georgia, USA, produced 61 kg dissolved organic matter per ha per year. This is only 1 or 2% of the primary production in this area.

3.1.7 *Photosynthetic bacteria*

In most estuaries photosynthetic bacteria are not very important as primary producers, but they are interesting enough to warrant a short discussion. The sulphur bacteria are obligate anaerobes which means that they grow only in habitats without oxygen. In estuaries they occur in anaerobic stagnant water masses as well as in the reduced part of the sediment. Since they need light as a source of energy these photosynthetic bacteria occur as close to the light as possible. In sediments this is at the boundary of oxidized and reduced layers. Blackburn *et al.* (1975) demonstrated that H_2S oxidation by photosynthetic bacteria in estuarine sediments is restricted to the top three millimetre of the sediment. In such sediments a purple coloured layer of sulphur bacteria may be discovered when the green upper layer of algae is scratched away.

As biotic transformers photosynthetic bacteria are usually unimportant. No exact figures are available for estuaries. For Japanese fresh water lakes Takahashi and Ichimura (1968) report that these bacteria are responsible for 3–5 per cent of the total primary production in most lakes, although values up to 25 per cent were found for lakes rich in H_2S.

3.2 Chemosynthesis

Chemosynthesis occurs only among bacteria. These chemosynthetic bacteria reduce CO_2 by means of energy obtained by the oxidation of simple inorganic compounds. Examples are the oxidation of sulphide to sulphur and/or sulphate, ammonia to nitrite and nitrite to nitrate.

Several of these bacteria are important in geochemical cycles, e.g. those of nitrogen and sulphur. In the nitrogen cycle *Nitrosomonas* converts ammonia to nitrite and *Nitrobacter* nitrite to nitrate. In general the ammonia results from the mineralization of organic matter and therefore these bacteria are often considered to be decomposers, but strictly speaking they are autotrophs.

In the sulphur cycle several colourless sulphur bacteria may take part. Especially the species of the genus *Thiobacillus* are fairly well known. All species are obligate aerobes, except for *T. denitrificans* and *Thiomicrospira denitrificans* (Timmer-ten Hoor, 1975), which use nitrate as a terminal electron acceptor. The various species oxidize either sulphide, sulphur or thiosulphate to sulphate. These sulphur bacteria are especially important in the top few centimetres of estuarine sediments. Here their activity results in a form of primary production which is usually neglected. Since this primary production is dependent on the sulfide production after the mineralization process of sulphate reduction by *Desulfovibrio* it has been called 'secondary primary production'. This form of primary production rescues energy which otherwise would be lost to the ecosystem. Its importance in estuaries is not well-known.

3.3 The role of animals

3.3.1 Introduction

Animals are heterotrophic organisms, i.e. they are dependent on other organisms as a source of food supplying them with energy and material for growth and maintenance. Many animals consume plants. Because of their place in the food web they are known as primary consumers or secondary producers. Other animals eating primary consumers are known either as secondary consumers or as tertiary producers, and so on. However, this picture is too straightforward, since in estuarine food webs the interrelationships are mostly diffuse. One of the few general principles is that relatively large animals tend to eat relatively small organisms.

3.3.2 Zooplankton

The relatively large summer zooplankton population in estuaries probably is the major difference between estuarine and oceanic waters. This summer peak results from a strong development of the holoplankton, i.e. the organisms living all their life suspended in the water, as well as from an influx of meroplankton, i.e. the pelagic larvae of bottom animals and the pelagic sexual stages of hydroids and jellyfishes.

Estuarine zooplankton are considered to feed mainly on planktonic algae, although planktonic predators do occur. Until recently bacteria seemed less important as a food source, mainly because the filtering apparatus of several common plankton animals is too coarse to retain bacteria. However, aggregates of detritus and bacteria have been demonstrated to serve as a source of food for estuarine copepods (Heinle *et al.*, 1977). Planktonic ciliates appeared to be an important link in the food chain between bacteria and these planktonic copepods (Berk *et al.*, 1977).

Zooplankton in estuaries probably will not consume more than 50–60% of the phytoplankton production (Riley, 1967). This is partly due to the important role of filter-feeding zoobenthos in estuaries and partly to the fact the zooplankton development in spring cannot keep pace with the phytoplankton production.

Production figures for estuarine zooplankton are rather scarce. Heinle (1966) records $77 \, \text{mg} \, \text{C} \, \text{m}^{-2} \, \text{day}^{-1}$ for the copepod *Acartia tonsa* in the period July–September. Its daily production: biomass ratio was 0.50.

3.3.3 *Zoobenthos*

In general the zoobenthos is in estuaries more important than either in the open sea or in fresh water. This is due to the shallow nature of estuaries as well as to the tidal currents causing high turbulence. These two factors ensure that in estuaries a higher proportion of the phytoplankton is available to the zoobenthos than is the case in other water bodies (Wolff *et al.*, 1976a).

For methodological reasons three size categories of zoobenthos are distinguished. Microbenthos are animals smaller than about $100 \, \mu\text{m}$, macrobenthos are species retained by a sieve with meshes of 1 mm diameter and meiobenthos are the animals in between.

The microzoobenthos comprises especially the ciliated Protozoa. These are unicellular animals living aerobically as well as anaerobically in the sediments. Various species of bacteria are important as their food.

The meiozoobenthos consists of such animal groups as nematodes, turbellarians, gastrotrichs, harpacticid copepods, and oligochaetes. Also in this group anaerobic metabolism occurs although aerobically living species are more numerous. Bacteria as well as benthic algae seem to be important food categories.

The macrozoobenthos includes especially molluscs, polychaetes, and crustaceans. All species are dependent on oxygen, at least temporarily, for their metabolism.

The biomass of the various groups diverges widely, the macrobenthos usually showing much larger biomass values than either the meio- or the microbenthos. However, this is not reflected by their role in the community since the metabolic rates of micro- and meiobenthic animals are considerably larger than those of the macrobenthic species. Vernberg and Coull (1974) show for three different stations studied by Fenchel (1969) that each of the three benthic size groups is able to play the dominating role in benthic metabolism. However, under strongly tidal conditions the macrobenthos and especially the suspension-feeding species tend to become the most important species.

Macrobenthic biomasses locally may reach values up to 2,000 g ash-free dry weight m^{-2} and more, but as an average for a whole estuary values in the range 10–25 g m^{-2} have been found (Wolff, in press). Similarly, production values can be high as well. Individual species may reach values up to 200–300 g ash-free dry weight m^{-2} $year^{-1}$, but this occurs only in local aggregations of animals, e.g. mussel-beds or oyster reefs. Average production values for estuaries are up to about 50 g ash-free dry weight m^{-2} $year^{-1}$ (Wolff, in press).

Such high secondary production values are chiefly based on suspension feeding species, such as oysters, mussels, and cockles. These species utilize phytoplankton and possibly detritic aggregates as food, brought to them by the tidal currents. The deposit-feeding species which take benthic algae and possibly also detritus, show less impressive production rates.

The production of meio- and microbenthos in estuaries is not well-known. Since all species are deposit-feeders they probably cannot match the high production figures of filter-feeding macrobenthos.

With an assumed food chain efficiency of 10% estuarine filter-feeding benthos will need an impressive share of the primary production. For the Grevelingen estuary, The Netherlands, Wolff (1977) gives a benthos production of 30 g C m^{-2} $year^{-1}$, indicating a consumption of about 300 g C m^{-2} $year^{-1}$, and a primary production of about 160–220 g C m^{-2} $year^{-1}$. This means that either the food chain efficiency of the benthos will be much higher than 10% or that detritus from sources outside the estuary plays a significant role as food.

Another effect of benthic organisms on the chemistry of estuaries is caused by their reworking of the sediment. By digging burrows, and by eating sediment which is afterwards defaecated the top layer of the sediment is loosened and turned upside down. The thickness of this annually re-worked top layer varies according to type of organism, sediment, and locality, but figures of 10–30 cm per year are not uncommon (Cadée, 1976). Some larger figures do not seem to be completely trustworthy. This activity of benthic organisms has important effects on the chemistry of the overlying water as well as on the turbidity of the water through resuspension of mud (Rhoads, 1974).

3.3.4 *Fishes, birds, and mammals*

With the exception of the mullets, which take benthic algae and meiofauna, all fishes are predators of either bottom fauna or pelagic animals. Kuipers (1977) estimated that young plaice (*Pleuronectes platessa*) took 1.7–5.0 g ash-free dry weight m^{-2} year^{-1} of the benthic fauna of a tidal flat area in the Wadden Sea, The Netherlands. Gobiid fishes may consume another 1.4 g m^{-2} year^{-1} in the same area (Van Beek, 1976). These figures are important for the benthic invertebrates, but will hardly result in an influence on the chemistry of the area. Data from other estuaries lead to the same conclusion.

Bird predation is of the same order of magnitude. Consumption of benthic invertebrates by birds resulted in losses of 3–37% of the benthic production in several areas (Nilsson, 1969; Hibbert, 1976; Swennen, 1976; Wolff *et al.*, 1976). Since the high value was found for a small area only, it may be assumed that bird predation also has no measurable impact on the chemistry of estuaries.

It is believed that the same is true for the consumption by marine mammals, e.g. manatees, seals, and dolphins. Swennen (1976) presents some data for the Harbour Seal in the Dutch Wadden Sea from which may be derived that these seals take less than 0.5 g ash-free dry weight of fishes m^{-2} year^{-1}.

3.4 The role of detritus

Detritus has been defined in various ways. Here it is all particulate dead organic matter still containing chemical energy which may be used by organisms. This excludes dissolved organic matter, although it has to be admitted that usually the difference between detritus and dissolved material is based on the pore size of the filters employed. In practice therefore detritus has a minimum size of about 0.20–0.45 μm.

Detritus is usually more abundant than living organic matter in sea and freshwater (Fenchel and Jørgensen, 1977). Manuels and Postma (1974) show for the estuarine Dutch Wadden Sea that on average only 5% of the particulate organic matter is contained in living organisms. In summer, up to 80% may locally be occurring in living material, but the average in summer is only 20% and in winter less than 0.1%. In nearshore areas, on tidal watersheds, and in the brackish Ems estuary the share of detritus is even higher. The concentrations of detritus in estuarine waters range from 0.1 to over 125 mg l^{-1} (Fenchel and Jørgensen, 1977; Manuels and Postma, 1974). Estuarine sediments also show very different amounts of organic matter, e.g. from less than 0.1% in sands (Wolff, 1973) up to nearly 100% in sapropels and peat.

Detritus derives from several sources, e.g. dead animals, exuviae of crustaceans, faecal pellets of zooplankton and zoobenthos and dead plants. Especially macrophytes seem to be important contributors to the detritus pool. In many cases 90% and more of the macrophyte primary production ends up as detritus (Mann, 1972; Nienhuis and Van Ierland, 1978).

This particulate plant material is colonized by bacteria and fungi, which use the detritus as a source of energy and carbon for their metabolism, while conserving the nitrogen and other nutrients. Newell (1965) studied the consumption of fine-grained sediment, containing detritus by the mud snail *Hydrobia ulvae*. Prior to ingestion this material contained about 10% organic carbon and 0.2–0.3% nitrogen. The faeces of the animal contained only 0.02% of nitrogen, but three days later the nitrogen content was 1.7%. This material was fed again to the snails and the second set of faeces also had a nitrogen content less than 0.1%. After three days it had increased again to 1.5%. Apparently, microorganisms improve the quality of dead organic matter and animal consumers mainly strip off the microorganisms again and again. This continuing process results in a decrease of the size of the material and finally may lead to its disappearance. Kofoed (1975) fed sterile, ^{14}C-labelled hay particles, similar particles on which bacteria were growing, and pure ^{14}C-labelled bacteria to the snail *Hydrobia ventrosa*. The assimilation efficiencies for the three types of food were 34, 56, and 70%, respectively, whereas 15, 34, and 33% of the assimilated material was used for growth.

It is now fairly well accepted that detritus feeders mainly use the bacterial cell material of the bacteria–detritus aggregates (Fenchel and Jørgensen, 1977). Such detritus feeders are found among the Protozoa, Mollusca, Nematoda, Oligochaeta, Polychaeta, Crustacea, and fishes. Well-known fishes capable of digesting bacteria from detritus are mullets and the milkfish *Chanos chanos*. In fact, most living animals will be able to digest bacteria, but the typical detritus-feeders have developed some mechanism to concentrate these tiny sources of food. Whether a digestive apparatus comparable to that of ruminant mammals has evolved in aquatic invertebrates is still unclear (Fenchel and Jørgensen, 1977).

Detritus plays an important role in aquatic ecosystems by stabilizing the source of food created by seasonally fluctuating primary production.

3.5 The role of dissolved organic matter

In sea and fresh water dissolved organic matter usually is the largest pool of organic matter present. In estuarine waters particulate dead organic matter may be of a comparable importance, both categories showing concentrations in the range 1–40 mg l^{-1}. Interstitial waters of sediments may contain much higher concentrations of dissolved organic matter. The exact nature of dissolved organic matter is largely unknown, although sugars, amino acids, and acetate have been demonstrated to occur (Fenchel and Jørgensen, 1977; Sepers, 1977).

Sources of dissolved organic matter are manifold. Many plants excrete part of the assimilated carbon as dissolved organic matter. Estimates of this excretion range between 5 and 40% depending on species, age, and circumstances (compare Section 3.1.6). Autolysis of dead organisms is another important source. Also animals may excrete a large part of assimilated food as dissolved organic matter, as do actively metabolizing bacteria (Fenchel and Jørgensen, 1977).

Part of the organic matter no doubt is nearly undigestable by hetertrophic organisms, but another part is readily used. This results in a very rapid turn-over and very low concentrations. Because of these extremely low concentrations our knowledge of the fate of such substances is still very limited.

It seems that the capacity for uptake of dissolved organic matter is widely spread among marine invertebrates. It has been demonstrated to occur among Protozoa, Porifera, Coelenterata, Rhynchoela, Sipunculoidea, Bryozoa, Chaetognatha, Annelida, Mollusca, Echinodermata, Hemichordata, and Pogonophora, but it seems to be absent among Crustacea. An important nutritional role of dissolved organic matter is only likely to occur in the marine Pogonophora, but estuarine animals seem to rely largely on particulate organic matter. The main reasons for this conclusion are that all estuarine animals have well-developed systems for digestion of particulate organic matter, that uptake of organic matter seems incompatible with osmoregulation, which process occurs among many estuarine animals and that aquatic animals will be outcompeted by bacteria when sharing the same limited resource of readily available dissolved organic matter (Sepers, 1977; Fenchel and Jørgensen, 1977).

The conclusion is that dissolved organic matter will be mainly metabolized by bacteria, although some unicellular algae under special circumstances may live heterotrophically on high concentrations. The important role of bacteria is supported by many observations reviewed by Sepers (1977) and Fenchel and Jørgensen (1977). Crawford *et al.* (1974) demonstrated for an American estuary that bacterial production based on the uptake of amino acids alone amounts to 10% of the algal production during summer. This figure does not include the processes within the sediment, although it seems that bacterial mineralization in the sediment is much more intensive than in the water of estuaries (Vosjan and Olanczuk-Neyman, 1977).

3.6 Decomposition and mineralization

Autotrophic organisms, viz. photosynthetic plants and bacteria as well as chemosynthetic bacteria, produce organic matter continuously. This organic matter is used as a source of energy by all heterotrophic organisms, viz. animals and the other groups of bacteria. Also respiration by autotrophic organisms is a heterotrophic process. The respiration of all these organisms is responsible for the breakdown or decomposition of organic material into much simpler compounds such as monosaccharides, simple amino acids and other organic acids and eventually into inorganic compounds such as ammonia and phosphate. In a way the chemosynthetic bacteria (Chapter 3.2.) may also be considered as belonging to the large group of decomposers, since they use the oxidation of the inorganic products of the former group as a source of energy. The transformation of organic material into inorganic compounds is also known as mineralization. Mineralization may be considered as the counterpart of photosynthesis, since it transforms organic matter again into CO_2.

The chapters 3.4 and 3.5 already presented data indicating that bacteria play a major role in mineralization of detritus and dissolved organic matter. A reliable quantification of the share of various groups of estuarine organisms in mineralization processes is still far ahead. Pamatmat (1968) used oxygen consumption to estimate the respiration of benthic organisms on an intertidal flat in False Bay, Washington, USA. He estimated that macrofaunal organisms were responsible for 8–34% of the total respiration of the bottom organisms. Benthic algae respired another 32–73%, and bacteria plus micro- and meiofauna were responsible for only 17–54% of the total respiration.

However, the oxygen consumption method does not represent total community metabolism since anaerobic bacteria and some animals living in deeper sediment layers use respiration processes not dependent on oxygen (Pamatmat, 1977). However, Jørgensen (1977) demonstrates that 90% of the sulphide produced by anaerobic sulphate reduction escapes to the surface of the sediment, where it will be included in the measurement of oxygen uptake. A promising method is measurement of the activity of the respiratory electron transport system by means of 2,3,5-triphenyl-tetrazolium chloride, although this method has also its limitations (Packard, 1971; Pamatmat, 1977; Olanczuk-Neyman and Vosjan, 1977). Vosjan and Olanczuk-Neyman (1977) found with the latter method that the rate of bacterial respiration in the sediment was one or two orders of magnitude larger than that in the overlying water of the estuarine Dutch Wadden Sea and that 90% of the mineralization in the sediment took place in the anaerobic layers deeper than 1 cm. Jørgensen (1977) calculated for the Limfjorden, Denmark, that 53% of the mineralization is due to anaerobic sulphate reduction. Similarly, Jørgensen and Fenchel (1974) found in a laboratory model of a marine sediment that more than 50% of the organic matter was mineralized by bacterial sulphate reduction. It is concluded that bacterial sulphate reduction plays a major role in the mineralization process in estuarine areas.

These and other data point to a quantitatively much more important role of bacteria in estuarine mineralization processes than may be derived from Pamatmat's (1968) data.

In most estuaries aerobic respiration will take place in the top layer of the sediment and in the overlying watermasses. Only in stagnant and/or polluted estuaries anaerobic processes may occur in the water. Aerobic respiration is dependent on the presence of oxygen, which functions as an electron acceptor. Oxygen removed in respiration processes is replenished by exchange with the atmosphere, by advection of oxygen-rich water and by photosynthesis of plants. The top layer of the sediment is supplied with oxygen by molecular diffusion, but also by water movements caused by hydrostatic pressure changes under the influence of waves, currents and tides, and by the activity of burrowing animals. The end products of aerobic respiration are CO_2 and H_2O.

Oxygen respiration occurs among the majority of animals and plants, and also among aerobic bacteria. Oxygen respiration releases more energy than anaerobic respiration but aerobic estuarine organisms are often confronted with conditions in which oxygen is not available to them. Several animal species switch over to anaerobic respiration during low tide, when their gills cannot function due to the absence of sufficient free water. In sich situations fumarate functions as a electron acceptor and succinate will be formed. When with the next incoming flood oxygen becomes available again, the succinate will be oxidized. Such adaptations have been demonstrated to occur in mussels and lugworms (Zebe, 1975; De Zwaan *et al.*, 1976).

Bacteria may also switch over to anaerobic respiration processes. One possibility is fermentation in which an organic compound, e.g. pyruvate, acts as an electron acceptor. In fermentation the electron acceptor thus is provided by the organic substrate itself. The process results in the formation of lactate or other simple organic compounds. Another possibility is nitrate respiration in which nitrate acts as a electron acceptor. The end product is gaseous nitrogen. The process is known as denitrification.

Aerobic mineralization is the major process for plants and animals in estuaries, but its importance to the chemistry of estuaries is possibly less than that of the anaerobic mineralization. The magnitude of the aerobic mineralization will in general be limited by the supply of oxygen, since organic matter in estuaries usually is overabundant. The supply of oxygen on its turn will be dependent on the exchange and flow rates of water, which will be closely related to the tidal amplitude.

Anaerobic respiration occurs in estuaries in those places where oxygen is absent. In all estuaries such conditions are found in the deeper sediment layers, but in some polluted estuaries or in estuaries with little water movement these circumstances also may occur in the water overlying the sediment. Temperature and salinity stratifications may play a role to cause such conditions.

In anaerobic conditions the facultative anaerobic bacteria discussed above occur also. Exclusively occurring under anaerobic conditions are the obligate anaerobic bacteria. Among these are the sulphate reducing bacteria and the methane forming bacteria. Also micro- and meiofauna organisms, e.g. Protozoa, Turbellaria, Nematoda and Gastrotricha, can live as anaerobes (Fenchel and Riedl, 1970; Boaden, 1974).

Contrary to the situation in fresh water sulphate occurs in relatively large quantities in sea and estuarine water. Sulphate reducing bacteria therefore occupy a prominent place in anaerobic sediment layers in estuaries. Sulphate reduction results in sulphide formation of which a part produces FeS, and another larger part escapes to the surface (Jørgensen, 1977). The black colour of many estuarine sediments below the first few milli- or centimetres is a demonstration of the occurrence of bacterial sulphate reduction. Usually the sulphate reducing bacteria are limited in their activity by their supply of organic compounds. Addition of lactate produces a higher rate of reduction (Vosjan, 1974). When in spite of the relatively high concentration of sulphate in estuarine sediments the sulphate supply becomes exhausted before the organic compounds are finished, methane producing bacteria may take over. Since the sulphate concentration is linearly correlated with salinity methane production will especially occur in low salinities. Also in sediment layers below the sulphate reduction zone methane formation may occur.

4 MAIN BIOTIC PROCESSES AFFECTING THE CHEMISTRY OF ESTUARIES

In the preceding chapters it has been demonstrated that organisms may influence the chemistry of estuaries in a multitude of different ways. Quantitatively many of these influences are rather unimportant and it is our aim to select here only the more important biotic processes in estuaries.

From Chapter 2 it may be concluded that in general biotic transports are rather unimportant. Only the concentration of particulate organic matter by filter-feeding benthos may deserve attention.

From Chapter 3.1, it may be concluded that photosynthesis certainly is a very important process and among the photosynthetic organisms unicellular algae, seagrasses, and saltmarsh and mangrove plants compete for the top position. The outcome depends on local topography, tidal conditions and climate. Photosynthetic and chemosynthetic bacteria (Chapter 3.2.) seem rather unimportant as producers of organic matter.

Zooplankton (Chapter 3.3.1) and zoobenthos (Chapter 3.3.2.) both may be important as consumers. Especially in estuaries with a large tidal amplitude filter-feeding benthic organisms may occupy a prominent place. Vertebrates have hardly any importance for the chemistry of estuaries (Chapter 3.3.3.).

Mineralization processes are the counterpart of photosynthesis and chemosynthesis, also quantitatively. Although all organisms take part in these processes, the bacteria appear to play the largest role. In estuaries especially the anaerobic sediment layers seem to be important as places of mineralization (Chapter 3.6).

As a connection between photosynthesis, heterotrophic consumption and mineralization dissolved and particulate dead organic matter (detritus) appear to play important roles (Chapter 3.4 and 3.5). Understanding of the behaviour of these compartments of the ecosystem may provide a key for the understanding of the complete ecosystem.

Due to various abiotic processes estuaries often import considerable quantities of particulate organic matter from neighbouring ecosystems (Wolff, 1977). This causes mineralization in estuaries to dominate over photosynthesis and may lead to undersaturation with regard to oxygen.

A biotic process which is difficult to localize with one particular ecological group is nitrogen fixation. It appears that atmospheric or dissolved nitrogen may be fixed by a range of microorganisms, such as autotrophic blue-green algae, chemosynthetic bacteria, and aerobic as well as anaerobic heterotrophic bacteria. The process may be of considerable importance for the estuarine nitrogen budget and more so because denitrifying bacteria liberate nitrogen. The widespread occurrence of nitrogen fixation may be related to the assumed nitrogen limitation of photosynthesis in coastal waters and estuaries (Ryther and Dunstan, 1971).

5 REFERENCES

Admiraal, W. (1977). Influence of light and temperature on the growth rate of estuarine benthic diatoms in culture. *Mar. Biol.*, **39**, 1–9.

Admiraal, W. (1977a). Salinity tolerance of benthic estuarine diatoms as tested with a rapid polarographic measurement of photosynthesis. *Mar. Biol.*, **39**, 11–18.

Admiraal, W. (1977b). Influence of various concentrations of orthophosphate on the division rate of an estuarine benthic diatom, Navicula arenaria, in culture. *Mar. Biol.*, **42**, 1–8.

Barsdate, R. J., Nebert, M., and McRoy, C. P. (1974). Lagoon contributions to sediments and water of the Bering Sea. In: D. W. Hood & E. J. Kelley (eds.). *Oceanography of the Bering Sea*, pp. 553–576.

Beek, F. A. van (1976). Aantallen, groei, produktie en voedselopname van de zandgrondel (P. minutus) en de wadgrondel (P. microps.) op het Balgzand. *Rapp. Versl. Ned. Inst. Onderzoek Zee*, 1976–9.

Bennekom, A. J. van, Krijgsman-van Hartingsveld, E., van der Veer, G. C. M., and van Voorst, H. J. F. (1974). The seasonal cycles of reactive silicate and suspended diatoms in the Dutch Wadden Sea. *Neth. J. Sea Res.*, **8**, 174–207.

Berk, S. G., Brownlee, D. C., Heinle, D. R., Kling, H. J., and Colwell, R. R. (1977). Ciliates as a food source for marine planktonic copepods. *Microb. Ecol.* **4**, 27–40.

Beukema, J. J. (1973). Migration and secondary spatfall of Macoma balthica (L.) in the western part of the Wadden Sea. *Neth. J. Zool.*, **23**, 356–357.

Blackburn, T. H., Kleiber, P., and Fenchel, T. (1975). Photosynthetic sulphide oxidation in marine sediments. *Oikos*, **26**, 103–108.

Boaden, P. J. S. (1974). Three new thiobiotic Gastrotricha. *Cah. Biol. Mar.*, **15**, 367–378.

Bousfield, E. L. (1955). Ecological control of the occurrence of barnacles (Crustacea, Cirripedia) in the Miramichee Estuary. *Bull. Nat. Mus. Canada*, **137**, 1–69.

Braarud, T. (1951). Salinity as an ecological factor in marine phytoplankton. *Physiologia Pl.*, **4**, 28–34.

Breteler, W. C. M. K. (1976). Migration of the shore crab, Carcinus maenas, in the Dutch Wadden Sea. *Neth. J. Sea. Res.*, **10**, 338–353.

Broekhuijzen, G. J. (1936). On development, growth and distribution of Carcinides maenas (L.). *Arch. neerl. Zool.*, **6**, 1–100.

Bunt, J. S. (1975). Primary productivity of marine ecosystems. In: H. Lieth and R. H. Whittaker (eds.). *Primary Productivity of the Biosphere*, Springer, New York, pp. 169–183.

Cadée, G. C. (1976). Sediment reworking by Arenicola marina on tidal flats in the Dutch Wadden Sea. *Neth. J. Sea. Res.*, **10**, 440–460.

Cadée, G. C., and Hegeman J. (1974). Primary production of phytoplankton in the Dutch Wadden Sea. *Neth. J. Sea. Res.*, **8**, 240–259.

Cadée, G. C., and Hegeman, J. (1974a). Primary production of the benthic microflora living on tidal flats in the Dutch Wadden Sea. *Neth. J. Sea. Res.*, **8**, 260–291.

Colijn, F., van Buurt, G. (1975). Influence of light and temperature on the photosynthetic rate of marine benthic diatoms. *Mar. Biol.*, **31**, 209–214.

Conover, S. A. M. (1975). Nitrogen utilization during spring blooms of marine phytoplankton in Bedford Basin, Nova Scotia, Canada. *Mar. Biol.*, **32**, 247–261.

Crawford, C. C., Hobbie, J. E., and Webb, K. L. (1974). The utilization of dissolved free amino acids by estuarine micro-organisms. *Ecology*, **55**, 551–560.

Creutzberg, F., Eltink, A. T. G. W., and Noort, G. J. (1978). The migration of plaice larvae into the western Wadden Sea. In McLusky, D. S. and Berry, A. J. (eds.) *Physiology and behaviour of marine organisms*. Pergamon Press, Oxford, pp. 243–251.

Fenchel, T. (1969). The ecology of marine microbenthos. IV. Structure and function of the benthic ecosystem, its physical factors and the microfauna communities with special references to the ciliated Protozoa. *Ophelia*, **6**, 1–182.

Fenchel, T. M., and Barker Jørgensen, B. (1977). Detritus food chains of aquatic ecosystems: the role of bacteria. *Adv. Microbiol. Ecol.*, **1**, 1–58.

Fenchell, T. M., and Riedl, R. J. (1970). The sulfide system: a new biotic community underneath the oxidized layer of marine sand bothoms. *Mar. Biol.*, **7**, 255–262.

Gallagher, J. L., and Daiber, F. C. (1974). Primary production of edaphic algal communities in a Delaware salt marsh. *Limnol. Oceanogr.*, **19**, 390–395.

Gallagher, J. L., Pfeiffer, W. J., and Pomeroy, L. R. (1976). Leaching and microbial utilization of dissolved organic carbon from leaves of Spartina alterniflora. *Estuarine Coastal Mar. Sci.*, **4**, 467–471.

Grindley, J. R. (1972). The vertical migration behaviour of estuarine plankton. *Zool. Africana*, **7**, 13–20.

Grøntved, J. (1960). On the productivity of microbenthos and phytoplankton in some Danish fjords. *Medd. Danmarks Fisk. of Havunders*, N.S., **3**, 55–92.

Haines, E. B. (1977). The origins of detritus in Georgia salt marsh estuaries. *Oikos* 29: 254–260.

Hall, C. A. S., and Moll, R. (1975). Methods of assessing aquatic primary productivity. In: H. Lieth and R. M. Whittaker. *Primary production of the biosphere.* Springer New York, pp. 19–53.

Haperen, A. A. M. van (1974). De plaats van de vogels in de koolstofkringloop van het Grevelingenbekken. *Unpubl. Rep.*, Delta Inst. Hydrobiol. Res., Yerseke, 31 pp.

Hartsuijker, L. (1966). Daily tidal migrations of the shrimp Crangon crangon L. *Neth. J. Sea Res.*, **3**, 52–67.

Heinle, D. R. (1966). Production of a calanoid copepod, Acartia tonsa, in the Patuxent River estuary. *Chesapeake Sci.*, **7**, 59–74.

Heinle, D. R., Harris, R. P., Unstads, J. P., and Flemer, D. A. (1977). Detritus as food for estuarine copepods. *Mar. Biol.*, **40**, 341–353.

Hellebust, J. A. (1967). Excretion of organic compounds by cultures and natural populations of marine phytoplankton. In: Lauff G. H. (ed.) *Estuaries*. Publ. Am. Ass. Advanc. Sci., **83**, 361–366.

Hibbert, C. J. (1976). Biomass and production of a bivalve community on an intertidal mud flat. *J. Exp. Mar. Biol. Ecol.*, **25**, 249–261.

Hoek, C. v.d., Admiraal, W., Colijn, F., and de Jonge, V. N. (1979). The role of algae and seagrasses in the Wadden Sea a review. In Wolff, W. J. (ed.) *Flora and Vegetation of the Wadden Sea*. Balkema, Rotterdam, pp. 9–118.

Hulscher, J. B. (1975). Het wad, een overvloedig of schaars gedekte tafel voor vogels? In: Swennen, C., de Wilde, P. A. W. J., and Haeck, J. (eds.), *Symposium waddenonderzoek*, Amsterdam, pp. 57–82.

Joenje, W., and Wolff, W. J. (1979). The functional role of salt marshes in the Wadden Sea. In: W. J. Wolff (ed.). *Flora and vegetation of the Wadden Sea*, Balkema, Rotterdam, pp. 161–170.

Jørgensen, B. B. (1977). The sulfur cycle of a coastal marine sediment (Limfjorden, Denmark). *Limnol. Oceanogr.*, **22**, 814–832.

Jørgensen, B. B., and Fenchel, T. (1974). The sulphur cycle of a marine sediment model system. *Mar. Biol.*, **24**, 814–201.

Keefe, C. W. (1972). Marsh production: a summary of the literature. *Contr. Mar. Sci.*, **16**, 165–181.

Kofoed, L. H. (1975). The feeding biology of Hydrobia ventrosa (Montagu.). *J. exp. mar. Biol. Ecol.*, **19**, 233–242, 243–254.

Kuipers, B. R. (1975). Experiments and observations on the daily foodintake of juvenile plaice (Pleuronectes platessa (L.). In Barnes, H. (Ed.). *Proc. 9th Eur. Mar. Biol. Symp.*, Aberdeen University Press, pp. 1–12.

Kuipers, B. R. (1977). On the ecology of juvenile plaice on a tidal flat in the Wadden Sea. *Neth. J. Sea Res.*, **11**, 56–91.

Leach, J. H. (1970). Epibenthic algal production in an intertidal mudflat. *Limnol. Oceanogr.*, **15**, 514–521.

Lelek, A. (1976). Veränderungen der Fichfauna in einigen Flüssen Zentraleuropas Donau, Elbe and Rhein). *Schriftenreihe für Vegetationskunde*, **10**, 295–308.

Mann, K. H. (1972). Macrophyte production and detritus food chains in coastal waters. *Mem. Inst. Ital.*, 29 Suppl., 353–383.

Mann, K. H. (1973). Seaweeds: their productivity and strategy for growth. *Science*, **182**, 975–981.

Manuels, M. W., and Postma, H. (1974). Measurements of ATP and organic carbon in suspended matter of the Dutch Wadden Sea. *Neth. J. Sea Res.*, **8**, 292–311.

Marshall, N., Oviatt, C. A., and Skanen, D. M. (1971). Productivity of the benthic microflora of shoal estuarine environments in Southern New England. *Int. Revue Ges. Hydrobiol.*, **56**, 947–956.

Newell, R. (1965). The role of detritus in the nutrition of two marine deposit-feeders, the prosobranch Hydrobia ulvae and the bivalve Macoma balthica. *Proc. Zool. Soc. Lond.*, **144**, 25–45.

Nienhuis, P. H., and van Ierland, E. T. (1978). Consumption of eelgrass, *Zostera marina*, by birds and invertebrates during the growing season in Lake Grevelingen (SW-Netherlands). *Neth. J. Sea Res.*, **12**, 180–194.

Nilsson, L. (1969). Food consumption of diving ducks wintering at the coast of South Sweden in relation to food resources. *Oikos*, **20**, 128–135.

Nixon, S. W., and Oviatt, C. A. (1973). Ecology of a New England salt marsh. *Ecol. Monogr.*, **43**, 463–498.

Nordli, E. (1957). Experimental studies on the ecology of Ceratia. *Oikos*, **8**, 200–265.

Odum, W. E., and Heald, E. J. (1975). Mangrove forests and aquatic productivity. In: A. D. Hasler (ed.). *Coupling of Land and Water Systems*. Springer Verlag, New York, pp. 129–136.

Olanczuk-Neyman, K. M., and Vosjan, J. H. (1977). Measuring respiratory electron transport system activity in marine sediment. *Neth. J. Sea Res.*, **11**, 1–13.

Packard, T. T., (1971). The measurement of respiratory electron transport activity in marine phytoplankton. *J. Mar. Res.*, **29**, 235–244.

Pamatmat, M. M., (1968). Ecology and metabolism of a benthic community on an intertidal sandflat. *Int. Revue Ges. Hydrobiol.*, **53**, 211–298.

Pamatmat, M. M., (1977). Benthic community metabolism: a review and assessment of present status and outlook. In: B. C. Coull (ed.), *Ecology of Marine Benthos*, Belle W. Baruch Library. *Mar. Sci.*, **6**, pp. 89–111.

Platt, T., (1971). The annual production by phytoplankton in St. Margarets Bay, Nova Scotia. *J. Cons. Petm. Int. Explor. Mer.*, **33**, 324–333.

Pomeroy, L. R., (1959). Algal productivity in salt marshes of Georgia. *Limnol. Oceanogr.*, **4**, 386–397.

Postuma, K. H., and Rauck, G. (1979). The fishery in the Wadden Sea. In: N. Dankers, W. J. Wolff & J. J. Zijlstra. Fishes and fisheries of the Wadden Sea. Balkema, Rotterdam, pp. 139–157.

Qasim, S. Z., (1973). Productivity of backwaters and estuaries. In B. Zeitschel, *The biology of the Indian Ocean*, 143–154.

Reimold, R. J. (1972). The movement of phosphorus through the salt marsh cord grass, Spartina alterniflora Loisel. *Limnol. Oceanogr.*, **17**, 606–611.

Rhoads, D. C. (1974). Organism-sediment relations on the muddy sea floor. *Oceanogr. Mar. Biol. Ann. Rev.*, **12**, 263–300.

Rhoads, D. C., and Young, D. K. (1970). The influence of deposit-feeding organisms on sediment stability and community trophic structure. *J. Mar. Res.*, **28**, 150–178.

Riley, G. A. (1956). Oceanography of Long Island Sound, 1952–1954. IX. Production and utilization of organic matter. *Bull. Bingham Oceanogr. Coll.*, **15**, 324–334.

Riley, G. A. (1967). The plankton of estuaries. In: G. H. Lauff, (ed.)—*Estuaries. Publ. Am. Ass. Advanc. Sci.*, **83**, 316–326.

McRoy, C. P. (1970). Standing stocks and other features of eelgrass (Zostera marina) on the coast of Alaska. *J. Fish Res. Bd. Can.*, **27**, 1811–1821.

McRoy, C. P., Barsdate, R. J., and Nebert, M. (1972). Phosphorus cycling in an eelgrass (Zostera marina L.) ecosystem. *Limnol. Oceanogr.*, **17**, 58–67.

Ryther, J. H., and Dunstan, W. M. (1971). Nitrogen, phosphorus, and eutrophication in the coastal marine environment. *Science*, **171**, 1008–1013.

Sand-Jensen, K. (1975). Biomass, net production and growth dynamics in an eel-grass (Zostera marina L.) population in Vellerup Vig., Denmark. *Ophelia*, **14**, 185–201.

Sepers, A. B. J. (1977). The utilization of dissolved organic compounds in aquatic environments. *Hydrobiologia*, **52**, 39–54.

Sieburth, J. M., and Jensen, A. (1969). Studies on algal substances in the sea. II. The formation of gelb stoff by. exudates of Phaeophyta. *J. exp. mar. Biol. Ecol.*, **3**, 275–289.

Steemann Nielsen, E. (1975). *Marine photosynthesis with special emphasis on the ecological aspects.* Elsevier, Amsterdam, 141 pp.

Swemmen, C. (1976). Wadden Seas are rare, hospitable and productive. In: M. Smart (ed.), *Proc. Int. Conf. Conserv. Wetlands Waterfowl*, Heiligenhafen, 1974, pp. 184–198.

Takahashi, M., and Ichimura, S. (1968). Vertical distribution and organic matter production of photosynthetic sulfur bacteria in Japanese lakes. *Limnol. Oceanogr.*, **13**, 644–655.

Teal, J. M. (1962). Energy flow in a salt marsh ecosystem of Georgia. *Ecology*, **43**, 614–624.

Thomas, J. P. (1971). Release of dissolved organic matter for natural populations of marine phytoplankton. *Mar. Biol.*, **11**, 311–323.

Timmer-ten Hoor, (1975). A new type of thiosulphate oxidizing, nitrate reducing microorganism: *Thiomicrospira denitrificans* sp. nov. *Neth. J. Sea Res.* **9**, 344–350.

Turner, R. E. (1976). Geographic variations in salt marsh macrophyte production: a review. *Contrib. Mar. Sci.*, **20**, 47–68.

Vader, W. J. M. (1964). A preliminary investigation into the reactions of the infauna of the tidal flats to tidal fluctuations in water level. *Neth. J. Sea Res.*, **2**, 189–222.

Vegter, F. (1968). Measurements of primary production by means of radioactive carbon. *Neth. J. Sea Res.*, **4**, 111.

Vegter, F. (1977). The closure of the Grevelingen estuary: its influence on phyto-plankton primary production and nutrient content. *Hydrobiologia*, **52**, 67–71.

Vernberg, W. B., and Coull, B. C. (1974). Respiration of an interstitial ciliate and benthic energy relationships. *Oecologia (Berl.)*, **16**, 259–264.

Verwey, J. (1952). On the ecology of the distribution of cockle and mussel in the Dutch Waddensea, their role in sedimentation and the source of their food supply. *Arch. Néerl. Zool.*, **10**, 171–239.

Vosjan, J. H. (1974). Sulphate in water and sediment of the Dutch Wadden Sea. *Neth. J. Sea Res.*, **8**, 208–213.

Vosjan, J. H., and Olanczuk-Neyman, K. M. (1977). Vertical distribution of mineralization processes in a tidal sediment. *Neth. J. Sea Res.*, **11**, 14–23.

Vosjan, J. H., and Siezen, R. J. (1968). Relation between primary production and salinity of algal cultures. *Neth. J. Sea Res.*, **4**, 11–20.

Wilde, P. A. W. J. de, and Berghuis, E. M. (1978). Laboratory experiments on the spawning of Macoma balthica; its implication for production research. In: McLusky, D. S., and Berry, A. J. (eds.), *Physiology and Behaviour of Marine Organisms.* Pergamon Press, Oxford, pp. 375–384.

Witte, J. Y. and Zijlstra, J. J. (1979). The species of fish occurring in the Wadden Sea. In: Dankers, N., Wolff, W. J., and Zijlstra, J. J. (eds.). *Fishes and Fisheries of the Wadden Sea.* Balkema, Rotterdam, pp. 10–19.

Wolf, P. de (1973). Ecological observations on the mechanism of dispersal of barnacle larvae during planktonic live and settling. *Neth. J. Sea Res.*, **6**, 1–129.

Wolff, W. J. (1973). The estuary as a habitat. An analysis of data on the soft-bottom macrofauna of the estuarine area of the rivers Rhine, Meuse and Scheldt. *Zoöl. Verh., Leiden*, **126**, 1–242.

Wolff, W. J. (1977). A benthic food budget for the Grevelingen estuary, the Netherlands, and a consideration of the mechanisms causing high benthic secondary production in estuaries. In B. C. Coull (ed.), *Ecology of Marine Benthos*. University of South Carolina Press, Columbia, 267–280.

Wolff, W. J. (1978). The degradation of ecosystems in the Rhine. In: M. W. Holdgate & M. J. Woodman (eds.). *The Breakdown and Restoration of Ecosystems*. Plenum Press, New York, pp. 169–191.

Wolff, W. J. (in press). Estuarine benthos. In: B. H. Ketchum (ed.), *Estuaries and Enclosed Seas*. Elsevier, Amsterdam.

Wolff, W. J., Eeden, M. J. v., and Lammers, E. (1980). Primary production and transport of particulate organic matter between a salt marsh and the adjacent estuary in the Netherlands. *Neth. J. Sea Res.*

Wolff, W. J., Haperen, A. M. M. v., Sandee, A. J. J., Baptist, H. J. M., and Saeijs, H. L. F., (1976). The trophic role of birds in the Grevelingen estuary. The Netherlands, as compared to their role in the saline Lake Grevelingen. In Persoone, G., and Jaspers, E. (eds.), *Proc. 10th Eur. Symp. Mar. Biol.*, Ostend, vol. 2, 673–689.

Wolff, W. J., Vegter, F., Mulder, H. G., and Meijs, T. (1976a). The production of benthic animals in relation to the phytoplankton production. Observations in the saline Lake Grevelingen, the Netherlands. *Proc. 10th Eur. Symp. Mar. Biol.*, *Ostend*, vol. 2, 653–672.

Wood, E. J. F., Odum, W. E., and Zieman, J. C. (1969). Influence of sea grasses on the productivity of coastal lagoons. In Lagunas Costeras, un Simposio. UNAM–UNESCO, Nov. 28–30, 1967, Mexico, D. F., pp. 495–502.

Wood, B. J. B., Tett, P. B., and Edwards, A. (1973). An introduction to the phytoplankton, primary production and relevant hydrography of Loch Etive. *J. Ecol.*, **61,** 569–585.

Woodwell, G. M., Whitney, D. E., Hall, C. A. S., and Houghton, R. A. (1977). The Flax Pond ecosystem study: exchanges of carbon in water between a salt marsh and Long Island Sound. *Limnol. Oceanogr.*, **22,** 833–838.

Zebe, E. (1975). In vivo Untersuchung über den Glucosev Abbau bei Arenicola marina. *J. Comp. Physiol.*, **101,** 133–145.

Zwaan, A. de, Kluijtmans, J. H. F. M., and Zandee, D. J. (1976). Facultative anaerobiosis in molluscs. *Biochem. Soc. Symp.*, **41,** 133–168.

Chemistry and Biogeochemistry of Estuaries
Edited by E. Olausson and I. Cato
Copyright © 1980 by John Wiley & Sons Ltd.

E. OLAUSSON

Marine Geological Laboratory,
University of Göteborg

9

The Carbon Dioxide—Calcium Carbonate System in Estuaries

1 INTRODUCTION

Calcium and carbon dioxide are of fundamental importance for organisms. Photosynthetic reactions transform CO_2 and water into organic compounds which are the base for all life. CO_2 and Ca are essential in many living systems. Calcium and carbon dioxide are major component of organic hard parts. Many organisms secrete hard parts which are mode up exclusively or in large part of calcium carbonate. The CO_2-$CaCO_3$-system regulate in the short term the pH of the environment and affect thus strongly on the estuarine conditions.

2 CaCO$_3$

Calcium carbonate occurs in two crystal forms in the sea: calcite (rhombohedral form) and the unstable aragonite (orthorhombic form). Vaterite, a third modification of $CaCO_3$, has not been found in natural carbonate sediments so far (Lippmann 1973, p. 67) and are therefore here left aside. Organisms are known to secrete both calcite and aragonite and both mineralogical and chemical composition differ from species to species. However, calcite is the most abundant form in cool water and in deep-sea deposits, aragonite occurs mainly in warm water and on the shelf. High pH and elevated Mg in solution favour also aragonite precipitation. Magnesium is enriched in the hard parts of many organisms. Molluscs with calcite shells can contain up to 5 per cent $MgCO_3$, echinoderm test up to 15% and many algae, particularly the red algae, up to 20% $MgCO_3$ in their calcite (Lowenstam, 1963). There is a tendency that the $MgCO_3$ content decreases with temperature. (Chave, 1954).

Calcium carbonate secreting organisms are e.g. many members of the algae, foraminifera, calcareous sponges, corals, bryozoa, molluscs, echinoderms, and ostracods (detailed list at Revelle and Fairbridge, 1957; Lowenstam, 1963; Millimann, 1974).

The formation of dolomite, $CaMg(CO_3)_2$, is favoured by elevated pH, probably also high CO_2 pressure, and might be a result of diagenetic reactions (Berner, 1971).

The pH is a measure of the degree to which the water is acid (pH $<$ 7) or alkaline (pH $>$ 7). The pH values of estuarine waters are changing. It drops during the day due to photosynthetic reactions, and rises at night. River waters are seldom buffered, the free carbon dioxide concentration and pH vary in that part of the estuary dominated by stream influences. Colloidal organic substances from land may be slightly acid. This will settle out in the estuary. pH values between 7.5 and 9 are common. In the seaward region

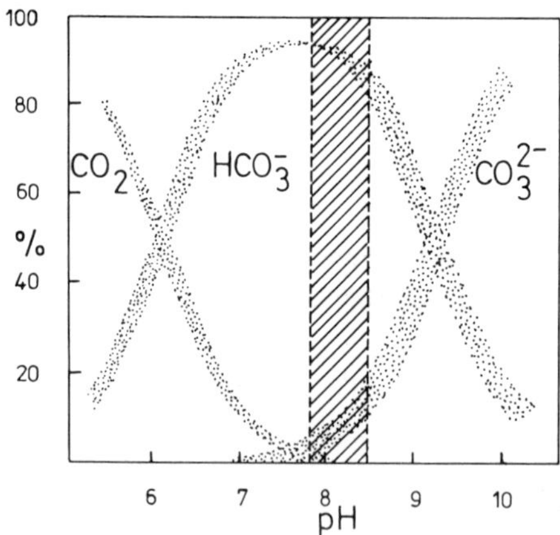

Figure 1. Percentage of total carbon dioxide in each of its three forms in water as a function of pH. The vertical broken lines indicate the approximate pH range of sea water. The leftest part of the wavy band represents 25 °C and 0 °C is to the right of the broad curve, based on Lyman's apparent constants.

pH will fall within the normal marine range (7.8–8.2) because of the buffering action of seawater.

CO_2 concentration in an estuary changes during the day due to biological activity. Dissolved CO_2 combines with water to form carbonic acid and this dissociates as appears in Figure 1. The amount of CO_2 in simple solution plus that in the form of H_2CO_3 is called free CO_2 and constitute the main forms at lower pH than about 6. Bicarbonate (HCO_3^-) constitutes the dominate form up to pH 9, and at still higher pH carbonate ions (CO_3^{2-}) dominates in seawater. HCO_3^- and CO_3^{2-} are called combined CO_2. Addition or removal of CO_2 (e.g. photosynthesis and respiration) will, as stated above, affect pH; conversely, any other factor changing the pH will affect the CO_2 equilibrium. However, HCO_3^- and CO_3^{2-} and other anions of weak acids form a buffer system tending to resist changes in pH.

2.1 Stability of calcareous matter

The thermodynamic equilibrium constant for the reaction

$$CaCO_c \rightleftarrows Ca_{aq}^{2+} + CO_{3_{aq}}^{2-}$$

has a value of $10^{-8.3}$ for calcite and $10^{-8.1}$ when the solid phase (c) is aragonite (at 25 °C and 35‰ salinity).

The dissolution intensity depends primarly on the amount of CO_2 present and the salinity.

Carbonic acid is the predominating acid dissolving calcium carbonate in nature and the dissolution can be written in two ways:

$$CaCO_3 + H_2CO_3 \rightarrow Ca^{2+} + 2HCO_3^- \tag{1}$$

$$CaCO_3 + CH_2O + O_2 \rightarrow Ca^{2+} + 2HCO_3^- \tag{2}$$

The normal atmospheric partial pressure of carbon dioxide is $10^{-3.5}$ atm and surface water exhibit a range of CO_2 activities from about 10^{-4} to as large as $10^{-2.6}$ depending upon the exposure to the atmosphere (see also Chapter 7, this volume), photosynthesis and the amount of decaying organic products. As average a litre of seawater contains approximately 0.0024 moles of CO_2 in various forms and its solubility increases with decreasing temperature being about twice as large at 0° than at 20 °C (details in Skirrow, 1975).

The solubility of calcareous matter increases with salinity being about 10 times higher in normal sea water than in distilled water. Unlike most substances the solubility increases with decreasing temperature. The apparent solubility product of calcite, defined as:

$$K'_{sp} = (Ca^{2+})(CO_3^{2-}),$$

is found to be $(4.59 \pm 0.05) \times 10^{-7}$ moles^2/(kilogram of seawater)^2 at 25 °C

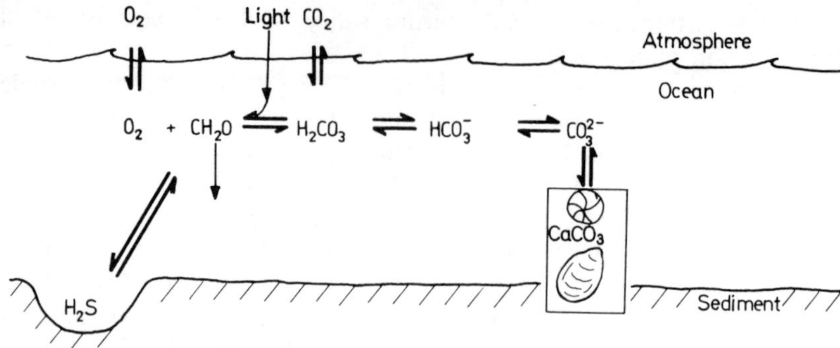

Figure 2. The CO_2–$CaCO_3$ cycle.

with a temperature coefficient of -0.0108×10^{-7}/°C between 2 and 25 °C (Ingle *et al.*, 1973; see also Morse *et al.*, 1980).

Calcium carbonate is stable in normal seawater (pH 8). This is usually the case in the surface water. Further down in the water column the water is undersaturated with respect to carbonates and they began to dissolve. This is also and particularly the case at and below the sediment surface. Calcite is less soluble than aragonite as appears from the thermodynamic equilibrium constants given above. As long as calcareous matter is present it acts as a buffer against changes in pH. If we only consider the molar (m) concentration of the components in equation (1) two theoretical possibilities exists for estuarine sediments:

$$\begin{array}{lll} & pH & Sediment \\ ^{m}CaCO_3 > {}^{m}CH_2O & \sim 8 & \text{Calcareous mud} \\ ^{m}CaCO_3 < {}^{m}CH_2O & \sim 6\text{-}7 & \text{Inorganic mud} \end{array}$$

$$^{m}CH_2O > {}^{m}O_2$$

The possibility that the amount of oxygen in estuarine sediments is larger than the concentration of organic matter does probably not exist. Consequently, whether or not the carbonate is completely dissolved and the pH is lowered in the bottom by one or two units depends as long as oxygen is available, on the ratio in the supply of carbonate/organic matter. When O_2 is used up both carbonate and organic matter can be buried together. Carbonate producing organisms found in estuaries are predominantly infaunal filter-feeders (such as bivalves) and bulk-detritus feeders (such as echinoids). In sandy areas, molluscs and echinoid fragments constitute most of the carbonate; in silty and clayey sediments are benthic foraminifera also important.

The production of calcareous matter is generally higher at lower latitudes than at higher ones and rapid shallow-water carbonate sedimentation generally restricts to tropical and subtropical waters. Genus such as *Mytilus*, *Littorina*, and *Balanus* are common at moderate latitudes.

The stable isotope composition of the calcium carbonate depends on temperature and salinity of the water in which it is formed. Since both vary the $^{18}O/^{16}O$ ratio in the hard parts of the organisms also change in estuaries. A temperature drop by 5 °C causes a $\delta^{18}O$ increase in calcareous tests by 1 per mil (relative to the standard PDB). There is a negative correlation between the $\delta^{18}O$ of the precipitation and the air temperature (Dansgaard, 1961). Therefore the $\delta^{18}O$ of the river water decreases with temperature of the drainage area, becoming particularly low in meltwater from ice sheets. A discharge of river and meltwater will, thus, cause a low $\delta^{18}O$, the more its proportion is and the cooler the adjoining land is (see Martin and Letolle 1979).

Shallow-water molluscs and foraminifera have normally a $\delta^{18}O$ between +1 and −4‰. In the upper estuaries where river and/or meltwater contribution are large still lower (more negative) values can occur (down to −10 per mil). The ^{18}O content of aragonite at 25 °C is 0.6‰ higher than in coprecipitated calcite.

3 FORMATION OF ORGANIC MATTER

The organic matter in an estuary is either formed within the area by photosynthesis or tranported as particulate or dissolved matter into the estuary. The former source dominates in the lower and middle estuary while in the upper regions the river input of organic matter dominate.

Plant utilize sunlight and inorganic nutrients to produce organic molecules (carbohydrates, amino acids, and fat). This photosynthetic reaction can be represented by the following equation:

$$106\,CO_2 + 16\,NH_3 + H_3PO_4 + 106\,H_2O$$

$$\rightarrow (CH_2O)_{106}(NH_3)_{16}H_3PO_4 + 106\,O_2 \quad (3)$$

The magnitude of this production of organic materials depends primarly on the concentrations of nutrients necessary for the organisms.

The organic matter transported from offshore sources into estuaries is of less importance. Estuaries receive high amount of inorganic and organic material from terrestrial runoff. The carbon cycle within an estuary is illustrated in Figures 2–3. The amount of particulate organic carbon (POC) can, in an upper estuary e.g. Chesapeake Bay, be greater than 3 mg C l^{-1} and in a middle region less than 1 mg C l^{-1}. The total amount of organic carbon brought yearly to the sea by rivers is estimated to $(3–18) \times 10^{13}$ g. (Williams, 1975). This is a comparatively small amount on a global scale, some 1 part per mil of the net yearly primary production in the ocean.

The marine plankton of temperate regions have a $\delta^{13}C$ of about −18 per mil (relative to PDB standard) with no apparent differences between phyto- and zooplankton. Common land plants have $\delta^{13}C$ values about 10 per mil lower (Degens 1969). The $\delta^{13}C$ of estuarine samples tended to increase with

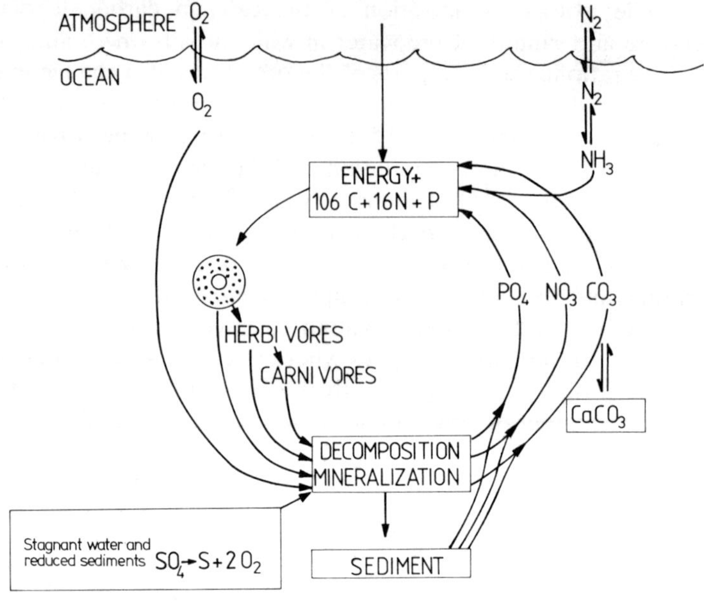

Figure 3. The biological carbon cycle. This biogeochemical cycle is essential for the maintenance and continuation of life. Solar radiation provides the energy. The decomposition and mineralization occurs both in waters as well as in the sediments.

radial distance from the river mouth. (see Hunt, 1968; Shultz and Calder, 1976).

3.1 Decomposition of organic matter

After the death of an organism the breakdown begins. This process can be expressed as follows:

$$(CH_2O)_{106}(NH_3)_{16}H_3PO_4 + 106\,O_2 \rightarrow 106\,CO_2 + 16\,NH_3 + H_3PO_4 + 106\,H_2O \tag{4}$$

If oxygen is available the released ammonia is oxidized, through nitrite, to nitrate:

$$NH_3 + 2O_2 \rightarrow HNO_3 + H_2O$$

The products of the mineralization of organic matter leads (equation (4)) to 6 times more acid products than alkaline ones. Therefore the pH should be lowered. However, the calcareous matter buffer as long as $CaCO_c$ are avaliable. The production of acid compounds stops either when CH_2O or $O_2 + SO_4^{2-}$ are used up, and therefore a minimum pH exists in marine sediments. Thus, there is to some extent also a negative correlation between the content of $CaCO_3$ and carbon in sediments.

In order to oxidize a mole of carbon (12 g) there is a theoretical need for 22.4 litre O_2. The oxygen content of sea water is ≤ 8 ml/1 which is sufficient to oxidize ≤ 4 mg carbon. This amount of carbon (as living or dead material) or higher (up to 10 mg) is as stated above found in the upper regions of several estuaries. In the middle part of an estuary less than 1 mg C/l is the common concentration. From a theoretical point of view such a high concentration of organic matter as in the upper part of an estuary could remove all the dissolved oxygen from the water. However, the oxidation goes on slowly and only in basins with restricted water exchange, with the open sea oxygen deficiency enters in the water (e.g. in fjords of estuary type). Large deposition of organic matter makes the sediment strongly reduced in the estuary, particularly in its upper part. Oxygen deficiency can enter already at the sediment–water interface in the upper part of an estuary, while further seawards the redox discontinuity occurs a few millimetres of centimetre below the sediment surface.

A simplified redox potential equation of a system can be expressed as

$$Eh = E^0 + \frac{k}{n} \log \frac{(ox)}{(red)}$$

where Eh is the oxidation potential, E^0 is the voltage of the reaction when all substances involved are at unit activity, and (ox) and (red) stand for the oxidizing and reducing agents. For a sediment, a heterogeneous system which is slowly brought into equilibrium, we may use the relation

$$Eh \sim k' + \log (ox) - \log (red)$$

Since oxygen is the most important oxidizing substance and organic matter is the dominating which consumes oxygen we may illustrate Eh of a sediment as

$$Eh \sim {}^m O_2 - {}^m CH_2O$$

If ${}^m O_2 > {}^m CH_2O$ the environment is oxidized (Eh > 0) but if ${}^m O_2 < {}^m CH_2O$ the sediment is reduced (Eh < 0).

4 EARLY DIAGENETIC REACTIONS

The sediment formed underwent changes. These alterations are called diagenesis. The diagenesis of organic matter means a degradation of organic molecules to the final product graphite under the release of CO_2, CH_4, NH_3, H_2O and small organic molecules. This diagenesis proceeds through several stages when also new compounds are produced and only the early stages happen within the loose sediment. The main factors affecting these stages are the sediment accumulation rate, the amount and composition of the trapped organic matter and its consumption by organisms, temperature, concentration of oxygen and sulphate and interactions between organic compounds and between organic compounds and metals and minerals.

The decrease of organic substances after burial may be considerable if oxygen is available, less intense in reduced environments. Most sediments are devoid of free oxygen a few centimetres below the sediment-water interface even though they are overlain by oxygenated water. The marine benthos generally recycles the upper 2–10 cm of sediment, thereby exposing to oxygenated conditions material which already had reached reducing condition. Amino acids, carbohydrates, and fatty acids constitute together with the more stable humic and fulvic acids the bulk of the active organic constituents in sediments. Both free and combined sugars decrease with sediment depth, particularly in oxidizing environments. The amino acid concentration decrease to zero within the first few metres (Degens, 1965). The total and particularly the undersaturated fatty acids show a marked decrease with depth (Farrington and Quinn, 1971).

When O_2 is used up as hydrogen acceptor SO_4^{2-} is used by anaerobic bacteria and nitrogen regeneration leads to the formation of NH_3, N_2, and N_2O (Vaccaro, 1965). The overall reaction may be represented by the following equation:

$$(CH_2O)_{106}(NH_3)_{16}H_3PO_4 + 53\,SO_4^{2-}$$

$$\rightarrow 106\,CO_2 + 106\,H_2O + 53\,S^{-2} + 16\,NH_3 + H_3PO_4 \quad (5)$$

Since there is a linear relationship between sulphate ion concentrations and salinity, the number of sulphate ions in the interstitial water decreases with decreasing salinity and the water content of sediments. In sediments rich in organic matter the sulphate ions used up and the aforementioned reaction stops. Then, methane formation follows by bacterial fermentation:

$$2\,CH_2O \rightarrow CH_4 + CO_2$$

There is some evidence that sulphate reduction (equation (5)) causes a slight increase in the pH (Emery and Rittenberg, 1952; Kaplan *et al.*, 1963).

5 REFERENCES

Berner, R. A. (1971). *Principles of Chemical Sedimentology. McGraw-Hill Book Company*, 240 pp.

Chave, K. E. (1954). Aspects of the biogeochemistry of magnesium. *Jour. Geol.*, **62,** 266–283.

Dansgaard, W. (1961). The isotopic composition of natural waters, with special reference to the Greenland ice cap. *Medd. om Grönland,* **165,** (2), 120 pp.

Degens, E. T. (1965). *Geochemistry of Sediments.* Prentice-Hall, Englewood Cliffs, 342 pp.

Degens, E. T. (1969). Biogeochemistry of stable carbon isotopes. In *Organic Geochemistry* (G. Eglinton and M. T. J. Murphy, eds.). Springer Verlag, Berlin, 304–329.

Emery, K. O., and Rittenberg, S. C. (1952). Early diagenesis of California basin sediments in relation to origin of oil. *Bull. Amer. Assoc. Petr. Geol.,* **36,** 735–752.

Farrington, J. W., and Quinn, J. G. (1971). Comparison of sampling and extraction techniques for fatty acids in recent sediments. *Geochim. Cosmochim. Acta,* **37,** 259–268.

Hunt, J. M. (1968). The significance of carbon isotope variations in marine sediments. In *Carbon in Biochemistry* (G. D. Hobson ed.), Pergamon, 27–35.

Ingle, S. E., Culbertson, C. H., Hawley, J. E., and Pytkowicz, R. M. (1973). The solubility of calcite in seawater at atmospheric pressure and 35‰ salinity. *Marine Chemistry,* **1,** 295–307.

Kaplan I. R., Emery, K. O., and Rittenberg, S. C. (1963). The distribution and isotopic abundance of sulphur in recent marine sediments off southern California. *Geochim. Cosmochim. Acta,* **27,** 297–332.

Lippmann, F. (1973). *Sedimentary Carbonate Minerals.* Springer-Verlag. 229 pp.

Lowenstam, H. A. (1963). Biological problems relating to the composition and diagenesis of sediments. In *The Earth Sciences* (Donnelly, T. W., ed.). Univ. Chicago Press, 137–195.

Martin, J. M., and Letolle, R. (1979). Oxygen 18 in estuaries. *Nature,* **282,** 292–294.

Milliman, J. D. (1974). *Marine Carbonates.* Springer-Verlag. 379 pp.

Morse, J. W., Mucci, A., and Millero, F. J. (1980). The solubility of calcite and araonite·in seawater of 35‰ salinity at 25°C and atmospheric pressure. *Geochim. Cosmochim. Acta,* **44,** 85–94.

Olausson, E. (1972). Water sediment exchange and recycling of pollutants. In *Marine Pollution and Sea Life* (Ruivo, M. ed.). FAO. Fishing News (Books) Ltd, London, 158–161.

Revelle, R., and Fairbridge, R. (1957). Carbonates and carbon dioxide. In: Treatise on Marine Ecology and Paleontology, 1, *Geol. Soc. Amer. Memoir,* **67,** 239–295.

Shultz, D. J., and Calder, J. A. (1976). Organic carbon $^{13}C/^{12}C$ variations in estuarine sediments. *Geochim. Cosmochim. Acta,* **40,** 381–385.

Skirrow, G. (1975). The dissolved gases—carbon dioxide. In *Chemical Oceanography* (Riley, J. R., and Skirrow, G., eds.). Academic Press, 2, 1–192.

Vaccaro, R. F. (1965). Inorganic nitrogen in sea water. In *Chemical Oceanography* (Riley, J. P., and Skirrow, G., eds.). Academic Press. **1,** 365–408.

Williams, P. J. LeB. (1975). Biological and chemical aspects of dissolved organic material in sea water. In *Chemical Oceanography* (J. P. Riley and G. Skirrow eds). Academic Press, **2,** 301–363.

Chemistry and Biogeochemistry of Estuaries
Edited by E. Olausson and I. Cato
Copyright © 1980 by John Wiley & Sons Ltd.

U. FÖRSTNER
Institut für Sedimentforschung,
Universität Heidelberg,
Heidelberg

10

Inorganic Pollutants, Particularly Heavy Metals in Estuaries

1 INTRODUCTION

The outbreak of a mysterious, non-infectious neurological illness among the inhabitants (especially fishermen and their families who mainly subsist on sea foods) living around Minamata Bay in southwestern Kyushu, Japan, marked the first case of an acute poisoning by heavy metal contamination in marine coastal areas. This disease accounted for 52 deaths and permanent damage to the health of hundreds of other residents. In 1959 it was revealed that the 'Minamata disease' was caused by methylmercury in fish and other foodstuffs. Bioaccumulation of mercury in the aquatic food chain of the bay resulted in 4.1–8.3 mg Hg per kg for fish (Fujiki, 1972); a value about 100 times greater than the respective values in unpolluted areas.

The source of methylmercury contamination was traced to effluents from various plants of the chemical firm, Chisso Co., manufacturers of plastics

                      U. Förstner

(PVC). For several years before 1953, these plants had discharged methyl-mercury formed from acetaldehyde and inorganic mercury (used as a catalyst) into the drainage channel, which leads into the bay. Investigations performed by Kitamura (1968) some years after these events indicated that more than 2000 mg mercury per kg dry sediment are still accumulated at the channel inflow to Minamata Bay; the sediments of the central part of the bay contained mercury concentrations between 40 and 60 mg/kg, exceeding normal mercury levels of unpolluted sediments 100-fold. At great expense — estimated at 40,000,000 dollars attempts have been made to dredge areas of greatest contamination, but problems associated with land disposal of the still highly toxic muds are practically not to be solved (Buffa, 1976). One of the greatest difficulties is that dissolution of a part of the heavy metals in the dredge material can occur and may threaten the supply of drinking water. Since Minamata, the potential danger of such metal contaminations has found an echo of interest in the public. A large number of investigations were carried out to find the origin, extent and the consequences of these substances, and to introduce appropriate counter measures. The estuaries have proved thereby to be a particular problem area especially at the outer reaches of heavily populated and industrialized areas where large amounts of waste materials transported by rivers meet the near-coast marine systems, regions that traditionally function as an important source of human nutrition.

The following contribution considers the distribution of inorganic pollutants in estuary areas; the group of heavy metals and artificial radioactive isotopes make up the major field of study. Since such a profusion of literature has been published in the last 10 years in this field, it seems reasonable to refer to only a few, but characteristic, examples.

2 TOXIC METALS AND RADIONUCLIDES

Viewed from the standpoint of environmental pollution, metals may be classified according to three criteria (i) non-critical, (ii) toxic but very insoluble or very rare, and (iii) very toxic and relatively accessible. Such a classification has been made by Wood (1974) and is listed in the following table (Table 1). Here, special importance must be attached to arsenic and selenium, besides the heavy metals mercury, cadmium, thallium, and lead.

A complication of metal pollution is that some metals considered to be dangerous are in fact essential to all living organisms even though normally present in only trace amounts. It has been shown, for example, by Venugopal and Luckey (1975) that the metal is toxic when concentration levels exceed those required for proper nutrition by factors ranging from 40 to 220.

Table 1.   Classification of elements according to toxicity and availability (from Wood, 1974).

Non-critical			Toxic but very insoluble or very rare		Very toxic and relatively accessable		
Na	C	F	Ti	Ga	Be	As	Au
K	P	Li	Hf	La	Co	Se	Hg
Mg	Fe	Rb	Zr	Os	Ni	Te	Tl
Ca	S	Sr	W	Rh	Cu	Pd	Pb
H	Cl	Al	Nb	Ir	Zn	Ag	Sb
O	Br	Si	Ta	Ru	Sn	Cd	Bi
N			Re	Ba		Pt	

One of the factors determining toxicity concerns the chemical form of the metal to which the organism is exposed. With the exception of mercury vapour, the free metals generally present no toxicity problems. What really matters is the particular species of the metallic compound or ion, its availability, solubility, binding affinity, and biological activity. It has been generally established that heavy metal species in solubilized form are more hazardous than their solid counterparts found in suspended material or sediment particles. However, certain metals and their compounds can be converted to a more toxic form by the action of microorganisms. For instance, some microorganisms living in the bottom sediments of aquatic systems are capable of transforming inorganic mercury to the monomethylmercury, CH_3Hg^+, which is the most toxic and common form of mercury in the biological environment. Mercury released to the aquatic environment as methylmercury is rapidly assimilated by water organisms such as fish and *via* the food chain ultimately by man; these conversion processes are enhanced by elevated temperatures and a high concentration of nutrients in the affected water.

The ability of microorganisms to detoxify their environment through organometallic transformations is obviously not restricted to mercury, although these processes are particularly characteristic and important with regard to the widespread pollution of that element. In the meantime it has been established that the mechanism of biological methylation is also effective in the formation of volatile compounds of As, Pb, and Se, such as alkyl arsines (Edmonds and Francesconi, 1977), tetramethyl lead (Wong *et al.*, 1975) and dimethyl selenide (Francis *et al.*, 1974). On the other hand, most of the compounds of the transition series metals are not decomposed by microbial activity, as is true for many organic pollutants; on the contrary, these substances tend to accumulate in the environment, especially in sediments by association with organic and inorganic matter through processes of adsorption, precipitation and complex formation. Recent investigations indicate highly complicated interactions of the biological cycles of elements, as exemplified by the detoxifying effects of selenium salts in mercury contaminated organisms (Ganther *et al.*, 1972; Wood *et al.*, 1977).

The chemical properties of radioactive isotopes, the second group of inorganic pollutants treated here, are essentially identical to those of the corresponding non-radioactive isotopes, having the same atomic number (Sayre *et al.*, 1963). They are acted upon, in their passage through the aquatic environment, by the same processes which affect stable elements (Burton, 1975). When taken internally, alpha emitters such as ^{226}Ra, ^{238}U, and ^{239}Pu—due to their high specific ionization—are considered to be about 20 times as damaging as either beta or gamma emitters such as ^{60}Co, ^{90}Sr, and ^{137}Cs (U.S. Nat. Bureau of Standards, 1955). In a given time interval, long-lived nuclides must be present in relatively larger quantities in order to give off as much radiation as shorter lived radionuclides.

3 SOURCES AND ACCUMULATION OF INORGANIC POLLUTANTS

There are five major sources of heavy metals in aquatic systems (Wittmann and Förstner, 1975):

(i) *Geological weathering.* This is the source of 'background levels.' It is to be expected that in areas characterized by metal-bearing lithological formations, these metals will also occur in higher levels in the water of the area.

(ii) *Industrial processing of ores and metals.* During the processing of ores, metal-bearing dust particles are formed, which may only be partially filtered out by air purification systems. Appreciable quantities of metals go to waste during chemical metal refinement processes (e.g. galvanizing and pickling).

(iii) *The use of metals and metal compounds.* Chromium salts are used in processes in tanneries, copper compounds as plant protection agents, mercury in chlorine–alkali production, zinc in water pipes, and tetramethyl lead as an anti-knock agent in gasoline.

(iv) *Burning of fossil fuels, production of cement and bricks.* Fossil-fuel mobilization is particularly high for arsenic, zinc, cadmium, copper (coal), nickel, and vanadium (oil). Strong emissions of zinc, lead, selenium, and arsenic result from cement production (Goldberg, 1976); metal enrichment was found around brickworks (Brumsack, 1977).

(v) *Leaching of metals from garbage and solid waste dumps.* The contribution of this source to metal contamination of inland and coastal waters merits close attention. Mine dumps especially can be a serious source of pollution.

A comparison of natural metal enrichment of sediments by geological weathering (as evidenced by analysis of the lower section of sediment cores) with present-day enrichment is given in Figure 1 for the example of the lower Rhine in the F.R.G. The greatest increases in concentration levels are found for the dangerous heavy metals; more than 90% of the concentrations of copper, zinc, lead, mercury, and cadmium in recent sediments of the lower Rhine originate from man-made sources (Förstner and Müller, 1973). Similar results are given from a comparison of the consumption rates of heavy metals with the natural concentrations of the respective elements in rocks, soils and sediments. The *Index of the Relative Pollution Potential* (Förstner and Müller, 1973) and the *Technophility Index* (Nikiforova and Smirnova, 1975) show that particularly strong man-made enrichments can be expected from emissions of lead, mercury, copper, cadmium, and zinc in surface sediments and soils.

Three types of artificially produced radioactive species are introduced in the marine environment (Goldberg, 1976): (i) nuclear fuels such as uranium-235 and plutonium-238; (ii) fission products arising from the use of nuclear fuels, such as cesium-137; and (iii) activation products resulting from the interaction of nuclear particles with the components of nuclear reactors and weapons, such as zinc-65 and iron-55. Radium-226 is the critial radionuclide for mining and milling operations, uranium is critical for the production of UF_6 enrichment and fuel fabrication, and plutonium is the critical element in recycling fuel. For reactors the order of relative importance has been established from ^{137}Cs, ^{134}Cs and ^{58}Co, ^{60}Co, ^{54}Mn to ^{3}H. In reprocessing plants tritium is one of the major nuclides released to the aquatic

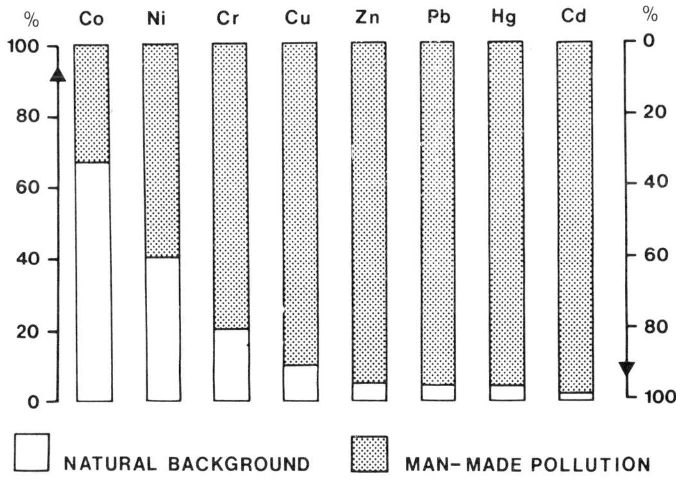

Figure 1. Geochemical and anthropogenic sources of heavy metals in pelitic sediments from the lower Rhine River (Förstner, 1980).

environment (Wrenn and Jinks, 1975). In view of their relative toxicities, radium is of particular importance as a potential contaminant of the environment.

4 POLLUTION ASSESSMENT BY ANALYSIS OF WATER, SEDIMENT, AND ORGANISMS

The most obvious medium of pollution assessment is surface *water*. For many heavy metals, in particular mercury, chromium, and lead, it has been established for a given sampling station monitored over a long period of time that values tend to vary by several orders of magnitude, even when the samples were collected over short time intervals. Such fluctuations are attributable to variables such as daily and seasonal variations in water flow, surreptitious local discharges of effluent, changing pH and redox conditions, salinity, temperature and biological activity. Thus, meaningful monitoring necessitates a closely-knit system of water sampling—preferably on a continuous-flow basis—which is both costly and time consuming. On the other hand, pollutant concentrations in water organisms and in *sediments* provide a more stable means of obtaining an indication of the state of associated waters.

Sediment profiles (cores) often uniquely preserve the historical sequence of pollution intensities; lateral distributions (quality profiles) are used to determine and evaluate local sources of pollution. Sediment analyses, however, are not only useful when selecting critical sites for routine water sampling, but they can reveal the fate fo contaminants under varying environmental conditions; by means of selective chemical extraction of different types of metal bonding, e.g. adsorbed, organically complexed, coprecipitated with Fe/Mn-oxides and carbonates, residually bonded trace metals (Förstner, 1978). Even for routine analysis, it is imperative to base the findings on a standardized procedure with regard to particle size, since there is a marked decrease in the content of trace metals (and also for artificial radionuclides) as sediment particle size increases. Several methods have been introduced in order to exclude (mostly coarser-grained) sediment fractions which are chemically inert (De Groot *et al.*, 1976; Förstner, 1977): (i) separation of characteristic grain fractions, e.g. the pelitic fraction of $<2 \mu$m and the clay-silt fraction of $<63 \mu$m, (ii) extrapolation from the grain size distribution, (iii) correction for the specific surface area, (iv) chemical treatment for the extraction of acid-soluble (hydrogenous) fractions, (v) mineral separation, as exemplified by the quartz correction method (Thomas, 1972) and (vi) comparison to conservative elements, such as aluminium, silicon, potassium, and sodium, which can be taken as a reference base for cultural metal inputs (Bruland *et al.*, 1974; Kemp *et al.*, 1976).

It was shown in the early seventies that the distribution of waste materials in aquatic systems can be determined from the amount of organic carbon and trace metals present. Klein and Goldberg (1970) and Applequist *et al.*, (1972) utilized the concentrations of mercury in sediment to estimate the rate of accumulation of sewage sludge and to trace the sources of pollution near marine coasts and in harbour areas. Of all the minor elements studied by Gross *et al.* (1970) in sediments of the New York Bight, lead proved to be the mose useful indicator of waste materials (see Figure 3). Rutherford and Church (1975) proposed the use of silver and zinc as well, to trace sewage sludge dispersal in coastal waters. In many estuaries, a characteristic lowering of metal levels in the suspended particulates and in the sediments can be observed seawards, particularly among those metal species strongly enriched by anthropogenic influences. From observations in the Elbe Estuary in Northern Germany, it was concluded that the mixing of heavily polluted river sediment with relatively 'clean' marine sediments is the main cause of the decrease of heavy metal concentrations within the solid phases (Förstner and Müller, 1974; Müller and Förstner, 1976), although processes of remobilization, e.g. desorption, complexing, dissolution (treated in more detail in Chapter 6) may also be effective in the estuarine environment. Similar effects have been observed in the Rhine Estuary by Duinker and Nolting (1976). A typical example of the behaviour of metal-polluted sediments, in this case from the Välen Estuary in southwestern Sweden (Figure 2) has been given by Cato (1977). Here, copper

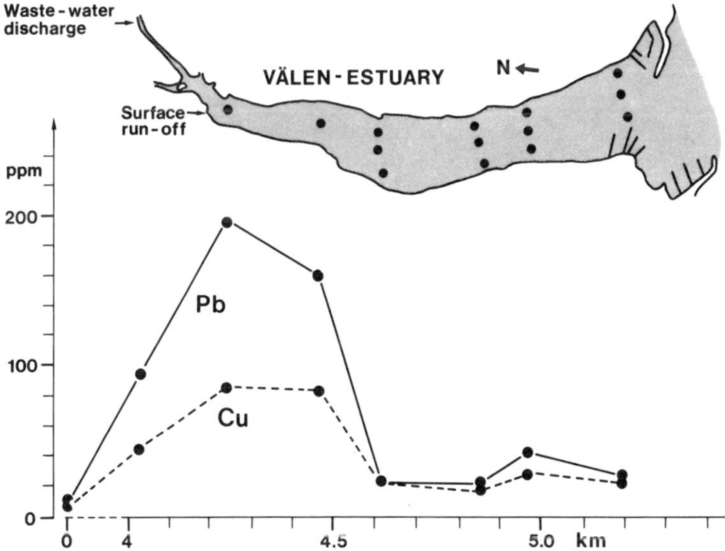

Figure 2. Concentration of Pb and Cu in sediments from the Välen Estuary, Sweden (Cato, 1977).

314 *U. Förstner*

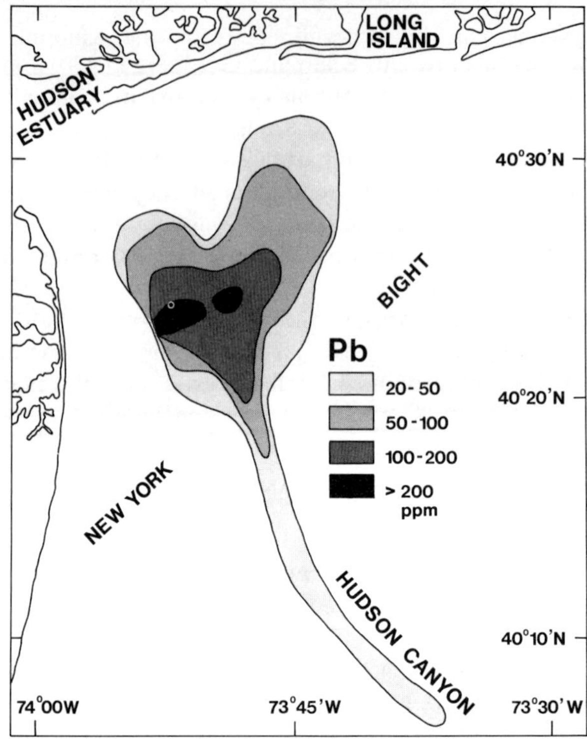

Figure 3. Distribution of lead in bottom sediments from
the New York Bight (Carmody *et al.*, 1973).

and—more strongly—lead are enriched in the sediments by inputs from
waste water effluents and surface runoff near the mouth of the Stora Ån
River. Although lead is usually more strongly associated to sediment particles than is copper, its decrease within the estuarine mixing zone is more
pronounced, most probably due to the higher contamination/background
ratio for lead in the inflowing material.

Bio-monitoring of pollution has been gaining attention, particularly with
regard to trace elements, since various marine organisms have the ability to
accumulate these substances. A review of the use of *indicator organisms* to
monitor trace metal pollution in marine and estuarine environments has
been provided by Phillips (1977); it is stressed, in particular, that data from
organisms such as macroalgae, bivalve molluscs and teleosts represent a
moving time-averaged value for the relative biological availability of metals
at each site studied. On the basis that molluscs are well-known concentrators of xenobiotics, Goldberg (1975) proposed a global monitoring programme utilizing *Mytilus edulis* and similar species for assessing marine

pollution. The International Environmental Programme Committee (IEPC) has endorsed the practical programme of *Mussel Watch* with the analysis of specimens from some 100 open ocean and coastal marine sites.

5 WASTE DISPOSAL IN ESTUARIES

The contamination of coastal areas including estuaries and marginal seas can generally be attributed to one of the following causes (Weichart, 1973): (i) direct input of effluents from industries and communities situated near the coast, (ii) effluents from communities via sewer outfalls, (iii) dumping of waste from ships, (iv) soluble and suspended load from rivers, (v) atmospheric fallout, (vi) waste derived from the extraction of raw materials from the sea, and (vii) pollution by the shipping industry.

Calculations from Goldberg (1972) on a global basis, indicate that approx. 140 million tons of wastes are annually dumped by ships into the sea. Furthermore, approx. 15 million tons of carbon and fly ash from fossil fuel combustion and 30 million tons of industrial particulates will reach the sea *via* atmospheric emissions. Although there is no accurate data available, it is assumed that wastes from input of rivers are discharged to the same extent, or, even more, that direct dumping occurs. However, since the dumped wastes tend more to concentrate in certain areas, they pose more immediate problems with regard to the effects on organic life than do the more diffuse influences from the atmosphere and also (in most cases) from the river discharges. In some areas, as described below, planned or municipal waste dumps often surpass the deposits caused by natural sedimentation in quantity and quality of waste materials. Gross (1972) has stated that 'waste-solid disposal by coastal cities modifies shorelines and cover adjacent ocean bottom with characteristic deposits on a scale large enough to be geologically significant; these activities change sediment distribution and dislocate sediment-transport processes in estuaries and along beaches'. A typical example is given from Long Island Sound near New York where waste discharges from dumping are ten times higher that the input of river-borne suspended sediment (5.1). In the context of pollution problems arising from trace metals and artificial radionuclides, particular interest is given to the emissions from metal-working industries (including disposal of mine tailings and smaller wastes, 5.2), as well as to the input of radioactivity from nuclear reactors and reprocessing plants situated in estuaries or in the lower reaches of inflowing waters (5.3).

5.1 Ocean disposal of dredge materials and sewage sludge

Estuarine areas have in the past received large volumes of waste solids; Table 2 is a compilation of U.S. data from the Atlantic coast, Gulf of Mexico and the Pacific coast (Gross, 1970).

Table 2. Ocean waste disposal sites of the U.S.A with estimated tonnage of wastes dumped in each region, in the 1960s (from Gross, 1970), 1 = Regional total.

Disposal sites	No. of active sites	Estimated tonnage (10^6 tons/yr)
Atlantic Coast	61[1]	14.3[1]
Gulf of Maine	13	1.4
Mid-Atlantic coast	29	7.8
South Atlantic	17	5.0
Puerto Rico	2	0.1
Gulf of Mexico	20[1]	29.3[1]
Eastern Gulf	9	3.6
Western Gulf	5	10.0
Mississippi R.	6	15.7
Pacific Coast	59[1]	12.6[1]
California	3	1.7
Oregon–Washington	38	10.8
For comparison: Sediment load Mississippi		
		310

Hudson Estuary, New York Bight; U. S. coast

The greatest amount of sediments entering the North Atlantic appears to come from the dumping of waste into the coastal near New York City. About 8.6 million tons were disposed of each year between 1964 and 1968 from this area. To visualize the volume, one must imagine that if the annual discharge of these waste solids were spread uniformly over Manhattan, it would form a layer 13 cm thick each year (Goldberg, 1976).

Details on the waste discharge in the New York Metropolitan Region are presented by Gross (1970): four waste disposal sites in the New York Bight have received large volumes of waste for many years. In addition, thirteen sites are actively used for disposal of waste solids in Long Island Sound. The types of wastes commonly dumped in the coastal ocean around the New York area are presented in Table 3 (in accordance with Gross *et al.*, 1971).

About 76 per cent of the wastes dumped in New York Bight in the 1960s was dredged wastes, fine-grained materials (for example, ash) and small stones. After waste solids, rubble from building construction and demolition is the second largest source; a third major type of waste dumped in the New York Bight is sewage sludge separated from municipal sewage during treatment, typically containing 3 to 6 per cent solids on a dry weight basis. Coal ash has been dumped in amounts averaging about 0.1 million tons per year between 1960 and 1968, but because of the reduction in the use of coal for fuel and increased use of ash for various purposes, less coal ash is expected to be dumped in the New York Bight in the future. Among the

Table 3. Discharge and composition of various wastes dumped into coastal waters near New York and their probable impact on these waters (from Gross *et al.*, 1971).

Waste solids	Discharge (10^6 tons/yr)	Composition	Environmental effect
Dredged wastes ('dredged spoils')	49	Silicate sediment (85–95%) mixed with various carbon-rich wastes (5–15%)	Shoaling of waterways, turbidity of waters, oxygen demand, release of materials
Sewage sludges	0.19	Silicate sediment (45%) Organic wastes (55%) Industrial wastes (?%)	Bacteria, viruses released Oxygen demand, floating materials Dissolved constituents
Coal ash	0.9	Silicates–quartz mullite glass	Turbidity, shoaling, adsorption
Waste chemicals			
Liquids	?	Variable	Toxic constituents
Solids	?	Variable	Oxygen demand, turbidity

various industrial wastes discharged at sea having substantial solid concentrations, the largest quantities are derived from titanium-pigment processes. These residues of extracted titanium ores suspended in dilute iron-rich sulphuric acid total approximately 3.5 million m^3/yr, 10% of which is solid content by weight.

Lead concentrations in sediments were used by Carmody *et al.* (1973) to assess major waste material in the New York Bight (Figure 3). The Pb concentrations in these deposits were twenty times higher than in normal shales and about fifty times greater than the concentrations in marine plants.

High concentrations of lead (as well as Cr, Cu, Ni, and Zn) were found in the areas to the east and west of the Hudson Shelf Valley. These areas correspond to the sewage sludge and dredge spoil disposal areas, respectively. The metal concentration decreases with the distance from the central areas of the two sites, but there is a broad section of contaminated sediment extending in a north-east direction towards Long Island as well as a narrower one stretching south-east along the axis of the Hudson Shelf Valley. In an area of approximately 60 km^2, sediments contained Cu, Cr, Pb, and Zn concentrations ten times higher than normal values, whereas a larger area encompassing some 170 km^2 was shown to contain a five-fold increase.

The different sources of metal pollution have been investigated in some detail by Klein *et al.* (1974) and Mueller *et al.* (1976) as part of the New York Bight Project, which was begun in 1973. The four major sources of pollutants, namely barge dumps, atmospheric fallout, runoff, and wastewater were comparatively evaluated. It was shown that the Bight proper receives direct discharges from the first two sources, whereas wastes from the latter

Table 4. Mass loads of suspended solids and metals into New York Bight by total
(metric tons per year) and source (percent). From Mueller *et al.*, **48**, (1976).

	Mass load (metric tons/yr)	Barge %	Direct Bight Atmospheric %	Wastewater Municipal	Industrial	Coastal Runoff Gauged	Urban	Groundwater
Suspended solids	8.8×10^6	63	5	4	0.2	16	12	0
Cadmium	876	82	2	5	0.6	5	5	<1
Chromium	1,825	50	1	22	0.8	10	16	<1
Copper	5,037	51	3	11	9	10	16	<1
Iron	83,950	79	3	5	0.5	6	6	<1
Mercury	110	9	—	71	2	13	5	—
Lead	4,636	44	9	19	3	6	19	<1
Zinc	12,045	29	8	8	2	21	22	<1

two originate in the coastal zones, i.e. Long Island and New Jersey. The
results of the work of Mueller *et al.* (1976) are displayed in Table 4.

Upon comparison with background loads of flow entering the Bight from
the ocean, it is clear that heavy metals, especially lead and chromium, are
the most significant of the man-made inputs entering the New York Bight.
Dredge spoils contribute the major portion (30–80%) of the heavy metal
input with the exception of mercury, for which wastewater contributes 70
per cent.

Data of Klein *et al.* (1974) have indicated that the percentage of trace
metal input to New York Bight from urban surface runoff is surprisingly
high in all cases and particularly for zinc. It could therefore be concluded
that even with perfect wastewater treatment and successive land disposal of
sewage sludge, there would still be significant pollution by trace metals from
city areas. Similarly, as has been noted by Klein *et al.*, the elimination of all
industrial discharges ('zero discharge') would reduce only the amount of
nickel to any great extent (nickel is mostly from electroplating industries);
high amounts of zinc, copper, and cadmium would still reach the harbour
area.

The second largest source area of waste materials on the Atlantic coast is
the *Chesapeake Bay* region. A pronounced population concentration centres
in the vicinity of the city of Baltimore, and, as a consequence, the human
influence on pollution of the northern part of the bay is most strongly felt in
the Patapsco subestuary of Baltimore Harbour. Of the 3.05×10^8 tons/yr of
total wastewater discharge from sewage treatment plants to the Chesapeake
Bay, 2×10^8 tons flow into Baltimore Harbour. Helz (1976) calculated that
minimum percentages of anthropogenic effects in this area are 74% for lead,
63% for chromium, 50% for cadmium and copper, and 30 per cent for zinc.

Cadmium presents the greatest potential for serious pollution because it appears to be readily remobilized from sediments; its input from wastewater exceeds fluvial input. (Helz *et al.*, 1975).

Along the *U.S. west coast,* the sewage discharges into the Southern California's marine regions are probably among the most remarkable examples of environmental pollution. Combined annual mass emission rates of Southern California's five largest municipal effluents have been calculated annually to be approximately 0.9 million tons of suspended particulates containing 1,500 tons of zinc, 650 tons of chromium, 520 tons of copper, 200 tons of lead, 50 tons of cadmium, 25 tons of silver and 2–3 tons of mercury; about two-thirds of these masses reach the sea *via* Hyperions Joint Water Pollution Control Plant near Los Angeles (Schafer, 1976). Sediment investigations around Hyperion's 7-mile outfall performed by Schafer and Bascom (1976) indicate strong enrichment of trace elements; near the outfall the cadmium content of the sediment is as much as 300 times higher than in the less contaminated sediments some distance away. Respective enrichment factors are found to be 140 for mercury, 85 for lead, 80 for copper, 60 for silver, and 15 for nickel and chromium. San Franscisco Bay mud samples have been analysed by Parsons *et al.* (1973); approximately 4,000 tons per year of gross heavy metals are dumped into the entire bay. High cadmium levels have been explained to be a product of a lead smelter and several petroleum refineries that are located nearby (Moyer and Budinger, 1974).

Sediment investigations from the Southern California coastal region also indicate characteristic influences from atmospheric pollution. Chow *et al.* (1973) demonstrated from their isotope ratios, that the lead accumulations in sediment cores from Baja California are probably derived from lead additives such as are present in the type of gasoline sold in Southern California. It has been estimated that on an annual basis, 180 tons of lead are deposited into the zone off California by atmospheric fallout, 200 are discharged through municipal waste effluents and 90 tons come from storm and river runoff.

In the southern part of Puget Sound (Wash.), smokestack emissions from the Tacoma Cu-smelter have resulted in significant accumulations of As, Sb, Cu, Pb, and Zn. Atmospheric input to Puget Sound was calculated by Schell and Nevissi (1977) as approximately 450 tons/yr for copper, 2,730 tons of lead, and 820 tons/yr of zinc; the respective values for the inputs from Metro's West Point Sewage Plant were 29 tons/yr (Cu), 9 tons/yr (Pb), and 56 tons/yr (Zn). At the waste outfall from a mercury cell chloralkali plant in Bellingham Bay (northern Puget Sound), an estimated 5–10 kg of mercury were discharged daily between 1965 and 1970 (Bothner and Carpenter, 1974).

Waste disposal along coastlines of Asia

A high degree of metal pollution observed in coastal areas of Japan is due to the strong increase of industrial production during the last decades. The amount of municipal and industrial wastes increased from 425×10^6 tons in 1967 to more than 1000×10^6 tons in 1975. According to Goto (1973), mercury losses from chlor-alkali production amounted to approximately 650 tons discharged to the environment in 1970 alone. Furthermore, the production of Pb stabilizers for vinyl chloride doubled between 1965 and 1970; in 1970, production of cadmium sulphide pigments amounted to 600 tons. Parallel to waste material, an increase could also be observed in the metal level of the marine coastal sediments. Values exceeding the normal concentrations of mercury and cadmium by more than 100-fold were measured in sediment samples from Tokyo Bay near Kawasaki Industrial District (Goto, 1973). Similar effects were found for other areas of Tokyo Bay (Matsumoto and Yokota, 1976), in Tokuyama Bay and Nagoya Harbour (Nisimura, 1974: in Buffa, 1976; Ito, 1975) as well as in lagoons and bays of Hokkaido (Saroma, 1972).

The sewage from the twin cities of Kowloon and Victoria, Hong Kong (more than 4 million live in this general area) passes into the harbour from 11 harbour-wall outlets; the volume of discharged sewage at present stands at about 2 million tons per year (dry weight; 560,000,000 l per day) and receives no treatment apart from screening to remove large solids. About 1 million tons of rubbish per year from Hong Kong Island, Kowloon and from the New Territories is either burned at one of two coastal incinerators or is dumped into Gin Drinkers Bay near Tsuen Wan (Morton, 1976).

Altogether, some 19,000 small factories are located in the densely populated urban areas of Hong Kong; most of the resulting effluents is discharged untreated *via* domestic drains and sewers into the sea—particularly into the harbour. Elevated heavy metal concentrations have been recorded in the receiving waters; lead has 160 times and cadmium 180 times the concentration found in the open ocean (Chan *et al.*, 1974). Analysis of 12 drain samples showed a lead content on the average 50 times higher and in one case 1,000 times higher than that found in the open sea; at the same site, zinc was registered with a content 330 times that of normal seawater. Fishing is prohibited in the harbour waters, though the sea bottom, comprised of a thick organic sludge, supports little or no life anyway (Morton, 1976).

From Taiwan, it is reported by Hung *et al.* (1977) that Kaohsiung Harbour has been thoroughly contaminated by heavy metals. The pollutants from Kaohsiung Harbour are probably the main cause of the complete extermination of the oyster cultur along the southwest coastline of Taiwan Island that has occurred during the last 10 years.

Heavy metal contamination from extensive disposal of waste materials has been reported from Port Phillip Bay in Australia, which receives the industrial and domestic effluents from the city of Melbourne (Talbot *et al.*, 1976a). A survey on metal accumulation in organisms revealed particularly strong enrichment of cadmium in oysters and mussels (Talbot *et al.*, 1976b).

Metal accumulation from waste disposal in marine areas of Europe

The pollution problems in the Mediterranean Sea are chiefly due to the high quantity of domestic sewage and the virtually total absence of control on toxic components. Among the rivers entering the Mediterranean, the Rhône, Po, and Tiber discharge by far the highest pollution load, which originates from both industrial and municipal inputs (Naeve, 1974). Characteristic increases of mercury concentrations have been reported from the French and Italian coast, e.g. the Gulf of Lion (Aubert *et al.*, 1974), the Ligurian coast and the Gulf of Genoa (Renzoni *et al.*, 1974), and from the industrial centres of Cagliari (Sardinia) and Naples (Sheppard and Bellamy, 1974). The Adriatic Sea is particularly susceptible to pollution, especially that from the industrial centres in the northern part of the region. Naeve (1974) noted 76 wastewater effluents from industrial plants in Venice; thus the concentration of certain heavy metals, e.g. mercury, in water and edible organisms lies at or even slightly above the acceptible safety levels. Ui and Kittamura (1971) have drawn attention to the possible hazard arising from effluents of a chemical factory at Ravenna, which produces acetaldehyde by the same process as was employed at Minamata. Grancini *et al.* (1976) found that concentrations of Cr, Ag, Sb, and Zn are noticeably higher in the sediments from the northern Adriatic Sea than those found in open seas; they suggest that this increase is related to the waste inputs from rivers. Sediment studies in the upper Saronikos Gulf, Greece (Papakostids *et al.*, 1975) revealed an 8 to 200-fold increase in the concentrations of Sb, As, Cr, Hg, Ag, and Zn of bottom deposits in the vicinity of the Athens sewage outfall.

In northern Europe there are five areas of particularly high amounts of waste disposal: Clyde Sea (Liverpool Bay), Thames Estuary, Rhine-Waal/Meuse/Scheldt estuaries, the German Bight and the Baltic Sea. During the last decade, many studies have been performed with regard to metal pollution, of which only a few can be mentioned here. Sewage sludge from the city of Glasgow and adjoining areas is dumped into a restricted area of the Firth of Clyde, at a rate of 1×10^6 tons per year (Mackay *et al.*, 1972). Metal analyses from bottom sediments indicate characteristic enrichment of mercury, lead, zinc, and copper. In particular the 'discrete association of mercury with sewage sludge' (Halcrow *et al.*, 1973) here serves as an indicator of sewage discharges: based on a background of approx. 0.1 mg/kg

Table 5. Enrichment of organic carbon and trace metal levels in the sediments of the sewage sludge disposal area off Garroch Head, compared with sediments from a control area at the Firth of Clyde (in mg/kg dry weight; Halcrow *et al.*, 1973).

	org. C	Ag	As	Cd	Cr	Cu	Mn	Mo	Ni	Pb	Zn
Disposal area	7.5%	4	21	6.4	122	269	407	7.0	77	361	631
Clyde control	0.9%	0.2	7	1.6	33	16	355	0.3	30	42	85

in the less polluted parts of the Firth of Clyde, an increase of mercury concentrations by a factor of more than 20 could be detected in the centre of the disposal area. Table 5 shows enrichments of trace metals, including rather significant increases in silver, copper, molybdenum, lead, and zinc (from Halcrow *et al.*, 1973). Considerably higher amounts of metals are discharged with sewage materials into Liverpool Bay than into the Firth of Clyde disposal site (Table 6); it must be noted, however, that the Liverpool discharges are distributed over a larger area. Even higher amounts of metals are dumped with the sludge from the London sewage works (Table 6): at present, more than five million tons (wet weight) are dumped each year in the Barrow Deep at the outer Thames Estuary. In contrast to the New York dumping grounds, however, the deposits at the Barrow Deep are generally swept away by strong tidal currents (Shelton, 1971). The remarkably high concentrations of copper (up to 2,500 mg/kg dry weight) and cadmium (30–70 mg/kg) in these materials should be nevertheless considered as a potential hazard to water organisms.

A particularly high input of pollutants is recorded in the southern North Sea. The annual supply of suspended matter to this region from the Rivers Rhine–Waal was measured at 3.2×10^6 tons, from the Meuse 0.7×10^6 tons (Terwindt, 1967) and from the Scheldt River contributes approx. 0.4×10^6 tons (McCave, 1973). A large portion of these materials is derived from sewage particles, as exemplified by the concentrations of organic carbon and trace metals (De Groot *et al.*, 1971; Wollast, 1976). The strong increase of the cadmium concentration in the solid matter of the Rhine River—which nearly doubled between 1960 and 1970—is particularly alarming, since due

Table 6. Discharges of Cu, Pb, and Zn in various coastal areas of the UK (in tons/year).

	Copper	Lead	Zinc
Firth of Clyde[a]	15–37	26–29	58–95
Liverpool Bay[b]	299	365	1606
Thames Estuary[a]	584–1460	—	1280–6205

[a] = metal dumped in sewage sludge only,
[b] = local authority outfalls not included (from Prater, 1975).

to these components, problems will arise in disposal of these dredge materials from the inner harbours of Rotterdam (De Groot *et al.*, 1976).

5.2 Atmospheric influences

The influence of atmospheric metal pollution has already been mentioned with the examples from the New York Bight and Southern California.

From a summary of the heavy metal input to the Clyde Sea given by Topping (*lit. cit.* Cambray *et al.*, 1975), it is evident that rainwater plays a significant role with regard to the input of cadmium, copper, zinc, and, in particular, lead (Table 7). A comparison of the atmospheric input by rainwater to the North Sea and the discharges of heavy metals and metalloids through the Rhine River into the same area was undertaken by Weichart (1973). Selected data in Table 8 indicates that the atmosphere is at least as important a source of pollution for the North Sea as the Rhine River (Cambray *et al.*, 1975).

Characteristic increases of metal inputs to the North Sea from atmospheric sources will be further discussed in Section 6.2. Sediment studies from the German Bight (southeastern North Sea) have shown significant increases of heavy metals, such as mercury, cadmium, lead, zinc, chromium, and copper, which may partly or dominantly be due to enhanced emissions from burning of fossil fuels, cement production and lead additives (Gadow and Schäfer, 1973, 1974; Förstner and Reineck, 1974; Nauke, in Caspers, 1975). Similar evidence has been gathered for the Baltic Sea by Erlenkeuser *et al.* (1974). and Suess and Erlenkeuser (1975). Suess (1978) writes of a typical 'coal residue assembly' for the heavy metals, cadmium, lead, zinc, and copper, which show in sediment profiles—in that order—an especially pronounced enrichment. Patchineelam and Förstner (1977) found relatively high lead contents in a residual compound—probably lead-oxide—and concluded that there must be a heavy atmospheric contribution of lead in sediments of the south-east North Sea.

Table 7. Annual quantities of metals (in tons/yr) discharged into the Clyde Sea from the most important anthropogenic sources 'sewage sludge' and 'rainwater' (from Topping, *lit. cit.* Cambray *et al.*, 1975).

	Sewage sludge	Rainwater
Cadmium	0.25	1.7
Copper	25	50
Lead	25	60
Zinc	60	250

Table 8. Annual inputs of metals to the North Sea from the Rhine River and atmospheric emissions (Weichart, 1973).

	Discharge through Rhine River (t/a)	Input into North Sea by rainwater
Mercury	100	<20
Cadmium	200	230
Chromium	1,000	840
Arsenic	1,000	1,500
Copper	2,000	6,900
Lead	2,000	8,100
Zinc	20,000	38,000

Effluents from metal-working industries (including mines and smelters)

Severe metal pollution in an estuarine environment was investigated by Stoffers *et al.* (1977) and is given here as an introductory example for the effects from predominantly industrial sources. During a study of sediment dispersal in the Massachusetts coastal zone, it was found that the bottom sediments of New Bedford Harbour on the estuary of the Acushnet River contain large amounts of heavy metals which appear to have an industrial origin. Figures 4(a, b) show the distribution of copper and cadmium in the clay-sized (smaller than <2 μm) fractions of 92 surface sediments collected from grab samples. Copper concentrations gradually decrease seaward from values of more than 6,000 mg/kg near the head of the harbour to less than 100 mg/kg at the edge Buzzards Bay; cadmium concentrations decrease from more than 100 mg/kg in the estuary of the Acushnet River to less than 0.5 mg/kg in the bay area. Figure 4(c) summarizes the metal data of four selected cores, taken in distances from the presumed industrial waste effluents of; (No. 9)—a few hundred meters, (No. 8)—2km, (No. 4)—6 km, and (No. 19)—12 km. Marked increases in the concentrations of heavy metals occur at burial depths of 100 cm in core 9, 50 cm in core 4 and at about 20 cm in core 19. The maximum amount of metal enrichment, compared with the data of a pre-cultural background taken from the analyses of the deeper portions of the Buzzard Bay sediment profiles, is 150 times for Cu, 100 for Cd, 30 for Pb and Cr, and 10 times for Zn. Since the sediments are organically enriched black silts and smell of hydrogen sulphide, it seems highly unlikely that a significant amount of the metal found near the surface results from diagenetic processes occurring within the sediment column; instead, it seems likely that the enrichment of metals reflects environmental contamination. Summerhayes *et al.* (1977) calculate that metal–rich mud has been accumulating at rates of about 3 cm/yr in the deeper parts, and at about 4 mm/yr in the shallower parts of the harbour. The history of waste metal accumulation covers a period of some 80 years, which coincides with the known history of Cu discharge from industrial plants in this area. Inspection of publicly available EPA records suggest that industrial output of Cu may have reached a level of approximately 30 tons in recent years, and that about 1,400 tons of copper have been discharged into the harbour area during the past 80 years. Conservative estimates by Summerhayes *et al.* (1977) indicate that at current prices for Cu, Cr, and Zn, the respective deposits in 4 km mud would be worth about 2 million dollars. The question, however, is whether the metals are in a form that is readily extractible and whether recovery would be economically feasible.

It is often impossible to associate elevated metal contents in sediment samples with certain industrial influences, especially if significant amounts of sewage sludge has been deposited in a certain area by means of dumping or *via* outfalls. From the first four examples in Table 9, the influence of mixed sewage inputs from domestic and industrial sources seems evident; a clear rise in cadmium values in sediments from Port Phillip Bay and in mercury concentrations in sediments from the Solent (Southampton) can most probably be attributed to industrial emissions. When dealing with the direct input of metal pollutants from the metal-working industry into the sea, mention must be made of the disposal of tailings from mines and smelters. Such studies have been mainly carried out on the coast of British Columbia (Thompson, 1977), although strong metal accumulation from mine waste has also been observed and studied in the Rio Tinto Estuary of Spain (Stenner and Nickless, 1975), in estuarine sediments near Moreton's Harbour, Newfoundland (Penrose *et al.*, 1975), off southwestern England (Bryan and Hummerstone, 1971), as well as in the Philippines (Lesaca, 1977), New Guinea (Brown, 1974) and Tasmania (Ayling, 1973; Thrower and Eustace, 1973). In Derwent Estuary (Tasmania), emissions from a zinc refinery were found to have caused a thousand-fold increase of Cd-, Cu-, Pb-, and Zn-levels in the sediments near the plant outlets (Table 9, Bloom and Ayling, 1977). Particular problems may arise when mine and other metalliferous wastes are released into fjords where the exchange of water and sediments with the open ocean is limited. Examples are given from the Sagueney Fjord on the St. Lawrence Estuary, which is injected with mercury from industrial effluents (Loring, 1975), and from the Ide Fjord on the Skagerrak where the mercury level of surface sediments increased by a factor of 100 in a time period of 50 years (Olausson, 1975). In the Sörfjord of western Norway, 45 tons of lead and 60 tons of zinc were discharged daily from a smelting plant, whereby levels of 3% Pb and 12% Zn by weight are reached in the bottom sediments in some parts of the fjord (see Table 9; Skei *et al.*, 1972).

Special problems arise when acid iron waste and alkaline red mud is dumped, here into the Mediterranean Ocean and the southern North Sea

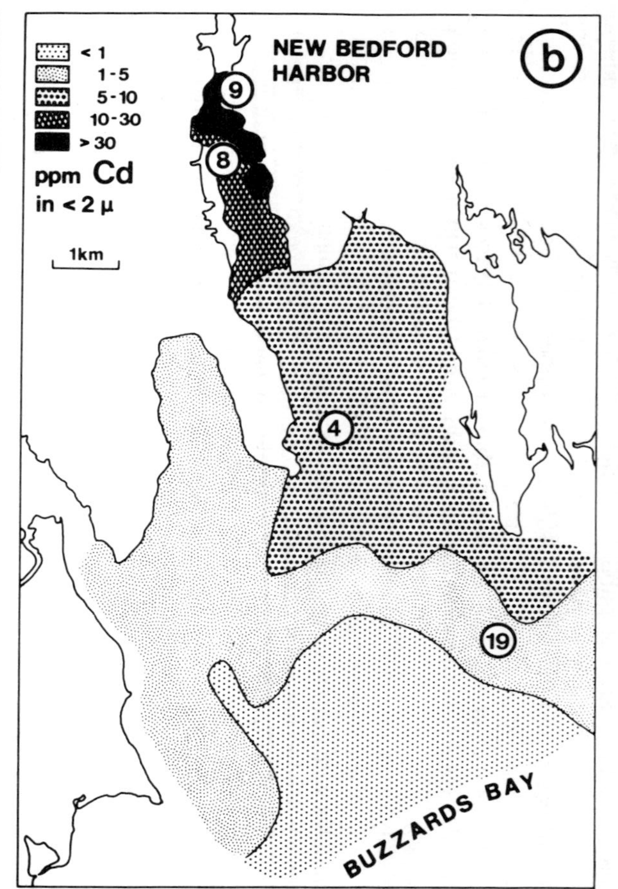
a
< 100
100 - 500
500 - 1000
1000 - 3000
3000 - 5000
> 5000
ppm Cu
in < 2 μ
1km
NEW BEDFORD
HARBOR
9
8
4
19
BUZZARDS BAY

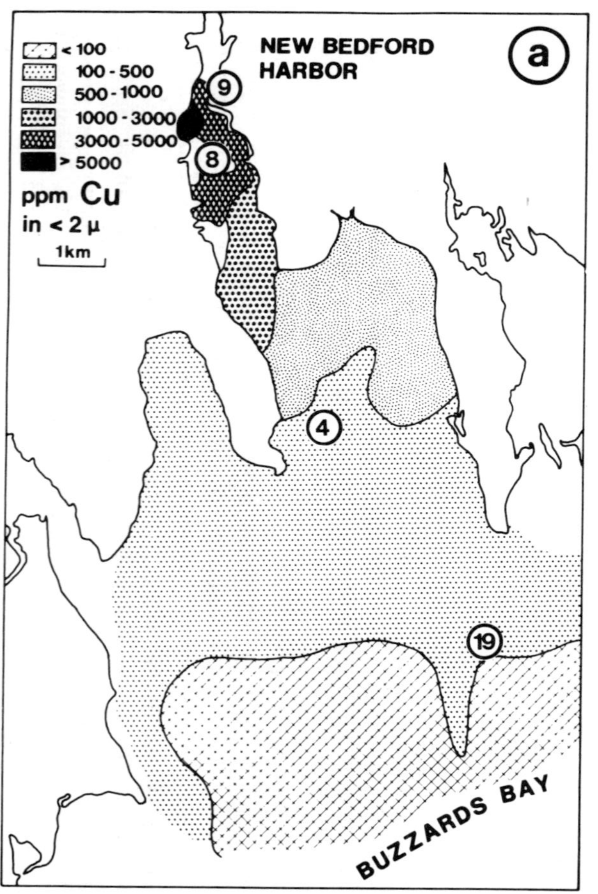
b
< 1
1 - 5
5 - 10
10 - 30
> 30
ppm Cd
in < 2 μ
1km
NEW BEDFORD
HARBOR
9
8
4
19
BUZZARDS BAY

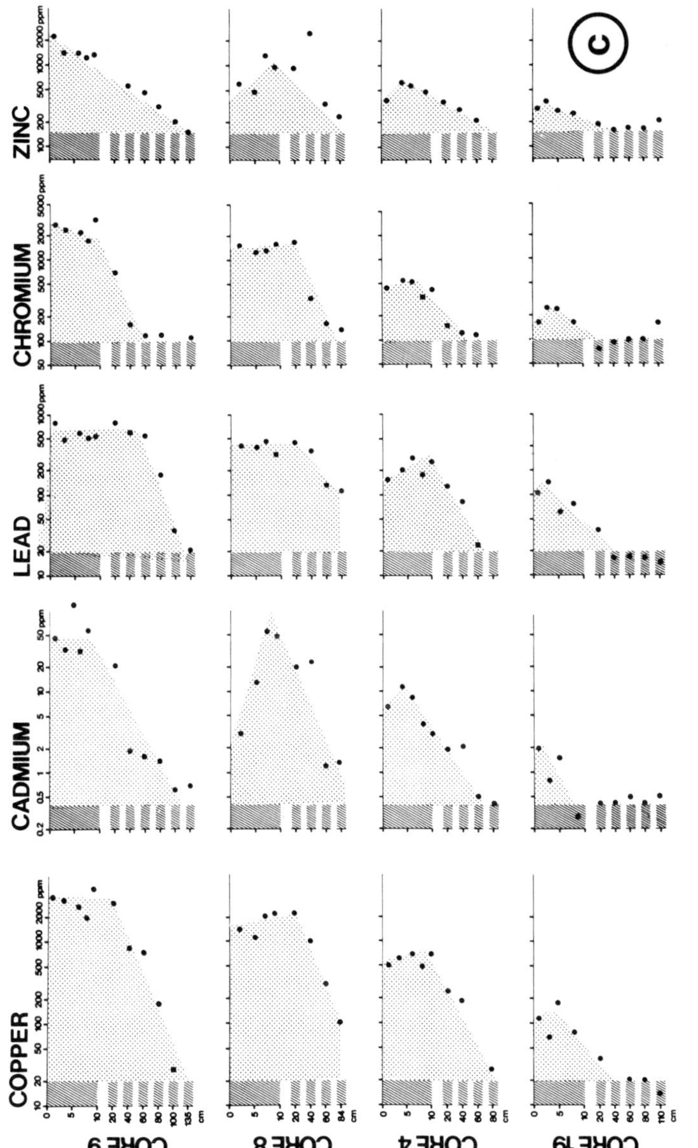

Figure 4. Distribution of heavy metals in the pelitic fraction of surface (a: Cu, b: Cd) and core sediments (c) from New Bedford Harbour and Buzzards Bay, Massachusetts (Stoffers *et al.*, 1977 and unpublished data, Institute of Sediment Research, Heidelberg University).

Table 9. Background and maximum metal concentrations (all values in mg/kg) in sediments from bays, estuaries, and harbour areas.

	Ag	Cd	Cr	Cu	Hg	Pb	Zn
	(0.1)[a]	(0.3)[a]	68[b]	14[b]	(0.4)[a]	13[b]	70[b]
Narragansett Bay[1] (Rhode Island, USA)	3.3[2]	3.9[2]	155	190	—	140	250
Port Phillip Bay (Victoria/Australia)[3]	—	9.9	—	85	—	183	278
Severn Estuary (U.K.)[4]	—	4.7	—	—	—	200	590
Solent (S'hampton, U.K.)[5]	—	4.2	—	—	5.6	—	800
Rio Tinto Est. (Spain)[6]	—	4.1	—	1,400	—	1,600	3,100
Restonguet Estuary (U.K.)[7]	7	12	1,060	4,500	—	1,620	3,000
New Bedford Harbour[8]	40	76	3,200	7,500	3.8	560	2,300
Corpus Christi (USA)[9]	—	130	—	—	—	—	11,000
Derwent Estuary (Tasmania/Australia)[10]	—	862 (1,400)	258	— (10,100)	1,130	1,000 (11,000)	10,000 (104,000)
Sörfjord (Norway)[11]	190	850	—	12,000	—	30,500	118,000

[a] = Average shale. Krauskopf. 1967: [b] = lower section—60–80 cm—of Narragansett core[1] [1]Goldberg *et al.*, 1977; [2]Eisler *et al.*, 1977; [3]Talbot *et al.*, 1976a, b; [4]Butterworth[4] *et al.*, 1972; [5]Leatherland and Burton, 1974; [6]Stenner and Nickless, 1975; Thornton *et al.*, 1975; [8]Summerhayes *et al.*, 1977; [9]Holmes *et al.*, 1974; [10]Bloom and Ayling, 1977, the numbers in parentheses indicate maximum concentrations near the refinery's outfall; [11]Skei *et al.*, 1972.

(Weichart, 1973; Naeve, 1974). Alkaline red mud is a waste product which increases steadily in areas where alumina is industrially extracted from bauxite; harmful affects were observed on various filter and suspension feeders. The dumping of red mud then, seems likely to be a danger to the stocks of bottom fish and fauna.

5.3 Nuclear wastes

The influence of artificial radionuclide wastes on the marine coastal zones is still a topic for lively discussion (Bowen, 1974). Particular concern arises from the fact that the amounts of transuranics produced in nuclear reactors and subsequently used as fuels or in nuclear weapons, are continually being stockpiled. It has been argued that 'an increasing number of nuclear power plants in the coastal areas of the world, with a large number of small leaks of waste materials, can eventually produce a dangerously radioactive ocean over long time periods' (Goldberg, 1976). On the other hand, the management of nuclear debris has become so effective that the amount of radioactive material reaching the marine environment per unit of fuel burnt and reprocessed is constantly being reduced. It is inferred 'that these practices, for example those in the U.K., will ensure that up to the end of this century the environmental waste disposal problem will not lead to any significant increase in public radiation exposure' (Preston *et al.*, 1971). With regard to the behaviour of artificial radionuclides in estuaries, typical examples are given here:

From the early 1940's until 1971 (when the discharges were stopped), the Columbia
reactors in the Hanford area near Richland, Wash. (Glenn, 1973). The principal isotopes found in the effluents were ^{51}Cr, ^{65}Zn, and ^{60}Co. These isotopes are neutron-capture products from cobalt and zinc, present in corroded condenser material and from chromium added as chromate to the cooling water in order to reduce corrosion (Duursma, 1976). Between 1961 to 1963, a period of particularly heavy output, the yearly estimates were 40,000 curies, of which 29,000 curies was ^{51}Cr and 11,000 curies was ^{65}Zn; these had been distributed downstream throughout the river system and the Columbia River Estuary to the Pacific Ocean. Concentrations of ^{51}Cr and ^{65}Zn in surficial sediments from the estuary are approximately 6.2 and 2.2 times, respectively, greater than the concentration of naturally occurring ^{40}K, which averages about 14 picocuries per gram of sediment (Hubbell and Glenn, 1973). Sediment investigations performed by Cutshall *et al.*, (1973) indicates that the ^{65}Zn in the Columbia River Estuary remains bound to the sediment, thereby acting as a tracer for the transport of river sediment to the sea.

The use of sediment cores shows the historical development of the impact of artificial readionuclides; these could originate either from nuclear reactor effluents or from atmospheric fallout. Table 10 shows the results of a study performed by Simpson *et al.* (1976) on the Hudson Estuary. At 'mile point 43' the test results are of a sediment sample taken near a commercial reactor; 'mile point 53.8' is located upstream from the power plant, and 'mile point 0.1' is situated where the Hudson River flows into New York Bight. The sharp increase of the Cs-137 values in the MP 53.8 sample— reaching a maximum between 1962 and 1965—is attributed to radioactive

Table 10. Anthropogenic radionuclides in sediments of the Hudson River Estuary (Simpson *et al.*, 1976).

Measurement station	Sediment depth	Activity in pCi/kg			
		Cs-137	Cs-134	Co-60	Pu 239, 240
MP 53.8	0–5 cm	2475	98	69	62.2
	5–10 cm	1825	17	19	52.1
	10–25 cm	210	−8	15	5.7
	20–25 cm	35	0	7	—
MP 43	0–10 cm	2700	345	400	25.0
MP 0.1	0–5 cm	1260	230	190	32.7
	10–15 cm	1995	480	245	43.3
	20–25 cm	1030	175	125	26.4

fallout from atomic bomb testing. In core MP 53.8, the contents of Pu-239 and Pu-240 tend to register much the same as the Cs-137 values, but it inferred that in the sea, Pu-239, 240 is more quickly and durably bound to solid substances than is Cs-137. The increase of Cs-134 and Co-60 values is accounted for by the periodical and relatively insignificant discharges from the nuclear power plant.

Significant enrichment of Cs-137 was detected by Patel *et al.* (1975) from discharges of low level liquid radioactive waste into Bombay Harbour, a tidal estuary. Maximum deposition of Cs-137 activity was around the zone off Bombay, reaching more than 400 pCi/g dry sediment during 1970–71; it decreases with distance from discharge point and is 100 times lower towards the mouth of the habour in the south.

Large volume low activity aqueous radioactive wastes were discharged from the Windscale (Cumberland, U.K.) nuclear processing plant to the Irish Sea by means of two pipelines, 2.5 km long. Throughout the period 1955 to 1963, the effluent was characterized by a high percentage of Ru-106. The introduction of a new radiochemical reprocessing plant in 1964 has resulted in a sharp increase in the amount of Zr-95 and Nb-95. Measurement of the radioactivity in surface silt from the Ravenglass Estuary, approx. 10 km south of the effluent pipline, indicated gamma dose-rates of approx. 750pCi/g dry sediment for Zr-95+Nb-95, 750 pCi/g for Ru-106 and approx. 300 pCi/g for Ce-144 in the study period 1965/1966 (Jefferies, 1968). The accumulation factors of Zr-95, Nb-95, and Ru-106 in the surface silt layer are approximately 1.5×10^4 compared with the respective values found in seawater, and are about 30 times greater than those for beach sand. Surface sediment samples taken from seven estuarine areas 12 to 120 km away from the Windscale pipeline indicate decreasing concentrations with increasing distance, which can be described by the equations (Preston, 1972);

$$C \; (\text{pCi } {}^{106}\text{Ru/g dry sediment}) = 1{,}250 \, D^{-1.7},$$

$$C \; (\text{pCi } {}^{144}\text{Ce/g dry sediment}) = 2{,}500 \, D^{-2.0},$$

$$C \; (\text{pCi } {}^{95}\text{Zr} + {}^{95}\text{Nb/g dry sediment}) = 3{,}150 \, D^{-2.3},$$

where D = distance in km from the Windscale outlet. The increasing values of the exponents reflect the radio-active half-lives of the three nuclides (${}^{95}\text{Zr} < {}^{144}\text{Ce} < {}^{106}\text{Ru}$), the values of the constituents reflect the relative affinities of the nuclides concerned for the sediments (${}^{106}\text{Ru} > {}^{144}\text{Ce} > {}^{95}\text{Zr}$), i.e. they are a measure of the relative concentration factors. With these data, it has been suggested by Preston (1972) that it would be possible 'to derive estimates of the contamination which may arise through the cumulative effect of several geographically separate contributions—an approach of undoubted value in semi-enclosed areas of international waters such as the southern North Sea.'

6 POLLUTION OF ESTUARINE WATERS BY ARTIFICIAL RADIONUCLIDES AND TRACE METALS

At the end of the sixties, in another estuary in the U.K., Preston and co-workers performed a large number of investigations, which have affected our knowledge of the distribution of artificial radionuclides and trace metals in estuarine areas, as well as giving an insight into the intake of pollutants by the local organisms: the Blackwater Estuary, which received low-level radioactive effluent from the nuclear power station at Bradwell-on-Sea, Essex is an example (Preston, 1968). Two groups of radionuclides (and also their stable counterparts) can be distinguished: those which are largely conservative with respect to water and tend to be flushed from the estuary, and those which are readily adsorbed to sediment and remain trapped in the estuary for long periods of time. The radionuclides of zinc, iron, manganese, cobalt, ruthenium, cerium and zirconium are very readily transferred to estuarine sediment. Cesium, in marked contrast to its behaviour in fresh water, where it completes with sodium and potassium for exchangeable positions, is removed to a much smaller extent, like silver; chromium and antimony show negligible involvement with sediments or biological material. Many of the metals removed to sediments subsequently show high concentrations in filter-feeding bivalve molluscs (Preston, 1972; see also Section 7 of the present article).

6.1 Artificial radionuclides in estuarine water

As compared with other relatively mobile artificial radionuclides, Cs-^{137} is specially considered as the tracer of other pollution inputs in freshwater systems (Plato and Jacobson, 1976). Kautsky (1973) has shown that measurements of the spatial and temporal distribution of Cs-137 provide detailed evidence about water mass transport and current velocities in coastal areas. Figure 5 (after Jefferies *et al.*, 1973 and Murray and Kautsky, 1977) shows the travel routes of Cs-137 emanating from the Windscale nuclear fuel processing plant: The contours illustrate that the outflow from the Irish Sea is through the North Channel; Cs-labelled water flows close to the coastline of Scotland and through the Minch. From Cape Wrath in North Scotland, the concentration of Cs-137 falls only slowly with distance along the north coast of Scotland towards the Pentland Firth and into the northern North Sea. (There is an additional point source influx of Cs^{137} into this coastal system from the Experimental Reactor at Dounreay, but discharges are only approx. 2 per cent of the Windscale discharges; Jefferies *et al.*, 1973). The Cs-labelled water-masses are then transported along the English east coast down to the southern North Sea and then, together with the water-masses which come from the English Channel, continue along the Norwegian coast back into the Atlantic (Kautsky, 1973). Elevated concentrations of Cs-137

                           U. Förstner

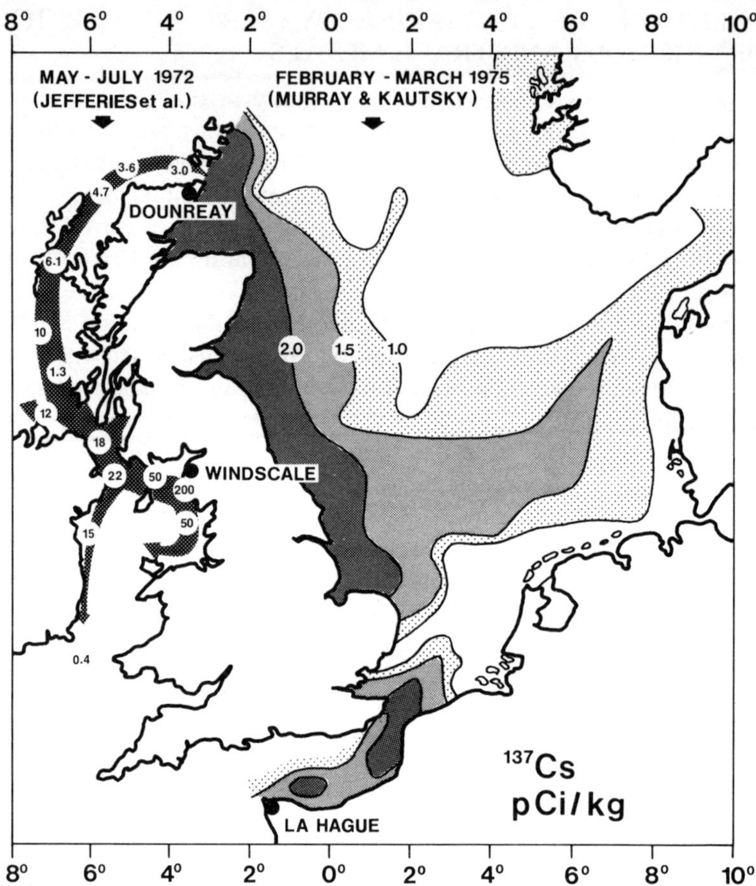

Figure 5. Distribution of ^{137}Cs (pCi/kg) in surface water from the Irish Sea (Jefferies *et al.*, 1973) and the North Sea (Murrray and Kautsky, 1977).

off the Belgian/Netherlands coastline (Figure 5) can be traced back to the single ejection of a rather large amount of active waste solution from the nuclear fuel reprocessing plant at La Hague, near Cherbourg (Kautsky, 1973). In this area, particularly high concentrations (approx. 1×10^{-2} pCi/kg water) of $^{239+240}$Pu has been found by Murray and Kautsky (1977), probably introduced from nuclear wastes, whereas in other parts of the North Sea, most of the plutonium (concentrations being 10 to 15 times smaller) appear to be due only to fallout from nuclear weapons testing.

6.2 Metal pollution in estuarine water

Over the years, many research groups have thoroughly investigated the distribution of trace metals in the British Isles. Some of their results are given here, to illustrate the origins and distribution patterns of dissolved heavy metal components in estuaries, with special emphasis on possible mercury pollution.

Burton and Leatherland (1971) analysed coastal waters off Netley and Calshot (North Sea) and the English Channel and found values of 0.013 to 0.021 μg of mercury per litre water; Smith *et al.* (1971) detected between 0.006 and 0.013 μg Hg/l for tidal waters of the Thames River. The data of dissolved mercury do not differ regionally from normal seawater levels ($\sim$0.011 μg/l; Gardner, 1975), and it was suggested that this behaviour might be due to the short residence times of mercury, which is caused by its affinity for suspended matter (Gardner and Riley, 1973). Recent investigations by Baker (1977) however, clearly show the major sources of mercury emissions, which are probably caused by sewage sludge dumping. 'High spots' of mercury contamination are apparent in the Thames Estuary, Liverpool Bay, the area off the Humber Estuary and to a less extent, the Bristol Channel; samples from the North Sea indicate elevated values at the mouths of the Rhine and Elbe rivers. With regard to other 'anthropogenic' metals, significant enrichment has been observed in water samples from the coastal zone west of England, Wales, and Scotland. Table 11 summarizes analytical data for zinc, copper, lead, and cadmium from various authors.

The data indicate particular strong enrichment of dissolved metals in the Firth of Clyde, Conway Bay, Liverpool Bay, Cardigan Bay, and Bristol Channel, more than 10-fold for Zn, Cd, Pb, and 5-fold for Cu, compared to the normal values from the open ocean or even from the Irish Sea. The distribution of these trace metals in the latter three coastal areas has been studied by Abdullah *et al.* (1972). Figure 6 shows the examples of cadmium and zinc: The concentration of cadmium (Figure 6(a)) in the Bristol Channel is the highest encountered in the Irish Sea coastal water, levels over 4 μg/l (about 50 times that of open sea levels) being found in

Table 11. Mean values of zinc, copper, lead, and cadmium in coastal waters off Great Britain.

	Zn μg/l	Cu μg/l	Pb μg/l	Cd μg/l	
Surface world ocean	1.4	0.8	—	0.07	Chester and Stoner (1974)
Atlantic Ocean (Iceland-Faroe Ridge)	1.9	0.5	—	0.07	Kremling and Petersen (1977)
Irish Sea (offshore)	3.0	0.59	0.19	0.11	Preston *et al.* (1972)
Firth of Clyde	18.0	3.0	9.0	0.5	Halcrow *et al.* (1973)
Liverpool Bay	11.9	1.45	1.74	0.27	Abdullah *et al.* (1972)
Conway Bay (near-shore)	8.8	3.0	3.6	0.76	Elderfield *et al.* (1971)
Cardigan Bay	7.5	1.72	2.24	1.11	Abdullah *et al.* (1972)
Bristol Channel	8.1	2.24	1.38	1.94	Abdullan and Royle (1974)
English Channel	2.0	0.46	0.17	0.06	Preston *et al.* (1972)
North Sea (near-shore)	6.3	1.4	—	0.5	Dutton *et al.* (1973)

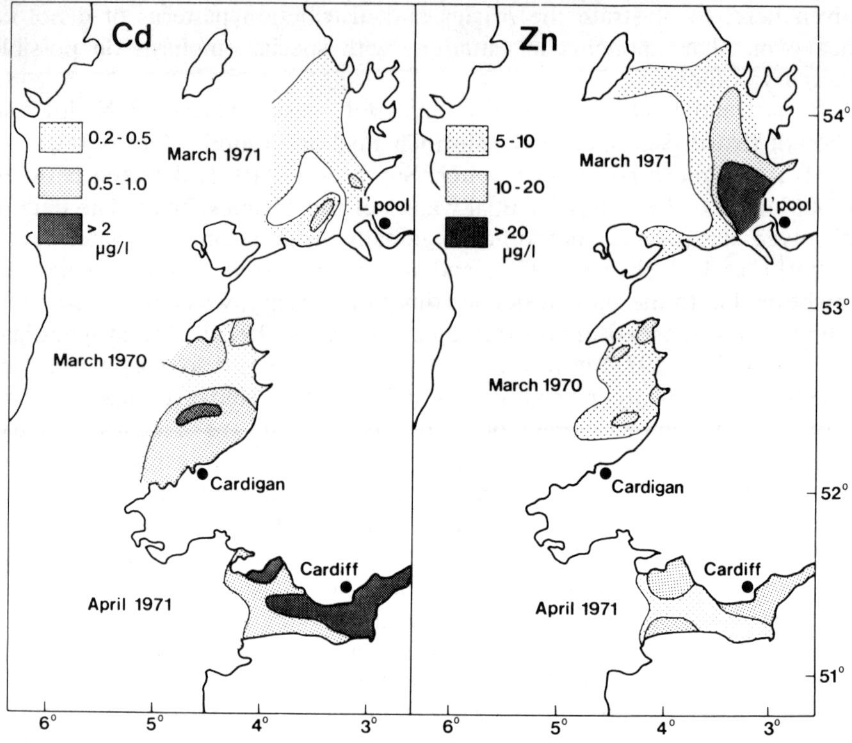

Figure 6. Zinc and cadmium enrichment in waters of the Irish Sea derived from industrial, domestic, and mine waste effluents (Abdullah *et al.*, 1972).

the eastern part of the Channel. The cadmium is possibly derived entirely from industrial effluents entering the Bristol Channel from the Avon and Severn Estuary. In Cardigan Bay, which is relatively free from industrial effluent and where, because of a low population density, little domestic waste is present, the runoff from the mineralized zones and sites of former mining activity is the main source of trace metals, in particular cadmium. On the other hand, the dominant sources of material affecting trace metal levels in Liverpool Bay are domestic and industrial effluents; highest values of zinc (Figure 6(b)) are found in the Mersey Estuary, being approximately 30 times those of open sea levels.

7 BIOACCUMULATION OF INORGANIC POLLUTANTS

It has been emphasized by many investigators of marine pollution problems (among others, Bryan, 1976, Bergmann and Müller-Hoberg, 1977, Patchineelam and Förstner, 1977) that conditions for the mobilization of metals may be particularly favourable in estuaries in view of the fluctuating salinity and the disturbance of the sediments by currents or by dredging, and that as a result, contaminated sediments may persist as sources of metals even when the original source has been removed. At the same time, as has been shown in the previous section, metal concentrations either in dissolved or complex form are particularly enriched in the estuarine waters, mainly as a result of polluted river inflow. Estuaries, therefore, are characteristic areas of bioaccumulation of trace metals and of other inorganic and organic pollutants.

Exemplary investigations performed by Butterworth *et al.* (1972) in the Severn Estuary are chosen here to demonstrate the effects of pollution on the concentrations of metals in aquatic organisms. Coastal waters bordering the southern shore of the Bristol Channel have been shown to contain abnormal amounts of cadmium, zinc, and lead, which are probably introduced from the Bristol area via the River Avon. In the water samples, the effects of the pollution have been traced as far away as Hartland Quay, some 150 km to the west from Avonmouth into the Bristol Channel. Table 12 indicates that the contamination in the water (by cadmium, similar effects have been observed for zinc) is obviously transmitted to the living material inhabiting this shore—at relatively low levels in seaweed *Fucus* (the producer), at higher levels in the limpet *Patella* (a primary consumer) and greatest concentrations in the dog whelk *Thais* (the secondary consumer of polluted organisms). Although this type of amplification of metals at higher trophic levels has generally not been confirmed by the studies of Butterworth *et al.* (1972), there is clear evidence that the highest concentration, both in the water and in the biological material, occur near the source of pollution and that the metal content in the estuary decreases nearer the open sea.

Investigations on the time sequence of metal accumulation in organisms in polluted estuaries have been performed by Stenner and Nickless (1974) in the same area. Both limpets and dog whelks were transferred from a relatively unpolluted coastal area near Beer in Dorset to St. Andrews Head on the Bristol Channel (see: Table 13). Although only a small number of

Table 12. Cadmium concentrations in water, seaweeds and shore animals of four collecting stations on the southern side of Severn Estuary and Bristol Channel (from Butterworth *et al.*, 1972).

Collecting point	Distance from Avonmouth	Seawater μg Cd/1	*Fucus* mg Cd/kg	*Patella* mg Cd/kg	*Thais* mg Cd/kg
Portishead	4 km	5.8	220	550	—
Brean	25 km	2.0	50	200	425
Minehead	60 km	1.0	20	50	270
Lynmouth	80 km	0.5	30	50	65

Table 13. Time-dependent absorption of cadmium, copper, and zinc by limpets and dog whelks in the Bristol Channel (from Stenner and Nickless, 1974).

Sample	Date (1973)	(n)	Patella Cd mg/kg	Patella Cu mg/kg	(n)	Thais Cd mg/kg	Thais Cu mg/kg
Before transplantation	April 8	(11)	11	7	(10)	36	70
Transplanted specimen	April 18	(4)	10	11	(6)	40	140
	May 6	(3)	14	17	(10)	63	175
	June 2		—	—	(6)	92	120
	July 29	(2)	200	30	(6)	184	200
Local specimens	Dec. 1972	(9)	220	30	(10)	460	740

Patella specimen could be analysed the data indicate that limpets rapidly absorb both copper and cadmium. On the other hand, the concentrations of copper and cadmium show only a small increase in the transplanted specimen of the dog whelk during the four month study period. Such effects must be considered when biological indicator organisms are used to monitor trace metal pollution in estuaries (e.g. Phillips, 1977).

In a review on *Heavy Metal Contamination in the Sea*, Bryan (1976) has summarized the factors affecting the adsorption, excretion, and storage of metals in organisms. From a large number of investigations in marine coastal areas on heavy metals in organisms, a compilation of geometric mean values was made, pertaining to less-contaminated conditions: selected metal examples and organism species are presented in Table 14. It can be seen that an increase in metal concentration in the higher trophic levels will not occur (at least in the range of normal metal contents), although the opposite has often been postulated. However, certain species of organisms seem to be capable of a selective enrichment by metals, exemplified by the data of copper, cadmium, and zinc in oysters.

If the normal contents in the algae samples are compared with those of seawater, enrichment factors of 2.5 to 10×10^3 for Ni, Cd, and As; 15–20×10^3 for Hg, Cu, and Ag; 60×10^3 for Zn and 130×10^3 for lead, are arrived at. On the other hand, enrichment factors in fish, with the exception of mercury, where a clear increase in respect to the algae is measured, are 1/2 the size of the algae values. Seen from the sediment aspect—in this case, the shale standard—there is an insignificant enrichment of As and Cd; Ag, Zn, and Hg levels are about the same in both mediums; Cu, Pb and especially Ni levels in the organism samples are clearly lower than the corresponding inorganic matrices.

Table 14 gives other examples of trace elements in mussels—the most important indicator organisms—for local pollution, especially in estuarine areas. Crucial estuary pollution areas include the English west coast to the

Irish Sea, especially the Severn Estuary, then those of Tamar and Derwent, areas of large sewer outlets such as those on the California coast and areas of restricted water circulation. Maximum enrichments in respect to the normal values were registered especially in areas affected by waste outputs from mines and smelters. The increase of metal contents in a number of organisms is highest where maximum enrichment have been observed in sediments; for example, in Sör Fjord, an arm of Hardanger Fjord, which is polluted by waste effluents from a Pb–Zn smelter (Stenner and Nickless, 1974). Here, in green algae, lead was 300 times greater, cadmium (in mussels) 4–70 times greater and zinc 25 times greater than normal levels. The strain on mussels and oysters from cadmium and mercury is also quite pronounced in the estuarine districts of Tasmania and in Port Phillip Bay, near Melbourne (Australia). With regard to the accumulation of artificial radionuclides in estuarine organisms, maximum concentrations have been reported by Preston *et al.* (1971) from the Blackwater Estuary. Oysters kept in cages close to the cooling water outfall of the Bradwell nuclear power station contained 43.1 pCi Zn-65 per gram wet weight; specimen taken from the nearest commercial bed (0.5 km from the outfall) exhibited values of more than 4 pCi of Zn-65. Investigations of Preston (1972), however, indicate that an average consumption rate of 75 g/day by the critical consumers, local fishermen and their families, would contribute less than 0.1 per cent of the maximum permissible dose rate. Bioaccumulation of strontium-90 and phosphorous-32 in fish appear to be of greater significance. An example is given from the Columbia River, where the incorporation of phosporous-32 in fish results in about 10% of the acceptable dose limit found in bone material of the fish-eating population (Wrenn and Jinks, 1975).

In respect to an acute harmful effect of heavy metals on man, the manifold increased percentages of mercury and cadmium warrants concern. The concentrations of the metals are, however, not as important as the daily rate of intake as well as the form the metals take in food. In the UK, the average normal daily intake of mercury by an adult is about 7.5 μg, of which 2 μg are contributed by eating 24 g of fish, with an average Hg (content wet weight of 0.08 μg/g (Bryan, 1976).

In some areas a further increase in mercury contents has recently been noted. For deep water fishing as well as coastal fishing—the latter exemplified by the fishing region near the mouth of the Elbe, clearly greater Hg-enrichments in young fish is accurately traced to an abnormally high contamination of this region (Krüger *et al.*, 1975). Similar effects in fish have been observed in dutch coastal waters, the mouth of the Thames and the Irish Sea (Gerlach, 1976), and in mussels from these same locations (de Wolf, 1976).

Increased cadmium values are presently most obvious in oysters; according to Ratkowsky *et al.* (1974), high concentrations of Cd and Zn in oysters

Table 14. Metal concentrations in marine fauna (ppm) (b = brown algae, g = green algae; m = mammals, s = seals; + = p.p.m. wet weight; other values in p.p.m. dry weight.

		Phyto-plankton	Algae	Mussels	Oysters	Gastropods	Crustaceans	Fish	Seals Mammals
Arsenic	Geometric mean [1]	—	20	15	10	20	30	10	
	Newfoundland [2]	—	9.8–17(b)	1.6–5	—	4.0–11.5	3.8–7.6	0.4–0.8	
	England [3]	—	26.0–54(b)	1.8–15	2.6–10	8.1–38	16	1.7–8.7	
	Greenland [4]	—	36(b)	14–17	—	—	63–80	14.7–307	
Cadmium	Geometric mean (1)	2	0.5	2	10	6	1	0.2	
	Spain [5]	—	0.8–4(b)	0.5–8	2.9–3.5	1.1–9	0.7–32	<0.4–4.3	
	England [3][6–9]	—	0.2–53(b)	3.7–65	6–54	3.5–1120	2.8–33	0.06–3.96+	2.2–11.6+
	Australia [10–12]	—	—	4.2–83	9–174	2.8–30	—	0.05–0.4+	—
	Norway [13]		1.0–13(b)	1.9–140	—	0–51	1.9–7	<0.01–0.03+	
Copper	Geometric mean [1]	7	15	10	100	60	70	3	
	Spain [5]	—	5–26 (g)	6–14	120–435	5–50	110–435	<0.6–10	
	England [6–8][14–15]	—	4–141 (b)	7–15	20–6480	0–1750	6–64+	0.5–14.6+	
	California [16–21]	—	—	7–77	10–2100	3–177	(4–150)	(16–29.3)	14.5–386 (m)
	Norway [13]		9–170 (g)	3–120	—	17–190	2–90	—	
Lead	Geometric mean [1]	4	4	5	3	5	1	3	
	Spain [5]	—	4–20 (g)	2–15	4–11	10–27	<1.2–11	<1.2–22	
	England [6][8][15][21, 22]	—	16–66 (g)	7–19	5–17	0.2–0.8	8	14–28	0–4+
	California [18][24–26]	—	—	0.3–42	—	0.6–21	—	<0.001–5.3	0.3–34.2 (s)
	Norway [13]	—	3–1200 (g)	2–3100	—	0–39	8.3	—	—

Mercury	Geometric mean [1]	0.17	0.15	0.4	0.4	0.2	0.4	0.4	
	Hawaii [27, 28]	—	—	—	—	0–0.03	0.03–0.12	0.02–23	0.6–103 (m)
	California [21][29]	—	—	—	—	<0.01–0.07+	0.02–0.04+	0.02–0.2	0.1–700 (s)
	Atlantic [30–35]	0.2–5.3	<0.01–0.07+	<0.01–0.13+	0.02–180+	—	<0.05–0.6+	0.1–9.0	—
	Mediterranean [36, 37]	—	<0.5–0.7+	0.25–0.4	—	0.1–3.5+	0.3–4.5+	0.1–29.8+	—
	Australia [11, 12]	—	—	0.05–0.23+	1.5–8.2	0.32–0.65	—	0.3–16.5	
	Norway [38][39][40]	0.5–25.2	—	0.24–0.84	—	0.61	0.31–0.39	0.14–7.3	0.1–106 (m)
	England [3, 8, 9, 22, 41–43]	—	<0.01–25.5 (g)	0.64–1.86	0.56–1.2	0.02–1.84	−0.98	0.02–1.8	0.4–225 (s)
Nickel	Geometric mean [1]	3	3	3	1	2	1	1	
	England [3][6][9]	—	4–33 (g, b)	5–12	2–174	8.8–12.3	1.1–12.3	0.5–10.6	
	California [18][21]	—	—	3.3–20	—	1.8–18.5	—	—	
Silver	Geometric mean [1]	0.2	0.2	0.3	—	1	0.4	0.1	
	California [18, 44, 21]	—	—	0.7–46	—	0.4–10.7	—	—	0.1–1.2 (m)
Zinc	Geometric mean [1]	38	90	100	1,700	200	80	80	
	South Africa [45, 46]	0.6–710	5.6+(g)	73–113	400–886	12+	17+	3.2–7.2+	
	California [17, 18, 44, 47]	0.1–725		46–244	70–8430	1.7–288	—	—	78–875
	Spain [5]	—	63–345 (g)	190–370	310–920	60–120	79–330	21–220	
	Australia [11, 12]	—	—	170–1350	3,740–38,700	56–1050	—	4–375	
	England [3, 6, 8, 9]	—	28–1240 (g)	12–779	1,830–99,200	9.7–4500	36–82	2–342	
	Norway [13]	—	20–2,310 (g)	105–2,370	—	87–2,900	12–32	—	

(1) Bryan. 1976; (2) Penrose *et al.*, 1975; (3) Leatherland & Burton, 1974; (4) Bohn, 1975; (5) Stenner and Nickless, 1975; (6) Boyden, 1975; (7) Stenner and Nickless, 1974a; (8) Peden *et al.*, 1973; (9) Wright, 1976; (10) Talbot *et al.*, 1976a,b; (11) Bloom and Ayling, 1977; (12) Mackay *et al.*, 1975; (13) Stenner and Nickless, 1974b; (14) Bradfield *et al.*, 1976; (15) Boyden and Romeril, 1974; (16) Young and McDermott, 1975; (17) Ruddell and Rains, 1975; (18) Schwimer, 1973; (19) Parsons *et al.*, 1973; (20) Fletcher *et al.*, 1975; (21) Martin *et al.*, 1976; (22) Hardisty *et al.*, 1974; (23) Roberts *et al.*, 1976; (24) Chow *et al.*, 1976; (25) Chow *et al.*, 1974; (26) Braham, 1973; (27) Klemmer *et al.*, 1976; (28) Schultz *et al.*, 1976; (29) Eganhouse and Young, 1976; (30) Windom *et al.*, 1973; (31) Johnson and Braman, 1975; (32) Freeman *et al.*, 1974; (33) Koper, 1974; (34) Greig *et al.*, 1976; Gibbs *et al.*, 1974; (36) Renzoni *et al.*, 1974; (37) Vucetic *et al.*, 1974; (38) Skei *et al.*, 1976; (39) Andersen and Neelakantan, 1974; (40) Havre *et al.*, 1973; (41) Jones *et al.*, 1972; (42) Raymont, 1972; (43) Pentreath, 1976; (44) Alexander and Young, 1976; (45) As *et al.*, 1975; (46) Watling and Watling, 1976a,b; (47) Knauer and Martin, 1972.

from the Derwent Estuary in Tasmania probably caused nausea and vomiting. The admissible daily intake of cadmium is considered to be $100\,\mu g/70\,kg$ of body weight (Ohnesorge, 1974). This amount is reached with approx. 50 grams of oyster (wet weight) from a moderately polluted area, and 10 grams of oysters from some parts of Derwent Estuary.

It has been suggested by Bryan (1976) that a certain risk exists from metal contamination of marine origin for those who eat large amounts of fish or shell fish from estuarine or coastal industrial areas. For those who eat fish more sparingly, there seems to be no real present danger. On the other hand, sub-lethal toxicity experiments have shown that with a further increase in heavy metal contents in estuary zones, increasingly unfavourable ecological consequences—as well as chronic toxic effects on man—will have to be expected.

REFERENCES

Abdullah, M. I., and Royle, L. G. (1974). A study of the dissolved and particulate trace elements in the Bristol Channel. *J. Mar. Biol. Ass. U.K.*, **54,** 581–597.

Abdullah, M. I., Royle, L. G., and Morris, A. W. (1972). Heavy metal concentration in coastal waters. *Nature*, **235,** 158–160.

Alexander, G. V., and Young, D. R. (1976). Trace metals in Southern California mussels. *Mar. Pollut. Bull.*, **7,** 7–9.

Andersen, A. T., and Neelakantan, B. B. (1974). Mercury in some marine organisms from the Oslofjord. *Norw. J. Zool.*, **22,** 231–237.

Anon. (1955). U.S. National Bureau of Standards: Maximum permissible amounts of radioisotopes in the human body and maximum permissable concentrations in air and water. *Handbook*, **52,** 45p.

Applequist, M. D., Katz, A., and Turekian, K. K. (1972). Distribution of mercury in the sediments of New Haven (Conn.) *Harbor. Environ. Sci. and Technol.*, **6,** 1123–1124.

As, D. van, Fourie, H. O., and Vleggaar, C. M. (1975). Trace element concentrations in marine organisms from the Cape west coast. *S. Afr. J. Sci.*, **71,** 151–154.

Aubert, M., *et al.* (1974). Use of a trophodynamic chain of the neritic type with molluscs for the study of the transfer of metallic pollutants. *Rev. Int. Oceanogr. Méd.*, **33,** 7. cit. JWPCF, **47** (6), p. 1625.

Ayling, G. M. (1973). Uptake of cadmium, zinc, copper, lead, and chromium in the Pacific oyster, *Crassostrea gigas*, grown in the Tamar River, Tasmania. *Water Res.*, **8,** 729–738.

Baker, C. W. (1977). Mercury in surface waters of seas around the United Kingdom. *Nature*, **270,** 230–232.

Bergmann, H., and Müller-Hoberg, C. (1977). Spurenmetalle in Küstengewässern (Literaturstudie). Bundesanstalt für Gewässerkunde, Koblenz, 91p.

Bloom, H., and Ayling, G. M. (1977). Heavy metals in the Derwent Estuary. *Environm. Geol.*, **2,** 3–22.

Bohn, A. (1975). Arsenic in marine organisms from west Greenland. *Mar. Pollut. Bull.*, **6,** 87–89.

Bothner, M. H., and Carpenter, R. (1974). Rate of mercury loss from contaminated estuarine sediments in Bellingham Bay, Washinton, *Proc. First Ann. NSF Trace Contam. Conf.*, 1973, 198–210.

Bowen, V. T. (1974). Transuranic elements and nuclear wastes. *Oceanus*, **18,** 43–54.

Boyden, C. R. (1975). Distribution of some trace metals in Poole Harbour, Dorset.

Mar. Pollut. Bull., **6**, 180–187.

Boyden, C. R., and Romeril, M. G. (1974). A trace metal problem in pond oyster culture. *Mar. Pollut. Bull.*, **5**, 74–78.

Bradfield, R. E. N., Kingsbury, R. W. S. M., and Rees, C. P. (1976). An assessment of the pollution of Cornish coastal waters. *Mar. Pollut. Bull.*, **7**, 187–193.

Braman, H. W. (1973). Lead in the California sea lion (*Zalophy californianus*). Environ. Pollut., **5**, 253–258.

Brown, M. J. F. (1974). A development consequence, disposal of mining waste on Bougainville, Papua, New Guinea. *Geoforum*, **18**, 19–27.

Bruland, K. W., Bertine, K., Koide, M., and Goldberg, E. D. (1974). History of metal pollution in Southern California, coastal zone. *Environ. Sci. Technol.*, **8**, 425–432.

Brumsack, H. J. (1977). Potential metal pollution in grass and soil samples around brickworks. *Environ. Geol.*, **2**, 33–41.

Bryan, G. W. (1976). Heavy metal contamination in the sea. In: Johnston, R. (ed.) *Marine Pollution*, pp. 185–302, Academic Press, London.

Bryan, G. W., and Hummerstone, L. G. (1971). Adaptation of the polychaete *Nereis diversicolor* to estuarine sediments containing high concentrations of heavy metals. I. General observations and adaption to copper. *J. Mar. Biol. Ass. U.K.*, **52**, 845–863.

Buffa, L. (1976). Review of environmental control of mercury in Japan. Canada Environmental Protection Service. *Econ. and Technol. Review Rept.*, EPS 3-WP-76-7, 81 pp.

Burton, J. D. (1975). Radioactive nuclides in the marine environment. In: Riley, J. P., and Skirrow, G. (eds.). *Chemical Oceanography*, Vol. 3, 2nd ed., pp. 91–191, Academic Press, London–New York–San Fransisco.

Burton, J. D., and Leatherland, T. M. (1971). Mercury in a coastal marine environment. *Nature*, **231**, 440–441.

Butterworth, J., Lester, P., and Nickless, G. (1972). Distribution of heavy metals in the Severn Estuary. *Mar. Pollut. Bull.*, **3**, 72–74.

Cambray, R. S., Jeffries, D. F., and Topping, G. (1975). An estimate of the input of atmospheric trace elements into the North Sea and the Clyde Sea (1972–1973). *U.K. Atomic Energy Authority Harwell Rept. AERE-R* 7733, 30 pp.

Carmody, D. J., Pearce, J. B., and Yasso, W. E. (1973). Trace metals in sediments of New York Bight. *Mar. Pollut. Bull.*, **4**, 132–135.

Caspers, H. (1975). Pollution in coastal waters: An interim report on the results of the priority programme from 1966–1974. *A Report of the German Research Society*, Boppard, Boldt-Verlag.

Cato, I. (1977). Recent sedimentological and geochemical conditions and pollution problems in two marine areas in southwestern Sweden. *Striae Upsaliensis pro Geologia Quaternaria*, **6.**

Chan, J. P., Cheung, M. T., and Li, F. P. (1974). Trace metals in Hong Kong waters. *Mar. Pollut. Bull.*, **5**, 171–174.

Chester, R., and Stoner, J. H. (1974). The distribution of zinc, nickel, manganese, cadmium, copper, and iron in some surface waters from the world ocean. *Mar. Chem.*, **2**, 17–32.

Chow, T. J., Bruland, K. W., Bertine, K., Soutar, A., Koide, M., and Goldberg, E. D. (1973). Lead pollution, records in Southern California coastal sediments. *Science*, **181**, 551–552.

Chow, T. J., Snyder, H. G., and Snyder, C. B. (1974). Mussels (mytilus sp.) as an indicator of lead pollution. *Sci. Total Environ.*, **6**, 55–63.

Chow, T. J., Snyder, C. B., Snyder, H. G., and Earl, J. L. (1976). Lead content of some marine organisms. *J. Environ. Sci. Health*, **A11**, 33–44.

Cutshall, N., Renfro, W. C., Evans, D. W., and Johnson, V. (1973). Zinc-65 in Oregon-Washington continental shelf sediments. In: Nelson, D. J. (ed.) Radionuclides in Ecosystems. *Proc. 3rd Nat. Symp. Radioecol. Rept. Conf.*, 710501, U.S. AEC, Washington D. C., pp. 694–702.

Duinker, J. C., and Nolting, R. F. (1976). Distribution model for particulate trace metals in the Rhine Estuary and the Southern Bight. *Mar. Pollut. Bull.*, **8,** 56–71.

Dutton, J. W. R., Jefferies, D. F., Folkard, A. R., and Jones, P. G. W. (1973). Trace metals in the North Sea. *Mar. Pollut. Bull.*, **4,** 135–138.

Edmonds, J. S., and Francesconi, K. A. (1977). Methylated arsenic from marine fauna. *Nature,* **265,** 436.

Eganhouse, R. P., and Young, D. R. (1976). Mercury in mussels. Coastal Water Res. Proj. El Segundo, Calif., *Ann. Rept.*, 105–109.

Eisler, R., Lapan, R. L., Telek, G., Davey, E. W., Soper, A. E., and Barry, M. (1977). Survey of metals in sediments near Quonset Point, Rhode Island. *Mar. Pollut. Bull.*, **8,** 260–264.

Elderfield, H., Thornton, I., and Webb, J. S. (1971). Heavy metals and oyster culture in Wales. *Mar. Pollut. Bull.*, **2,** 44–47.

Erlenkeuser, H., Suess, E., and Willkomm, H. (1974). Industralization affects heavy metal and carbon isotope concentration in recent Baltic sea sediments. *Geochim. Cosmochim. Acta,* **38,** 823–842.

Fletcher, G. L. *et al.* (1975). Copper, zinc, and total protein levels in the plasma of sockeye salmon (*Oncorhynchus nerka*) during their spawning migration. *J. Fish. Res. Board Can.,* **32,** 78. *cit. JWPCF,* **48,** (6), 1444/45.

Förstner, U. (1977). Metal concentrations in freshwater sediments—natural background and cultural effects. In Golterman, H. L. (ed.), *Interactions between Sediments and Freshwater,* pp. 94–103. Pudoc/Junk B. V. Publ. Wageningen/The Hague.

Förstner, U. (1978). Sources and sediment associations of heavy metals in polluted coastal regions. In Ahrens, L. H. (ed.), *Proc. 2nd Symp. on the Origin and Distribution of the Elements,* pp. 849–866. Pergamon Press, Oxford.

Förstner, U. (1980). Cadmium in polluted sediments. In: Nriagu, J. O. (ed.), *Biogeochemistry of Cadmium.* Wiley, New York (in press).

Förstner, U., and Müller, G. (1973). Heavy metal accumulation in river sediments, a response to environmental pollution. *Geoforum,* **14,** 53–62.

Förstner, U., and Müller, G. (1974). *Schwermetalle in Flüssen und Seen.* Springer-Verlag, Berlin–Heidelberg–New York, 225 pp.

Förstner, U., and Reineck, H-E. (1974). Die Anreicherung von Spurenelementen in den rezenten Sedimenten eines Profilkerns aus der Deutschen Bucht. *Senckenbergiana Marit.,* **6,** 175–184.

Förstner, U., and Wittmann, G. T. W. (1979). *Metal Pollution in the Aquatic Environment.* Springer-Verlag, Berlin–Heidelberg–New York 486 pp.

Francis, A. J., Duxbury, J. M., and Alexander, M. (1974). Evolution of dimethylselenide from soils. *Appl. Microbiol.,* **28,** 248–250.

Freeman, H. C. *et al.* (1974). Mercury in some Canadian Atlantic coast fish and shellfish. *J. Fis. Res. Board Can.,* **31,** 369. *Cit.: JWPCF,* **47,** (6), p. 1626/27.

Fujiki, M. (1972). The transitional condition of Minamata Bay and the neighbouring sea polluted by factory waste water containing mercury. *6th Int. Water Pollut. Res.,* June 18–23, Paper No. 12.

Gadow, S., and Schäfer, A. (1973). Die Sedimente der Deutschen Bucht: Korngrößen, Tonmineralien und Schwermetalle. *Senckenbergiana Marit.,* **5,** 165–178.

Gadow, S., and Schäfer, A. (1974), Schwermetalle in den Sedimenten der Jade. *Senckenberg. Marit.*, **6**, 161–174.

Ganther, H. E., Goudie, C., Sunde, M. L., Kopecky, M. J., Wagner, O., Sang-Hwan, O., and Hoekstra, W. G. (1972). Selenium, relation to decreased toxicity of methylmercury added to diets containing tuna. *Science*, **175**, 1122–1124.

Gardner, D. (1975). Observations on the distribution of dissolved mercury in the ocean. *Mar. Pollut. Bull.*, **6**, 43–46.

Gardner, D., and Riley, J. P. (1973). The distribution of dissolved mercury in the Bristol Channel and Severn Estuary. *Estuar. Coast. Mar. Sci.*, **1**, 191–192.

Gerlach, S. A. (1976). *Meeresverschmutzung—Diagnose und Therapie.* Springer-Verlag, Berlin–Heidelberg–New York, 145 pp.

Gibbs, R., Jarosewich, E., and Windom, H. L. (1974). Heavy metal concentrations in museum fish species: Effects of preservatives and time. *Science*, **184**, 475–477.

Glenn, J. L. (1973). Relations among radionuclide content and physical, chemical, and mineral characteristics of Columbia River sediments. *U.S. Geol. Surv. Prof. Paper*, 433-M, 51 pp.

Goldberg, E. D. (1972). Man's role in the major sedimentary cycle. In: Dyrssen, D. and Jagner, D. (eds.) *The Changing Chemistry of the Oceans.* Nobel Symposium, pp. 267–288. Almquist and Wiksell, Uppsala (Sweden).

Goldberg, E. D. (1975). The mussel watch. A first step in global marine monitoring. *Mar. Pollut. Bull.*, **6**, 111.

Goldberg, E. D. (1976). *The Health of the Oceans.* UNESCO Press: Paris.

Goldberg, E. D., Gamble, E., Griffin, J. J., and Koide, M. (1977). Pollution history of Narragansett Bay as recorded in its sediments. *Estuar. Coast. Mar. Sci.*, **5**, 549–561.

Goto, M. (1973). Inorganic chemicals in the environment—with special reference to the pollution problems in Japan. *Environ. Qual. and Safety*, **2**, 72–77.

Grancini, G., Stievano, M. B., Firardi, F., Guzzi, G., and Pietra, R. (1976). The capability of neutron activation for trace element analysis in sea water and sediment samples of the northern Adriatic *Sea. J. Radioanal. Chem.*, **34**, 65–72.

Greig, R. A., Nelson, B. A., and Nelson, D. A. (1975). Trace metal content on the American oyster. *Mar. Pollut. Bull.*, **6**, 72–73.

Greig, R. A., Wenzloff, D. R., and Pearch, J. B. (1976). Distribution and abundance of heavy metals in finfish, invertebrates and sediments collected at a deep water disposal site. *Mar. Pollut. Bull.*, **7**, 185–187.

de Groot, A. J., Goeij, J. J. M., and Zegers, C. (1971). Contents and behaviour of mercury as compared with other heavy metals in sediments from the Rivers Rhine and Ems. *Geol. en Mijnbouw*, **50**, 393–398.

de Groot, A. J., Salomons, W., and Allersma, E. (1976). Processes affecting heavy metals in estuarine sediments. In: Burton, J. D. and Liss, P. S. (eds.). *Estuarine Chemistry*, pp. 131–157. Academic Press, London–New York–San Francisco.

Gross, M. G. (1970). Analysis of dredged wastes, fly ash, and waste chemicals. *Mar. Sci. Res. Centre Technol. Rept. 7.*

Gross, M. G. (1972). Geologic aspects of waste solids and marine waste deposits, New York metropolitan region. *Geol. Soc. Am. Bull.*, **83**, 3163–3176.

Gross, M. G. (1976). New York Bight II: Problems of research. *Oceanus*, **19**, 11–14.

Gross, M. G., Black, J. A., Kalin, R. J., Schramel, J. R., and Smith, R. N. (1970). Survey of marine waste deposits, New York metropolitan region. Mar. Sci. Res. Centre, State Univ. New York, *Technol. Rept.*, **8**, 72 pp.

Halcrow, W., Mackay, D. W., and Thornton, I. (1973). The distribution of trace metals and fauna in the Firth of Clyde in relation to the disposal of sewage sludge. *J. Mar. Biol. Assoc. U.K.*, **53**, 721–739.

   U. Förstner

Hardisty, M. W., Huggins, R. J., Kartar, S., and Sainsbury, M. (1974). Ecological implications of heavy metal in fish from the Severn Estuary. *Mar. Pollut. Bull.*, **5**, 12–15.

Havre, G. N., Underdal, B., and Christiansen, C. (1973). Cadmium concentration in some fish species from a coastal area in southern Norway.   **24**, 155–157.

Helz, G. R. (1976). Trace element inventory for the northern Chesapeake Bay with emphasis on the influence of man. *Geochim. Cosmochim. Acta*, **40**, 573–580.

Helz, G. R., Huggett, R. J., and Hill, J. M. (1975). Behavior of manganese, iron, copper, zinc, cadmium, and lead discharged from a waste-water treatment plant into an estuarine environment. *Wat. Res.*, **9**, 631–636.

Holmes, C. W., Slade, E. A., and McLerran, C. J. (1974). Migration and redistribution of zinc and cadmium in marine estuarine systems. *Environ. Sci. Technol.*, **8**, 255–259.

Hubbell, D. W., and Glenn, J. L. (1973). Distribution of radionuclides in bottom sediments of the Columbia River estuary. *U.S. Geol. Surv. Prof. Paper*, 433-L, 50 pp.

Hung, T. C., Li, T. H., and Wu, D. C. (1977). The pollution of heavy metals in the Kaohsiung Harbour, Taiwan. Proc. Int. Conf. *Heavy Metals in the Environment, Toronto*, 1975, Vol. II/2, pp. 809–820.

Ito, K. (1975). Heavy metal pollution in the sediments of Nagoya Harbour. *Kogai To Taisaku*, **11**, 650–659. *Cit.:* CA: 208520r.

Jefferies, D. F. (1968). Fission-product radionuclides in sediments from the northeast Irish Sea. *Helgolander Wiss. Meeresunters*, **17**, 280–290.

Jefferies, D. F., Preston, A., and Steele, A. K. (1973). Distribution of cesium-137 in British coastal waters. *Mar. Pollut. Bull.*, **4**, 118–122.

Johnson, D. L., and Braman, R. S. (1975). The speciation of arsenic and the content of germanium and mercury in the pelagic *Sargassum* community. *Deep Sea Res.*, **22**, 503–507.

Jones, A. M. *et al.* (1972). Mercury in marine organisms of the Tay region. *Nature*, **238**, 164–165.

Kautsky, H. (1973). The distribution of the radio nuclide cesium-137 as an indicator for North Sea watermass transport. *Dtsch. Hydrogr. Zeitschr.*, **26**, 241–246.

Kemp, A. L. W., Thomas, R. L., Dell, C. I., and Jaquet, J. M. (1976). Cultural impact on the geochemistry of sediments in Lake Erie. *J. Fish. Res. Board Can.*, **33**, 440–462.

Kitamura, S. (1968). Determination of mercury content in bodies of inhabitants, cats, fishes, and shells in Minamata District and in the mud of Minamata Bay. *Minamata Disease*, 257–266.

Klein, D. H., and Goldberg, E. D. (1970). Mercury in the marine environment. *Environ. Sci. Technol.*, **4**, 765–768.

Klein, L. A., Lang, M., Nash, N. and Kirschner, S. L. (1974). Sources of metals in New York City waste-water. *JWPCF*, **46**, 2653–2662.

Klemmer, H. W., Unninayer, C. S. and Okubo, W. I. (1976). Mercury content of biota in coastal waters in Hawaii. *Bull. Environ. Contam. Toxicol.*, **15**, 454–457.

Knauer, G. A., and Martin, J. H. (1972). Mercury in a marine pelagic food chain. *Limnol. Oceanogr.*, **17**, 868–876.

Kopfler, F. C. (1974). The accumulation of organic and inorganic mercury compounds by the Eastern Oyster (*Crassostrea virginica*). *Bull. Environ. Contam. Toxicol.*, **11**, 275–280.

Krauskopf, K. (1967). *Introduction to Geochemistry*. McGraw-Hill Book Co., New York-St. Louis-San Francisco.

Kremling, K., and Petersen, H. (1977). The distribution of zinc, cadmium, copper, and iron in seawater of the Iceland-Faroe Ridge area. *'Meteor' Forsch. Ergebn.*, **A19**, 10–17.

Krüger, K. E., Nieper, L., and Auslitz, H.-J. (1975). Bestimmung des Quecksilbergehaltes der Seefische auf den Fangplätzen der deutschen Hochsee- and Küstenfischerei. 1. Mitteilung. *Archiv f. Lebensmittelhyg.*, **26**, 210–240.

Leatherland, T. M., and Burton, J. D. (1974). The occurrence of some trace metals in coastal organisms with particular reference to the Solent region. *J. Mar. Biol. Ass. U.K.*, **54**, 457–468.

Lesaca, R. M. (1977). Monitoring of heavy metals in Philippine rivers, bay waters, and lakes. Proc. Int. Conf. *Heavy Metals in the Environment*, Vol. II/1, Toronto, 1975. pp. 285–307.

Loring, D. H. (1975). Mercury in the sediments of the Gulf of St. Lawrence. *Can. J. Earth Sci.*, **12**, 1219–1237.

Mackay, D. W., Halcrow, W., and Thornton, J. (1972), Sludge dumping in the Firth of Clyde. *Mar. Pollut. Bull.*, **3**, 7–11.

Mackay, N. J., Williams, R. J., Kacprzac, J. L., Kazacos, M. N., Colins, A. J., and Auty, E. H. (1975). Heavy metals in cultivated oysters (*Crassostrea commercialis* = *Saccostrea cucullata*) from the estuaries of New South Wales. *Aust. J. Mar. Freshwater Res.*, **26**, 31–46.

Martin, J. M., Meybeck, M., Salvadori, F., and Thomas, A. (1976). Pollution chimique des estuaires: état actual des connaissance. *Rapp. Scient. Techn.* CNEXO, Paris, 286 pp.

Matsumoto, E., and Yokota (1976). History of heavy metal pollution in Tokyo Bay and Osaka Bay. *Kaguku*, **46**, 182–184.

McCave, I. N. (1973). Mud in the North Sea. In Goldberg, E. D. (ed.) *North Sea Science*, pp. 75–100. MIT Press, Cambridge, Mass.

Morton, B. S. (1976). The Hong Kong seashore—an environment in crisis. *Environ. Conservation*, **3**, 243–254.

Moyer, B. R. and Budinger, T. F. (1974). Cadmium levels in the shoreline sediments of San Francisco Bay. In Hemphill, D. D. (ed.) Trace Substances in Environmental Health, Vol. 8, pp. 127–135. University of Missouri.

Mueller, J. A., Anderson, A. R., and Jeris, J. S. (1976). Contaminants in the New York Bight. *JWPCF*, **48**, 2309–2326.

Müller, G., and Förstner, U. (1976). Schwermetalle in den Sedimenten der Elbe bei Stade: Veränderungen seit 1973. *Naturwissenschaften*, **63**, 242–243.

Murray, C. N., and Kautsky, H. (1977). Plutonium and americium activities in the North Sea and German coastal regions. *Estuar. Coast. Mar. Sci.*, **5**, 319–328.

Naeve, E. (1974). Sources of pollution in the Mediterranean and its effects on living resources and fishing. *Rev. Intern. Oceanogr. Méd.*, **XXXV–XXXVI**, 5–20.

Nikiforova, E. M. and Smirnova, R. S. (1975). Metal technophility and lead technogenic anomalies. Intern. Conf. *Heavy Metals in the Environment*, Toronto. Abstr. C-94/96.

Nisimura, H. (1976). *Hg in Fish and Sediments in Tokuyama Bay*. University of Tokyo, Feb. 1974.

Ohnesorge, F. K. (1974). Toxizität von Cadmium und Quecksilber unter dem Aspekt möglicher Gesundheitsschäden bei der Allgemeinbevölkerung. In *Problems of the Contamination of Man and his Environment by Mercury and Cadmium*, Comm. Europ. Communities, Luxembourg, 3–5 July, 1973, p. 409–419.

Olausson, E. (1975). Man-made effect on sediments from Kattegat and Skagerrak. *Geol. Fören. Stockholm*, **97**, 3–12.

Papakostides, G., Grimanis, A. P., Zafiropoulos, D., Griggs, G. B., and Hopkins, T. S. (1975). Heavy metals in sediments from the Athens sewage outfall. *Mar. Pollut. Bull.*, **6**, 136–138.

Parsons, T. R. *et al.* (1973). Preliminary survey of mercury and other metals contained in animals from the Fraser River mudflats. *J. Fish. Res. Board Can.*, **30**, 1014. *Cit.: JWPCF* **46** (6), p. 1444.

Patchineelam, S. R., and Förstner, U. (1977). Bindungsformen von Schwermetallen in marinen Sedimenten. *Senckenberg. Marit.* **9**, 75–104.

Patel, B., Mulay, C. D., and Ganguly, A. K. (1975) Radioecology of Bombay Harbour—a tidal estuary. *Estuar. Coast. Mar. Sci.* **3**, 13–42.

Peden, J. D., Crothers, J. H., Waterfall, C. E., and Beasley, J. (1973). Heavy metals in Sommerset marine organisms. *Mar. Pollut. Bull.*, **4**, 7–9.

Penrose, W. R., Black, R., and Hayward, M. J. (1975). Limited arsenic dispersion in sea water, sediments, and biota near a continuous source. *J. Fish. Res. Board Can.*, **32**, 1275–1281.

Pentreath, R. J. (1976). The accumulation of inorganic mercury from sea water by the plaice *Pleuronectes platessa* L. *J. Exp. Mar. Biol. Ecol.*, **24**, 103–117.

Phillips, D. J. H. (1977). The use of biological indicator organisms to monitor trace metal pollution in marine and estuarine environments—a review. *Environ. Pollut.*, **13**, 281–317.

Plato, P. A., and Jacobson, A. P. (1976). Cesium-137 in Lake Michigan sediments: areal distribution and correlation with other manmade material. *Environ. Pollut.*, **10**, 19–34.

Prater, E. E. (1975). The metal content and dispersion characteristics of steelwork's effluents discharging to the Tees Estuary. *Wat. Pollut. Control*, **74**, 63–78.

Preston, A. (1968). The control of radioactive pollution in a North Sea oyster fishery. *Helgoländer Wiss. Meeresunters.*, **17**, 269–279.

Preston, A. (1972). Artificial radioactivity in freshwater and estuarine systems. *Proc. Royal Society (London)*, **180B**, 421–436.

Preston, A., Fukai, R., Volchok, H. L., and Yamagata, N. (1971). Report of the panel on radioactivity. In *Report of the Seminar on Methods of Detection, Measurement, and Monitoring of Pollutants in the Marine Environment*, p. 87–99, Rome, FAO (*Marine Fisheries Reports*, No. 99, Suppl. 1).

Preston, A., Jefferies, D. F., Dutton, J., Harvey, B., and Steele, A. K. (1972). The concentration of selected heavy metals in sea water. *Environ. Pollut.*, **3**, 69–82.

Ratkowsky, D. A., Thrower, S. J., Eustace, I. J., and Olley, J. (1974). A numerical study of the concentration of some heavy metals in Tasmania oysters. *J. Fish. Res. Board Can.*, **31**, 1165–1171.

Raymont, J. E. G. (1972). Some aspects of pollution in Southampton water. *Proc. R. Soc. London*, **B180**, 451–468

Renzoni, A., Bacci, E., and Falcia, L. (1973). Mercury concentration in the water sediments and fauna of an area of the Tyrrhenian coast. *Rev. Intern. Oceanogr. Méd.* **XXXI–XXXII**, 17–45.

Renzoni, A. *et al.* (1974). Mercury concentration in the water, sediments and fauna of an area in the Tyrrhenian coast. *Rev. Int. Oceanogr. Méd.*, **XXXII**, cit. *JWPCF*, **47** (6), p. 1626/27.

Roberts, W. P., and Pierce, J. W. (1976). Deposition in upper Patuxent estuary, Maryland, 1968–1969. *Est. and Coast. Mar. Sc.*, **4**, 167–280.

Ruddell, C. L., and Rains, D. W. (1975). The relationship between zinc, copper, and the basophils of two crassostreid oysters, *C. gigas* and *C. virginica*. *Comp. Biochem. Physiol.* **51A**, 585–591.

Rutherford, F., and Church, T. (1975). Use of silver and zinc to trace sewage sludge dispersal in coastal waters. In: Church, T. T. (ed.) Marine Chemistry in the Coastal Environment, pp. 440–452. *Amer. Chem. Soc. Symp. Ser.* **18.**

Saroma, A. M. (1972). Heavy metal distribution in bottom sediments of lagoons and bays in Hokkaido. *Chikyu Kaguku,* **25,** 149–159.

Sayre, W. W., Guy, H. P. and Chamberlain, A. R. (1963). Uptake and transport of radionuclides by stream sediments. *Geol. Surv. Prof. Paper* **433-A,** A1–A33.

Schafer, H. A. (1976). Characteristics of municipal wastewater discharges, 1975. Southern California Coastal Water Res. Proj. El Segundo, *Ann. Report,* 57–60.

Schafer, H. A., and Bascom, W. (1976). Sludge in Santa Monica Bay. Southern California Coastal Water Res. Proj., El Segundo, *Ann. Report,* 77–82.

Schell, W. R., and Nevissi, A. (1977). Heavy metals from waste disposal in Central Puget Sound. *Environ. Sci. Techol.* **11,** 887–893.

Schultz, C. D., Crear, D., Pearson, J. E., Rivers, J. B., and Hylin, J. W. (1976). Total and organic mercury in the Pacific blue marlin. *Bull. Environ. Contam. Toxicol.,* **15,** 230–234.

Schwimer, S. R. (1973). Trace metal levels in three subtidal invertebrates. *Veliger,* **16,** 95–103.

Shelton, R. G. J. (1971). Sludge dumping in the Thames Estuary. *Mar. Pollut. Bull.,* **2,** 24–27.

Sheppard, C. R., and Bellamy, D. J. (1974). Pollution of the Mediterranean around Naples. *Mar. Pollut. Bull.,* **5,** 42–44.

Simpson, H. J., Olson, C. R., Trier, R. M., and Williams, S. C. (1976). Man-made radionuclides and sedimentation in the Hudson River Estuary. *Science,* **194,** 197–183.

Skei, J. M., Price, N. B., Calvert, S. E., and Hogdahl, E. (1972). The distribution of heavy metals in sediments of Sörfjord, West Norway. *Water, Air, Soil Pollution,* **1,** 452–461.

Skei, J. M. *et al.* (1976). Mercury in plankton from a polluted Norwegian fjord. *Mar. Pollut. Bull.,* **7,** 34. *Cit.: JWPCF,* **49** (6), p. 1319.

Smith, J. D., Nicholson, R. A., and Moore, P. J. (1971). Mercury in water of the tidal Thames. *Nature,* **232,** 393–394.

Stenner, R. D., and Nickless, G. (1974a). Absorption of cadmium, copper, and zinc by dog whelks in the Bristol Channel. *Nature,* **247,** 198–199.

Stenner, R. D., and Nickless, G. (1974b). Distributions of some heavy metals in organisms in Hardangerfjord and Skjerstadfjord, Norway. *Water, Air, Soil Pollution,* **3,** 279–291.

Stenner, R. D., and Nickless, G. (1975). Heavy metals in organisms of the Atlantic coast of southwest Spain and Portugal, *Mar. Pollut. Bull.,* **6,** 89–92.

Stoffers, P., Summerhayes, C., Förstner, U., and Patchineelam, S. R. (1977). Copper and other heavy metal contamination in sediments from the New Bedford Harbour, Mass.: A preliminary note. *Environ. Sci. Technol.,* **11,** 819–821.

Suess, E. (1978). Distinction between natural and anthropogenic materials in sediment. In Goldberg, E. D. (ed.), *Biogeochemistry of Estuarine Sediments,* pp. 225–237. UNESCO Press, Paris.

Suess, E., and Erlenkeuser, H. (1975). History of metal pollution and carbon input in Baltic Sea sediment. *Meyniana,* **27,** 63–75.

Summerhayes, C. P., Ellis, J. P., Stoffers, P., Briggs, S. R., and Fitzgerald, M. G. (1977). Fine-grained sediment and industrial waste distribution and dispersal in New Bedford Harbour and Western Buzzard Bay, Massachusetts. *WHOL-Rept.* 76–115. Woods Hole Oceanographic Institution, 110 p.

Talbot, V., Magee, R. J., and Hussein, M. (1976a). Distribution of heavy metals in Port Phillip Bay. *Mar. Pollut. Bull.,* **7,** 53–55.

Talbot, V., Magee, R. J., and Hussein, M. (1976b). Cadmium in Port Phillip Bay mussels. *Mar. Pollut. Bull.*, **7**, 84–86.

Terwindt, J. H. J. (1967). Mud transport in the Dutch delta area and along the adjacent coast line. *Neth. J. Sea Res.*, **3**, 505–531.

Thomas, R. L. (1972). The distribution of mercury in the sediment of Lake Ontario. *Can. J. Earth Sci.*, **9**, 636–651.

Thompson, J. A. J. (1977). Copper in marine waters—effects of mining wastes. *Proc. Int. Conf. Heavy Metals Environment*, Toronto, 1975, Vol. II/1, 273–284.

Thornton, I., Watling, H., and Darracott, A. (1975). Geochemical studies in several rivers and estuaries for oyster rearing. *Sci. Total Environ.*, **4**, 325–345.

Thrower, S. J., and Eustace, I. J. (1973). Heavy metal accumulation in oysters grown in Tasmanian waters. *Food Technol. Australia*, **25**, 546–554.

Ui, J., and Kitamura, S. (1971). Mercury in the Adriatic. *Mar. Pollut. Bull.*, **2**, 56–58.

Venugopal, B., and Luckey, T. D. (1975). Toxicology of non-radioactive heavy metals and their salts. In: Luckey, T. D., Venugopal, B., and Hutcheson, D. (eds.). *Heavy Metal Toxicity, Safety, and Hormology*, pp. 4–73. Thieme Publishers, Stuttgart.

Vucetic, T. *et al.* (1974). Long-term annual fluctuations of mercury in the zooplankton of the east central Adriatic. *Rev. Int. Oceanogr. Méd.*, **33**, 75. *Cit.: JWPCF*, **47** (6), p. 1627.

Watling, H. R., and Watling, R. J. (1976a). Trace metals in oysters from Knysna Estuary. *Mar. Pollut. Bull.*, **7**, 45–48.

Watling, H. R., and Watling, R. J. (1976b). Trace metals in *Chromytilus meridionalis*. *Mar. Pollut. Bull.*, **7**, 91–94.

Weichart, G. (1973). Pollution of the North Sea. *Ambio*, **2**, 99–106.

Windom, H., Stickney, R., Smith, R., White, D., and Taylor, F. (1973). Arsenic, cadmium, copper, mercury, and zinc in some species of north Atlantic fishes. *J. Fish. Res. Board Can.*, **30**, 275–279.

Wittmann, G. T. W., and Förstner, U. (1975). Metal enrichment of sediments in inland waters—the Hartbeespoort Dam. *Water SA*, **1**, 76–82.

de Wolf, P. (1975). Mercury content of mussels from west European coasts. *Mar. Pollut. Bull.*, **6**, 61–63.

Wollast, R. (1976). Transport et accumulation de pollutants dans l'estuaire de L'Escaut. In Nihoue, J. C. J. and Wollast, R. (eds.), *Projet Mer, Vol.* 10, *L'Estuaire de L'Escaut*, pp. 191–218. Progr. Polit. Scient., Bruxelles, Belgium.

Wong, P. T. *et al.* (1975). Methylation of lead in the environment. *Nature*, **253**, 263–264.

Wood, J. M. (1974). Biological cycles for toxic elements in the environment. *Science*, **183**, 1049–1052.

Wood, J. M., Segall, H. J., Ridley, W. P., Cheh, A., Chudyk, W., and Thayer, J. S. (1977). Metabolic cycles for toxic elements in the environment. *Proc. Int. Conf. Heavy Metals Environment*, Toronto, 1975, pp. 49–68.

Wrenn, M. E., and Jinks, S. M. (1975). The dosimetric implications of releases to the aquatic environment from the nuclear power industry. In Miller, M. W., and Stannard, J. N. (eds.). *Environmental Toxicity of Aquatic Radionuclides—Models and Mechanisms*, 229–265. Ann Arbor Sci. Publ., Ann Arbor, Mich.

Wright, D. A. (1976). Heavy metals in animals from the Northeast coast. *Mar. Pollut. Bull.*, **7**, 36–38.

Young, D. R., and McDermott, D. J. (1975). Trace metals in harbour mussels. *Southern California Coastal Water Res. Proj.*, *El Segundo*. Ann. Report, 139–144.

Chemistry and Biogeochemistry of Estuaries
Edited by E. Olausson and I. Cato
Copyright © 1980 by John Wiley & Sons Ltd.

L. REUTERGÅRDH

National Swedish Environment Protection Board,
Special Analytical Laboratory,
University of Stockholm

11

Chlorinated Hydrocarbons in Estuaries

1 INTRODUCTION

The problem of environmental pollution is usually considered to be a consequence of the industrialization and urbanization processes which made progress in the late 19th and early 20th centuries. Fertilizers and herbicides were introduced into the agriculture, the chemical industry developed new products and the large scale use of fossil fuels increased rapidly. A concentration of large population to cities gave rise to the problem of garbage disposal and domestic effluents into surrounding waters and the contribution of industrial discharges directly into the environment grew rapidly.

During the last decade, however, both scientists and society have realized the growing pollution problems. This includes the fact that some of the man-made chemicals were shown to be extremely persistent in the environment and accumulate in fatty tissues of animals. At the same time, both insects, herbs, and fungi developed new forms that were resistant to the chemicals used to combat them.

Industries grew up where the population densities were high and transportation facilities were good. These places were in many cases identical to river banks, bays, and coast lines. The phenomenon thus created the issue of pollution in estuaries. The present chapter will briefly consider the influence of some classes of chlorinated organic pollutants on the specific ecosystems that are represented by the estuaries. Emphasis will be put on substances which are persistent and which show bioaccumulation properties in food webs. Within this group of substances the majority is chlorinated compounds.

2 ORGANOCHLORINE COMPOUNDS FOUND IN ESTUARIES

The stable chlorinated compounds encountered in estuaries are generally brought there, intentionally or unintentionally, through man's activities. Insecticides, herbicides, and fungicides are introduced into the environment through their use to control pests, while polychlorinated biphenyls (PCB) and other industrial compounds leak into the environment from manufacturing sites or *via* their use and disposal routes. So far, the efforts have been concentrated to map down the substances dealt with under this headline. Today, there exists limited or no literature on whether other substances like polychlorinated terphenyls (PCT), polychlorinated naphthalenes (PCN) and both low- and high-molecular weight polychlorinated paraffins (PCA), are present in estuaries or not. Generally, the persistent organochlorine compounds are characterized by low water solubilities and moderate or low vapor pressures.

2.1 Pesticides

DDT or *p,p*-DDT (1,1,1-trichloro-2,2-bis(4-chlorophenyl) ethane, (Figure 1(a)) was first described in 1874, but its insecticidal properties were not proved until 1939 by Paul Müller who received the Nobel prize 1948 for his findings.

The discovery of DDT started the modern pesticide era and the compound was used during World War II with great success. DDT has primarily been used to defeat tropical diseases spread by insects, but has also been used in agriculture and forestry. The water solubility is about 1.2 mg/m³ at 20 °C, but decreases with increasing salinity. Technical DDT contains 10–20% of the 2-chlorophenyl isomer (*o,p*-DDT) which is more easily degraded than DDT itself. The two most important metabolites of *p,p*-DDT are *p,p*-DDE (1,1-dichloro-2,2-bis(4-chlorophenyl) ethene, (Figure 1(b)) and *p,p*-DDD (1,1-dichloro-2,2-bis(4-chlorophenyl) ethane, (Figure 1(c)). In a number of organisms and environmental materials, like soils and sediments, the latter two compounds constitute the major residues of DDT, and DDE is generally more stable than DDT itself (Fishbein, 1974; Metcalf, 1973). Aldrin (Figure 1(d)), dieldrin (Figure 1(e)), and endrin belong to a group of pesticides collectively named cyclodienes. These compounds are used mainly

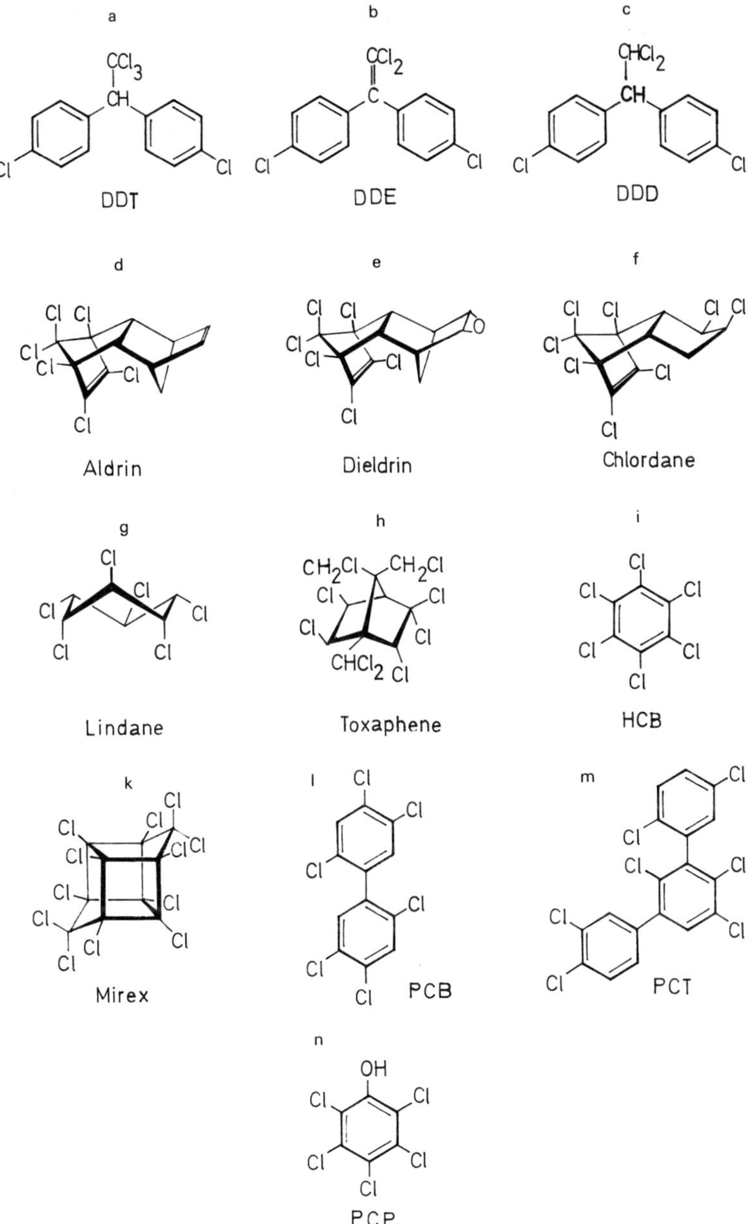

Figure 1. Representative structures of organochlorine compounds mentioned in the text.

in the agriculture, but are also used to combat swarms of locusts in tropical areas of the world. Fish are known to be very susceptible to cyclodiene insecticides, but also higher animals have been adversely affected by the use of aldrin and dieldrin, i.e. in the US and in Great Britain. The technical preparations of aldrin and dieldrin are not chemically pure, but contain 85–95% of the desired components. In most animals aldrin is metabolized to dieldrin which in fact means that a compound is 'de-toxified' to a more toxic compound. Dieldrin in turn is metabolized further to water-soluble components which are excreted.

Chlordane (Figure 1(f)) also belongs to the cyclodiene pesticide group and has been used since 1940 to control termites, ants, and other insects in domestic areas as well as in agriculture pest control. Chlordane is a multi-component mixture of mainly polychloromethanoindenes (Cochrane and Greenhalgh, 1976; Sovocool *et al.*, 1977). At least 45 components, some still awaiting characterization, have been indicated in the technical product. Chlordane like other cyclodienes is very toxic for aquatic organisms. The chlorination of benzene in UV-light gives rise to a product containing at least 9 isomeric components. One of these, γ-hexachlorocyclohexane (γ-HCH, Figure 1(g)) is mainly responsible for the insecticidal properties of the product and constitute in turn 99% of the pesticide lindane. The most persistent isomer, however, is the β-HCH isomer which is resistent both to metabolic and chemical breakdown. Lindane has been used in agriculture to combat insects and together with DDT in forestry and for household use against various insects. The HCH compounds are metabolized mainly *via* dehydrohalogenation reactions mainly to chlorophenols and chlorinated benzenes. Toxaphene (Figure 1(h)) was introduced as an insecticide in the middle of the forties, and by 1948 it was being used commercially for the control of a variety of insect pests (West and Campbell, 1952). Toxaphene is a mixture of more than 170 components produced by chlorination of terpene derivatives, mainly camphene. The product contains an average of 67–69% chlorine and is one of the most dominant insecticides used in the US (Pollock and Kilgore, 1978). It is also known to be used in East Europe and in the USSR. Like the cyclodienes, toxaphene is very toxic against water-living organisms, but very few data can be found in the literature on the residue levels of toxaphene in the environment, despite its large-scale use (Jansson *et al.*, 1978). Hexachlorobenzene (HCB, Figure 1(i)) is used as a fungicide, but is also found as a by-product in a number of industrial processes. Thus, it is known that HCB is formed during electrolysis of brine, during the use of carbon electrodes e.g. for aluminium production and by incineration of chlorine-containing polymers.

Mirex (dechlorane, Figure 1(k)) was prepared in 1946 and introduced for pesticidal use in 1969. This compound has its widest use in combating some species of ants which can be serious pests in agriculture. Mirex is also claimed to be used as a flame retardant additive to polymers and in coatings for various materials. Mirex and its degradation product kepone (mirex with one methylene group exchanged for a keto group) is highly toxic to fish.

2.2 Industrial chemicals

The polychlorinated biphenyls (PCB, Figure 1(l)) were known as chemicals before the 20th century and have been produced industrially since 1929. Because of the chemical stability of PCB and its excellent dielectrical properties it has been widely used as capacitor fluid, fire retardant, heat transfer media, and additives to hydraulic oils, paints, adhesives, etc. (Hutzinger *et al.*, 1974). PCBs are prepared by chlorination of biphenyl which, depending on degree of chlorination, gives complicated mixtures of up to 80 different components. The presence of PCB residues in the environment is very well documented and has been traced back to 1942 in Swedish biota (Jensen, 1972). PCB shows a low acute toxicity to most organisms, but tends to increase the insecticidal properties of DDT (Lichtenstein *et al.*, 1969) and lindane (Risbec, 1954).

Polychlorinated terphenyls (PCT, Figure 1(m)) are claimed to be used for the same purpose as the PCBs. The presence of PCT in biota has been confirmed in a few studies (see references in Renberg *et al.*, 1978), but only at high trophic levels.

Pentachlorophenol (PCP, Figure 1(n)) has a variety of usage areas, e.g. as a wood and textile preservative (fungicide), a slimicide in paperpulp industry, and as an insecticide for the control of termites and wood-boring insects. PCP is manufactured by hydrolysis of hexachlorobenzene (HCB) and PCP also is a major metabolite of HCB. PCP is highly toxic to aquatic organisms and its toxic properties have been summarized by Becker and Thatcher (1973).

3 TRANSPORT MECHANISMS, DISTRIBUTION, AND ACCUMULATION

Many of the compounds blamed to cause environmental pollution problems have in common a poor water solubility. They possess a high affinity to particles or any solid surface and adsorb on lipophilic surfaces. This in turn means that solubility data given in the literature are ambiguous and not directly applicable to environmental conditions, since the character of natural waters is highly dependent on the amount and type of particles, organisms, etc., present in the water (Gunther *et al.*, 1968; Tulp and Hutzinger, 1978).

The interface between the air and the sea is the stage for important processes in pollution chemistry. Long-chain fatty acids, hydrocarbons, alcohols, and chlorinated hydrocarbons are enriched in surface slicks (Garrett, 1967; Williams, 1967; Boehm and Quinn, 1974). During suitable wind and water conditions, the slicks can be concentrated with nutrients and organic matter and become an area of high biological activity where hydrophobic pollutants can enter the biosphere (Seba and Corcoran, 1969; Ogner and Schnitzer, 1970).

In rivers and other fresh water tributaries containing humic substances and other organic matter, these compounds act as carriers for lipophilic substances into estuaries (Ogner and Schnitzer, 1970; Matsuda and Schnitzer, 1971; Poirrier *et al.*, 1972). Owing to the increase in ionic strength during the mixing of fresh water and sea water, the organic components will be salted out.

Since subsurface currents are generally directed landwards, material transport to the deep sea is restricted. Lipophilic compounds adsorbed on settling material are therefore carried back to the estuaries (Strakhov, 1967; Meade, 1969; Arrhenius, 1963). Depending on particle size, tidal action, wind, and water exchange, organic pollutants reside in the water body for different periods of time before being deposited in sediments (Long and McDevit, 1952; McDevit and Long, 1952; Paul, 1952). In estuaries with a pronounced halocline, there is often a subsidiary maximum of particles near this strong density gradient which also retards the settling (Jerlov, 1959).

Both organic and inorganic material in the water column will eventually sink to the bottom and add to the sediment. However, deposition in recent sediments may not be final. Depending on bottom conditions, sediment particles with attached compounds can be eroded by tidal currents to accumulation areas. During these transports, the lipophilic compounds are recycled with the suspended particulate matter and once more exposed to free-water living organisms in the estuaries (Moor and Harriss, 1972). Another way of recirculation for organic pollutants is by stripping effects caused by bubbles leaving the sediment. When organic pollutants are deposited, benthos and microbiological activities are involved in the complex routes of these compounds in the environment.

Marine phytoplankton rapidly take up DDT residues from the aquatic environment. In natural populations virtually all of the DDT residues which are available for uptake, i.e. DDT bound to particles larger than 1–2 μm in diameter, are incorporated into phytoplankton cells and similarly sized detrital particles (Cox, 1972). However, the amounts incorporated still only represent 10% of the DDT residues recoverable from the whole sea water. Cox (1972) has suggested two mechanisms for the uptake of DDT by phytoplankton. One mechanism involves phase partitioning of DDT between sea-water and the lipid portion of algal cells. The other involves adsorption of DDT onto cell surfaces. The latter mechanism was proposed after it was discovered that a saturation value of ^{14}C-DDT accumulated by *Dunaliella salina*, was independent of the ambient concentration, the saturation value was estimated to 5×10^{-15} pg ^{14}C-DDT/μm^2.

The uptake of DDT by phytoplankton has been studied by Södergren (1973) who exposed algal unicells to ^{14}C-DDT and noted that the uptake was rapid and almost quantitative, and that incorporated residues were not released into DDT-free medium.

Two different pathways have been proposed for the transfer of biomass through marine food webs. One leads from large phytoplankton by way of a one to three step food chain to fish that can be harvested by man (Greve and Parsons, 1977). The other leads from smaller phytoplankton through about five trophic levels to various gelatinous predators (Ryther, 1969). Many zooplankters are selective grazers, often choosing food on the basis of size, shape, and species (Parsons and Le Brasseur, 1970). If algal cells are too small, they may not be grazed efficiently by common estuarine and coastal copepods (Nival and Nival, 1976). This implies that if changes in composition of the phytoplankton community occur, due to the impact of, for example, chlorinated hydrocarbons, the whole nourishment web or stages of it can be adversely affected.

The figures for bioconcentration ratios of organochlorine compounds in phytoplankton are scattered. From the determinations of Cox (1972) the levels of DDT residues in phytoplankton range from 1.2–2.7 μg/g organic carbon (dry weight). Harvey *et al.*, (1971) reported 0.2–0.5 ng/g wet weight for DDT residues in Sargassum weed. The average concentration of DDT residues in all marine phytoplankton has been put at 0.01 μg/g (Goldberg, 1971). The corresponding figures for PCB were set to 0.007–0.64 μg/g (net weight) by Harvey *et al.* (1973). Ware and Addison (1973) found that PCB concentration was inversely related to particle size and that this trend persisted in spite of a 34-fold difference in residue levels between composite samples. They also indicated a linear relation between PCB concentration and cumulative rainfall which should indicate that PCBs are transported from the atmosphere to the sea surface and then to the plankton.

Risebrough and Vreeland (1972) suggested in an investigation of PCB residues in Atlantic zooplankton that levels found should be in physical–chemical equilibria with those in the ambient aqueous environment and could be used as an indicator of the contamination of the water. This opinion is shared by Clayton *et al.* (1977) who made an effort to elucidate factors that control bioaccumulation of PCB in lower trophic levels of near-shore ecosystems. In a study of the upper Chesapeake Bay Munson *et al.* (1977) showed that only a very small percentage of chlordane, DDT, and PCB in the water column were associated with zooplankton (average 12 pg/g PCB in the water column on suspended sediment *versus* 0.072 in zooplankton). This means that there is a potential reservoir of chlorinated hydrocarbons that will flow into the biological system when zooplankton blooms occur.

Several monitoring studies have been performed within the United States on different media to obtain knowledge about the distribution and concentration of organochlorine compounds in estuarine biota. Rowe *et al.* (1971) found, when investigating dieldrin and endrin concentrations in a Louisiana estuary that the major pesticide influx to the sampling area was *via* land drainage. They also concluded that the concentration of pesticides had decreased since the monitoring began in 1964. Water, sediment, and oysters were analysed, the levels in water was less than 0.2 ng/g for both dieldrin and endrin. The highest levels of dieldrin in sediments were 4 ng/g and for endrin 5 ng/g. The concentrations in oysters ranged from the detection limit (1 ng/g) to 3.4 ng/g with a mean value of 1.4 and for endrin the figures were from the detection limit to the highest value 2.3 ng/g with a mean of 1.8 ng/g.

Butler and Schutzman (1978) have presented residue levels of pesticides and PCBs in estuarine fish from 144 United States estuaries. In their opinion, juvenile fish are satisfactory tools for gauging pesticide pollution trends in estuaries, provided 25 individuals (6–12 months old of the same species) are sampled annually at a specific location. The magnitude and frequency of biotic residues of DDT, dieldrin, endrin, and toxaphene declined substantially between 1965–1970 and 1972–1976. This tendency was also verified by monitoring estuarine molluscs during the period 1972–1977 (Butler *et al.* 1978). In 85 of the 87 estuaries investigated, no PCB could be determined with a detection limit of 50 ng/g. Eleven pesticides were monitored, but only DDT could be detected at low levels in two of the estuaries in 1977. In 1965, when the project started, DDT was found in almost all samples. Dieldrin was the next most commonly detected pesticide, and residues of endrin, mirex, toxaphene, and PCB were found occasionally.

Chlordene epoxide, a metabolite of chlordene, was detected in the range of 0.04–0.27 ppm in catfish. The metabolite was found in three out of 36 fishes and constituted 46% of the total amount of chlordane (Castillo *et al.* 1978).

Levels of the cis- and trans-isomers of nonachlor and chlordane in aquatic fauna collected between 1971–1977 on the east coast of Canada have been presented by Zitko (1978). The levels were generally below 0.5 μg/g lipid for herring (*Clupea harengus*) both for the cis- and trans-isomers of the two compounds, while the PCB concentrations were 10 to 100 times higher. Chlorinated norbornadiene derivatives, two of which are used as intermediates in the production of endrin, have been identified and surveyed in fish (Yrawecz and Roach, 1978). The concentrations found ranged from trace levels to 3.22 μg/g for fish caught in the lower Mississippi river, and corroborates with results presented of Barthel *et al.* (1969).

4 EFFECTS OF ORGANOCHLORINE COMPOUNDS ON DIFFERENT TROPHIC LEVELS

Bourquin and Cassidy (1975) made 85 bacterial isolates from various estuarine environments to investigate if PCBs had any inhibitory effect upon growth. In a disk assay, 26 isolates were inhibited to varying extent by the PCB formulation applied (Aroclor 1242). From these sensitive bacteria, representative cultures were tested for growth inhibition in a liquid culture, and a greatly extended log phase was observed. This should, according to Bourquin and Cassidy, imply that since a substantial portion of the bacteria was adversely affected, the turnover of carbon in estuarine systems could be severely reduced.

Keil *et al.* (1972) reported a stimulatory effect of PCB on the growth of *Esherichia coli*, a bacterium from human intestinal microflora used as an indicator of water quality. Sayler *et al.* (1978) found an increase in the growth rate of estuarine *Pseudomonas* during a degradation study of PCB by this strain. However, there was a 1,000-fold difference in concentrations between the inhibition and stimulation experiments described.

Various studies have been performed to evaluate the influence of different organochlorine compounds on the primary producers of the estuaries (Moore and Harriss, 1972; Hardling, 1976; Mosser *et al.*, 1972a, b; Fisher, 1973, 1975; Fisher *et al.*, 1974). Fisher and Wurster (1973) found that the marine diatome *Rizosolenia setigera* was affected by PCB concentrations as low as 0.1 ng/l, and that the sensitivity was temperature dependent. This led to the conclusion that phytoplankton living under suboptimal conditions may be more vulnerable to stress than those growing under optimal conditions. The alteration of phytoplankton community composition has been studied by Powers *et al.* (1975, 1976, 1977a, b). One litre dialysis bags containing the cultures together with the organochlorine compound to be investigated were put in natural waters (Powers *et al.*, 1976). In the presence of PCB (10 μg/l), there was a decrease in biomass, and particles larger than 9 μm (equivalent spherical diameters) were observed during an exposure time of four days (Powers *et al.*, 1977b). There was no loss of photosynthesis function per unit of chlorophyll *a*, but carbon fixation was nevertheless inhibited due to reduced levels of chlorophyll *a* per PCB-treated cell. This implies that the PCB should interfere in the synthesis of chlorophyll *a*.

A similar study has been made for chlordane and PCB by Biggs *et al.*, (1978). Chlordane gave the same effects as PCB although the inhibition of cell division, chlorophyll *a* synthesis, and ^{14}C-uptake was more pronounced for PCB. Chlordane gave no alterations of community composition as compared to controls, while PCB devasted algae larger than 8μm. The marine dinoflagellate *Exuviella baltica* Lohmann was exposed to 50 μg/l of either heptachlor or chlordane in cultures (Magnani *et al.*, 1978). The effect on cell size distribution was severe for the chlordane treated cells, but the heptachlor treated culture differed only slightly from the control. Both compounds were found to reduce chlorophyll *a* per unit volume of culture, ^{14}C-uptake per cell, and carbon fixation per unit of chlorophyll *a*. Chlordane was the more toxic compound and caused disintergration of cells. Other chlorinated insecticides, including DDD, DDE, dieldrin, and endrin have also been shown to inhibit marine phytoplankton growth (Wurster, 1968; Mentzel *et al.*, 1970; Bowes, 1972). In a study where the siliat protozoans *Tetrahymena vorax* and the chlodoceran *Daphnia pulex* were exposed to Aroclor 1242 and DDT, there was no effect on the growth or cell density for the protozoan, but *Daphnia* was quite sensitive to as low concentrations of PCB as 0.02 μg/g (Morgan, 1972).

Bioassay studies to determine chronic and sublethal effects have been carried out for various chlorinated hydrocarbons on a diversity of organisms (Parrish *et al.* 1976). Togatz *et al.*, (1977) found, when they allowed bentos organisms to develop in sand from planctonic larvae in a flow-through system, that pentachlorophenol decreased the number of individuals. Mollusks were markedly fewer at a concentration of 7 μg/l, annelids and arthropods at 76 μg/l. Almost no animals occurred at 622. μg/l. Sensitivity differed among species of the same phylum, so that pentachlorophenol altered the experimental communities, changing the relative abundance of animals by species and phylum. For a review on the toxicity of PCP to aquatic biota, see Becker and Thatcher (1973).

The toxicity of HCH-isomers to different phyla of estuarine organisms has been investigated in flow-through and static bioassays (for references see Schimmel *et al.*, 1977a). In a flow-through system Schimmel *et al.*, (1977b) assessed 96-hr lindane LC50 values to: mysid (*Mysidopsis bahia*), 6.3 μg/l; pink shrimp (*Peaeus duorarum*), 0.19 μg/l; grass shrimp (*Palaemonetes pugio*), 4.4 μg/l; sheepshead minnow (*Cyprinodon variegatus*), 104 μg/l; and pinfish (*Lagodon rhomboides*), 30.6 μg/l. The accumulation of HCH-isomers was shown to be similar for all isomers in the species investigated and no evidence was found for isomer specific degradation. The technical HCH turned out to be less toxic to pink shrimp and pinfish, 0.34 and 86.4 μg/l respectively, than pure lindane (Schimmel *et al.*, 1977b).

The toxicity of PCB has been investigated for several estuarine organisms by Nimmo *et al.* (1975). Population growth rate for the ciliate protozoan *Tetrahymena pyriformis* was significantly depressed at a level of 1.0 μg PCB/l. This concentration was also revealed to be toxic to shrimps and fish. An effort was made to find out if there was any minimum level, below which shrimp could not accumulate PCB. Grass shrimps were tested against concentrations of 0.04, 0.09, and 0.62 μg PCB/l, and corresponding levels in shrimp 0.2, 1.0, and 10 mg/kg were reached within five weeks. No threshold level could be determined. It was shown by Butler (1966), that shell deposition in oysters was disturbed at concentrations of 0.1 to 0.5 mg PCB/l in short-term tests. When *Crassostrea virginica* was exposed to PCB for 30 weeks at a level of 5 μg/l there was a decrease in growth rate, but not at 1 μg/l. Oysters exposed to 5 μg PCB/l (Nimmo *et al.*, 1975) for half a year developed pathological changes. The normal vesicular connective tissue lost its regular cell distribution and was infiltrated by leukocytes, degeneration of the tissue was also observed. There was an atrophy of digestive gland tubules epithelium, and enlarged lumen in exposed oysters. When the organisms were moved to uncontaminated water for several weeks a partial or complete recovery was observed. Tissue changes associated with chronic PCB exposure were also observed in shrimp. The normal digestive gland heptophanchreas epithelical cells had inclusions of crystalloids. In some cases the nuclear membrane was ruptured as the crystalloid grew.

An investigation of chronic toxicity of Arochlor 1254 and the subsequent elimination has been made on pin-fish (*Lagodon rhomboides*) and spot *Leiostomus xanthurus* (Nimmo *et al.*, 1975). Pinfish and spot died when exposed to 5 p.p.b. for 14–45 days while spot appeared unaffected by exposure to 1 p.p.b. for a period of 56 days. The liver concentrated the greatest amount of PCB, followed by the gills, heart, brain, and muscle in decreasing order. The elimination study revealed that after 84 days of flushing in PCB-free water, the PCB-amount decreased 61%.

5 DEGRADATION

Organochlorine compounds can be eliminated from estuaries by a few principal ways, microbial or chemical decomposition, photodegradation, metabolisation, and vaporization. Studies of metabolic pathways and degradation products for organochlorine compounds in estuarian ecosystems are in its infancy, and almost no literature exists in this important field. The photochemical transformations of pollutants that can take place in water systems have been reviewed by Sundstrom and Ruzo (1978).

The uptake and metabolism of DDT in marine phytoplankton have been reviewed by Cox (1972). The conversion of DDT to DDE ranged from a few percent to ten, during the experimental time of about three weeks. Several phytoplankton species were investigated without finding major differences in residual concentrations of DDT and DDE. Metcalf *et al.* (1973) studied the environmental fate of six organochlorine pesticides (aldrin, dieldrin, endrin, mirex, lindane, hexachlorobenzene) in a model ecosystem. Radiolabelled compounds were introduced into the model which represented several food chains, algae—snail, plankton—water flea—mosquito—fish, and the experiment continued for 33 days. It was concluded

that there was an inverse correlation between levels of ecological magnification and the solubility of the compounds investigated. The concentration
of the pesticides in fish and snail were calculated to be from 200 to 84,000
times larger than those in the ambient water. Using the same model (Metcalf
et al., 1975a), and the same technique, tri-, tetra-, and pentachlorobiphenyls
were compared to DDE. The conclusion from the first study was confirmed,
pentachlorobiphenyl having an environmental bioaccumulation rate comparable to DDT. Several metabolic products were indicated though not identified.

The evaluation of environmental distribution and fate of hexachlorocyclopentadiene, chlordene, heptachlor and heptachlorepoxid have also been
studied (Metcalf *et al.*, 1975b) in the same laboratory model ecosystem, as
described above.

Bacterial degradation of PCB was studied by Ahmed and Focht (1973a, b)
who exposed Achromobacter to different PCBs. Although complete
mineralization was not observed, oxygenation, and cleavage of the benzene
rings were reported, with *p*-chlorobenzoic acid as one of the major metabolites. Hydroxylation of biphenyl molecules followed by meta cleavage of the
aromatic ring, have been reported (Lunt and Evans, 1970). This is a
common mechanism of degradation by pure cultures of gram negative
bacteria from soil. The bacterial genus *Pseudomonas* was isolated and
allowed to grow both on a commercial PCB and a ^{14}C-hexachlorobiphenyl
(Sayler *et al.*, 1977). A stimulatory effect on growth was observed at a
concentration of 10 μg PCB/l, as well as the oxygen uptake. The degradation of PCB was determined both with radio-isotope assay and gas
chromatography.

Zoro *et al.* (1974) investigated the transformation of *p,p*-DDT to *p,p*-
DDD under different conditions, to elucidate the mechanism.
Five different environments were studied: (1) sewage sludge, (2) baker's
yeast, (3) iron salts and iron porphyrins, (4) bacterial cultures, and (5)
annelid worms. It was concluded that the DDT to DDD conversion was
affected in any situation in which DDT and reduced iron porphyrins were
brought together in solution. Iron porphyrins are actually abundant in living
species, bound to proteins in complex molecules, for instance catalases,
cytochromes, and haemoglobin. These porphyrins can be released during
decay processes which prevail in decomposing organic material, and catalyse
the turnover of DDT to DDD. That iron redox pairs are essential for
degradation in estuarine sediments were shown by Williams and Bidleman
(1978) who studied toxaphene degradation in four different systems (1)
sterilized sediment, (2) unsterilized sediment, (3) sterilized sand containing
$Fe_3(OH)_8$, (4) sterilized sand plus distilled water. Toxaphene was present in
a concentration of 5 μg/g. Degradation products which had shorter retention times than toxaphene were confirmed with gas chromatography. The

fact that break down took place also in case (3) indicated that observed changes were due to chemical rather than biological processes. Though toxaphene components could be detected in oysters, there was no evidence for those compounds in sediments.

6 SUMMARY

The literature that has been reviewed is not considered to cover the whole subject, but is rather an introduction into specific areas of the literature that has not been reviewed before, or areas of special interest. Much of the knowledge that is needed to fully understand the impact of organochlorine compounds on the estuarine environments, still remains to be gathered through scientific work. Very little is known about air transport and fall out, and few approaches have been made to obtain a mass balance of these chemicals. Much effort has to be put into the study of the synergistic effects where the literature is limited today. Investigations of juvenile stages, where the adverse effects of chlorinated hydrocarbons seem to be most pronounced will have to be intensified, as these stages are the most important for the life cycles in the estuaries.

7 REFERENCES

Ahmed, M., and Focht, D. D. (1973a). Oxidation of polychlorinated biphenyls by *Achromobacter* pCB, *Bull. Environ. Contam. Tox.*, **10**, 70–72.

Ahmed, M., and Focht, D. D. (1973b). Degradation of polychlorinated biphenyl by two species of *Achromobacter, Can. J. Microbiol.* **19**, 47–52.

Arrhenius, G. (1963). Pelagic Sediments, in *The Sea*, (ed. M. N. Hill), pp 655–727, Interscience 3, New York.

Barthel, W. F., Hawthorn, J. C., Ford, J. H., Bolton, G. C., McDowell, L. L., Grissiner, E. H., and Parsons, D. A. (1969). Pesticides in water, *Pestic. Monit. J.*, **3**, 8–66.

Becker, C. D., and Thatcher, T. O. (1973). Toxicity of power plant chemicals to aquatic life, *Report No. WASH-1249 U.S. Atomic Energy Comission.* Richland, Washington: Battelle, Pacific Northwest Laboratories.

Biggs, D. C., Rowland, R. G., O'Conners Jr., H. B., Powers, C. D., and Worster, C. F. (1978). A comparison of the effects of chlordane and PCB on the growth, photosynthesis, and cell size of estuarine phytoplankton, *Environ. Pollut.*, **15**, 253–263.

Boehm, P. D., and Quinn, J. G. (1974). Solubility behaviour of No. 2 fuel oil in sea water, *Mar. Pollut. Bull.*, **5**, 101–105.

Bourquin, A. W., and Cassidy, S. (1975). Effect of polychlorinated biphenyl formulations on the growth of estuarine bacteria, *Appl. Microbiol.*, **29**, 125–127.

Bowes, G. W. (1972). Uptake and metabolism of 2,2-bis-(p-chlorophenyl)-1,1,1-trichloroethane (DDT) by marine phytoplankton and its effect on growth and chloroplast electron transport, *Pl. Physiol.*, Lancaster, **49**, 172–176.

Butler, P. A. (1966). Pesticides in the marine environment, *J. appl. Ecol.*, **3**, 253–259.

Butler, P. A., and Schutzmann, R. L. (1978). Residues of pesticides and PCBs in estuarine fish, 1972–1976—National pesticide monitoring program, *Pestic. Monit. J.*, **12**, 51–59.

Butler, P. A., Kennedy, C. D., and Schutzmann, R. L. (1978). Pesticide residues in estuarine mollusks, 1977 versus 1972—National pesticide monitoring program, *Pestic. Monit. J.*, **12**, 99–101.

Castillo, G. D., Jeffus, M. T., and Kenner, C. T. (1978). Gas–liquid chromatographic determination of chlordane epoxide in catfish, *J. Ass. Off. Anal. Chem.*, **61**, 1–4.

Clayton, J. R., Pavlov, S. P., and Breitner, N. F. (1977). Polychlorinated biphenyls in coastal marine zooplankton: Bioaccumulation by equilibrium partitioning, *Environ. Sci. Technol.*, **11**, 676–682.

Cochrane, W. P., and Greenhalgh, R. (1976). Chemical composition of technical chlordane, *J. Ass. Off. Anal. Chem.*, **59**, 696–702.

Cox, J. L. (1972). DDT residues in marine phytoplankton, *Residue Rev.*, **44**, 23–38.

Fishbein, L. (1974). Chromatographic and biological aspects of DDT and its metabolites, *J. Chromatog.*, **98**, 177–251.

Fisher, N. S., Graham, L. B., Carpenter, E. J., and Wurster, C. F. (1973). Geographic differences in phytoplankton sensitivity to PCBs, *Nature*, **241**, 548–549.

Fisher, N. S. (1975). Chlorinated hydrocarbon pollutants and photosynthesis of marine phytoplankton: A reassessment, *Science, N. Y.*, **189**, 563–564.

Fisher, N. S., and Wurster, C. F. (1973). Individual and combined effects of temperature and polychlorinated biphenyls on the growth of three species of phytoplankton, *Environ. Pollut.*, **5**, 205–212.

Fisher, N. S., Carpenter, E. J., Remsen, C. C., and Wurster, C. F. (1974). Effects of PCB on interspecific competition in natural and gnotobiotic phytoplankton communitites in continuous and batch cultures, *Microb. Ecol.*, **1**, 39–50.

Garrett, W. D. (1967). The organic chemical composition of the ocean surface, *Deep–Sea Res.*, **14**, 221–227.

Goldberg, E. D., Butler, P., Meier, P., Menzer, D., Risebrough, R. W., and Stickel, L. F. (1971). Chlorinated hydrocarbons in the marine environment. A report prepared by the *Panel on Monitoring Persistent Pesticides in the Marine Environment*, National Academy of Science Washington D.C.

Greve, W., and Parsons, T. R. (1977). Photosynthesis and Fish production: Hypothetical effects of climatic change and pollution. *Helgol. Wiss. Meeresunters.*, **30**, 666–672.

Gunther, F. A., Westlake, W. E., and Jaglan, P. S. (1968). Reported solubilities of 738 pesticide chemicals in water, *Residue Rev.* **20**, 1–148.

Hardling, L. W. (1976). Polychlorinated biphenyl inhibition of marine phytoplankton photosysthesis in the Northern Adriatic Sea, *Bull. Environ. Contam. Toxicol.*, **16**, 559–565.

Harvey, G. R., Bowen, J. T., Backus, R. H., and Grice, G. D. (1971). Chlorinated hydrocarbons in open-ocean Atlantic organisms, Proc. Nobel Symposium *The Changing Chemistry of the Oceans*, Gothenburg, Sweden, Aug. 16–20.

Harvey, G. R., Miklas, H. P., Bowen, V. T., and Steinhauser, W. G. (1973). Observations on the distribution of chlorinated hydrocarbons in Atlantic Ocean Organisms, *J. Marine Res.*, **32**, 103–118.

Hutzinger, O., Safe, S., and Zitko, U. (1974). The Chemistry of PCB's, *CRC Press, Inc.*

Jansson, B., Vaz, R., Blomkvist, G., Jensen, S., and Olsson, M. (1979). Chlorinated terpenes and chlordane components found in fish, guillemot, and seal from Swedish waters, *Chemosphere*, **4**, 181–190.

Jensen, S. (1972). The PCB story, *Ambio*, **1**, 123–131.

Jerlow, N. G. (1959). Maxima in the vertical distribution of particles in the sea, *Deep-Sea Res.*, **5**, 173–184.

Keil, J. E., Sandifer, S. H., Graber, C. D., and Preister, L. E. (1972). DDT and polychlorinated biphenyl (Aroclor 1242) effects of uptake on E. coli growth, *Water Res.*, **6**, 837–841.

Lichtenstein, E. P., Schultz, K. R., Fuhremann, T. W., and Liang, T. T. (1969). Biological interaction between plasticizers and insecticides, *J. Econ. Entomol*, **62**, 761–765.

Long, F. A., and McDevit, N. F. (1952). Activity coefficients of nonelektrolyte solutes in aqueous salt solutions, *Chem. Rev.*, **51**, 119–169.

Lunt, D., and Evans, W. C. (1970). Microbial metabolism of biphenyl, *Biochem. J.*, **118**, 54–55.

McDevit, W. F., and Long, F. A. (1952). Activity coefficient of C_6H_6 in aqueous salt solutions, *J. Am. Chem. Soc.*, **74**, 1773–1777.

Magnani, B., Powers, C. D., Wurster, C. F., and O'Connors, Jr, H. B. (1978). Effects of chlordane and heptachlor on marine dinoflagellate, *Exuviella baltica* Lohmann, *Bull. Environ. Contam. Toxicol.*, **20**, 1–8.

Matsuda, K., and Schnitzer, M. (1971). Reactions between fulvic acid, a soil humic material, and dialkyl phtalates, *Bull. Environ. Contam. Toxicol.*, **6**, 200–204.

Meade, R. H. (1969). Landward transport of bottom sediment in estuaries, *J. Sedim. Petrol.*, **39**, 222–234.

Menzel, D. W., Andersson, J., and Randke, A. (1970). Marine phytoplankton vary in their response to chlorinated hydrocarbons, *Science*, **167**, 1724–1726.

Metcalf, R. L. (1973). A century of DDT, *J. Agric. Food Chem.*, **21**, 511–519.

Metcalf, R. L., Kapoor, I. P., Lu, P-Y., Schuth, C. K., and Sherman, P. (1973). Model ecosystem studies of the environmental fate of six organochlorine pesticides, *Environ. Health Persp.*, **4**, 35–44.

Metcalf, R. L., Sanborn, J. R., Lu, P-Y., and Nye, D. (1975a). Laboratory ecosystem studies of the degradation and fate of radiolabelled tri-, tetra-, and pentachlorobiphenyl compared with DDE, *Arch. Environ. Contam. Toxicol.*, **3**, 151–165.

Metcalf, R. L., Lu, P-Y., Hirwe, A. S., and Williams, J. W. (1975b). Evaluation of environmental distribution and fate of hexachlorocyclopentadiene, chlordane, heptachlor, and heptachlor epoxid in a laboratory model ecosystem, *J. Agric. Food Chem.*, **23**, 968–973.

Moore, S. A., and Harriss, R. C. (1972). Effects of polychlorinated biphenyls on marine phytoplankton communities, *Nature*, London, **240**, 356–357.

Morgan, J. R. (1972). Effects of Aroclor 1242 (a polychlorinated biphenyl) and DDT on cultures of an alga, protozoan, daphnic, ostracod, and guppy, *Bull. Environ. Contam. Toxicol.*, **8**, 129–137.

Mosser, J. L., Fisher, N. S., Teng, T.-C., and Wurster, C. F. (1972a). Polychlorinated biphenyls: Toxicity to certain phytoplankters, *Science*, N.Y., **175**, 191–192.

Mosser, J. L., Fisher, N. S., and Wurster C. F. (1972b). Polychlorinated biphenyls and DDT alter species composition in mixed cultures of algae, *Science*, N.Y., **176**, 533–535.

Munson, T. O., Palmer, H. D., and Forns, J. M. (1976). Transport of chlorinated hydrocarbons in the upper Chesapeake Bay, Natl. Conf. Polychlorinated biphenyls 1975 (EPA-506/6-75-004), 218–229.

Nimmo, D. R., Hansen, D. J., Couch, J. A., Cooley, N. R., Parrish, P. R., and Lowe, J. J. (1975). Toxicity of Aroclor 1254 and its physiological activity in several estuarine organisms, *Arch. Environ. Contam. Toxicol.*, **3**, 22–39.

Nival, P., and Nival, S. (1976). Particle retention efficiences of an herbivorous copepod, *Acartia clausi* (adult and copepodite stages): Effects on grazing, *Limnol. Oceanogr.*, **21**, 24–38.

Ogner, G., and Schnitzer, M. (1970). Humic substances: fulvic acid–dialkyl phtalate complexes and their role in pollution, *Science*, **170**, 317–318.

Parsons, T. R., and LeBrasseur, R. J. (1970). The availability of food to different trophic levels in the marine food chain, in *Marine Food Chains* (ed J. H. Steele, 1st ed.), pp. 325–343, Oliver and Boyd, Edinburgh.

Parrish, P. R., Schimmel, S. C., Hansen, D. J., Patric, J. M., and Forster, J. (1976). Chlordane: Effects on several estuarine organisms, *J. Toxicol. Environ. Health.*, **1**, 485–494.

Paul, M. A. (1952). The solubilities of naphthlene and biphenyl in aqueous solutions of electrolytes, *J. Am. Chem. Soc.*, **74**, 5274–5277.

Poirrier, M. A., Bordelon, B. R., and Laseter, J. L. (1972). Adsorption and concentration of dissolved carbon-14 DDT by coloring colloids in surface waters, *Environ. Sci. Technol.*, **6**, 1033–1035.

Pollock, G. A., and Kilgore, W. W. (1978). Toxaphene, *Residue Rev.*, **69**, 87–140.

Powers, C. D., Rowland, R. G., Michaels, R. A., Fisher, N. S., and Wurster, C. F. (1975). The toxicity of DDE to marine dinoflagellate, *Environ. Pollut.*, **9**, 253–262.

Powers, C. D., Rowland, R. G., and Wurster, C. F. (1976). Dialysis membrane chambres as a device for evaluating impacts of pollutants on plankton under natural conditions, *Water Res.*, **10**, 991–994.

Powers, C. D., Rowland, R. G., and Wurster, C. F. (1977a). Dieldrin-induced destruction of marine algal cells with concomitant decrease in size of survivors and their progeny, *Environ. Pollut.*, **12**, 17–25.

Powers, C. D., Rowland, R. G., O'Connors, H. B., and Wurster, C. F. (1977b). Response to polychlorinated biphenyls of marine phytoplankton isolates cultured under natural conditions, *Appl. Environ. Microbiol.*, **34**, 760–764.

Renberg, L., Sundström, G., and Reutergårdh, L. (1978). Polychlorinated terphenyls (PCT) in Swedish white-tailed eagles and in grey seals, a preliminary study, *Chemosphere*, **6**, 477–482.

Risbec, J. (1954). Conjoint action of pentachlorobiphenyl and the γ-isomer of hexachlorocycklohexane on *Musca domestica*, *Phytiatr. Phytopharm. Spec. issue*, **573**.

Risebrough, R. W., Vreeland, V., Harvey, G. R., Miklas, H. P., and Carmignani, G. M. (1972). PCB residues in Atlantic Zooplankton, *Bull. Environ. Contam. Toxicol.*, **8**, 345–355.

Rowe, D. R., Canter, L. W., Snyder, P. J., and Mason, J. W. (1971). Dieldrin and endrin concentrations in Luisiana estuary, *Pestic. Monit. J.*, **4**, 177–183.

Ryther, J. H. (1969). Photosynthesis and fish production in the sea, *Science*, **166**, 72–76.

Sayler, G. S., Sohn, M., and Colwell, R. R. (1977). Growth of estuarine *Pseudomonas* sp. on polychlorinated biphenyl, *Microb. Ecol.*, **3**, 241–255.

Sayler, G. S., Thomas, R., and Colwell, R. R. (1978). Polychlorinated biphenyl (PCB) degrading bacteria and PCB in estuarine and marine environments, *Estuarine and Coastal Marine Science*, **6**, 553–567.

Schimmel, S. C., Patrick, Jr, J. M., and Forester, J. (1977a). Uptake and toxicity of toxaphene in several estuarine organisms, *Arch. Environ. Contam. Toxicol.*, **5**, 353–367.

Schimmel, S. C., Patrick, J. M., and Forester, J. (1977b). Toxicity and bioconcentration of BHC and Lindane in selected estuarine animals, *Arch. Environ. Contam. Toxicol.*, **6**, 355–363.

Seba, D. B., and Corcoran, E. F. (1969). Surface slicks as concentrators of pesticides in the marine environment, *Pest. Monit. J.*, **3**, 190–193.

Sovocool, G. W., Lewis, R. G., Harless, R. L., Wilson, N. K., and Zehr, R. D. (1977). Analysis of technical chlordane by gas chromatography/mass spectrometry, *Anal. Chem.*, **49**, 734–740.

Strakhov, M. N. (1967). *Principles of lithogenesis,* New York, Consultants Bureau, **1,** 245 pp.

Sundström, G., and Ruzo, O. (1978). Photochemical transformation of pollutants in water, in *Aquatic pollutants: Transformation and biological effects.* (Eds. O. Hutzinger, J. H. van Lelyveld, and B. C. J. Zoeteman), pp. 205–222, Pergamon Press, Oxford.

Södergren, A. (1973). Transport distribution, and degradation of chlorinated hydrocarbon residues in aquatic model ecosystems, *Oikos,* **24,** 30–41.

Togatz, M. E., Ivey, J. M., Moore, J. C., and Tobia, M. (1977). Effects of pentachlorphenol on the development of estuarine communities, *J. Toxicol. Environ. Health,* **3,** 501–506.

Tulp, M. Th. M., and Hutzinger, O. (1978). Some thoughts on aqueous solubilities and partition coefficients of PCB, and the mathematical correlation between bioaccumulation and physico-chemical properties, *Chemosphere,* **10,** 849–860.

Ware, D. M., and Addison, R. F. (1973). PCB residues in plankton from the gulf of St. Lawrence, *Nature* **246,** 519–521.

West, T. F., and Campbell, G. A. (1952). '3956' (TOXAPHENE), in *DDT and newer persistent insecticides, Ch,* 17. ed. Chemical Publishing Co., New York.

Williams, P. M. (1967). Sea surface chemistry: organic carbon and organic and inorganic nitrogen and phosphorus in surface films and subsurface waters, *Deep-Sea Res. Oceanogr. Abstr. 14,* 791–800.

Williams, R. R., and Bidleman, T. F. (1978). Toxaphene degradation in estuarine sediments, *J. Agric. Food Chem.,* **26,** 280–282.

Wurster, C. F. Jr. (1968). DDT reduces photosynthesis by marine phytoplankton, *Science,* **159,** 1474–1475.

Yrawecz, M. P., and Roach, J. A. G. (1978). Gas–liquid chromatographic determination of chlorinated norbornene derivatives in fish, *J. Ass. Off. Anal. Chem.,* **61,** 26–31.

Zitko, V. (1978). Nonachlor and chlordane in aquatic fauna, *Chemosphere,* **1,** 3–7.

Zoro, J. A., Hunter, J. M., Eglinton, G., and Ware, G. C. (1974). Degradation of *p,p*-DDT in reducing environments, *Nature,* **247,** 235–237.

Chemistry and Biogeochemistry of Estuaries
Edited by E. Olausson and I. Cato
Copyright © 1980 by John Wiley & Sons Ltd.

S. R. CARLBERG

National Board of Fisheries,
Hydrographic Department,
Göteborg

12

Oil Pollution of the Marine Environment—With an Emphasis on Estuarine Studies

1 INTRODUCTION

Estuaries and other coastal waters represent areas of prime interest for a multitude of human activities such as, for example, navigation, fisheries, disposal of sewage from communities and industries, extraction of sand and gravel, recreation, and so forth. However, these waters also represent the most productive zones of the world oceans not only in terms of their primary production by phytoplankton but also by their important role as reproduction areas and nursery grounds for a great variety of fish species, of which many are of great importance for the global supply of protein.

Many of man's activities in marine areas, or close to them, involve use or handling of great quantities of petroleum or petroleum products. Consequently, marine areas receive a heavy burden of oil pollution which mainly affects coastal waters. This is in obvious contrast to our interest and obligation to keep and maintain the coastal waters as the valuable resources they indeed are.

As the global oil consumption, and hence also the transportation, steadily increases the numbers of major accidents and oil spills in coastal waters grow as well.

1967	*Torrey Canyon*	117,000 tons of crude oil hitting the coasts of Wales and Brittany (France)
1970	*Arrow*	9,500 tons of Bunker C into the Chedabucto Bay, Nova Scotia
1976	*Argo Merchant*	26,000 tons of No. 6 fuel oil off the coast of Rhode Island, USA
1976	*Monte Urquiola*	110,000 tons of light crude oil of which about 40,000 tons were spread into the coastal waters off La Coruña, Spain, polluting large estuarine areas
1978	*Amoco Cadiz*	216,000 tons of crude oil were released into the waters off the coast of Brittany

are but a few examples of large oil spills during recent years, although the most spectacular ones have been included. While this chapter is being written, *Andros Patria* loaded with 200,000 tons crude oil has started to leak during a storm in the Bay of Biscay. According to reports some 50,000 tons have been released.

In connection with different oil spills in coastal waters a great number of studies have been carried out concerning the effects and the fate of the oil. These studies are of varying complexity as well as more or less extensive in time and space. In order to understand and explain many of the observations, different processes have been studied in the laboratory or in model systems.

It was not meaningful to limit this review to studies carried out in true estuaries. The knowledge we have about oil pollution today is derived from work carried out in different kinds of coastal waters and in laboratories. In some cases the geographical description of the studied area does not tell the reader whether the study concerns an estuary or not. Quite often the different authors do not draw any conclusions as to whether the results obtained are significant for an estuary only, or if they are valid for other types of marine areas as well.

It was not possible to present in a few pages a complete review of the literature in this field, let alone the biological effects of oil pollution which can largely be regarded as being beyond the scope of this chapter. It has rather been the intention to present some relevant material in such a way that a person who is not active in this field will be stimulated to further reading.

2 PETROLEUM AND BIOGENIC HYDROCARBONS—WHAT THEY ARE AND HOW THEY ARE CLASSIFIED

2.1 Petroleum

From a strict point of view petroleum is equivalent to the crude (not refined) oil that is obtained mainly from wells drilled deeply into the earth's interior. In some cases the word petroleum is by mistake or ignorance also used for the different products obtained from crude oil by different refining operations. A better common name is oil, preferably with added distinctions such as crude oil, diesel oil, fuel oil, etc. The major part of all these oils are hydrocarbons, components consisting only of hydrogen and carbon. Logically the term petroleum hydrocarbon should thus be reserved for the hydrocarbons in crude oils. However, terms like diesel oil hydrocarbons, etc., were never considered logical either, simply because many of the hydrocarbons in one product appear also in other refined products as well as in the crude oil itself. Thus the term petroleum hydrocarbons refer to all hydrocarbons in crude oils and their refined products.

Crude oils are the results of geochemical processes. Organic material which sedimented to the bottom of the sea millions of years ago were covered by subsequent sedimentary layers and deeply buried. As a result of elevated temperature and pressure during very long periods of time an immensely complex mixture of hydrocarbons was formed. A great deal of this material accumulated in reservoirs from which crude oil can be obtained.

Crude oil is probably the most complicated natural mixture on earth. The hydrocarbons dominate (50 to 98 per cent) and the remaining part consists mainly of non-hydrocarbons i.e. compounds containing oxygen, sulphur, nitrogen, or metals, e.g. nickel, vanadium, iron, and copper.

Dependent on how the atoms are bound together in the molecules, the compounds can be classified into different chemical classes. The names, and to some extent also the subdivision, used in this nomenclature varies

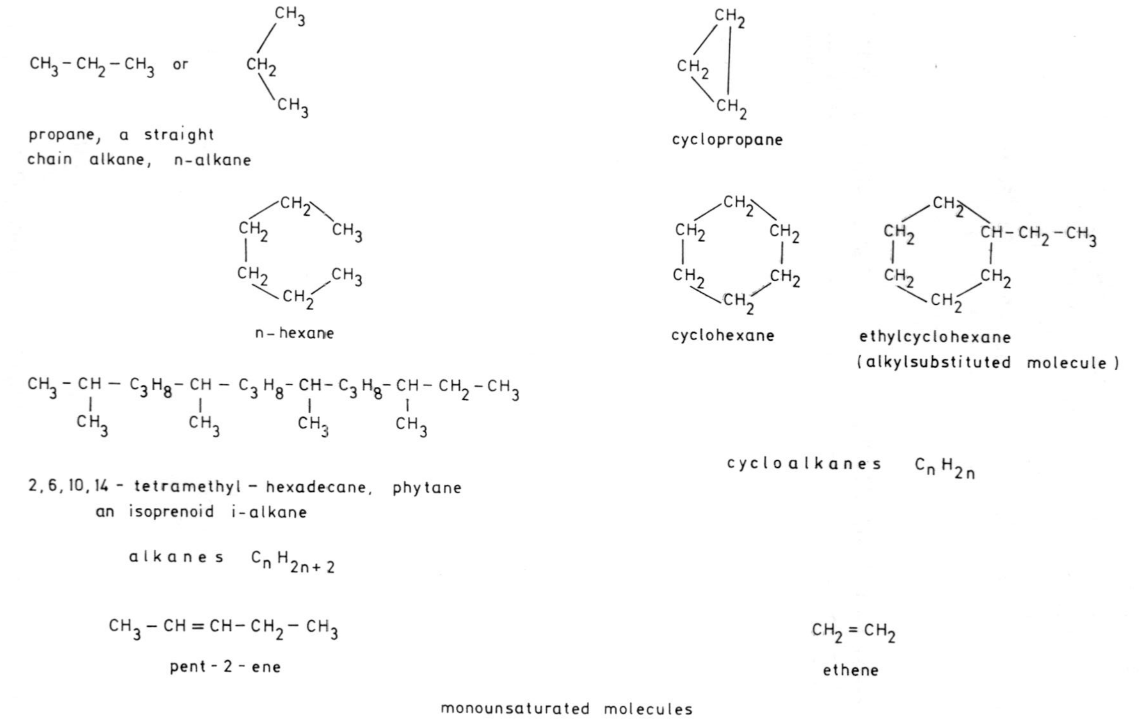

Figure 1. Examples of hydrocarbons and their basic structures.

somewhat from time to time reflecting what is the current use in textbooks in organic chemistry. The following system, which largely is adopted from Goldberg (1976), is valid (older or alternative names are given in brackets). See also Figure 1.

2.1.1 *Alkanes (paraffins)*

All compounds in this group have the general composition $C_n H_{2n+2}$. The group ranges from methane (CH_4) and ethane to compounds with sixty carbon atoms or more, such as n-hexacontane ($C_{60}H_{122}$, which is a wax at room temperature). They can be straight chained or branched, but the former class predominates.

The expression straight chained should be understood as a contrast to ringformed and branched. In fact the straight molecule is not straight, but rather bent or twisted. The straight chained molecule is called normal and the branched one is called iso. The compounds are frequently referred to as n-alkanes and i-alkanes.

2.1.2 *Cycloalkanes (cycloparaffins, naphthenes)*

This is a group of compounds with the general formula $C_n H_{2n}$. Most members of the group do not carry more than five or six carbon atoms in the ring that the molecule constitutes. Some polycyclic substances of this group are also known. Quite frequently alkyl groups are substituted like tails to the cycloalkane rings.

These two first groups are in some textbooks called with the older name aliphates and in other books they are named aliphates and alicyclics respectively.

2.1.3 *Alkenes (olefins)*

It is important to notice that *alkenes are normally not present to any significant extent in crude oils* (Posthuma, 1977). This is a major difference between the crude oils and the biogenous hydrocarbons. The alkenes are formed in some refining processes and therefore these compounds are common in many petroleum products.

The compounds have the general formula $C_n H_{2n}$ just as the cycloalkanes. However, the molecules are not shaped into rings but straight chained or branched as the alkanes. The difference is that the alkanes and cycloalkanes are saturated, whereas the alkenes are unsaturated. Thus in a saturated molecule each carbon atom is bonded to four other atoms, two carbon and two hydrogen (or one carbon and three hydrogen as is the case at the ends of the chained molecules). In the unsaturated molecule two of the carbon atoms are joined by a double bond. Each of these two atoms is then bonded to one carbon and one hydrogen. This is schematically shown in Figure 1 above.

2.1.4 *Aromatic compounds*

The basic constituent of an aromatic compound is the benzene ring, which contains three double bonds (Figure 2). In crude oils the aromatic content is

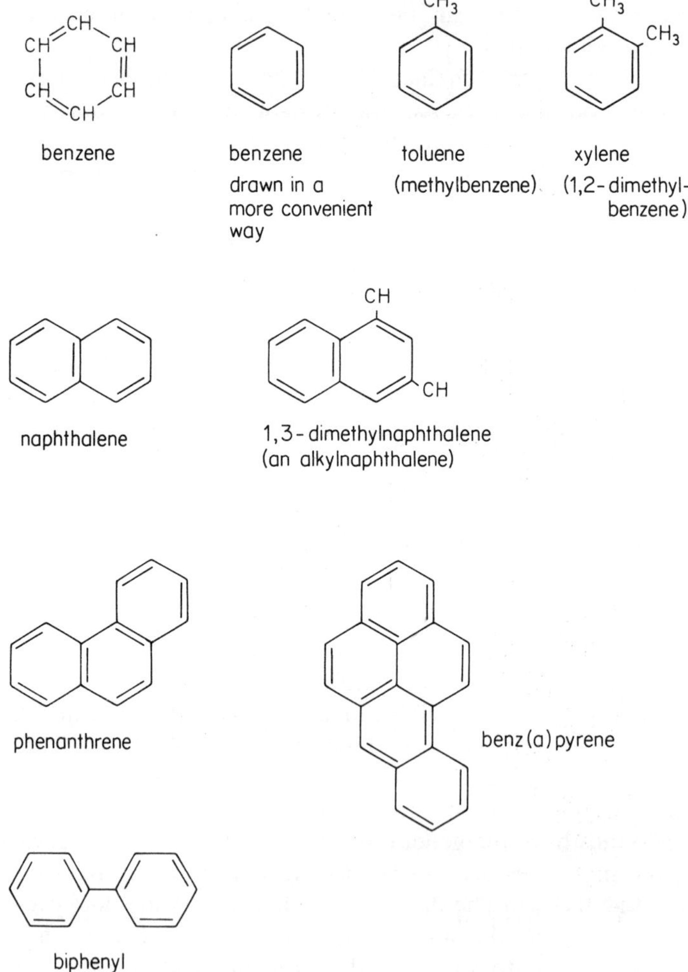

Figure 2. Examples of aromatic hydrocarbons and their basic structures.

very varying and seldom exceeds 40 per cent. In various refined products the aromatic content is often higher (up to 60 per cent according to Hargrave and Phillips, 1975).

The most simple examples of the group is benzene and the methylbenzenes such as toluene and xylene. In addition there are the polynuclear aromatics (consisting of several benzene rings) in the higher boiling fractions and they encompass two general types: the fused-ring compounds such as the alkyl naphthalenes, phenanthrene, and benz(a)pyrene and the linked rings such as the biphenyls. The former are usually more abundant.

In this group is usually included the *naphtheno-aromatics*, molecules part aromatic and part cycloalkane.

2.1.5 *Other compounds in oil*

The oxygen, nitrogen, and sulphur in a crude oil is mainly found in the *heterocyclic compounds*, which are aromatics or naphtheno-aromatics where one or more carbon atom in the molecule has been replaced with oxygen, nitrogen or sulphur. These compounds are mainly found in the *asphalthene residue* of the oil.

The total oxygen content of a crude oil may attain 2 per cent. The principal species include phenols (cresols and higher boiling point alkylphenols) and carboxylic acids, both straight chained and linked, such as hexanoic acid and 3-methylpentanoic acid. Nitrogen varies between 0.05 and 0.8 per cent in crude oils primarily in pyridine and quinolines. Sulphur exists both as the element and in compounds from trace amounts to 5 per cent by weight in crudes. These compounds include hydrogen sulphide, mercaptans, and aliphatic and cyclic sulphides.

The metallo-organics encompass the nickel and vanadium porphyrin complexes, which allow these metals to reach concentrations of 5–40 p.p.m. In addition, there are very small concentrations of other metals such as iron, sodium, and zinc, whose associations with other elements are still undetermined.

2.1.6 *Possible toxic compounds in oil*

Among the large number of possible toxic constituents of (crude) oil the following groups are included (Speers and Whitehead, 1969):

(a) Low-molecular weight aromatics such as benzene, toluene, xylene and other monocyclic aromatics.
(b) Acids and phenols such as acids, derived from alkanes or cycloalkanes (mainly carboxylic acids) and phenols, cresols, xylenols, and naphthols.
(c) Sulphur compounds such as thiols, sulphides, and thiophenes.

(d) Polynuclear aromatic hydrocarbons (PAH) such as 3,4-benz(a)pyrene (BaP), 1,2-benzanthracenes, 1,2-benzphenanthrene (chrysene), diphenyl-methane, fluorene, phenanthrene and dibenzthiophene.

2.2 Biogenous hydrocarbons

Hydrocarbons are found in terrestrial and marine organisms. According to a literature survey (Zsolnay, 1977) about 0.01 per cent is the mean content of hydrocarbons in marine organisms. This is one or two orders of magnitude lower than the values reported for the fatty acids. The following review is derived from NAS (1975) and Whittle (1977) unless otherwise stated.

Normal alkanes, primarily with odd-numbered carbon chains, are synthesized by both marine and terrestrial organisms. Often one or two compounds, usually with an odd number of carbon atoms, predominate. In marine phytoplankton, alkanes with 15, 17, 19, and 21 carbon atoms are most abundant, while in marsh grasses and *Sargassum* C_{21} to C_{29} normal alkanes have the highest concentrations. Some bacteria have equal amounts of even and odd numbered normal alkanes with 25 to 32 carbon atoms.

Branched alkanes are also to be found in organisms. The most notable compound is an isoprenoid called pristane (2,6,10,14-tetramethylpentadecane). Pristane is widespread in the food web and occurs as a major component of some copepod lipids and fish oil hydrocarbons. Although it is known to be derived from the phytyl moiety of chlorophyll present in the algal diet of copepods, it is not certain whether the copepods or the microflora in their gut achieve the conversion. As a branched alkane it is chemically far more stable than the polyenes (could be described as alkanes into which two or more double bonds have been introduced). If it is assumed that the copepods are the major source, pristane is also stable in the food web. Studies after release into the water column are much more difficult since *in situ* concentrations are low and analyses are prone to contamination. Besides pristane, several monomethyl branched alkanes have been identified.

Pristane is not only commonly produced by marine organisms, but it is also a constituent of crude oil and several refined products. Consequently, the presence of pristane in an environmental sample does not confirm incorporation of petroleum hydrocarbons. However, according to Anderson *et al.* (1974) and Ehrhardt and Heinemann (1974) the related isoprenoid phytane (2,6,10,14-tetramethylhexadecane) has not been found as a natural component of marine organisms. Its presence in samples therefore suggest oil incorporation. There are representatives of alkenes (olefins) in the biosphere, but as mentioned previously, normally not in crude oils. Many species of algae, including phytoplankton, have been analysed either from culture or from nature, but they are not truly representative of the algal biomass or the variations to which it is subject. Careful analyses and the use of labelled substrates suggest that the algae biosynthesize a relatively small number of compounds, perhaps not more than 25 in all, most of which are alkenes. Indeed, a few polyenes could be the major components. If indeed the polyenes represent the major algal input, their chemical structure suggests that turnover times would be rapid since they are particularly prone both to chemical degradation, especially to oxidative processes, and to biodegradation.

Squalene (a hexaene; polyene with six double bonds) is the primary hydrocarbon in shark and cod livers. Isoprenoid C_{19} and C_{20} mono-, di- and trienes occur in copepods and some fish. In addition, straight chain mono- to hexaenes have been found in organisms of the sea. The NAS (1975) report cites no definite reports of occurrences of cycloalkanes being synthesized by marine organisms.

There is some uncertainty whether aromatic hydrocarbons are biosynthesized in the marine ecosystem or not. Whittle (1977) state that it is not certain whether marine algae or bacteria are capable of such a production. Similarly, Broen and Huffman (1976) cite evidence suggesting that marine organisms do not produce aromatic hydrocarbon mixtures. Blumer and Youngblood (1975) describe forest fires and prairie fires as a likely source of many polynuclear aromatic hydrocarbons in marine sediments. Anderson *et al.* (1974) indicate that although aromatic hydrocarbons have not been isolated from plankton, they may be produced by some organisms. Despite these possibilities, elevated levels of polynuclear aromatic hydrocarbons are widely thought to indicate the presence of petroleum hydrocarbons. As an example Hargrave and Phillips (1975) use the presence of triaromatic and greater substituted aromatic compounds as a useful indication of oil pollution. The naphtheno-aromatic hydrocarbons have not been reported as constituents of marine organisms so far.

3 HOW TO DISTINGUISH BETWEEN PETROLEUM HYDROCARBONS AND BIOGENOUS HYDROCARBONS

For most studies of hydrocarbon pollution, it is necessary to be able to distinguish between recent biogenous hydrocarbons and those released to the environment through man's use of different oils. In addition, it would be valuable to identify substances synthesized by land plants which in one way or the other reach the oceans. There are at present no definitive criteria for distinguishing between those hydrocarbons introduced from recent biogeochemical processes and those from man's activities.

NAS (1975) indicates some guidelines to the differentiation of petroleum and biogenous hydrocarbons, noting that the following differences may not apply to all organisms, crude oils or refined products:

Petroleum contains a much more complex mixture of hydrocarbons with much greater ranges of molecular structure and molecular weight.

Petroleum contains several homologous series, with adjacent members usually present in nearly the same concentration. The approximate unit ratio for even and odd numbered alkanes is an example, as are the homologous series of C_{12}–C_{22} isoprenoid alkanes. As previously mentioned, marine organisms have a strong predominance of odd-numbered C_{15} through C_{21} alkanes.

Petroleum contains more kinds of cycloalkanes and aromatic hydrocarbons. Also, the numerous alkyl-substituted ring compounds have not been reported in organisms. Examples are the series of mono-, di-, tri- and tetramethyl benzenes and the mono-, di-, tri- and tetramethyl naphthalenes.

Petroleum contains numerous naphtheno-aromatic hydro-carbons that have not been reported in organisms. Petroleums also contain numerous heterocompounds containing sulphur, nitrogen, oxygen, and metals, and the heavy asphaltic compounds.

In addition to that, as was pointed out above, the isoprenoid compound phytane has not been found as a natural constituent in marine organisms.

4 INPUT OF OIL TO ESTUARINE AREAS

There is no statistics available, regarding the input of oil to estuarine areas as separated from other marine areas. However, attempts have been made to estimate the total annual input of oil to the oceans. From such estimates some guidance can be obtained as to the input to estuaries.

Three estimates are given in Table 1, which appeared in the NAS (1975) Report. The NAS Workshop produced the only one with figures for all the types of sources given in the table. In this paper the NAS estimate is thus regarded as complete.

The estimated figure for the natural seeps is not further considered below, because it is not known at all how much of that amount is significant for coastal waters. The same is true for the wildly differing estimates of atmospheric rainout. It is obvious, however, that the natural sources contribute to a very minor degree to the coastal oil pollution.

If we summarize the sources: offshore production, coastal oil refineries, industrial waste, municipal waste, urban runoff and river runoff (No. 2–7 in the table) we obtain a figure of 2.78 mta (millions of tons per annum) for the sources localized along the shores and discharging into coastal waters. Offshore oil production is here regarded as a coastal source as well, because most of that production takes place not far from the coasts and an oil spill

Table 1. Comparison of estimates of petroleum hydrocarbons annually entering the oceans, *circa* 1969–1971.

| | Authority (Millions of Tons per Annum) | | |
| | MIT SCEP Report (1970) | USCG Impact Statement (1973) | NAS Workshop (1973) |
Source			
1 Marine transportation	1.13	1.72	2.133
2 Offshore oil production	0.20	0.12	0.08
3 Coastal oil refineries	0.30	—	0.2
4 Industrial waste	—	1.98	0.3
5 Municipal waste	0.45	—	0.3
6 Urban runoff	—	—	0.3
7 River runoff[a]	—	—	1.6
SUBTOTAL of the sources No 2–7[c]	3.15	4.5	2.78
8 Natural seeps	?	?	0.6
9 Atmospheric rainout	9.0[b]	?	0.6
TOTAL	11.08	?	6.113
GRAND TOTAL[c]	15.96	11.24	6.113

[a] Petroleum hydrocarbon input from recreational boating assumed to be incorporated in the river runoff value.

[b] Based upon assumed 10 per cent return from the atmosphere.

[c] For calculation of the grand total the bars and question marks in the MIT SCEP and USCG columns were replaced by the corresponding estimates in the NAS column.

For further explanation see the text.

(Reproduced from *Petroleum in the Marine Environment* (1975), page 6, with the permission of the National Academy of Sciences, Washington, D. C.)

from a production site is likely to be a threat to a coastal water. There are of course examples of more remote offshore production, e.g. in the North Sea. However, at the time the three estimates, for North Sea production was very modest.

To the 2.78 mta above we must add an unspecified but great deal of the 2.133 mta given for the marine transportation. The reasons for stating a great deal of the figure are as follows. Each oil cargo is handled at least twice; at loading and unloading. These terminal operations are carried out in the coastal zones, and leakages are frequent at such operations. Despite international and national regulations much of the oil remainings from tank cleaning operations are discharged in coastal waters. Reality tells us that the supervision of these regulations are not effective enough. Furthermore, when the crude oils have been refined, very much of the products are then transported by coast tankers. Not all of these tankers have the load-on-top system that the big tankers have and which reduces the amount of oil released into the water.

As a result of this our figure of 2.78 mta might increase to some 4 mta. It is not possible to say directly how much of this, reaches the estuaries.

However, not less than 1.6 mta of that amount is estimated as the river runoff and that affects estuarine areas. Furthermore, much of man's activities are concentrated to estuaries where we find many big cities and important harbours, etc., and consequently a considerable amount of oil pollution. This discussion leads us to the not very unrealistic conclusion that some estuaries receive a very heavy burden while others are spared, because a large number of estuaries are situated in non-populated areas.

From this estimation it is realistic to state that the overall input of oil to estuaries may well exceed 2 mta and that figure may prove to be on the conservative side. This is then to be compared with the total figure of 6.113 mta that NAS gives for the entire input to the oceans. The estuaries and other coastal waters indeed receive a heavy burden.

In the two other studies in Table 1 the contribution from several sources were never estimated. If we want to use these studies for comparison with that from NAS, figures from the latter one may be used where the others display dashes or question marks. Summarized in that way the study from SCEP (Study of Critical Environmental Probe) yields 3.15 mta for the sources No 2–7 and 15.96 mta for the total input and the USCG (US Coast Guard) study yields 4.5 mta and 11.24 mta respectively.

In contrast to the discussion about global input one must always bear in mind that such figures give us the dimensions of the problem but they do not tell anything about the input to a specific estuary. It is the input to the estuary under consideration and the stress tolerance and capacity for recovery of that estuary that in any given case is of outmost importance for the impact of any oil pollution.

For a specific area of interest it can be possible to obtain a better estimate of the possible input of oil. Using local (national) statistics on transportation, river runoff, etc., combined with the principles for evaluation as given in the NAS report, it will be possible to analyse as above and reach a reliable figure for the area under investigation.

5 THE ENVIRONMENTAL FATE OF A POLLUTING OIL

In this section will be presented a generalized model and a likely course of events for an oil which is released into the marine environment. The different processes involved in transformation, and sometimes also elimination, of the oil will be described. In the following section some of the processes (those given in italics) will be further discussed from a chemical point of view. Much of the discussion here is derived from Parker *et al.* (1971) and Nelson-Smith (1972) unless otherwise indicated.

Oil in industrial and domestic effluents as well as in river runoff may be present either as dissolved substances, emulsions, or particles. The fate of such oils depends partly on the form in which the oils enter the water. The different hydrocarbons may be subjected to any of the process below.

For example, when an oil is released from a stranded ship it reaches the water as a separate phase, which largely is immiscible with the water. What happens to the oil during the first period of time it spends on the water depends partly on the properties of the oil—its composition and viscosity etc.—partly on some environmental factors such as temperature, wind and wave action, currents, etc.

At first the oil will form a surface slick which can be of considerable thickness. Depending on the viscosity of the oil, the temperature and the intensity of wind and wave actions the oil will in shorter or longer time spread into a thinner slick which covers a larger area (see Figure 3). Wind will force this surface slick to move and generally it will travel with a speed of about 2–4 per cent of the wind speed.

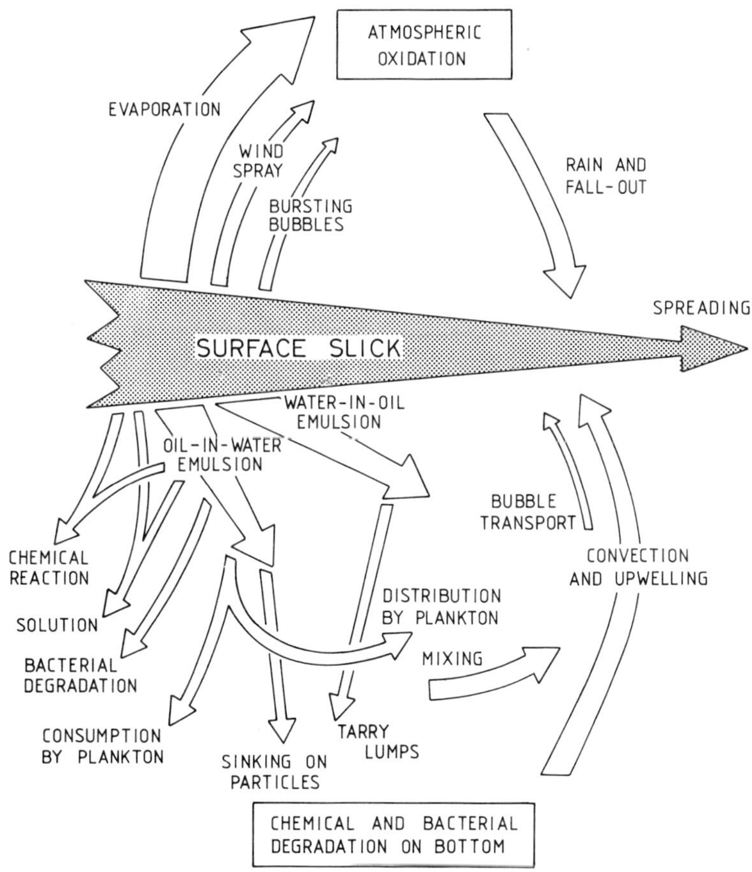

Figure 3. The main physical, chemical and biological processes governing the fate of a polluting oil (from Nelson-Smith (1972). Reproduced with permission of the author.

During this initial phase *evaporation* to the atmosphere is the most important modifying process, while *chemical* and *biological* processes will be subordinate. This evaporation will of course be speeded up by strong winds. As the oil looses its volatile compounds, its density increases. This process itself, as well as adsorption to denser particles or incorporation of water into the oil give the oil a greater density than the surrounding water and the oil will sediment to the bottom.

In tidal waters, particularly in tidal estuaries, there are large quantities of suspended silt and other particles. As the slick is agitated, these become incorporated in the emulsion, or are collected by dispersed droplets and make them more liable to sink. Quartz particles have a density of about 2.65 g/cm^3. The spreading of pulverized material of some sort on to the floating oil is a method that is sometimes used in dealing with spilled oil.

Salt water disturbs the ionic charge that surrounds dispersed particles. Because of this flocculation occurs in coastal waters, where fresh water from rivers or sewage water is mixed with sea water. The settling of particles occurs more rapid than in fresh water. Consequently, extensive banks of soft mud are to be found in such areas. Bottom deposits of oil, loosely bound by fine sediments, might be expected to occur in sluggish estuaries and shallow coastal waters. In most large estuaries, the net circulation (disregarding tidal oscillations) consists of a seaward movement at the surface, balanced by an upriver movement at the bottom; sedimented oil is thus more likely to be concentrated than dispersed.

As a result of the interaction between the properties of the oil and the environmental factors, oil and water will be mixed to a minor or major degree. According to NAS (1975) a Kuwait crude can take up 50 per cent water in just two to four hours, while a Tia Juana crude will absorb only 3 per cent during a period of two days. If sufficient amount of water is entrained in the oil it will form a water-in-oil emulsion and attain a gel-like consistency, which is often referred to as chocolate mousse. In this the water content can be as high as 70–80 per cent. Because of the low surface to volume area, chemical and biological processes (with the exception of *microbial degradation* by possibly occluded bacteria) can be expected to be slow. These may be more pronounced as the mousse breaks down into smaller pieces and particles as a consequence of wave action. This water-in-oil emulsion, which can be extremely stable, is widely regarded as the precursor for the more or less hard lumps of oil known as tar balls, pelagic tar, or beach tar. That these particles are capable of travelling far has been demonstrated, for example, by Forrester (1971). Following the grounding of the tanker Arrow in Chedabucto Bay, Nova Scotia, tar particles with sizes ranging from 5 μm to 1 or 2 mm were distributed down to at least 80 m within an area of 25 km times 250 km.

As a contrast to the chocolate mousse, the oil itself can be subdivided into fine particles, forming an oil-in-water emulsion. As long as this is agitated the particles will remain separated, but if the agitation ceases, as may be the situation in a calm estuary, the particles may coalesce thus reforming a surface slick. This formation of oil-in-water emulsion is mediated by the presence of natural emulsifiers in the water or by the addition of detergents. This is a method which has been widely advocated in dealing with oil spills. However, the resulting emulsions have often showed a synergistic effect (e.g. see Kühnhold, 1969; Lindén, 1976) and the use of detergents for these purposes have been restricted in many countries.

The oil-in-water emulsion has a very large surface to volume ratio which will facilitate several processes: *dissolution* will increase and *chemical oxidation* as well as microbial degradation will gain speed. Solar irradiance may penetrate the upper layer of the water when the oil slick has been emulsified, and this will lead to *photochemical oxidation* of the oil in the surface water.

The oil particles in the emulsion will to some extent adhere to different marine organisms. The most important pathway is probably the interaction with zooplankton. In a study following the Arrow accident Conover (1971) found that as much as 10 per cent of the Bunker C oil in the water was associated with zooplankton and that their faeces contained up to 7 per cent Bunker C, which had been ingested. These pellets, combined of faeces and oil, have higher density than water and this leads to sedimentation. Conover estimated that perhaps 20 per cent, or more, of the oil was sedimented to the bottom as zooplankton faeces. Maurer (1977) found, in a study after the Argo Merchant spill, that oil was ingested into zooplankton which were collected at several of the stations in the study area.

The processes described above play important roles in the modification and elimination of oil. During the course of these processes the properties of the oil, that remains in any given moment, are different from those of the original oil. With a collective term the processes are known as weathering and the oil as weathered oil. Obviously, the differences between weathered oils can be at least as big as the differences between the original oils. Sometimes the term is a bit more precise; e.g. chemical weathering, biological weathering.

6 WEATHERING—MECHANISMS FOR MODIFICATION AND ELIMINATION OF OIL

6.1 Evaporation

Evaporation is by far the single most important natural process for decreasing the amount of oil polluting a given sea area. The single most important factor governing the evaporation is the content of volatile compounds in the oil. Crude oil may contain compounds with boiling points ranging from well below room temperature and up to above 500 °C. The lower boiling fractions of a crude oil are comparatively rich in the simpler saturated alkanes and cycloalkanes as well as low molecular weight aromatics, whereas the higher boiling fractions, e.g. heavy fuel oils, contain more of the aromatic hydrocarbons.

An average crude oil may contain up to nearly 50 per cent of low molecular weight hydrocarbons (from C_4 to C_{12}) according to Golberg (1976). Thus evaporation will lead to rapid losses of hydrocarbons and the extent of the process may be quite considerable. Parker *et al.* (1971) state that 25–30 per cent can be lost in 2–3 days. According to the data from Golberg evaporation alone would be capable of removing about 50 per cent of an average crude from the sea surface. A Bunker C or other heavy oil, on the other hand, will probably not lose more than 10 per cent. The evaporation can lead to the situation that the remainings of a crude oil may resemble a Bunker C oil as far as hydrocarbon distribution is concerned. Only traces of low boiling products like gasoline or kerosene will remain in the water. A somewhat more extreme example is the Bravo blow out which occurred from a production platform in the North Sea in April 1977. A mixture of oil and gas at a temperature of about 75–90 °C was ejected up in the air to a height of about 50 m. Consequently the initial losses were great and part of the oil may even never have reached the sea surface. After a period of maximum 10 days about 60 per cent of the oil had evaporated (Bratberg, 1977).

In simulated evaporated experiment in the laboratory Kreider (1971) showed that hydrocarbons containing less than about C_{15} (boiling point <250 °C) may be lost from the surface of the sea in 10 days. Hydrocarbons in the C_{15}–C_{25} range (boiling point 250–400 °C) showed limited volatility and will therefore be retained to the greatest extent within the slick. Above C_{25} (boiling point >400 °C) there will be no substantial loss (NAS, 1975).

The evaporation is increased not only by high ambient temperature, but also by wind and wave action. From the wave crests at sea or waves striking the shore the wind can whip up an aerosol of fine droplets of oil. Similarly the bursting of air bubbles at the sea surface will transport oil up into the atmosphere.

The fate in the atmosphere of these hydrocarbons is not known. Their potential of returning to marine environment is yet to be determined.

6.2 Solution

In contrast to earlier considerations oils are not insoluble in water although the solubilities are low as far as the concentration figures are concerned. However, it must be remembered that a solution, containing as an example 1 mg/l, and covering one square kilometre and to a depth of 1 metre represents 1 ton of oil. Thus solution can play an important role for the modification of an oil in the marine environment.

As will be demonstrated below, generally the most soluble compounds of an oil are those which display the greatest volatility as well. Thus these two processes largely will remove the same kind of compounds, but an important difference remains between the processes; by evaporation, which is the quantitatively more important process, the compounds are removed from the marine habitat, whereas by solution they remain in the water. As the compounds are in the water they are readily available to marine organisms. Several of the most soluble compounds are very toxic but a large number of sublethal effects can occur as well.

The extent and the speed of the solution process is governed both by the environmental factors such as temperature, wind, and wave action, as well as the properties of the oil itself (chemical composition, solubility, viscosity, pour point, etc.). Because of the environmental conditions, the solution will remain a pure process only in very calm weather. The wind and waves will contribute not only by speeding up the solution, but by dispersing the oil as well.

The solution from a spilled oil will start immediately, but the process has a long term perspective also. Oxidation and microbial degradation produces more polar substances (fatty acids, fatty alcohols, etc.) which are by far more soluble than their parent compounds.

Between the different groups of compounds the solubility decrease in the order: aromatics—naphthenes, i-alkanes, n-alkanes. Within each group solubility is inversely related to the molecular weight. Thus, in practice, the highest concentrations to be found in relation with a spilled oil are those of aromatics and low molecular n-alkanes.

For the n-alkanes the solubilities are roughly related to their vapour pressures (Rossini *et al.*, 1953). This is valid for the low molecular weight compounds, but as solubility and vapour pressure is related to the chemical potential of a compound in a solution, this relation is expected to hold for the higher boiling compounds of a homologous series as well.

A few examples will be cited from NAS (1975) in order to demonstrate the principles discussed above. Of the n-alkane hydrocarbons, n-C_5 has a distilled water solubility of about 40 mg/l (all figures given in the same units), n-C_6 about 10, n-C_7 about 3, n-C_8 about 1, n-C_{12} about 0.01 and n-C_{30} about 0.002. For aromatic hydrocarbons, C_6 (benzene) has a distilled water solubility of about 1,800, C_7 (toluene) about 500, C_8 (xylenes) about 175, C_9 (alkylbenzenes) about 50, C_{14} (anthracene) about 0.075 and C_{18} (chrysene) about 0.002. Hydrocarbon solubilities in sea water are less than those in distilled water; a decrease by some 12–30 per cent can be estimated (Harned and Owen, 1958).

As a complement to the laboratory figures above, a study by Boylan and

Tripp (1971) gives data for the compounds released from crude oil and oil products into sea water. In their study they used filtered sea water and the figures given are not true solubilities, but rather reflects the concentrations obtained at equilibrium when the oil and water had been slowly stirred and the phases had been allowed to separate. Thus, the presence of some finely dispersed droplets cannot be fully ruled out. The water phases from Kuwait crude oil and kerosene contained mostly benzenes and naphthalenes, i.e. aromatic compounds in the low boiling range. The total aromatic content released from the Kuwait crude oil was 1,453 μg/l with the concentration of naphthalenes reaching 65 μg/l. From the kerosene the total aromatics reached 610 μg/l, whereas the content of naphthalenes comprised not less than 362 μg/l of that total amount. This clearly indicates a positive correlation between the higher toxicity to fish of kerosene than Kuwait crude and the content of naphthalenes in sea water extracts from these oils.

As was stated above, microbial activity, as well as chemical oxidations, yield polar components of higher solubility than their parent compounds. The resulting effect is that these processes, which will be further discussed below, will increase the solubility losses from the oil into the water. For example, naphthalene (solubility 32 mg/l) can be oxidized to α-naphthol (solubility 740 mg/l) and n-octanol (solubility 1 mg/l) yields n-octanic acid (solubility 600 mg/l).

6.3 Chemical degradation of oil at the sea surface

The chemical degradation process is not known in detail, but some guidance can be obtained from the literature.

For initiating and maintaining most of the chemical reactions which are likely to occur, more energy is required than for evaporation or solution. Energy is at hand in the form of sunlight, especially the energy rich ultraviolet radiation. The penetration of light into water is rather limited, and consequently most of the chemical reactions involving the petroleum hydrocarbons occur in the surface water layer and in the surface oil film. The reactions in tar balls and dispersed oil will be of subordinate magnitude, partly because oil in these forms is only to a minor degree present in the surface water and partly because the unfavourable surface to volume ratio of the particles limit light absorption. The reactions occurring in the oil are of oxidative nature. Reduction reactions will not occur unless the oxygen content in the surrounding water is extremely low. Auto-oxidation of oil by atmospheric oxygen is possible, but according to Hansen (1977) who studied photodegradation of crude oil in a model system, this did not occur in darkness at temperatures below 30 °C. It can be concluded that for initiating (and maintaining) the chemical reactions, the presence of light is essential. The photolysis thus induced will continue *via* a chain of free radical reactions. The reactions can be catalysed by the presence of metal ions but be inhibited by sulphur atoms in reacting molecules.

Many of the compounds formed in the oxidations are by far more soluble in water than the original ones. However, polymerization of simple molecules into larger aggregates may occur as well. Aldehydes or ketones may undergo condensation reactions with phenols, and esterification may occur between alcohols and carboxylic acids. The resin and asphaltene content of the oil may increase. Aromatic hydrocarbons of medium and high molecular weight probably have a basic part in this process. This increase is most noticeable in the case of petroleum products that have been subjected to heating, as is the case, for example, with fuel oils (according to the *GESAMP Report*, 1977; for reference see *IMCO*, etc.).

One crucial property of the available light is its wavelength. The speed of the photolytical reactions is dependent of the wavelength and this in turn has a quantitative aspect for the modification of the oil. Parker *et al.* (1971) reported that when thin films of oil were irradiated with light of wavelength shorter than 300 nm, i.e. shorter than the ultraviolet limit of sunlight, the photooxidation was found to be rapid. When the light was restricted to the wavelengths present in sunlight the processes were slower. The most effective region appeared to be 300–350 nm. This is in good agreement with the results obtained by Hansen (1977), who reported that the UV-spectrum of the crude oil he used for experiments showed an increase of absorption starting at 260 nm, due to the presence of aromatics and sulphur compounds. Hansen found that when the oil was irradiated with light and the spectral range was widened from 300 nm to 200 nm, the photooxidation processes increased by at least two orders of magnitude.

The type and extent of the chemical modification of the oil is not only governed by the availability of energy. Obviously the type of the oil and its composition is crucial. For example alkyl derivatives of either naphthalenes or cycloalkanes, as well as the isoprenoids, are more rapidly oxidized than n-alkanes. The gum produced in degraded gasoline consists largely of esters of higher molecular weight, while the degradation products of lubricating oils contain acids, alcohols, esters, and carboxyl compounds (Parker *et al.*, 1971). Hansen (1977) found that irradiation of a Libyan crude in the laboratory produced a great number of products. From the sea water beneath the oil were extracted carboxylic acids (20 per cent aromatic, 28 per cent aliphatic), alcohols (23 per cent) and ketones (18 per cent, aliphatic only). The aromatics comprised 20 per cent of the reaction products but only 3 per cent of the original oil. The preferential reaction by aromatics is obvious from this experiment, which shows the importance of the aromatics as initiators of photooxidation.

The mechanism of free radical oxidation is conveyed *via* several steps which include formation of peroxides or hydroperoxides as intermediate products. These intermediates are then converted into carboxylic acids, aldehydes, ketones, and alcohols. The aldehydes and alcohols can then be readily oxidized to carboxylic acids by reaction with peroxides (except tertiary alcohols, which are inert).

Carboxylic acids are surface active materials because of their hydrophilic and hydrophobic nature. They are readily dissolved from the surface film into the water, where they are subject to very limited oxidation. The formation of surfactants obviously affects the physical dispersion of the surface oil film. Hansen found a hydrocarbon concentration of 300 μg/ml under the surface film, which is far above the water solubility of the hydrocarbons. Additionally, the amount of water in the surface film increased to several times the weight of hydrocarbons, both effects due to the formation of oil-in-water and water-in-oil emulsions respectively, enhanced by the formation of surfactants.

As a result of the chemical oxidations, a number of compounds will be removed from the oil. The remaining part of the oil usually turns darker and its density and viscosity increases. Solid or semisolid particles are formed and as a result sedimentation may increase.

The quantitative role of the chemical reactions is estimated in different ways by different authors. Hansen reported from his laboratory studies that during a 19-day period, when the Libyan crude was maintained at 26 °C and irradiated with light of wavelengths down to 200 nm, the decomposition rate was some 0.7 per cent per day. Experimental estimates were made by Freegarde and Hatchett (1970) concerning the actual removal rates of slicks attributable to photolysis. Using light sources resembling the spectrum of sunlight, within a factor of two in internal relative intensities, the total rate of decomposition corresponded to the destruction of a 2.5 μm slick in 100 hours. A slick of this thickness has about 2 tons/km^2. Assuming an effective 8 hour day of sunlight, photolysis can initiate sufficient oxidation reaction to remove the slick in a few days. This might be a conservative removal estimate since it assumes fairly complete oxidation, whereas the formation of carboxylic or phenolic acids increases the solubility and emulsification rates, and perhaps the biodegradation rate as well (NAS, 1975).

6.4 Microbial degradation

The microbial degradation is the single most important process for modification, and possible elimination, of that part of the oil which remains when the evaporation has ceased. The statement of possible elimination is sometimes misinterpreted in that sense that the oil is believed to be completely degraded to carbon dioxide and water, thus no longer being a threat to the marine flora and fauna. However, along the path from oil to carbon dioxide and water a multitude of substances are produced, some of which can be harmful to marine life. In this connection it may be proper to remind of the old fact that a substance can never be regarded as being harmless; it is the actual concentration in any given situation that makes the difference between harm and harmlessness.

This section has a strong connection to a following one dealing with oil in the sediments, because a great part of the oil is likely to end up in the sediments, where the microbial activity is responsible for the further fate of the oil. Unless otherwise stated, the presentation below is largely based on the reviews by Karczewska (1972), NAS (1975), Goldberg (1976) and GESAMP (1977).

Many marine microorganisms are capable of degrading compounds of oil. Although other sources of carbon are usually preferred, the microorganisms can be adjusted to decomposition of oil. The process may be most efficient in the coastal zones, where the organisms are constantly adjusted. The oil can be so radically changed that, except for a possible content of metals like vanadium and nickel, it is not recognizable as an oil at all. However, also the opposite question is relevant to consider. The presence of oil in the ecosystem can influence the microbial decomposition of natural organic matter and thus influence the recycling of nutrients. In a laboratory system Chet and Mitchell (1976) studied the decomposition of organic substrates by motile marine bacteria. When kerosene was added to the system the decomposition rate decreased drastically but not because of decreased enzymatic activity but rather by the reduced motility of the organisms. When the system was carefully mixed the decomposition rate increased again because the contact between the bacteria and the substrate was facilitated. It might thus be concluded that the degradation is not only dependent on the composition of the oil itself, but also upon a number of factors in the physical and chemical environment.

The microbial degradation of oil is a multivariable system and the result of the system is not possible to predict on the basis of the preconditions known. A great number of studies have been reported in the literature. Because the processes involved are very difficult to study in nature separated from other influences most of the studies have been carried out in laboratories, using well-defined systems at optimized conditions. From these disparate studies it is not possible to draw firm conclusions which are directly applicable to what will happen with a given oil under natural conditions. However, based on the laboratory studies and the few reliable field measurements that have been carried out, some regulating factors and degradative mechanisms can be understood and discussed.

Studies *in vitro* have showed that petroleum hydrocarbons ranging from simple gas to complex solids are utilized by bacteria under proper conditions. However, the mechanisms involved and the resulting degradation rates are different for various compounds and thus the composition of the oil influences the degradation. Aliphatic hydrocarbons are oxidized more rapidly and by more microbial species than aromatics, however phenol and cresols can be readily oxidized by acclimatized microorganisms. Alkylbenzenes with short sidechains are rather resistant. Generally, the resistance to microbial attack increases through the series n-alkanes, i-alkanes, cycloalkanes and aromatics. Derivatives containing oxygen, nitrogen or sulphur are generally more readily oxidized than pure hydrocarbons. Some fractions of kerosene and naphtha are even reported to be bacteriocidal or bacteriostatic.

The state at which the oil is present is important. When the contact area between the oil and the environment containing the microorganisms increases, the degradation is facilitated. Thus, hydrocarbons present in naturally produced emulsions or associated to particles, can be expected to be more rapid degraded than those in solution.

Factors such as salinity stresses, wind and wave action, temperature and sunlight not only directly affect the growth and metabolism of biota, but also change the physical state and chemical nature (e.g. emulsification and oxidation) of the oil. When oil is buried in oxygen free sediments only little degradation occurs. Generally increasing factors are, for example, increasing temperature (oxidation is most rapid in the interval 15–35 °C) and availability of oxygen and nutrients, whereas insufficient organic substrate for development of viable microorganisms, presence of alternate carbon sources and presence of predators are limiting factors.

In some investigations yeasts have been used to mediate oil decomposition because of their greater tolerance to external stress, such as osmotic pressure and ultraviolet sun radiation.

More than 90 species of microorganisms capable of oil degration, bacteria as well as fungi, have been identified. Some examples of the most active species are *Pseudomonas, Mycobacterium, Proactinomyces, Micrococcus, Bacterium, Bacillus, Sarcina, Serratia, Penicillium and Aspergillus.*

A few examples will be mentioned in order to demonstrate the mechanisms (as found in model studies) which are involved in the degradation processes.

The first reaction involved in alkane oxidation is one requiring participation of molecular oxygen, whereas subsequent reactions involve the oxygen of water. Incorporation of ^{18}O into cell material of organisms grown in various n-paraffins has been found to decrease markedly as the length of the paraffin increased, which indicates that the longer molecules are more resistant. Oxidation of hydrocarbons by sulphate-reducing bacteria is slow and will take place in the absence of free or dissolved oxygen. The primary enzymatic attack occurs at the terminal carbon atom. Studies of how different organisms modified various alkanes, indicated at least three major pathways to the microbial utilization of alkanes. Certain short chain length alkanes, C_3–C_5, produce methyl ketones probably *via* arrangement of 1-alkyl free radicals. Intermediate length alkanes, C_6–C_{10}, are dehydrogenated to the corresponding 1-alkene. Long chain alkanes, C_{12}–C_{18}, are oxidized by the way of 1-alkyl hydroperoxides and the formation of n-alkohols which in turn are oxidized to n-fatty acids.

That the n-alkanes are more readily oxidized than their branched relatives are clearly shown, for example, Green *et al.* (1974). In their field study after an oiling of the Alert Bay in British Columbia (Canada), they followed the fate of the released oil during one year. In Figure 4 are shown gas chromatograms of the original oil and of an extract of oiled gravel one year later. When comparing the peaks of the n-C_{17} and n-C_{18} alkanes with those of the isoprenoids pristane and phytane, it becomes quite clear that the straight chain hydrocarbons are the most susceptible to microbial attack. This is characteristic of biological degradation of oil. Other natural degradative processes, such as photo-oxidation, normally oxidize the isoprenoids more quickly than the alkanes. Thus the ratio n-C_{17}/Pr or n-C_{18}/Ph can be used as an instrument to follow the microbial degradation process in such a study.

The only substances known with certainty to be produced in the marine environment as a result of microbial digestion, and whose concentrations have been measured in the field, are microbial tissue (Gunkel, 1968) and carbon dioxide (Robertson *et al.*, 1973). However, numerous intermediates and end products are known to accumulate in laboratory experiments. Some of these substances, for example aliphatic alcohols, acids, and equivalent aromatic derivatives, may be disruptive to chemotaxis by marine life (Mitchell *et al.*, 1972; Zafiriou, 1972). Some oil degrading organisms could produce toxins (Traxler, 1973).

Several of the intermediates and end products, such as acids, aldehydes, ketones, alcohols, peroxides, and sulphoxides are soluble in water and will thus be removed from floating oil or oil particles.

The rates for degradation of oil have been studied in a number of experiments. The results have been reported in a multitude of ways: $g/m^2/day$, $g/m^3/day$, mg/day/bacterial cell or percentage removed. Consequently the results from different studies are virtually impossible to compare directly. However, as most of the figures are derived from experiments carried out at more or less optimized conditions, they can only give indication of maximum rates. The often-quoted estimate of 36–350 $g/m^3/year$ (ZoBell, 1964), while it may apply to heavily polluted harbours, is many orders of magnitude too high for the open sea. For example, Robertson *et al.* (1973) measured oxidation rates of n-dodecane, one of the most readily oxidized hydrocarbons *in situ* to be a factor of 1,000 lower. In another work ZoBell (1969) summarizes work accomplished prior to 1969.

                    S. R. Carlberg

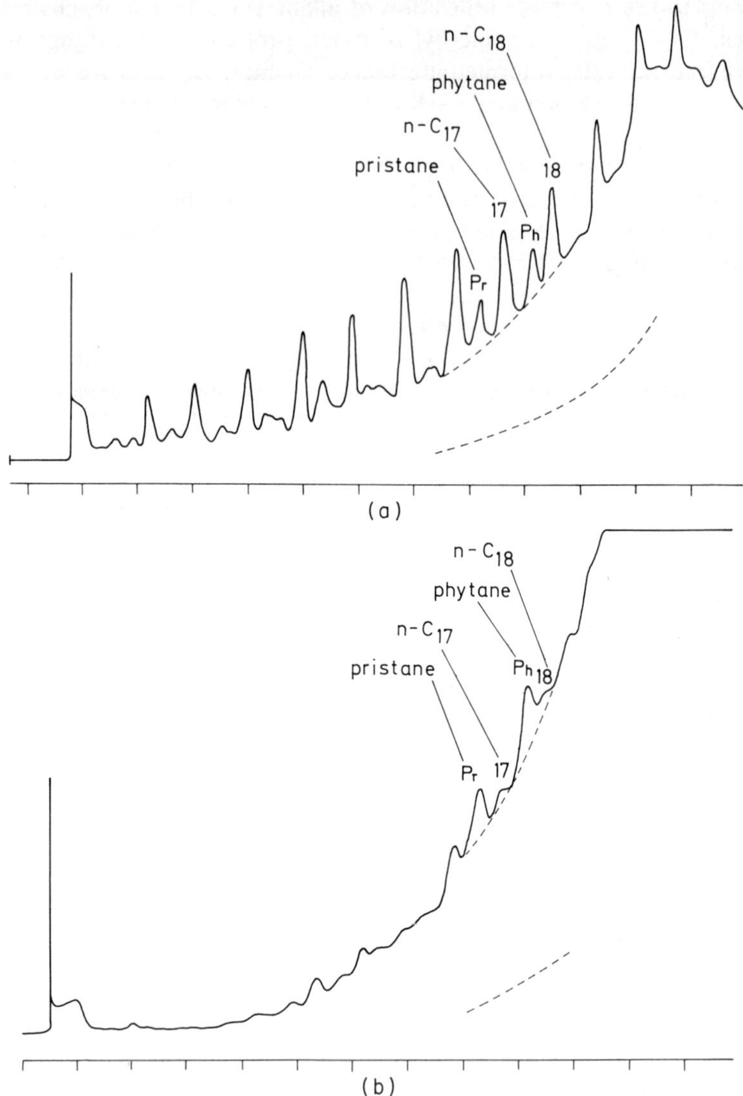

Figure 4. Biological degradation of oil. Both figure 4(a) and 4(b) show parts of gas chromatograms of the same No. 5 fuel oil.

Figure 4(a) shows the original oil. The ratio between n-C_{17} and pristane (C_{17}/Pr) is 1.9 and C_{18}/Ph is 2.2.

Figure 4(b) shows the same oil extracted from gravel one year after the spill. C_{17}/Pr is 0.3 and C_{18}/Ph is 0.2. This demonstrates the higher resistance of isoprenoids, as compared with normal alkanes, to biological degradation.

(After an unpublished manuscript by Green *et al.*, 1974. The final report was published by Cretney *et al.*, 1978.

Reproduced by permission of the authors).

The degradation rates of crude oils, lubricating oils, cutting oils, and oil wastes range from 0.02 to 2.0 g/m^2/day at 24 to 30 °C.

While weeks or months is the probable time duration for the degradation of n-alkanes, the more resistant fractions may require years to decades. This will be most pronounced in low-energy environments, where wind and wave actions are restricted, as pointed out by Owens (1978). In a study 6 years after the Arrow spill, Keizer *et al.* (1978) found isolated places within the upper high tide zone where the Bunker C oil remained in an 'pavement-like' consistency.

In laboratory studies it has been shown (Atlas and Bartha, 1972, Robertson *et al.*, 1973, and Traxler, 1973) that if special psychrophilic cultures develop the overall microbal degradation rate may not necessarily be significantly reduced by a temperature decrease to 5 °C. In *in situ* experiments it was shown, however, (Atlas *et al.*, 1978) that oil in arctic marine water was very slowly degraded indeed, and if the oil was on or below an ice cover the biodegradation was in practice inhibited during the winter months.

7 OIL IN SEDIMENTS

As has been demonstrated above, several processes will contribute to the removal of many of the constituents of a polluting oil. After the initial, and possibly rapid, evaporative phase the major part of the oil remains in the water and is subjected to microbial degradation and other processes which are comparatively slow. As a result of this, and furthermore aided by the sedimentary processes, the greater part of the oil is likely to end up in the sediments. This is most pronounced in estuaries and other coastal waters, where the water is shallow and sedimentation is strong.

The biological aspects of oil pollution—in sediments, as well as in other compartments—is by and large beyond the scope of this chapter. However, from some points of view the hazards arising from oil in sediments will be briefly discussed.

The microbial degradation of oil is much slower in sediments than in water (Jonston, 1970). Several reasons contribute to this fact. Oil which is dispersed in water displays a larger surface area, thus facilitating the bacterial attack. Oil which accumulates as a film or layer on the sediment surface has a much smaller effective surface area and this reduces the speed of degradation. The upper part of the sedimentary layer is not a static area, but rather a highly dynamic one. By the activity of the bottom epifauna and infauna—e.g. deposit-feeding polychaetes—the top layer is mixed. This is a process often referred to as bioturbation. As a result of this the oil will become incorporated in the sediment. When the oil in this way reaches the anaerobic layers of the sediment, the degradation rate will be considerably decreased presumably due to reduced bacterial activity. Consequently, the sediments will act as deposits. In areas where oil pollution is not very infrequent the inflow of oil to the sediment will exceed the degradation and hence the oil will be accumulated. In the literature there are numerous field studies supporting this (Scarratt and Zitko, 1972; Blumer and Sass, 1972; Thomas, 1973; Rudling, 1976; and Wennergren, 1976, 1978). In the studies referred to it was found that Bunker C and No 2 fuel oil persisted in shallow inshore and marsh sediments for more than two years. In the studies by Wennergren it was shown that following the great oil spill from the *Monte Urquiola* outside the La Coruña area, the oil content in the estuarine sediments increased considerably from May 1976 (shortly after the spill) to May 1977, whereas the oil content in the water decreased. In Table 2 are given some examples of petroleum hydrocarbon content of sediments from different areas.

In studies of oil pollution of aquatic areas, the chemical analysis of sediments offers a very valuable tool for assessment of the degree of contamination.

Depending on in which form (slick, chocolate mousse, tar balls, dispersion, etc.) the oil reaches the sediment, the oil will either be present as an absorbed film, a covering layer or as discrete particles. If we extend our discussion of sediments to include shores, especially the tidal beaches, we will find the oil deposited with varying degree of patchiness. On a rocky shore the oil will be caught in cavities and on other shores the oil may reside among stones and pebbles. On a sandy beach and also in shallow water, the oil may by the tidal activity be incorporated into the soft bottom material. The bioturbation, as mentioned above, will further facilitate this.

Considering the small scale distribution (metres), the variability might roughly be correlated to the particle size of the sediment. Hargrave and Phillips (1975) analysed replicate samples of different sediment types from Bermuda and from the Bedford basin (Nova Scotia). They found a large variation (62 per cent) for coarse sand (1,200 μm), intermediate (29 per cent) for fine sand (255 μm) and lowest (8 per cent) for mud sediments (16 μm). Variations in oil content could not be related to differences in organic carbon content or carbon:nitrogen ratios in the sediments examined. Based on a limited number of samples, the authors could not relate the proportion of organic carbon attributable to oil, to either the particle size or the amount of organic matter in the sediment.

Concerning the large scale distribution (kilometres or hundreds of metres), the variability will be governed by a number of factors such as the time duration of the oil spill, changes in direction and strength of currents, tidal activity, etc. In the study connected with the *Amoco Cadiz* disaster (Hess, 1978) samples from sediments were taken from the entire oiled coastline of Britany. From the estuary of l'Aber Wrac'h (not far from the stranded tanker) to Ile Grande (approximately 80 km) samples contained

Table 2. Examples of hydrocarbon levels in sediments

Location	Sample depth	Water depth (metres)	Concentration (p.p.m., dry weight)	No. of samples
Buzzard's Bay, Massachusetts, Wild Harbor river (over $2\frac{1}{2}$-year period)	Top 10 cm, sand and clay	<3	250–1,650	~180
Buzzard's Bay, Massachusetts, Station 37, subtidal unpolluted	Top 10 cm	11	38–70	10–20
Buzzard's Bay, Massachusetts, Silver Beach (over $2\frac{1}{2}$-year period)	Top 10 cm sand and clay	<2	500–12,000	~60
Mississippi coastal bog	—	0–1.5	350	1
Narragansett Bay, Rhode Island, mouth of Providence river (head of bay)	Top 8–10 cm	?	820–3,560	2
Middle of bay	Top 8–10 cm	?	350–440 cm	2
Mouth of bay	Top 8–10 cm	?	50–60	2
Vineyard Sound, Massachusetts	Top 7 cm	6–9	1.7 n-alkanes only	1
Chedabucto Bay, Nova Scotia (over 2-year period)	Top 5 cm	3	34–420	7
Chedabucto Bay, Nova Scotia (over 2-year period)	Top 5 cm	12	11–1,240	7
Coast of France, Le Havre vicinity	'Surface'	Subtidal	450	1
Seine estuary	'Surface'	Subtidal	33	1
Coast of France, Le Havre vicinity (different from the one above)	'Surface'	Subtidal	920	1
Bay of Veys	'Surface'	Subtidal	38	1
Port Valdez, Alaska	'Surface'	Subtidal	0.5–2.5 (C_{16}–C_{28} only)	?
Gulf of Batabano, Cuba	?	?	15–85	10
Orinoco Delta, Venezuela	?	?	27–110	10
Gulf of Mexico, open ocean	?	?	12–63	10
Mediterranean, open ocean	?	?	29	1
Cariaco Trench	?	?	56–352	16
South-east Bermuda to base of rise near Hudson Canyon	Top 5 cm of sediment	>3,000	1–4	5

(Reproduced from *Petroleum in the Marine Environment* (1975), page 57, with the permission of the National Academy of Sciences, Washington, D.C.)

from 742 μg/g to 4 μg/g (dry weight) but no consistent pattern of decreasing oil content with distance from the tanker could be observed.

Oil, residing in sediments or sandy beaches, poses a long-term threat to the environment. Several processes may lead to it resuspension into the water mass. Bioturbation has to be considered as a two-way mechanism; oil that has been covered by more recent layers may be brought up to the sediment surface again. By the action of bottom currents, density waves, and tides the oil can thus be forced into the water. The processes acting upon oil in sediments are not well understood yet. In some studies have been reported that oil contamination of sediments seem to grow geographically although the input has ceased. Blumer *et al.* (1971) found in the study of the West Falmouth oil spill that in a period of four to five months the contamination of the sediments grew tenfold into Buzzards Bay (Massachusetts), covering an area of 20 square kilometres. Although not proved, it seems reasonable that resuspension followed by spreading and resedimentation at least partly can explain such observations.

As was stated above, oil may remain in sediments for prolonged periods of time due to the slow degradation processes. In a study of aromatic hydrocarbons in intertidal sediments in Buzzards Bay, Teal *et al.* (1978) found that naphthalenes with up to three alkyl substitutions and phenanthrenes with up to two substitutions decreased in concentration with time in surface sediments, whereas the more substituted aromatics decreased relatively less and in some cases increased in absolute concentration. Keizer *et al.* (1978) made an extensive study of Chedabucto Bay (Nova Scotia) 6 years after the *Arrow* spill. Their chemical analysis included aliphatic as well as polynuclear aromatic hydrocarbons. Most samples revealed either a rather fresh contamination or highly weathered residues which could not be identified. However in some samples of intertidal and sublittoral sediments the Bunker C oil, originating from the *Arrow*, could still be identified after 6 years. Mayo *et al.* (1978) studied weathering characteristics of petroleum hydrocarbons which had been deposited in fine clay sediments in the area of Maine. The authors claimed that the physical weathering of the oil by evaporation and dissolution during the time between spillage and sedimentation largely determined the distribution of the hydrocarbons in the sediment. Once incorporated in the sediment, the hydrocarbons remained unchanged during a period of 5 years, as judged from the absence of significant changes in their gas-chromatographic profiles.

8 FATES IN ORGANISMS

Biological effects of oil pollution will not be considered in this chapter. Reviews of literature dealing with effect studies are presented, for example, in the reports by NAS (1975) and GESAMP (see IMCO, etc., 1977). Unless otherwise stated, this section is a synthesis of material presented in these two reports.

Although the biota may be severely affected by the presence of oil, it is not a major reservoir for spilled oil, nor is it a major factor in influencing the distribution pattern of spilled oil.

In the section dealing with the environmental fates of a polluting oil it was mentioned that the oil could be ingested by zooplankton. This is but one example of how oil may be taken up by organisms. Petroleum hydrocarbons are presented to pelagic organisms as dissolved or dispersed materials, adsorbed onto particulate material, or as small floating tar balls.

Lee and Benson (1973) have suggested that hydrocarbons enter the marine food web by several routes:
—adsorption onto particles, both living and dead, followed by ingestion of these particles
—active uptake of dissolved or dispersed petroleum
—passage into gut of fish which gulp or drink water.
A considerable portion of petroleum is adsorbed onto or dissolved in particulate matter, because the partition coefficient should favour the solution of hydrocarbons in the lipids of detritus. Detritus would then carry the adsorbed or dissolved hydrocarbons to the sea floor to be consumed by benthic organisms.

Dissolved hydrocarbons are taken up by the gill tissue in mussels and fish. It is presently believed that large floating aggregates (i.e., tar balls) are not taken up by marine organisms to any great extent, and are thus not an important introductory pathway into the marine ecosystem.

After petroleum hydrocarbons are taken up by an organism, they may be excreted unchanged, they may be metabolized, or they may be stored with possible elimination at a future date.

Insight into the fates and effects of petroleum in marine and estuarine organisms may be gained from different chemical and biological studies, which according to Andesson *et al.* (1974) may be classified in four general types:

1. Short-term toxicity studies—to determine the range of tolerance to the pollutant, to set exposure levels for sublethal studies, and to evaluate the significance of environmental concentrations.
2. Organism–environment transfer studies—to determine rates of accumulation and release of the pollutant by the organism.
3. Physiological studies—to determine the extent and nature of modification of metabolic parameters in response to exposure to sublethal concentrations of the pollutant.
4. Field studies—to compare responses of marine animals to natural and perhaps chronic exposures to pollutants with responses observed in the laboratory, and to determine the effects of pollutants on the metabolism and structure of marine communities.

In conjunction with (1) the reader is here referred to Table 3, which gives some examples of concentrations of petroleum hydrocarbons in marine

Table 3. Examples of hydrocarbon levels in marine macro-organisms.

Organisms	Area type[a]	Hydrocarbon type	Estimated hydrocarbon amount (μg/g)
Macroalgae			
Fucus	4	Bunker C[b]	40 dry
Enteromorpha	4	No. 2 fuel oil	429 wet
Sargassum	1	C_{14-30} range	1–5 wet
Higher plants			
Spartina	4	No. 2 fuel oil	15 wet
Molluscs			
Modiolus, mussel	4	No. 2 fuel oil	218 wet
Mytilus, mussel	4	No. 2 fuel oil[b]	36 dry
Mytilus	4	Bunker C[b]	10 dry
Mytilus	4	Bunker C, aromatics	74-100 wet
Mytilus	3	$n\text{-}C_{14-37}$[b]	9 dry
Mya, clam	4	No. 2 fuel oil	26 wet
Pecten, scallop	4	No. 2 fuel oil	7 wet
Littorina, snail	4	Bunker C, aromatics	46–220 wet
Mercenaria, clam	3	C_{16-32} range	160 dry
Crassostrea, oyster	2	Polycyclic aromatics	1 wet
Crustacea			
Hemigrapsus, crab	4	Bunker C[b]	8 dry
Mitella, barnacle	4	Bunker C[b]	8 dry
Lady crab	3	C_{14-30}	4 wet
Plankton	2	Benzopyrene	0.4 wet
Sargassum shrimp	1	C_{14-30}	3 wet
Lepas, barnacle	1	C_{14-30}	6 wet
Portunus, crab	1	C_{14-30}	34 wet
Planes, crab	1	C_{14-30}	11 wet
Fish			
Fundulus, minnow	4	No. 2 fuel oil	75 wet
Anguilla liver, eel	4	No. 2 fuel oil	85 wet
Smelt	3	Benzopyrene	0–5 dry
Flatfish	2	C_{14-20}	4 wet
Flying fish	1	C_{14-20}	0.3 wet
Sargassum fish	1	C_{14-20}	1.6 wet
Pipefish	1	C_{14-20}	8.8 wet
Triggerfish	1	C_{14-20}	1.7 wet
Birds			
Herring gull, muscle	4	No. 2 fuel oil	535 wet
Echinoderm			
Asterias, starfish	4	Bunker C, aromatics	20–147 wet
Luidia, starfish	2	C_{14-30}	3.5 wet

[a] 1, oceanic; 2, chronic pollution, coastal; 3, chronic pollution, harbour; 4, single spill.
[b] *n*-alkanes only.

(Reproduced from *Petroleum in the Marine Environment* (1975), page 62, with the permission of the National Academy of Sciences, Washington, D.C.)

macro-organisms. It should be pointed out, however, that it is not sufficient to study the concentrations in the biota and in the environment and whether these concentrations give rise to harmful effects after a certain exposure time. It is also essential to study the dynamics of the system. The dynamics of the different processes will tell us, in a specific case, whether the concentrations in the environment will reach harmful levels, and also whether such a situation will prevail after the source of the pollution eventually is eliminated.

The subject of (2) is discussed to some extent in the beginning at the end of this section. The matter in (3) will be further discussed here, because it is of special interest concerning the biogeochemistry of oil pollution. Finally, the subject in (4), being beyond the scope of this chapter, will not be discussed here.

Lee, Sauerheber and Benson (1972) found in particular that saturates and aromatics were taken up rapidly by mussels but were also discharged without metabolic breakdown after the mussels were returned to clean sea water. They found no evidence that mussels could not metabolize such aromatics as toluene, naphthalene, and benzpyrene. Metabolism of ingested petroleum hydrocarbons by marine organisms has also been shown to occur, followed by either incorporation into the body tissues or subsequent excretion. Marine fish tested by Lee, Sauerheber, and Dobbs (1972) were found to take up aromatic hydrocarbons *via* the gills. Metabolism then occurred in the liver, followed by transfer of the hydrocarbons and metabolites to the bile, and finally excretion. This suggests an efficient detoxification mechanism in the fish which allowed for the removal of polycyclic aromatics from the body tissues.

If some broad generalizations would be made it could be concluded that petroleum hydrocarbons become dissolved in the lipids and thus enriched in those tissues and organs which are rich in lipids. Saturated hydrocarbons as well as the aromatic ones are readily taken up, but the aromatics seem to be the most difficult to get rid of by depuration from the organisms. In a study of oil pollution in Dutch shellfish areas Kerkhoff (1974) found that polluted mussels lost 90 per cent of the gas oil (as measured by gas chromatography of the non-aromatic fraction) in one day when kept in clean sea water. The oily taste of the mussels remained for one or two months, however. Stegeman and Teal (1973) found that No 2 fuel oil was accumulated rapidly by oysters and dropped to a stable level of approximately 30 μg/g wet weight, where after two weeks of discharge it remained constant for at least $3\frac{1}{2}$ months. Blumer, Souza, and Sass (1970) found that in some cases, shellfish had a similar concentration of No 2 fuel oil eight months after a spill.

The metabolism by marine organisms other than bacteria is not well understood. The pathways involving oxidases and other enzymes, important in the degradation of aromatic and paraffinic hydrocarbons by mammalian systems, is well documented (Boyland and Soloman 1955; Daly, Jernis and Witkop, 1972; Falk *et al.*, 1962; Diamond and Clark, 1970; McCarthy, 1964). In the case of aromatic compounds, hydroxylation is followed by conjugation with sulphate or glucose and finally by excretion of the water-soluble product. Straight chain hydrocarbons are hydroxylated at the terminal end and further oxidized to the fatty acid which can be broken down by β-oxidation. Highly branched chain hydrocarbons, such as pristane and phytane, are probably oxidized to an acid (e.g. phytanic acid) which can be further oxidized by a combination of α- and β-oxidation (Mize *et al.*, 1969). Thus metabolism of petroleum hydrocarbons also by other organisms than bacteria can yield water-soluble compounds. This mechanism contributes to the removal of the oil as a visible and separate phase, and brings it into the water phase as dissolved compounds.

From the description it can be seen that the work on metabolism on hydrocarbons within marine organisms is as yet very patchy, but some patterns are beginning to emerge. However, observations are that fish appear to have much the same metabolic pathways as mammals, but the bivalves have a diminished capacity for hydrocarbon metabolism and that capacity differs from one species to another.

Whether all the pathways above are available for use in marine organisms is still being examined. Some information available for a few species of organisms from several marine areas can be given as examples. Degradation of sizable amounts (between 10 and 500 μg) of aromatic and paraffinic hydrocarbons occur in marine fish and some marine invertebrates (Stegeman and Teal, 1973; Lee *et al.*, 1972a, b). Other benthic marine invertebrates, phytoplankton, and some zooplankton, over a period of a month, were unable to oxidize either paraffinic or aromatic hydrocarbons. Several species of copepods were unable to metabolize hydrocarbons but could degrade paraffinic hydrocarbons. Hydroxylated products are found when fish and some crustacea are given aromatic and paraffinic hydrocarbons.

It should be kept in mind during this discussion that long chain paraffinic hydrocarbons (carbon chain lengths between C_{12}–C_{30}) are a common constituent of marine organisms, although usually accounting for only a few per cent of the lipid. However, most aromatic hydrocarbons present in petroleum are not known to be synthesised by marine organisms, though there are reports of biosynthesis of benzpyrenes by fresh water green algae (Borneff and Fischer, 1962).

There is no evidence for food web magnification (enrichment through the different steps of the food chain) in the case of petroleum hydrocarbons in the marine environment. On the contrary, evidence is strongest that direct uptake from the water and the sediments is more important than from the food chain, except in special cases.

9 REFERENCES

Ahearn, D. G., and Meyers, S. P. (editors) (1973). *The Microbial Degradation of Oil Pollutants.* Workshop held at Georgia State University, Atlanta, 1972. Publ. No. LSS-SG-73-01. Center for Wetland Resources, Louisiana State University, Baton Rouge, 322 pp.

Andersson, J. W., Clark, R. C., and Stegeman, J. J. (1974). Petroleum hydrocarbons. *Proceedings of Marine Bioassays Workshop.* Sponsored by API, EPA, and Marine Technology Society. Marine Technology Society, Washington D.C.

Andersson, J. W., Neff, J. M., Cox, B. A., Tatem, H. E., and Hightower, G. M. (1974). The effects of oil on estuarine animals: toxicity, uptake and depuration, respiration. In: *Pollution and Physiology of Marine Organisms.* Academic Press Inc., New York, San Francisco, London.

Atlas, R. M., and Bartha, R. (1972). Biodegradation of petroleum in sea water at low temperature. *Can. J. Microbiol.,* **18,** 1851–1855.

Atlas, R. M., Horowitz, A., and Busdosh, M. (1978). Prudhoe crude oil in arctic marine ice, water, and sediment ecosystems: degradation and interactions with microbial and benthic communities. *J. Fish. Res. Board Can.,* **35,** 585–590.

Blumer, Sanders, H. L., Grassle, J. F., and Hampson, G. R. (1971), A small oil spill, *Environment,* **13** (2), 2–12.

Blumer, M., and Sass, J. (1972). Oil Pollution: persistence and degradation of spilled fuel oil. *Science, N.Y.,* **176,** 1120–2.

Blumer, M., Souza, G., and Sass, J. (1970). Hydrocarbon pollution of edible shellfish by an oil spill. *Mar. Biol.,* **5,** 195–202.

Blumer, M., and Youngblood, W. W. (1975). Polycyclic aromatic hydrocarbons in soils and recent sediments, *Science,* **188,** 53–55.

Borneff, J., and Fischer, R. (1962). Canzerogene Substanzen in Wasser und Boden. Mitt. X: Untersuchungen von Phytoplankton eines Binnensees auf Polycyclische aromatische Kohlenwasserstoffe. *Arch. Hyg.,* **146,** 5.

Boylan, D. B., and Tripp, B. W., (1971). Determination of hydrocarbons in sea water extracts of crude oil and crude oil fractions. *Nature,* **230,** 44–47.

Boyland, E., and Soloman, J. B., (1955). Metabolism of polycyclic compounds, 8. Acid labile precursors of naphthalene produced as metabolites of naphthalene. *Biochem. J.,* **59,** 518–522.

Bratberg, E., (editor) (1977). The Bravo blow out, *Fisken og Havet,* Rapporter og meldinger fra Fiskeridirektoratets Havforskningsinstitutt, Bergen, Serie B. Nr. 5, 40 pp.

Brown, R. A., and Huffman, H. L., (1976). Hydrocarbons in open ocean waters, *Science,* **191,** 847–849.

Chet, I., and Mitchell, R., (1976). Petroleum hydrocarbons inhibit decomposition of organic matter in sea water, *Nature,* **267,** 308–309.

Conover, R. J., (1971). Some relations between zooplankton and Bunker C oil in Chedabucto Bay following the wreck of the tanker Arrow, *J. Fish. Res. Bd. Canada,* **28,** 1327–1330.

Cretney, W. J., Wong, C. S., Green, D. R., and Bawden, C. A. (1978). Long-term fate of a heavy fuel oil in a spill-contaminated B.C. coastal bay. *J. Fish. Res. Board Can.,* **35,** 521–527

Daly, J. W., Jernis, D. M., and Witkop, B. (1972). Arene oxides and the NIH shift. The metabolism, toxicity, and carcinogenicity of aromatic compounds. *Experientia,* **28,** 1129–1149.

Diamond, L., and Clark, H. F. (1970). Comparative studies on the interaction of benzo(a)pyrene with cells derived from poikilothermic and homothermic vertebrates. 1. Metabolism of benzopyrene. *J. Natl. Cancer Inst.*, **45**, 1005–1011.

Ehrhardt, M., and Heinemann, J. (1974). Hydrocarbons in blue mussels from the Kiel bight, *Marine Pollution Monitoring (Petroleum)*. NBS, Special Publication 409. Proceedings of a Symposium and Workshop held at National Bureau of Standards, Gaithersburg, Maryland, May 13–17.

Falk, H. L. *et al.* (1962). Intermediary metabolism of benzo(a)pyrene in the rat, *J. Natl. Cancer Inst.*, **28**, 699–745.

Forrester, W. D. (1971). Distribution of suspended oil particles following the grounding of the Tanker Arrow, *J. Mar. Res.*, **29(2)**, 151–170.

Freegard, M., and Hatchett, C. G. (1970). *The ultimate fate of crude oil at sea.* Interim Rep. Admiralty Materials Laboratory, U.K.

Goldberg, E. D. (1976). *The health of the oceans.* The Unesco Press, Paris, 172 pp.

Green, D. R., Bawden, C., Cretney, W. J., and Wong, C. S. (1974). The Alert Bay oil spill: a one-year study of the recovery of a contaminated bay. *Unpublished Manuscript.*

Gunkel, W. (1968). Bacteriological investigations of oil-polluted sediments from the Cornish coast following the Torrey Canyon disaster. In: Cowell, E. B. (ed.), *The Biological Effects of Oil Pollution on Littoral Communities.* Institute of Petroleum, London.

Hansen, H. P. (1977). Photodegradation of hydrocarbon surface films. In: *Petroleum hydrocarbons in the marine environment,* (eds. McIntyre and Whittle). *Rapp. P.-v. Réun. Cons. int. Explor. Mer.*, **171**, 101–106.

Hargrave, B. T., and Phillips, G. A. (1975). Estimates of oil in aquatic sediments by fluorescence spectroscopy, *Environ. Pollut.*, **8**, 193–215.

Harned, H. S., and Owen, B. B. (1958). *The physical chemistry of electrolytic solutions.* Reinhold, New York, pp. 531–534.

Hess, W. N. (editor) (1978). *The Amoco Cadiz oil spill: A preliminary scientific report.* NOAA/EPA Special Report, 348 pp. (U.S. Government Printing Office, Washington, D.C.).

IMCO/FAO/UNESCO/WMO/IAEA/UN. Joint Group of Experts on the Scientific Aspects of Marine Pollution (GESAMP) (1977). Impact of oil on the marine environment, *Rep. Stud. GESAMP*, **6**, 250 pp.

Johnston, R. (1970). The decomposition of crude oil residues in sand columns, *J. mar. biol. Ass. U.K.*, **50**, 925–37.

Karczewska, H. (1972). Microbial degradation of hydrocarbons in mineral oils—A literature survey, Report B-136, Swedish Water and Air Pollution Research Laboratory, Stockholm, 48 pp.

Keizer, P. D., Ahern, T. P., Dale, J., and Vandermeulen, J. H. (1978). Residues of Bunker C oil in Chedabucto Bay, Nova Scotia, 6 years after the Arrow Spill, *J. Fish. Res. Board. Can.*, **35**, 528–535.

Kerkhoff, M. (1974). Oil pollution of the shellfish areas in the Ooesterschelde estuary: December 1973. Unpublished manuscript.

Kreider, R. E. (1971). Identification of oil leaks and spills, In: *Proceeding, Joint Conference on Prevention and Control of Oil Spills,* American Petroleum Institute, Washington, D.C., 119–124.

Lee, R. F., and Benson, A. A. (1973). Fates of petroleum in the sea: biological aspects. In: *Proceedings of a workshop on inputs, fates, and effects of petroleum in the marine environment, 21–33 May 1973, Airlie, Virginia, Washington, D.C.*, National Academy of Sciences, Vol. 2, 541–551.

Lee, R. F., Sauerheber, R., and Benson, A. A. (1972). Petroleum hydrocarbons: uptake and discharge by the marine mussel Mytilus edulis. *Science*, **177**, 344–346.

Lee, R. F., Sauerheber, R., and Dobbs, G. H. (1972). Uptake, metabolism and discharge of polycyclic aromatic hydrocarbons by marine fish. *Mar. Biol.*, **17**, 201–208.

Lindén, O. (1976). The influence of crude oil and mixtures of crude oil dispersants on the ontogenic development of the Baltic herring. *AMBIO*, **5(3)**, 136–140.

Maurer, R. (1977). Unpublished manuscript.

Mayo, D. W., Page, D. S., Cooley, J., Sorenson, E., Bradley, F., Gilfillan, E. S., and Hanson, S. A. (1978). Weathering characteristics of petroleum hydrocarbons deposited in fine clay marine sediments, Searsport, Maine, *J. Fish. Res. Board. Can.*, **35**, 552–562.

McCarthy, R. D. (1964). Mammalian metabolism of straight chain saturated hydrocarbons, *Biochem. Biophys. Acta*, **84**, 74–79.

Mitchell, R., Togel, S., and Chet, I. (1972). Bacterial chemoreception: An important ecological phenomenon inhibited by hydrocarbons. *Water Res.*, **6**, 1137–1140.

Mize, C. E. *et al.* (1969). A major pathway for the mammalian oxidative degradation of phytanic acid. *Biochem. Biophys. Acta*, **176**, 720–739.

NAS. (1975). *Petroleum in the marine environment.* National Academy of Sciences, Washington D.C., 107 pp.

Nelson-Smith, A., (1972). *Oil Pollution and Marine Ecology.* Elek Science, London, 260 pp.

Owens, E. H. (1978). Mechanical dispersal of oil stranded in the littoral zone. *J. Fish. Res. Board Can.*, **35**, pp. 563–572.

Parker, C. A., Freegarde, M., and Hatchard, C. G. (1971). The effects of some chemical and biological factors on the degradation of crude oil at sea. In : Hepple, P. (ed.), *Water Pollution by Oil.* Institute of Petroleum. The Elsevier Publishing Company Ltd, Amsterdam, London, New York, 393 pp.

Posthuma, J. (1972). The Composition of petroleum. *Rapp. P.-v. Réun. Cons. int. Explor. Mer.*, **171**, 7–16.

Robertson, B., Arhelger, S., Kinney, P. J., and Button, B. K. (1973). Hydrocarbon biodegradation in Alaskan waters. In: Ahearn, D. G. and Meyers, S. P. (ed.), *The Microbial Degradation of Oil Pollutants.* Workshop held at Georgia State University, Atlanta, 1972. Publ. No. LSS–SG–73–01. Center for Wetland Resources, Louisiana State University, Baton Rouge, 322 pp.

Rossini, F. D., Mair, B. S., and Streiff, A. J. (1953). *Hydrocarbons from petroleum.* Reinhold Publishing Corporation, New York.

Rudling, L. (1976). Oil pollution in the Baltic Sea. A chemical analytical search for monitoring methods. Statens Naturvårdsverk (Sweden), *SNV PM* 783.

Scarratt, D. J., and Zitko, V. (1972). Bunker C oil in sediments and benthic animals from shallow depths in Chedabucto Bay, N.S. *J. Fish. Res. Board. Can.*, **29**, 1347–50.

Speers, G. C., and Whitehead, E. V. (1969). Crude petroleum. In: *Organic Geochemistry*, (ed. Eglington and Murphy). Springer-Verlag, New York, Heidelberg, Berlin, 638–675.

Stegeman, J. J., and Teal, J. T. (1973). Accumulation, release and retention of petroleum hydrocarbons by the oyster Crassostrea virginica. *Mar. Biol.*, **22**, 37–44.

Teal, J. M., Burns, K., and Farrington, J. (1978). Analyses of aromatic hydrocarbons in intertidal sediments from two oil spills of No 2 fuel oil in Buzzards Bay, Mussachusetts. *J. Fish. Res. Board, Can.*, **35**, 510–520.

Thomas, M. L. (1973). Effects of Bunker C oil on intertidal and lagoonal biota in

Chedabucto Bay, Nova Scotia. *J. Fish. Res. Board, Can.*, **30**, 83–90.

Traxler, R. W. (1973). Bacterial degradation of petroleum materials in low temperature marine environments. In: Ahearn, D. G., and Meyers, S. P. (ed.), *The Microbial Degradation of Oil Pollutants.* Workshop held at Georgia State University, Atlanta, 1972. Publ. No. LSS–SG–73–01. Center for Wetland Resources, Louisiana State University, Baton Rouge, 322 pp.

Wennergren, G. (1976). Kemisk-analytisk undersökning av oljespillet utanför la Coruña. Swedish Water and Air Pollution Research Lab. (unpublished manuscript).

Wennergren, G. (1978). Undersökning i La Coruña-området ett år efter förlisningen av 'Monte Urquiola' 1976–05–12. Stockholm (unpublished manuscript).

Whittle, K. J. (1977). Marine organisms and their contribution to organic matter in the ocean. *Marine Chemistry* **5**, 381–411.

Zafiriou, D. C. (1972). Response of Asterias vulgaris to chemical stimuli. *Mar. Biol.*, **17**, 100–107.

ZoBell, C. E. (1964). The occurrence, effects and fate of oil polluting the sea. In: *Proceedings International Conference on Water Pollution Research*, London 1962, Pergamon Press, London.

ZoBell, C. E. (1969). Microbial modification of crude oil in the sea. In: *Proceedings, Joint Conference on the Prevention and Control of Oil Spills.* American Petroleum Institute Washington, D.C.

Zsolnay, A. (1977). Inventory of non-volatile fatty acids and hydrocarbons in the oceans. *Marine Chemistry*, **5**, 465–475.

Chemistry and Biogeochemistry of Estuaries
Edited by E. Olausson and I. Cato
Copyright © 1980 by John Wiley & Sons Ltd.

I. CATO
*Marine Geological Laboratory,
University of Göteborg, Sweden*

I. OLSSON
*National Board of Fisheries,
Göteborg, Sweden*

and

R. ROSENBERG
*Institute of Marine Research,
Lysekil, Sweden*

13

Recovery and Decontamination of Estuaries

1 INTRODUCTION

Pollution of estuaries creates social, economical, and ecological problems. Political authorities are forced to make judgements concerning the future use of estuaries, and those are and have been mainly based on economical considerations. Lately, however, great emphasis has been placed also on social and human health aspects. Great costs are involved if decisions are made to clean up polluted areas or to install purification plants, leading to pollution abatement and, moreover, will such manipulations be followed by the establishing of a rather balanced ecosystem? The recovery capacity of estuaries is known from a few investigations of benthic macrofauna invertebrates, whereas such studies of meiofauna and sediments, are scarce. Macrofauna are considered a good indicator of environmental quality, since the benthic animals are fairly stationary and live in communities. Meiofaunal communities will probably respond more rapidly to environmental changes. Decontamination of sediments, on the other hand, are in general slow processes but are accelerated by the presence of burrowing benthic animals. The recovery of the bottom strata is, however, mainly due to the deposition of unpolluted sediments with a subsequent embedding of the contaminated sediment-cover. The structure of benthic communities is influenced by biotic and abiotic factors and will change under the influence of environmental stress. A reduced production and biomass of macrofaunal species due to pollution will have a direct impact on demersal fish.

2 FACTORS AFFECTING THE DECONTAMINATION OF SEDIMENTS

The recovery of estuary sediments is due to several factors which affect the content of pollutants in the sediments:

1. a decrease or cessation of the pollutants discharged,
2. a fairly good ventilation of the estuarine water-mass,
3. a mixing-in of uncontaminated sediments, as well as the action of other physical and biological forces re-working the sediments,
4. a physico–chemical and/or microbiological mobilization of the pollutants in the sediments with a subsequent release to the overlying water-column (recycling),
5. the persistence of the different pollutants,
6. a sedimentation of uncontaminated sediments with a subsequent embedding of the contaminated sediment cover.

These are all factors that tend to decrease, dilute or embed the amounts of the pollutants in contaminated sediments. In the following these factors will be elucidated and discussed.

A decrease or abatement of an effluent is of course a presumption for a recovery process, since a continued discharge of pollutants can only worsen the conditions within an estuary. Evidence of increased contamination has been given in many studies dealing with sediment cores as a historical record of the intensified metal pollution due to human activities. Up-to-date reviews concerning this subject, have been given by Förstner and Müller (1976), by Förstner (1976), by Cato (1977), and by Förstner in this volume.

The estuarine circulation system, in association with the tide and water changes due to wind and air pressure, generally result in a good ventilation of the water-mass and exchange with the adjacent sea. Exception from these conditions occurs in the bottom waters of the fjords, which are only occasionally replaced (see Bowden this volume). Consequently, a water-mass with pollutants is not able to maintain itself within an estuary. However, it should be pointed out that there are several mechanisms, which tend to hold the suspended load within or carry it back into the estuary e.g. the rapid flocculation and deposition of colloidal species and clay-size particles carried in suspension in the seaward freshwater flow (Duinker and Nolting, 1976) and of course the up-estuary flow (Pritchard, 1967; Bowden, 1967) which may to some extent carry suspended particles back into the estuary (see Meade, 1969; Cato, 1977; Duinker, this volume). In an attempt to predict the escape of suspended material from a dredging of the small estuary Välen (Sweden), Cederwall, and Svensson (1975) came to the conclusion that there here only 5–10% of the suspended material would be transported out into the adjacent sea. The residual 90–95% of the load would resettle within the estuary proper. This as well as other studies (e.g. Turekian, 1977; Duinker, this volume) points to the fact that an escape of suspended pollutants and/or pollutants adsorbed on or bound to suspended particulate matter probably is lower compared to pollutants dissolved in the water-mass.

Investigations of the sediments of the Elbe River and estuary have shown the importance of mixing-in of uncontaminated sediments in the interpretation of the sea-ward decrease of pollutants in estuarine sediments (Müller and Förstner, 1975). They found that the seaward decrease (65–98%) of the concentrations of heavy metals in the sediments of the Elbe tidal area was due to the mixing in of slightly contaminated North Sea sediments in the Elbe sediments by the up-estuary flow and the tide. Their interpretation is based on the fact that the reduction of heavy metals in the Elbe sediment started in the *suspended-load mixing zone* of the River Elbe and its estuary. Duinker and Nolting (1976) concluded that strong arguments can be adduced in support of Müller and Förstner's theory, since they found a corresponding seaward decrease of 20–60% of suspended, leachable, particulate Cu, Fe, and Zn in the Rhine estuary and the southern Bight.

Other evidence for the mixing-in theory has been given from North American estuaries by Meade (1969, 1972) and from the Swedish estuary Välen by Cato (1977). The seaward decrease of the C/N (carbon/nitrogen) ratios in the sediments of the Välen estuary shows a subsequent, increased influence of marine humus and a decreased influence of land-derived, organic matter with increased distance from the head of the estuary.

It is obvious from above that the mixing-in of uncontaminated sediments in estuaries results in a dilution of the pollutants and a subsequent reducing of the concentrations in the sediments. This is particularly true for the lower reaches of the estuaries.

Decontamination by mechanical transport, e.g. dune migration, has been reported by Rust and Waslenchuk (1976) from Ottawa River in Canada. They found that since the main sources of pollution were abated in 1971, the Hg level had declined by about 66% per year in the coarser sediment and by about 40% annually in the finer deposits. They thought that the differential rate was due to the greater influence of mechanical transport in areas with coarse sand, whereas chemical desorption was the principal pathway of mercury loss from the fine silty sand. Bioturbation of sediments increases the release and mixing of pollutants. Molluscs and tubificids are e.g. responsible for transferring methyl mercury down to a depth of burial of about 9 cm (Jernelöv, 1970). In a study of the release of Zn-65 from sediments by marine polychaete worms Renfro (1973) found that the release of Zn increased 3–7 folds, due to the burrowing of *Nereis*. However, the amount of bioturbation could decrease up the estuaries as the salinities become lower and the bottom-dwelling fauna changes in composition. This was shown in the Great Bay estuarine system of New Hampshire, USA, by Winston and Anderson (1971).

Diagenesis refers to changes that take place within a sediment during and after the burial of the sediment particles (see e.g. Berner, 1971; Olausson, 1974; Elderfield and Hepworth, 1975). Particularly the recycling of the elements from the sediments to the water column are important for the interpretation of the recovery capacity of polluted sediments. The break-down of organic matter buried within the sediments may release attached pollutants into the interstitial waters, which may then migrate upward through the interstitial waters by ionic or molecular diffusion, sediment compaction, etc., to the sediment-water interface. At the surface an enrich-ment may take place by adsorption, precipitation, etc., and/or the pollutant may diffuse out of the sediments and be lost to the overlying water column (recycled). Mobilization of heavy metals associated with the organic matter have been reported by de Groot *et al.* (1971) and de Groot (1973).

In the sediments the amount of organic matter is in excess compared with the heavy metals, and therefore the strong affinity of the heavy metals for organic matter (i.e. the large surface area) probably results in the immediate adsorption of most of the mobilized heavy metals (Håkanson, 1973; Kuijpers, 1974). Elderfield and Hepworth (1975) found in the Conway estuary (North Wales) that diagenetic remobilization contributes significantly to the metal deposition on the sediment surface. According to them a metal enrichment of about 10% at the sediment surface, compared with the background values of the sediment can be interpreted as a consequence of sediment diagenesis.

Under reducing conditions heavy metals mobilized in the interstitial water would probably be precipitated as sulphides (e.g. see Holmes *et al.*, 1974;

Loring, 1975), which are unable to migrate upwards if not released from the sediment by chelating substances (Elderfied and Hepworth, 1975). However the precipitation of simple sulphides cannot completely control the concentrations of heavy metals in the reduced sediments (Presley *et al.*, 1972; Elderfield and Hepworth, 1975).

In the deeper reducing part of the sediment iron and especially manganese hydrous oxides are dissolved and migrate upward through the sediment column (Eisma, 1973; Cline and Upchurch, 1973; Reinhard and Förstner, 1976) and are subsequently precipitated in the oxidizing sediment surface (Förstner and Müller, 1974) or lost to the overlying water. Co-precipitation and sorption of trace metals on hydrous oxides of iron and manganese is common (see e.g. Lee, 1970), but it is uncertain whether the trace metals mobilized by the dissolution of iron and manganese hydrous oxides also migrate upwards or are precipitated as insoluble sulphides within the sediment. The possibility of such a migration depends probably of the amount and the formation of soluble organo-metal complexes (Rashid and Leonard, 1973) and stable chloro-complexes (Feick *et al.*, 1972), which are easily lost from the interstitial waters and the sediments into the overlying water (Hallberg, 1973; Elderfield and Hepworth, 1975).

Lowering the pH increases the rates of solubility of most heavy-metal compounds, such as carbonates and hydrous oxides, and with the concurrence of hydrogen ions decreases the sorption of heavy metals on clay minerals, organic matter, and hydrous oxides of manganese and iron (Förstner and Patchineelam, 1976).

Changes in runoff as well as the tide causes the salt wedge to move up and down in an estuary (Postma, 1967). Sediment deposited within the fresh water regime of the estuaries may therefore occasionally be introduced to the sea water. During such occasions desorption of heavy metals present on clay minerals and ferric hydroxides by competing cations from the sea (Kharker *et al.*, 1968; Evans and Cutshall, 1973; Holmes *et al.*, 1974; Rohatgi and Chen, 1975) may take place in the upper reaches of estuaries.

Added to these chemical and physical processes must also be the microbiological processes (see review Summers and Silver, 1978). Certain microorganisms are able to methylate mercury (Jensen and Jernelöv, 1969; Fagerström and Jernelöv, 1971), lead (Wong *et al.*, 1975) and arsenic (McBride and Wolfe, 1971; Braman and Foreback, 1973) with a subsequent release of the methyl-products from the sediment. Pollutants bound to sediment particles may also be released from the sediment by direct accumulation in the benthic fauna, which serves as food for the more mobile aquatic organisms.

It is obvious from above that there are many processes working in the sediment in one or another direction which may give rise to a release of pollutants from the sediment to the overlying water. However, it should be remembered that all these physicochemical and microbiological processes are in general very slow. The diffusion coefficients, due to the concentration gradients in the interstitial water, and the biological diffusion coefficients, due to the reworking and mixing of bottom sediment by bottom organisms, are of the order of 10^{-11} to 10^{-12} cm^2 sec^{-1} and 10^{-6} to 10^{-7} cm^2 sec^{-1} respectively (Eisma, 1973). The methylation rate of mercury, for example, is about 0.1% of the total mercury present per year (Jernelöv, 1972). Consequently, it seems that the loss due to diffusion of trace elements in estuarine sediments, with a normal sedimentation rate, will in general almost negligible in the sedimentological discussion (cf. Aston *et al.*, 1973) but is, of course, relevant in the biological uptake and discussion of toxical effects.

Eisma (1973) points to another factor which is important for the preservation of particularly organically bound pollutants, namely, the biological degradation in relation to the rate of sedimentation. According to him the amount of bacteria decreases from 10^8 to 10^3 bacteria per gram of sediment from the sediment surface to a depth of burial of 1 m. The correspondingly great decrease in the breakdown of organic matter will therefore lead to a fairly good preservation of the pollutants.

The preservation of the pollutants is also due to the stability or persistence of the pollutants (see e.g. Jernelöv, 1971). This has a great influence on the recovery process, since the persistence determinate the degradation rate of the pollutants. Toxins with a low persistence are even if they are extremely toxic of rather low ecological importance, since the life-time is to short to give rise to accumulation in biota and sediments. Several elements as the heavy metals cannot be destroyed by biological degradation and therefore they tend to accumulate in the food chains (see e.g. Fagerström and Åsell, 1973), as well as in the bottom sediments. Also several man-made chemicals are extremely persistent in the environment e.g. pesticides as chlorinated hydrocarbons (DDT, chlordane, lindan, etc.) and polychlorinated biphenyls (PCB).

However, bacterial degradation of PCB have been reported, for example, by Ahmed and Focht (1973), but it is believed that these processes are slow, The latter may also be true for several oil-compounds (Johnston, 1970; Tea *et al.*, 1978), which finally end up in the sediments after the prior evaporation phase of hydrocarbons at the sea-surface (Goldberg, 1976). Several inorganic compounds are also degraded in their natures, e.g. mercury compounds as phenyl mercury and alkoxy–alkyl mercury, are converted to the divalent mercury ion which then are available to methylating microorganisms (see above).

Heavy metals and persistent chemicals, once incorporated in the sediments may consequently mainly remain in the bottom strata for prolonged times. Since there is a more or less continuous deposition of sediments in estuaries, the polluted sediment cover will subsequently be embedded with fresh unpolluted sediments if the effluents have ceased. As time goes by and, depending on the accumulation rate, the polluted sediment cover will be withdrawn below the sediment–water interface and the biological communities may occupy an uncontaminated sediment surface. The embedding of the polluted sediment layer is probably one of the most important factors in the recovery process.

However the bioturbation may prolong the recovery process, but may probably not really affect the total concentrations of the pollutants at depth in other ways than that the bioturbation homogenizes the uppermost sediment and its content of pollutants (see Jernelöv, 1970) during the first time span of the deposition phase of unpolluted sediments. After a time lapse of several years the effect of the bioturbation only will be a decreasing pollution gradient with increasing depth in the sediment which finally will end up in an unpolluted topmost bottom stratum.

3 RECOVERY OF BENTHIC COMMUNITIES

The recovery capacity of benthic animals in formerly polluted estuaries is based on several factors, e.g. currents, sedimentation, sediment structure, temperature, oxygen concentration, and type of pollutant. These factors vary from estuary to estuary and even within estuaries. Despite this, in general several ecological processes can be distinguished and related to pollution load. This has recently been described and reviewed by Pearson and Rosenberg (1976, 1978). In the following description examples of recovery capacity are taken from organic pollution abatements and oil pollution studies. Examples in the literature about recovery following pollution by heavy metals and chlorinated hydrocarbons are scarce, and results from the Välen estuary studied by the authors will be presented. Organically enriched sediments often contain increased concentrations of these other pollutants as well. This review is divided into two parts, the first dealing with shallow parts of estuaries and the second with deeper parts.

3.1 Recovery of shallow wake benthic communities

The resiliency after pollution in shallow waters (approx. 0–5 m) is generally more rapid than in deeper waters. The cleaning of shallow soft and hard bottom substrata from pollutants is accelerated by tides, wave action, and a high temperature. The organisms living in this zone are well adapted to changing conditions and several may be considered opportunists. Animals living in estuaries are probably especially fit for a rapid recolonization of defaunated areas, since they also possess tolerance towards changing salinities. They are eurytopic with a high general stress tolerance (Jernelöv and Rosenberg, 1976). Estuarine shallow water species generally reproduce several times annually with great numbers of larvae. This means that as soon as the bottom substrate is sufficiently cleaned to allow survival of the larvae, they will settle in this habitat.

This was illustrated in the Raritan River estuary in New Jersey, USA, where for example the barnacle *Balanus improvisus* coated all firm substrata in the previously uninhabited section shortly after the introduction of a trunk sewer system (Dean and Haskin, 1964). In a Swedish estuary, the Saltkällefjord, the shallow water habitats (0–1 m) were similarly rapidly repopulated after the cessation of a sulphite pulp mill (Rosenberg, 1971, 1976b). Data from 1965, when the pulp mill was still in operation, showed that hard-bottom species and soft-bottom species, typical of that area, were wiped out from the fjord or found in its mouth only (Figure 1). Following the closure of the mill in 1966 it took approximately 1 to 3 years, at least for the most conspicuous species of the littoral fauna, to invade the fjord to the inner distribution limit, now determined by resistance towards low salinities.

Discharge of oil refinery wastes in Los Angeles Harbour was prohibited by law in 1968 and from 1970 no such wastes were discharged. Reish (1971) reported that a more successful settlement and growth of a greater diversity of organisms occurred more rapidly on the boat floats than on the bottom, where the wastes had accumulated. The oil spill in West Falmouth, Massachusetts, USA, reduced the animal diversity to the greatest extent in intertidal and shallow water areas (Grassle and Grassle, 1974). A few months after the oil spill opportunistic polychaetes began the succession by building up a new benthic community, which illustrates their adaption to short-term selection. The dominant species in this recovery process was the polychate *Capitella capitata*, which first occupied the shallow water localities and, with several months' delay, also increased in density at deeper polluted stations (Sanders *et al.*, 1972).

The recovery of the Swedish estuary, Välen (Figure 2), situated south of Gothenburg on the Swedish west-coast, is being studied by the present authors (Cato *et al.* 1975, Cato 1977). The estuary is of the salt-wedge type and the bottom strata vary from clastic unconsolidated sand silt-clay in the inner part to more consolidated silty sand seawards. Discharge of mainly domestic and industrial sewage to the inner part of this shallow (0.2–1.5 m) estuary created organically enriched sediments (>80 mg carbon per gram dried sediment) contaminated by high concentrations of heavy metals such as Hg, Pb, Cu, and Zn. This serious condition was particularly obvious in the inner third of the estuary, where the enrichment factors for the different heavy metals in the surface sediments varied between 13–50, which are larger or about as large as those found e.g. in the heavily polluted, major rivers of West Germany (cf. Banat *et al.*, 1972).

Since the amount of dissolved oxygen has not been sufficient for continuous breakdown, the substantial effluents of oxygen-demanding substances (mainly sludge products) gave rise to reduced sediment–water–interface conditions (unlike the normal near-shore sediments; Borchert, 1965) in the upper reaches of the estuary. The condition implied an anaerobic, bacterial breakdown, including putrefaction and fermentation and a subsequent loss of hydrogen sulphide, methane, etc. resulting in conditions in which the benthos was unable to survive (Cato *et al.*, 1975).

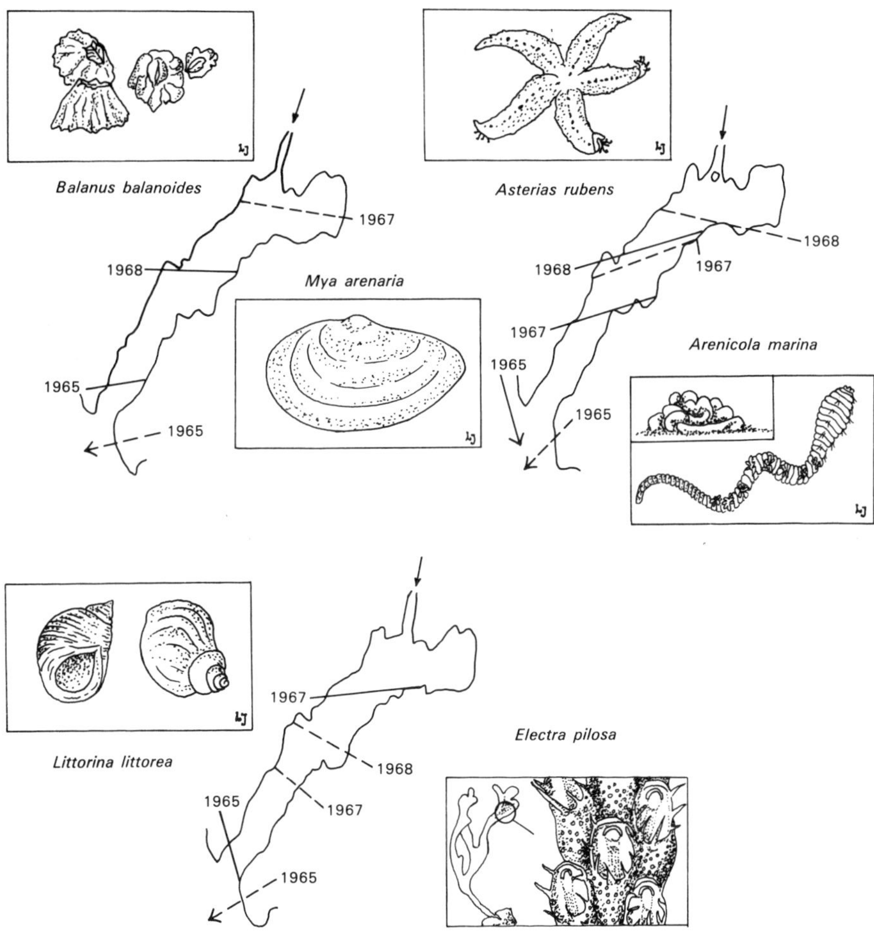

Figure 1. Inner distribution limits of some littoral species in the estuary Saltkällef-jord before and after the closure of the sulphite pulp mill in 1966. The wastes entered the fjord *via* the river indicated by an arrow (Redrawn from Rosenberg, 1976b). (Reproduced by permission of the International Union of Pure and Applied Chemistry.)

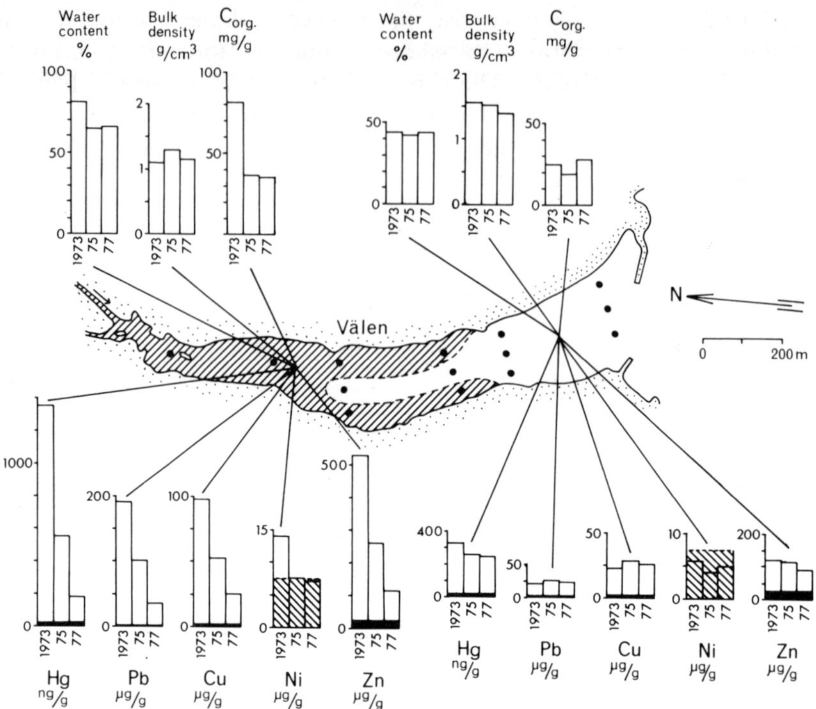

Figure 2. The recovery of the Välen estuary on the Swedish west coast during the years 1973 to 1977 illustrated by some superficial-sediment parameters. The parameters are expressed as means from the upper (hatched area) and the lower reaches of the estuary respectively, and show the polluted conditions 1973, as well as the conditions 1975 and 1977 one and three years respectively after pollution abatement in 1974. Filled circles indicate the sampled stations. The filled or hatched areas of the bars in the diagram show the background concentrations of the heavy metals found in the sediments of the Swedish west coast.

The pollution of the estuary was drastically abated in December 1974, when the treatment plant was closed down. One year later measurements of the redox potential (Eh) reflected that the relation between organic matter and oxygen (see Olausson, 1972) in the sediment–water interface had changed to one more favourable in the inner part of the recipient. In Figure 4 this is illustrated with the level of the redox-potential-discontinuity (RPD) zone (Eh = 0 mV), which between 1973 and 1975 had displaced from the sediment surface to an average depth of 0.5 cm. This probably was an effect of both sedimentation and depressing of the RPD-zone by feeding and respiratory activity of organisms (cf. Rhoads *et al.*, 1977). The improvement has then succeeded to a rather normal redox-condition in 1978 with the RPD-zone at a depth of 1 to 3 cm due partly to the effects given above and to the dredging operation in 1976, see below. Also the pH-situation was improved in the outer part of Välen in 1977. Generally the pH-values for 1973 compared to those from 1977 exhibited a larger range.

If the organic-carbon concentrations in 1975 are compared with the corresponding data from 1973 the results show that the organic carbon concentrations in the superficial sediments have decreased with fully 50% on the average in the upper reaches of the estuary and with slightly more than 20% in the lower reaches (Figure 2 and Table 1). The strong decrease

Table 1. The development of superficial sediment-parameters, after the pollution abatement of the Välen estuary in 1974. The changes are expressed as percentages of the 1973 values. Values in parentheses are not significant.

Regime	Year	No. of samples	Water con-tent	Bulk den-sity	Organic carbon	Mer-cury	Lead	Copper	Nickel	Zinc
Upper	1975	6	−19	+16	−54	−59	−46	−40	−46	−51
Välen	1977	6	−17	(+5)	−86	−82	−73	−49	−78	−56
Lower	1975	8	(−3)	(−1)	−23	−23	+13	−15	−26	(−3)
Välen	1977	8	(−1)	(−10)	+13	−24	(+4)	(+5)	(−7)	−23

of organic carbon (i.e. organic matter) in the upper Välen is also reflected in the corresponding decrease of 19% of the water content and in the increase of 16% of the bulk density. A decrease in the content of the organic matter (i.e., sludge products), with its considerable water-retaining capacity, also decreases the water content of the sediment, due to the relationship of these parameters (see Cato, 1977). The wet density, in turn, varies inversely with the water content. The heavy-metal concentrations in the upper Välen sediments decreased markedly between 1973 and 1975. The decreases of the different metals vary between 40 and 60% on average (Figure 2 and Table 1). In the lower reaches of the estuary the changes of the metal concentrations have been more scattered, an effect probably due to a stronger physical and biological reworking of the sediments in this part of the estuary also during the pre-abatement period.

The cessation of the sewage-and waste-water discharge also included an abatement of the heavy metals to the estuary. This means that the sediments deposited in the estuary since the discharge ceased can be considered as free from sludge products and heavy metals. However, the results show that the heavy-metal concentrations in the topmost sediments of the estuary still far exceeded their background concentrations. This has been interpreted as a consequence of bioturbation, since the recolonization of the bottom by organisms already had started (see below). The burrowing organisms homogenized the uppermost sediment layer, i.e. the freshly deposited unpolluted sediments were diluted with the older underlying and polluted sediments which retained high concentrations of heavy metals. This dilution effect is discernible in the unchanged relationships of heavy metals to organic carbon between 1973 and 1975 (Figure 3). The unchanged relationships also indicate that a strong breakdown of organic matter with a subsequent mobilization of the heavy metals must have been negligible between 1973 and 1975.

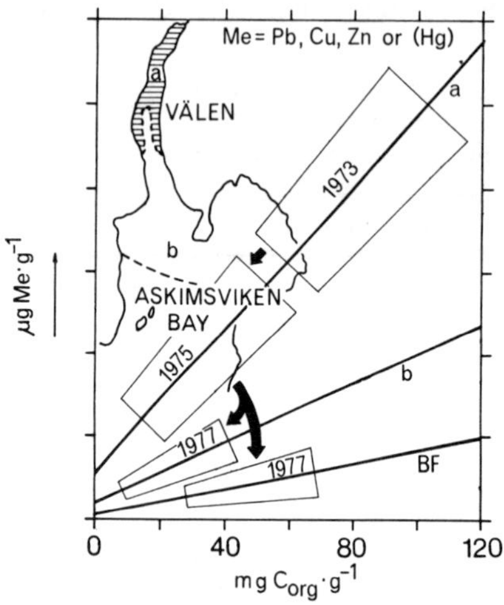

Figure 3. General outline of the pollution load of the Välen estuary, expressed as the relationships between heavy metals and organic carbon in the superficial sediments in 1973. (a) refers to the highly polluted upper Välen, (b) refers to the less polluted lower Välen, and *BF* refers to the natural load in an unpolluted regime of the Swedish westcoast (after Cato, 1977). The succession is given by the marked areas which show how the *x*/*y* values from the upper Välen (hatched area) were grouped one year before (1973), one year (1975) and two years (1977) after pollution abatement in 1974 respectively. The metal concentrations decreased between 1973 and 1975 but followed still the overloaded relation (a). Between 1975 and 1977 the relation changed to one more favourable to the recipient. Note, that several *x*/*y* values in 1977 follow the unpolluted relation *BF*.

The difference in the heavy-metal decrease in the sediments between the upper and the lower reaches of the estuary may both be a consequence of a decreasing bioturbation with a decreasing salinity (see Winston and Andersson, 1971) and the stronger action of waves and currents in the lower reaches of the estuary.

Since the embedding of the highly polluted sediments in an unpolluted sediment cover thick enough to prevent direct contact between the benthos and the highly polluted sediments would have taken about 20–30 years (Cato, 1977), the inner part of the estuary has been restored by dredging in

1976. The observations in 1977, three years after the pollution abatement and one year after the dredging operation, show that the improvement had succeeded. If the data from 1977 are compared with the corresponding data from 1973, one finds that the concentrations of the heavy metals in the upper reaches of the Välen have decreased between 50 and 80% on average, i.e. the concentrations in 1977 have decreased still more with 20–30% since 1975 (Figure 2 and Table 1). Exception is nickel, which already in 1975 reached its background concentration. In the lower reaches only zinc show a significant decrease. The other metal concentrations are rather unchanged compared with 1975. The latter is also valid for the organic–carbon concentrations within the whole estuary.

Since the polluted sediment layer was removed by the dredging operation the sediment samples analysed in 1977 contained freshly deposited sediments (accumulated after the dredging operation) diluted with the unpolluted sediments older than the polluted sediment layer removed. Therefore the relationships between the heavy-metal concentrations and organic carbon were expected to have changed to one more favourable to the estuary, particularly since the organic carbon content had not changed since 1975. This is also discernible in Figure 3, which shows how the gradient of the relationship of a certain heavy metal to organic carbon has decreased to a level corresponding to that in the adjacent water seawards to the estuary and in some cases even to a level corresponding to the normal background gradient for unpolluted regimes of the Swedish west coast.

The abundance of the meiofauna (0.1–1 mm) increased considerably at two sections in the middle part of the estuary from about 33 and 105 to 492 and $917 \times 10^3 \ m^{-2}$, respectively, between 1973 and 1975 (Figure 4). Four classes were identified in 1973, which were Foraminifera, Nematoda, Podoplea (Harpacticoida and Cyclopoidea) and Oligochaeta. In 1975 the number of classes had increased to eight. In addition to the four mentioned above Turbellaria, Halacarida, Bivalvia, and Polychaeta were then encountered. Thus, the diversity of the meiofauna had increased one year after the pollution abatement. The nematods dominated numerically at most stations both in 1973 (43–96%) and 1975 (34–95%). Podoplea was the second dominating group. In 1973, of the foraminiferan species, only the tolerant *Elphidium excavatum* was found in Välen, while *Ammoscalaria runiana* and *Miliammina fusca* were additionally observed in 1975.

The obvious recovery of the estuary has been shown also by the colonization of macrofaunal species (Figure 5). Three years after pollution abatement the abundance was high ($0.7–35 \times 10^3 \ m^{-2}$), especially in the outer parts. The most conspicuous colonizers were *Nereis diversicolor*, *Hydrobia ulvae*, and *Cardium* spp. The inner third part of the estuary was affected by dredging operations in 1976, and other environmental stress-factors in this shallow area were great seasonal variations in temperature and salinity.

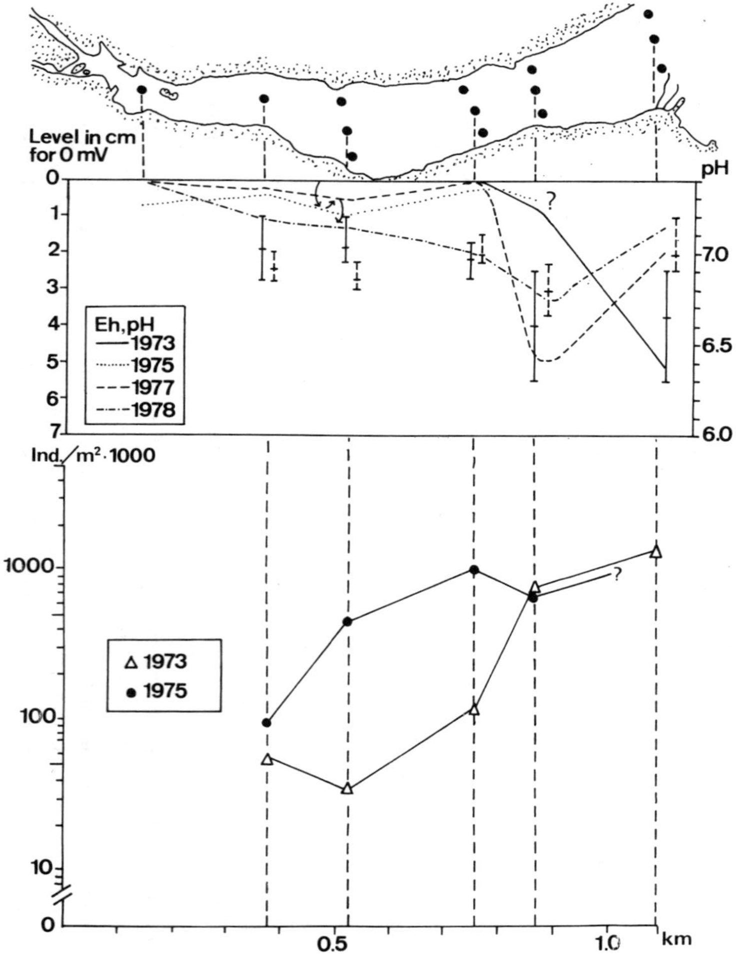

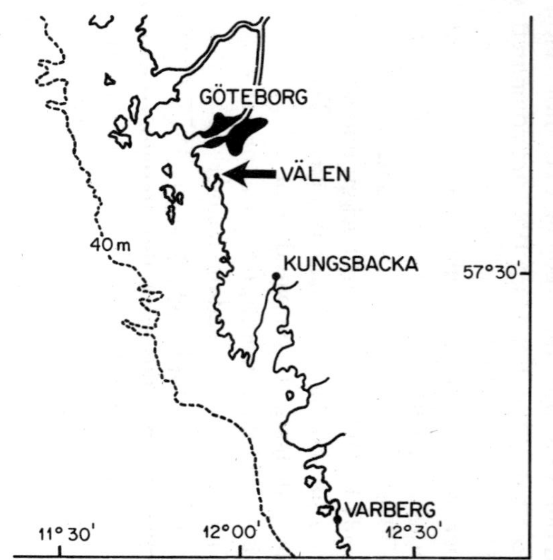

Figure 4. Upper left diagram: Level in cm in the sediment for the redox potential 0 mV 1973, 1975, 1977, and 1978 (curves), means and range of pH in the years 1973 and 1977 in the layer 0–5 cm. The small arrows give the succession of the redox-potential-discontinuity (RPD) zone. Lower left diagram: Abundance of the meiofauna 1973 and 1975 in the shallow Välen estuary. The sampled stations are indicated by filled circles. Lower right figure: Map showing the location of the Välen estuary on the Swedish west coast.

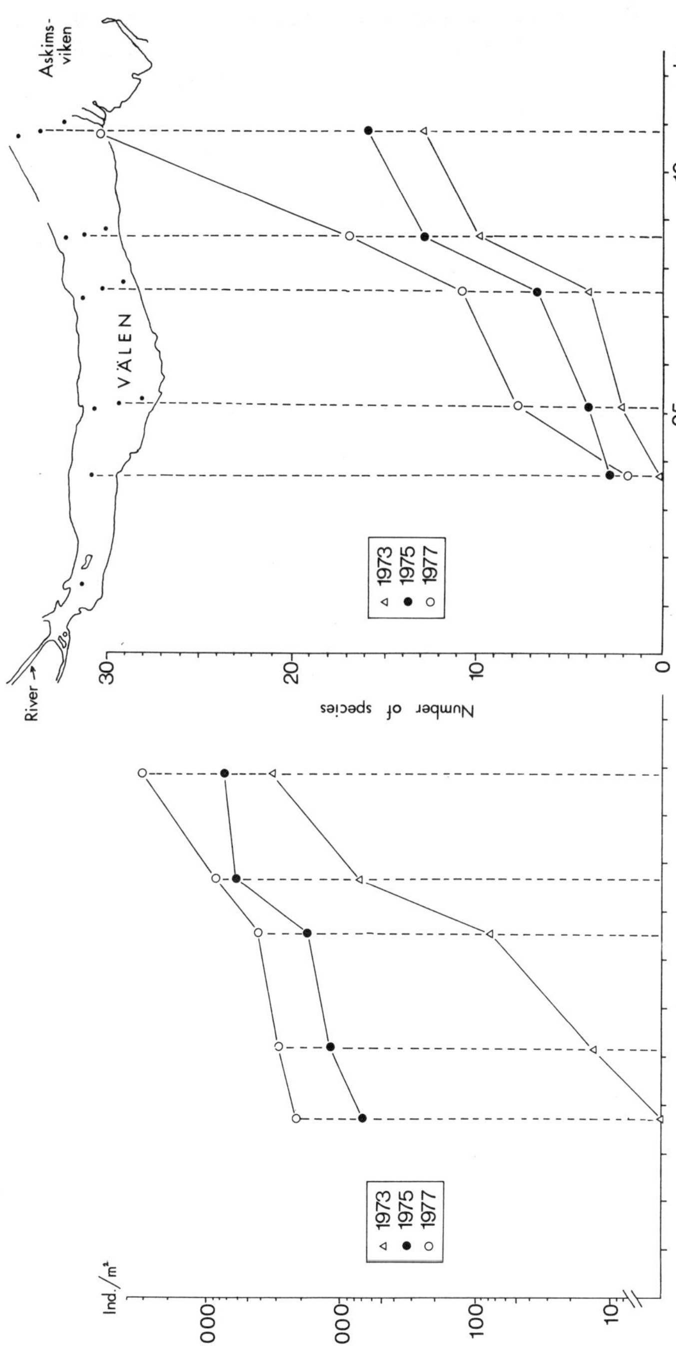

Figure 5. Number of species and individuals per 0.135 m² in the shallow estuary Välen on the Swedish west coast. In 1973 (△) the estuary was polluted by sewage and heavy metals via the river on the left-hand side. The waste water discharge was discontinued in 1974. Data from 1975 (●) represents one year of recovery and from 1977 (○) three years of recovery. The inner third of the estuary was disturbed by dredging operations in 1976, which especially affected the results from the two innermost set of stations. The sampled stations in the estuary are marked by filled circles.

The study of the sediments in the Välen estuary, have shown that the recovery capacity of the bottom strata in estuaries of this type is mainly due to the deposition of unpolluted sediments with a subsequent embedding of the polluted sediments. Since this process will take several years depending on the accumulation rate and the reworking of the sediments by the organisms, the tide, the waves, etc., the recovery process can be speeded up by removing of the polluted sediment bed. The two examples from the estuaries Raritan River and the Saltkällefjord suggest a recovery period for the benthic communities of approximately 3 years. Similar rapid recolonizations in shallow waters were found after defaunation by red tide in Florida, USA (Dauer and Simon, 1976), following the closure of a dam in the Netherlands (Wolff, 1974), and after construction of salt water ponds in Maryland, USA (Hanks, 1968).

3.2 Recovery of benthic communities in deeper parts of estuaries

Wastes accumulating in sediments in deeper parts of estuaries are normally of a greater longevity compared to wastes in shallower water. The degradation processes are slow due to low temperatures, and the sedimentation rate is usually low.

Succession in macrobenthic communities following pollution abatement in deeper water is a slower process than in shallow water (Rosenberg, 1972, 1973, 1976a). Several deep water species do not have recruitment every year, which may naturally delay the establishment of the 'normal' community. For example the brittle star *Amphiura chiajei*, a dominant in large areas of the North Sea, was reported to have no recruitment over a 5 year period off the Northumberland coast, U.K. (Buchanan, 1964).

The recovery capacity, also in deeper parts of estuaries, is probably greater than in similar non-estuarine regions because of the estuarine water circulation. This circulation has the ability to oxygenate the bottom water, to distribute the wastes over large areas, and to transport widely benthic larvae.

Göta River estuary

In the Göta River estuary outside the city of Göteborg the redox potential and pH in the sediment, together with the meio- and the macrobenthic fauna have been followed in an integrated study after mainly organic pollution abatement (Olsson, unpubl.). In 1972 treatment of the sewage water from Göteborg was started. During the period 1972–1976 the number of equivalents connected to the sewage plant was gradually increased and corresponded in 1976 to about 0.5 million inhabitants. The faunal reactions on the increasing pollution prior to 1972 have been studied by Bagge (1969) and Nyholm *et al.* (1977). The stations concerned, situated in the inner part of the estuary, cover the depth interval 5–12 m (Figure 6) and the bottom

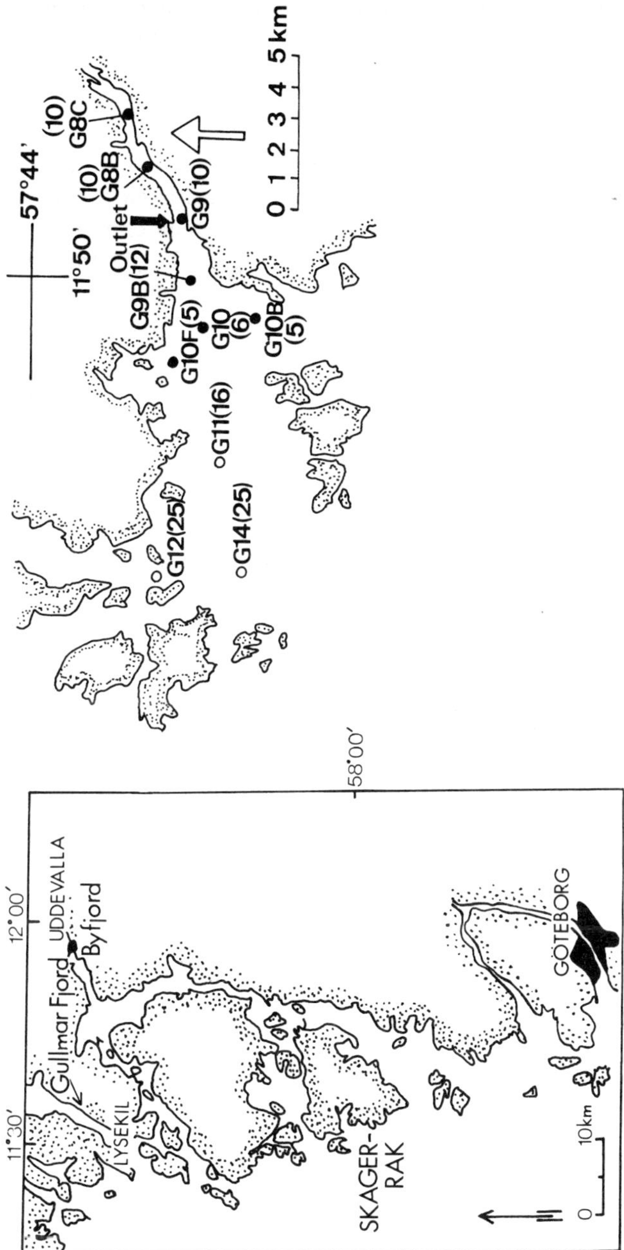

Figure 6. The benthic stations in the Göta River estuary outside Göteborg (filled circle = meio- and macrofauna, circle = macrofauna). Figures within brackets refer to depth in meters. The G10-stations form a transect. The arrow shows the outlet point of the sewage plant.

water may be designated as polyhaline (18–30‰ S). The temperature range at the shallower stations (~5 m) is 7.0–18 °C, while the corresponding extremes for the deeper stations (~10 m) amount to 7.0 and 15 °C.

Redox potential (EH) and pH

The redox potential integrates and reflects the effects of the decreased organic load on the estuarine sediments. The values from the station G8B recorded from July 1971 to November 1976 have been chosen to elucidate the improved redox conditions of the area (Figure 7). At the top of this figure the number of persons connected to the sewage plant is indicated. The curves of this and other stations as those from one with a more frequent sampling exhibits in principle a cyclic pattern with lower values in spring (May–June), when the redox zone approaches the mudwater interface and the vertical gradient turns more steep. In autumn (October–November) the redox values were higher and the isolines more separated. Generally during the period 1971–1976 the level in the sediment for negative Eh-values

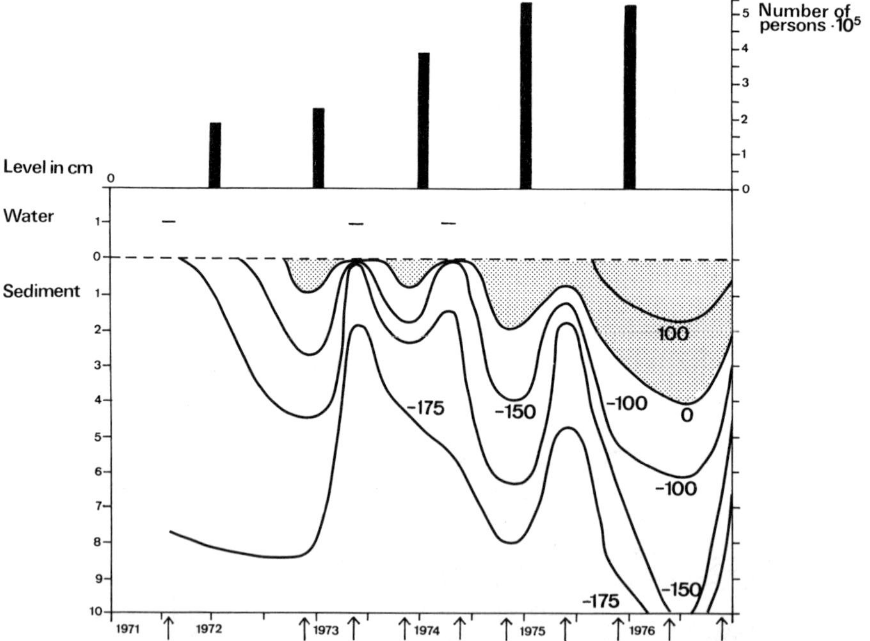

Figure 7. Number of persons connected to the sewage plant in the Göta River estuary 1971–1976 and the level in cm in the sediment for indicated redox isolines (Eh) at station G8B. Dotted area implies oxidized conditions. Horizontal lines in the level for the bottom water indicates negative *Eh*-values.

moved downwards 1–2 cm in the area. The Eh-values from the area also exhibit a horizontal gradient with more reduced conditions in the proper river-mouth. As a consequence of the decreased organic load the gradient has moved towards the proper mouth.

The total pH-range in the uppermost 10 cm at station G8B 1971–1976 was 6.5–7.9. In general the values were positively correlated to the redox potential above and negatively correlated to it below the RPD-zone (cf. Fenchel and Riedl, 1970). The pH values also exhibited a horizontal gradient in the area with more acid conditions in the proper river-mouth.

Abundance and biomass of the meio- and macrofauna

The meiofaunal (0.1–1 mm) abundances from 1971 to 1976 on seven stations in the Göta River estuary are summarized in Figure 8. The values for 1971 represent the situation prior to the changed outlet conditions and form a V-shaped curve. The values for the period November 1972–June 1974 represent the first phase after the pollution abatement. The general form of the curve was similar to that from 1971. During the next phase in the recovery process, October 1974–November 1976, the mean abundance curve changed its form principally due to considerably decreased densities in the innermost part of the estuary.

The changes of the macrofaunal abundance (>1 mm) in time and space after pollution abatement has recently been reviewed by Leppäkoski (1975) and by Pearson and Rosenberg (1978). Pearson and Rosenberg give a composite general outline of number of species, with abundance and biomass along an environmental gradient of organic enrichment. The abundance curve is characterized by four parts: one with practically no macrobenthic fauna, one with a peak of opportunists, one described as an ecotone point and one as a transition zone (cf. Figure 12). The ecotone point is designated by Leppäkoski (1975) as 'primary minimum'.

The macrofaunal as the meiofaunal abundances of the Göta River estuary (Figure 8) are divided into three periods. In July 1971 the two innermost stations were practically devoid of macrofauna. During the next period, November 1972–June 1974, the number of macrofauna was increased at all stations and the curve now exhibited two peaks. During the period from October 1974 to November 1976 the macro- and the meiofaunal abundance curves had a similar shape except in the innermost part, where the relation was inverse. The macrofaunal curve now showed two very distinct peaks corresponding to the primary and the secondary maxima according to Leppäkoski (1975).

To conclude the meio- and the macrofaunal abundance curves gradually exhibited about the same shape implying that the maxima and minima according to the terminology of Leppäkoski (1975) became common to both

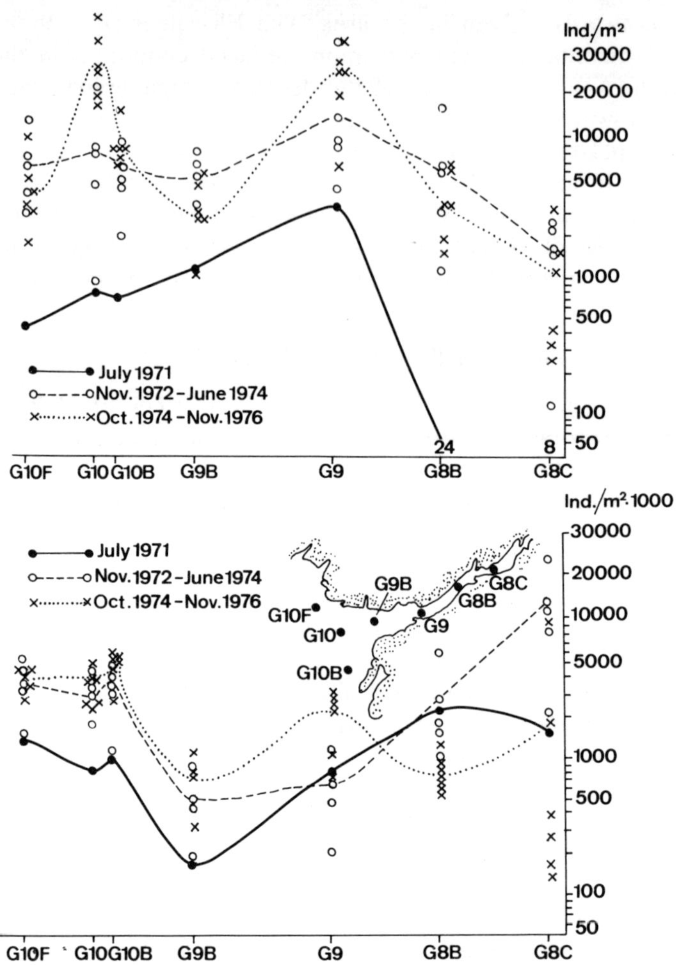

Figure 8. Mean abundance curves of the meio- and macrobenthic fauna (lower and upper diagram respectively) for indicated periods along the pollution gradient in the Göta River estuary.

faunal categories. However, inside the secondary maximum (the peak of opportunists) the two faunal curves generally were inversely related. Gerlach (1971) has generalized that meiofauna are quantitatively more important relative to macrofauna in brackish water than in marine areas. The greater importance might be valid also for pollution stressed sediments (Rosenberg *et al.*, 1977).

In July 1971 the meiofaunal biomass curve (range 7–10 g/m²) was similar to that for the abundance curve. During the next period, November 1972–

June 1974 there was a general increase of the wet weight (range 9–62 g/m^2). The biomass and the abundance curves for this period exhibited, as for the period October 1974–November 1976, a similar shape.

The macrofaunal biomass (wet weight) was in July 1971 between 0.2–27 g/m^2. The meiofaunal values exceeded the macrofaunal wet weights in the innermost part of the estuary at that time. During the next periods there was a general increase of the macrofaunal biomass. The highest values were recorded at the stations where the abundance peaks were observed.

Succession of the meiofauna

The abundance/m^2 and the numerically dominating main groups of the meiofanua (dominance value $\geq$1%) at the stations G8B, G9 and G10, are shown in Figure 9. As indicated above the total meiofaunal abundance and biomass during the period July 1971–November 1976 are assumed to represent three different periods: July 1971, November 1972–June 1974 and October 1974–November 1976.

The meiofauna of station G8B represents the first phase in the meiofaunal recovery process, with a fauna initially characterized by high abundances of nematods and larval forms of macroinvertebrates. Gradually the number of individuals of harpacticoids and foraminifers, mainly monothalamous forms, was increased at the same time as the total meiofaunal density decreased. Foraminifera exhibited its first colonizers in November 1973. Up to June 1974 only *Ammonia beccarii* and three monothalamous species to the Allogromiidae family one of them being *Allogromia crystalifera* were recorded here. The more diversified meiofauna of station G9 represents the next seral stage in the meiofaunal recovery process. Briefly the fauna are characterized by high percentage shares of harpacticoids and of the temporary meiofauna. The foraminifers exhibit an increased proportion and ostracods may appear with low proportions.

The next seral stage represented by station G10 was characterized by a meiofaunal community dominated by foraminifers and nematods up to more than 80%. At the end of the period of investigation the foraminifers share increased while the proportion of nematods decreased. The proportion of the temporary meiofauna was relatively stable. The foraminifers exhibited the highest dominance values at this station, ranging 32–60% (mean = 46%), with an increasing share of polythalamous forms. On almost all occasions aranaceous, polythalamous species were encountered. Totally 16 monothalamous, 8 polythalamous calcareous and 7 polythalamous aranaceous species were recorded during the period of investigation. The following species were constantly found: *Allogromia crystallifera, Psammosphaera bowmanni, Elphidium excavatum, Protelphidium anglicum,* and *Ammonia*

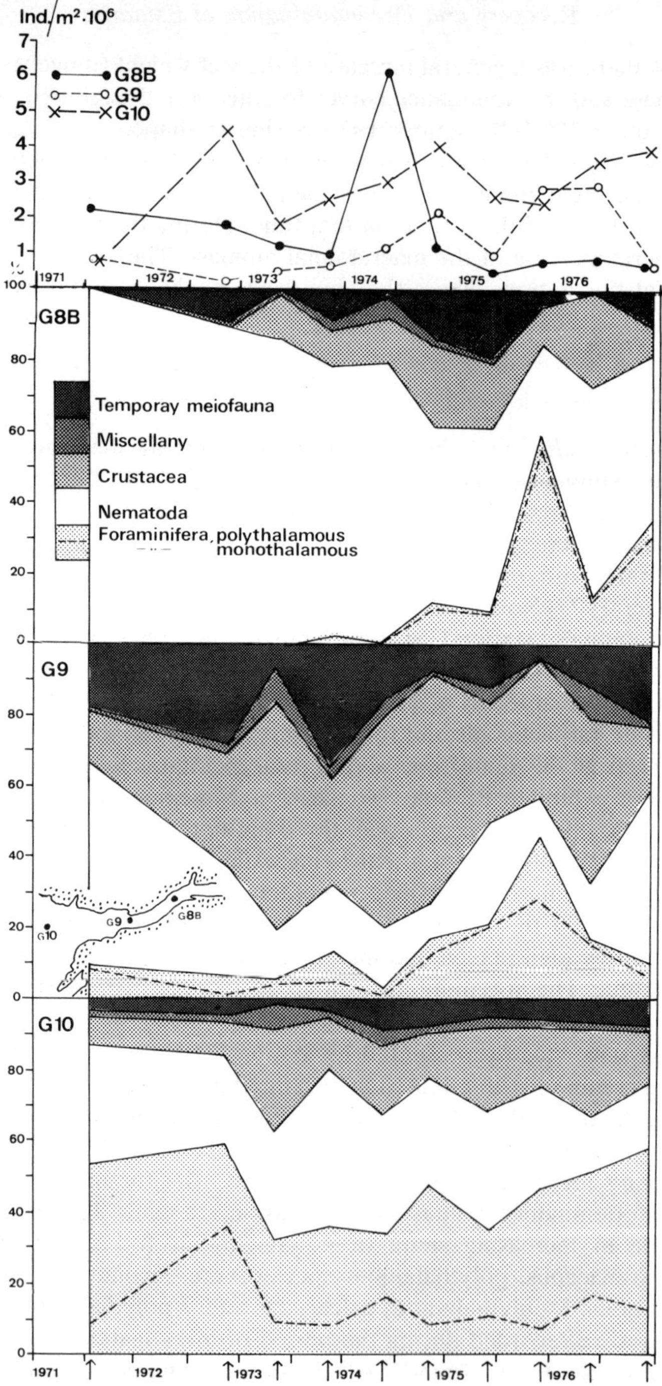

Ind./m²·10⁶
7
6 ●——● G8B
5 ○--○ G9
4 ×—×G10
3
2
1
100
1971   1972   1973   1974   1975   1976
G8B
80
60
40
20
0
Temporay meiofauna
Miscellany
Crustacea
Nematoda
Foraminifera, polythalamous
monothalamous
G9
80
60
40
20
0
G10
G9
G8B
G10
80
60
40
20
0
1971   1972   1973   1974   1975   1976

beccarii. The polythalamous aranaceous form *Ammotium cassis* was recorded on most occasions. Simultaneously as the foraminiferan density was increased the number of species per $30\,cm^2$ decreased, from about 15 in November 1972–June 1974 to about 10 in October 1974–November 1976.

Succession of the macrofauna

The numbers of macrofaunal species found in the Göta River estuary 1971–1976 are indicated in Table 2.

Curves based on the mean values in Table 2 drawn along the pollution gradient exhibit in general the same shapes as the abundance curves shown in Figure 8. The macrofaunal composition as to classes was changed after the pollution abatement in 1972. At station G8B there was a very high dominance of annelids, mainly polychaets, to June 1975. The oligochaets, however, gradually exhibited an increased share. Later on, the group Bivalvia generally increased its proportion.

Table 2. Number of macrofaunal species per $0.5\,m^2$ in the Göta River estuary 1971–1976.

Station	July 1971	November 1972– June 1974 Mean	October 1974– November 1976 Mean
G8C	1	6	12
G8B	3	12	22
G9	17	32	31
G9B	17	22	22
G10B	18	35	34
G10	15	28	30
G10F	13	37	28

The macrofaunal composition as to the dominating species was changed during the successions at the three stations G8B, G9, and G10. In July 1971 only three species were found at station G8B, one of them being *Capitella capitata* with rank number one. From November 1972 to June 1974 this

Figure 9. Abundance and composition of the meiobenthic fauna at the stations G8B, G9, and G10 in the Göta River estuary 1971–1976. Arrows indicate the sampling occasions. Foraminifera is divided into monothalamous and polythalamous forms. Crustacea mainly consists of Podoplea (harpacticoids and cyclopoids). 'Miscellany' includes Hydrozoa, Kinorhyncha, Halacarida, and Turbellaria. Juvenile forms of Gastropoda, Bivalvia, Polychaeta, Oligochaeta, and Echinodermata constitute the temporary meiofauna.

polychaet exhibited the same rank number, while the species *Hydrobia ulvae, Peloscolex benedeni, Polydora ciliata, Pontonema vulgaris* and *Nereis diversicolor* were recorded among the six most dominant species during this period. From October 1974 the faunal composition was considerably changed regarding the dominance of the species and higher proportion of filter-feeders than before was recorded. To conclude, the change of the faunal composition from October 1974 at station G8B, principally implied a shift of number one of the ranked species, followed by six to seven species more or less regularly found simultaneously as new species were added. At station G9, the peak of opportunists was progressively more pronounced; *P. benedeni* was the dominating species here with rank number 1 or 2 during the whole period, and *P. ciliata, P. vulgaris,* and *C. gibba* were also regularly found. *P. benedeni* also was the outstanding dominant species at station G10, followed by *C. gibba.* From 1974 there was a higher dominance of *P. elegans* and *C. capitata* at this station and in November 1976 also of *N. diversicolor.*

Due to inward migration of species (*Pholoë, Pygospio, Pelescolex, Corbula* and others) the fauna has been more homogeneous reflected by an increased macrofaunal similarity (cf. Pearson and Rosenberg, 1978).

Summary of the succession in the Göta River estuary

Major changes of the benthic fauna in relation to decreasing organic enrichment from 1972 in the inner part of the estuary (principally station G8B) are summarized as follows.

In the beginning of the investigation period with a strongly reduced sediment in July 1971 (cf. Figures 7 and 9) the abundance of the meiofauna was relatively high, 2.2×10^6 ind./m^2. The meiofauna was to almost 100% dominated by nematods. The macrofaunal number of species and density was extremely low. In November 1972 the sediment surface down to about 1 cm was oxidized, implying an increased number and proportion of the temporary meiofauna, mainly larval forms of *Capitella capitata* (about 190×10^3 ind./m^2). Furthermore, the abundance and number of species of the macrofauna increased and by May 1973 the numbers were 16×10^3 ind./m^2, with a high proportion of polychaets, mainly *Capitella capitata.* This implies that the survival of larvae recorded in November 1972 was good.

The meiofauna accounted for the most conspicuous change recorded in June 1974, when its abundance exhibited a peak of about 6×10^6 ind./m^2. The temporary meiofauna then reached a density of about 350×10^3 ind./m^2 and the harpacticoids an abundance of about 700×10^3 ind./m^2. The nematodan share amounted to about 80%. The macrofauna also exhibited an increased abundance in June 1974, about 6×10^3 ind./m^2 with a new raise of the polychaet share. As from October 1974 the level of the redox values 0 mV did not reach the sediment surface not even in spring situations. On the contrary a level of about 4 cm below the mud-water interface was recorded in June 1976. Generally the meio- and the macrofaunal abundance curves showed inversed relationships in the inner part from October 1974 implying an decreasing density of the meiofauna and an increasing macro-faunal number. In this month several major changes were observed. The proportion of the foraminifers was then increased (to about 10%) as was that of the harpacticoids. The temporary meiofauna now reached its highest share during the whole investigation period. Furthermore the polychaet proportion was drastically lowered, while the *Oligochaeta* exhibited its hitherto highest share (36%) with *Peloscolex benedeni* as the outstanding species.

Figure 10 summarizes in a correlation matrix the relations at station G8B between the percentage shares of the major classes i.e. Foraminifera, Nematoda, Podoplea, Polychaeta, Oligochaeta, and Bivalvia during the period July 1971–November 1976. The dominance values of these classes were ranked ($n = 10$) and Spearmans̀ rank correlation coefficient (r_s) was calculated (Elliott, 1971). In ten cases statistically significant, positive or negative r_s-values were obtained. Foraminifera was strongly positively and negatively correlated to Bivalvia and Nematoda respectively ($p < 0.001$). Rather strong, negative correlations ($p < 0.01$) were found between Foraminifera and Polychaeta, Bivalvia and Polychaeta, Bivalvia and Nematoda while Oligochaeta and Bivalvia exhibited an equal strong, posi-

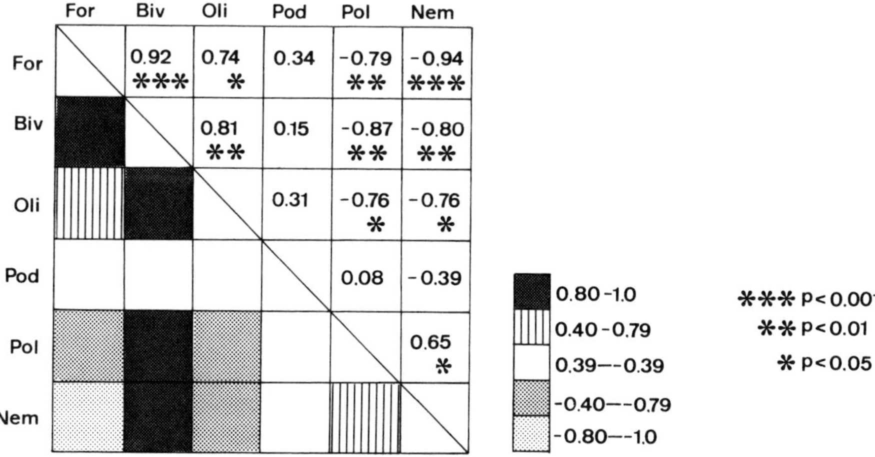

Figure 10. Matrix of rank correlation coefficients (r_s) calculated on the ranked percentage shares at station G8B for the classes Foraminifera (For), Nematoda (Nem), Podoplea (Pod), Polychaeta (Pol) Oligochaeta (Oli) and Bivalvia (Biv).

tive correlation. The positive and negative correlations mentioned indicate the complicate, faunal interactions connected to the recovery process after pollution abatement.

3.2.1 *Meiobenthic recovery*

The information in the literature on meiobenthic recolonization and recovery processes is scarce. The sediment structure has been demonstrated to be very important for the composition of the meiofauna (Scheibel, 1976). Organisms like foraminifers may exercise a stabilizing function and bind the sediment. The detrital layer is greatly reinforced by the agglutination of detritus between the pseudopodia (Nyholm, 1957). McEnery and Lee (1976) studied the life cycles of the monothalamous foraminiferan *Allogromia laticollaris* and three strains were followed. Each of the strains had a different, non-classical, and basically apogamic asexual life cycle. Sexual reproduction was rare. The species is considered to be a food gatherer and can survive many months in the absence of food organisms. This kind of species had been designated 'safety valve species', and organisms with a wide adaptational range dominating under marginal conditions. The species *Allogromia crystallifera* was found to be a dominating pioneer species in the Göta River estuary. Its life cycle is however not yet known (Dahlgren, 1962), but multinucleate forms and asexual reproduction by transverse division has been observed. Holland, Zingmark, and Dean (1974) experimentally found that diatom species secreted large quantities of mucilage thereby effectively stabilizing the sediment. The mucilage-secreting species significantly reduced resuspension and retarded laminar flow of the sediments. Many aquatic nematods continuously produce with their glands a slimy trace consisting of sticky, elastic threads. (Riemann and Schrage, 1978). By creeping repeatedly on their traces the nematods produce burrows and solid, stabilizing, branched concentrations in fine sediments. The authors suppose that the copious mucus secretion of nematodes is mainly involved in nutrition and they present a mucus-trap hypothesis. With the mucus threads the nematods entrap detritus particles, bacteria, and macromolecules.

It is essential for a recovering benthic community that the oxygen conditions in the sediments are improved, that is to reach the original complex structure with larger burrows and break up the laminar stratification with the oxygenated area confined to a narrow zone immediately below the sediment–water interface (Pearson and Rosenberg, 1978). It was shown by Winston and Anderson (1970) that the amount of bioturbation decreased up an estuary as the salinities became lower and the bottom dwelling fauna changed in composition. Species of the polychaete genus *Nereis* appeared to be an important bioturbation agent. Four different levels of bioturbation were observed, one of them being a nonturbate in a silty bottom/brackish water environment near the head of the estuary. As pointed out by Elmgren (1975) the meiofauna is a very useful indicator of the oxygen conditions in a water column since it extends deeper into an oxygen-poor zone than the macrofauna, and can thus show differences between depths and stations totally devoid of macrofaunal species. Representatives of the meiofauna are also characteristic for sediments free from oxygen (Fenchel and Riedl, 1970). Olsson (unpubl) found in an oxygen deficient estuary that the foraminiferan species *Elphidium incertum*, *Elphidium excavatum*, and *Ammonia beccarii* could endure oxygen concentrations below 0.5 mg/l.

Driscoll (1975) observed seasonal variations in the organic content of the sediment largely due to variations in the standing crop of benthic organisms. The standing crop of deposit feeders was, in part, controlled by the abundance of microorganisms. Bioturbation and fecal formation by deposit feeders result in increased surface area for colonization of microorganisms. Biodeposits represent a readily available nutrient source for such organisms. It was suggested that a feedback relationship exists, increasing microorganisms resulting in increasing deposit feeder abundance. The rate of this feedback is considered to be temperature dependant. Coprophagy of a large number of fecal pellets may involve 'seeding' of the pellets to increase colonization by microflora (Fenchel, 1970). Experiments on detrital utilization by the polychaete, *Capitella capitata* showed that there was an increase in the rate of net incorporation with increasing age of the detritus of all the different particle sizes (Tenore, 1975). Augustin and Anger (1974) experimentally found that the same polychaet did not react on water currents. Experiments on the oligochaet *Peloscolex benedeni* (Hunter and Arthur, 1978) indicated that this worm tolerated oxygen concentrations as low as 7.2% of air saturation with negligible mortality over 7 days. At the lowest level tested, 1.2% of air saturation, 37% of adult, and 52% of newly hatched worms died. In the Göta River estuary, as was indicated above, the species *C. capitata* and *P. benedeni* belonged to the key species during the recovery process.

Rhoads *et al.* (1977) studied the colonization of benthic macrofauna of a dredge-spoil dump. The colonization was divided into three stages. The first stage represented the initial recruitment of shallow burrowing surface deposit feeders, suspension feeders and meiofauna. The second stage was characterized by a phase of exponential recruitment of the first stage populations and by recruitment of deeper feeding infauna. The third stage implied a period of leveling off in population densities. Habitat modification related to the three colonization stages were observed. During the first stage (summer) the fecal pellet production started and the redox potential discontinuity was depressed by bioturbation and respiration activities. Stage two (autumn) exhibited destruction of the surficial layer of pellets by meiofaunal grazing. During the third stage (winter) the pelletal surface decayed and the microbial binding decreased. Gray and Johnson (1970) suggested that living micro-organisms are one of the key factors inducing fixation for many sediment living species. Meiofaunal animals by their activities and by excreting metabolic end products induce a bacterial productivity which would not be there without them, and feed on it (Gerlach, 1978). Chamroux, Boucher, and Bodin (1977) experimentally found that bacteria and nematod densities underwent a similar evolution, divided into three main periods.

Colonization experiments have been conducted in the Baltic by Scheibel (1974). Meiofauna samples were taken from submerged platforms. Already some days after submersion meiofaunal colonization began. The substrates were almost exclusively colonized by nematods and harpacticoids. No foraminifers were recorded. It was established that the main migration route to the platforms was up the securing lines. These experiments clearly show the ability of meiofaunal organisms to migrate and occupy new space. Wefer and Richter (1976) have studied, also in the Baltic, the colonization by foraminifers on artificial substrates. With few exceptions only, *Elphidium excavatum clavatum* was present. The authors assume that this species has an especially good passive distribution capability by advection. Schafer (1976) found that the mobility, or transportability of nearshore species, such as *Ammonia beccarii* and *Miliammina fusca* was relatively high. Schafer and Young (1977) summarize some experimental observations on the mobility of Foraminifera relevant to recolonization rates of certain species on substrates created by anthropogenic activities. It is stated that the equivalent diameters of certain, common, nearshore types of Foraminifera range between 0.10 and 0.14 mm. They would thus be within the size range of most easily eroded particles (Shephard, 1963), and therefore subject to continuous redistribution by wave turbulence. Gerlach (1977) has summarized the means of meiofaunal dispersal. During severe storms sea water turbulence erodes sediment not only in shallow water but also in the sublittoral region. The experiments by Schafer and Young (*ib.*) indicated that the transport susceptible calcareous, rotaloid Foraminifera also were the most active crawling species within the total population. The observed arenaceous species tended to be less active than calcareous forms.

The faunal processes found might support the view of multiple stable points and neighbourhood stability and so correspond to the theory of poly-climax (traditionally seral stages) in community development (cf. Gray, 1977). Communities under a fluctuating regime can adapt two alternative strategies. One is to conform with the external fluctuations as forcing functions in the system and to respond with corresponding variations. An alternative is to develop mechanisms dampening the oscillations of the forcing functions (Ott and Fedra, 1977). The interactions found in the Göta

River estuary might have changed the sediments in such a way that the effects of such oscillations have been gradually diminished. Losses of energy took place when the sediments, especially during early phases, were mixed through wave action or turbulence caused by the estuarine compensation current (cf. Fenchel and Riedl, 1970).

Arlt (1975) found in the Baltic that the abundance of some species of *Oligochaeta* and *Harpacticoidea* increased as there was an increase in the organic load. With decreasing salinity a higher temporary meiofaunal biomass and a lower ostracod biomass have been demonstrated (Elmgren *et al.*, 1979). Gerlach (1971) has generalized that meiofauna is quantitatively more important relative to macrofauna in brackish water than in marine areas. The greater importance might be valid also for pollution stressed sediments (Rosenberg *et al.*, 1977). It can be assumed that the meio- and macrobenthic recovery processes in the Göta River estuary described above, due to faunal interactions, will take about the same period of 5 to 10 years. The results indicate a spatial concentration of longdistance meiofaunal gradients.

3.2.2 *Macrobenthic recovery*

Pearson and Rosenberg (1978) have recently reviewed the literature about macrobenthic succession in relation to organic enrichment. The authors conclude that the successional process is similar all over the world, irrespective of an increasing or decreasing organic load. Naturally, the succession goes in opposite directions in these two cases. It is even the same species or genera that reoccur at the different seral stages, i.e. the last survivors in habitat under increasing organic enrichment are the pioneers in a defaunated region following decreased organic load. Rosenberg (1976a) goes one step further and describes from the literature, the same successional pattern in macrobenthic communities recovering from (1) oxygen deficiency, (2) temperature shock, (3) oil pollution, and (4) organic input. However, the species composition was somewhat modified with different salinities and geographic location in combination with temperature. It has recently been demonstrated in field experiments, by putting defaunated sediments on the sea bed, that also species other than the usual opportunists could be the first invaders (Brunswig *et al.*, 1976; McCall, 1977). In these experiments colonization can occur from the natural communities nearby and mobile species can occupy the new habitat immediately.

The speed of recovery in a formerly polluted benthic environment is dependent on the degree of pollution in the sediments and whether benthic larvae are available and hardy enough to settle and survive on that substrate (see, Mileikovsky, 1970; Rosenberg 1977). In the bottoms most polluted by organic wastes in the Saltkällefjord it took more than two years after pollution abatement before any macrobenthic species repopulated that area

(Rosenberg, 1972). The pioneer species are small polychaetes with opportunistic features, e.g. they have a reproduction strategy which allows for rapid colonization and development of extremely high population densities in a short time. It has recently been concluded that almost all opportunistic species are 'general opportunists', i.e. the initial colonizers of any denuded area irrespective of the richness of the substrate, and that only a few species are 'enrichment opportunists', i.e. initial colonizers only in organically rich areas (Pearson and Rosenberg, 1978). Opportunists have the ability to utilize the accumulated organic content in the sediments and their reworking (burrowing and feeding) of the sediments will later make it suitable for larvae of other species also to settle there. As the opportunists are poor competitors they will have to give way to the later immigrants. The portion of larger species in the macrobenthic community will increase during the succession and these species (e.g. echinoderms and molluscs) will rework the sediments vertically more deep and extensively (see Pearson and Rosenberg 1976): Figure 5; Rhoads *et al.*, 1977). This cultivation by the macrobenthos will increase the rate of mineralization and the loss of heavy metals (Renfro, 1973).

The speed of recovery in macrobenthic communities can be illustrated by a few examples. Organic pollution of the bottoms of a Swedish estuary, the Saltkällefjord, almost ceased in 1966 (Rosenberg, 1972, 1973, 1976a). Moderately polluted stations in the middle of the estuary showed the fastest recovery as the communities there had already passed some of the earliest seral stages of the succession (Figure 11). The macrobenthic communities of the most polluted area of the estuary (Figure 11: stations L6, L9) approached the final phase of the succession in 1972. Eight years after

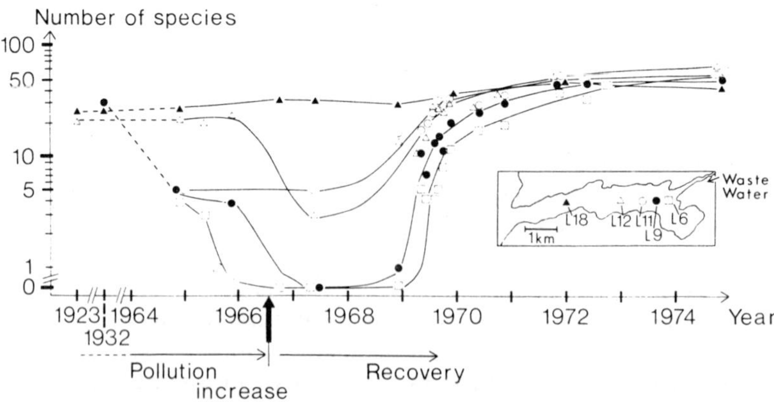

Figure 11. The succession of benthic species at five stations in the Saltkällefjord. The arrow indicates the time of pollution abatement. Semilog scale (From Rosenberg, 1976a). (Reproduced by permission of *Oikos.*)

pollution abatement, i.e. in 1974, the community structure was re-established. A key-species in the final organization of the community was *Amphiura filiformis*. In contrast to the smooth, successive increase in number of species (Figure 11), the population dynamics during succession along a temporal gradient of increasing or decreasing organic pollution is a messy and confusing process (Pearson, 1975; Rosenberg, 1976a). The temporal changes in abundance and dominance of different populations are most drastic with sudden shifts at the early stage of recovery, and graphically they show a typical bell-shaped dominance pattern. The Saltkällefjord is the only known example, where the whole process of benthic faunal recovery has been followed in an estuary. The tropical storm Agnes reduced the salinities drastically in the James and York estuaries in the lower Chesapeake Bay in USA. Irruptions of opportunistic species followed these perturbations and, 2.5 years after the storm, the deep mud bottoms in the lower York estuary had not yet recovered completely (Boesch *et al.*, 1976). The first phases of succession subsequent to the oil spill in West Falmouth were similar (Grassle and Grassle, 1974) to those reported from Sweden and an estimated recovery time in this case of 5 to 10 years seems reasonable.

The macrobenthic community structure along a spatial gradient of organic enrichment is illustrated in Figure 12. The succession in relation to time shows the same structural changes. Thus, if the macrofaunal species are wiped out by heavy pollution, the succession process will begin as illustrated on the left-hand side of Figure 12 and slide towards the right. The different phases of this succession have been described by Pearson and Rosenberg (1978) as follows: The first species encountered are small and few and will, within a short time or distance, rise to an extremely high abundance 'the peak of opportunists'. Later during succession the numbers will drop off and more species will invade the area, and this phase, the 'ecotone phase' will with time develop into a 'transitory community' including more and larger species. The transitory community will gradually develop into the 'normal' and more mature community. Leppäkoski (1975) used the term 'migrating communities' for such sequential changes.

The recovery capacity in relation to depth has been illustrated in a series of experiments. In defaunated boxes placed in an intertidal area at a depth of 10 m, and at a deep-sea station of 1760 m, the rates of colonization were tremendously different (Grassle, 1977). After 2 months, 43 individuals were found in the deep-sea box, whereas thousands of individuals were observed at 10 m. The experiment in the intertidal zone demonstrated high densities and the fastest response of settling larvae, many growing to adults within a month. This experiment is in agreement with what has been demonstrated by field data above. We may conclude that, in general, littoral ecosystems have the greatest recovery capacity and that the recovery time to reach fairly normal conditions is approximately 3 to 5 years. Sublittorial systems down

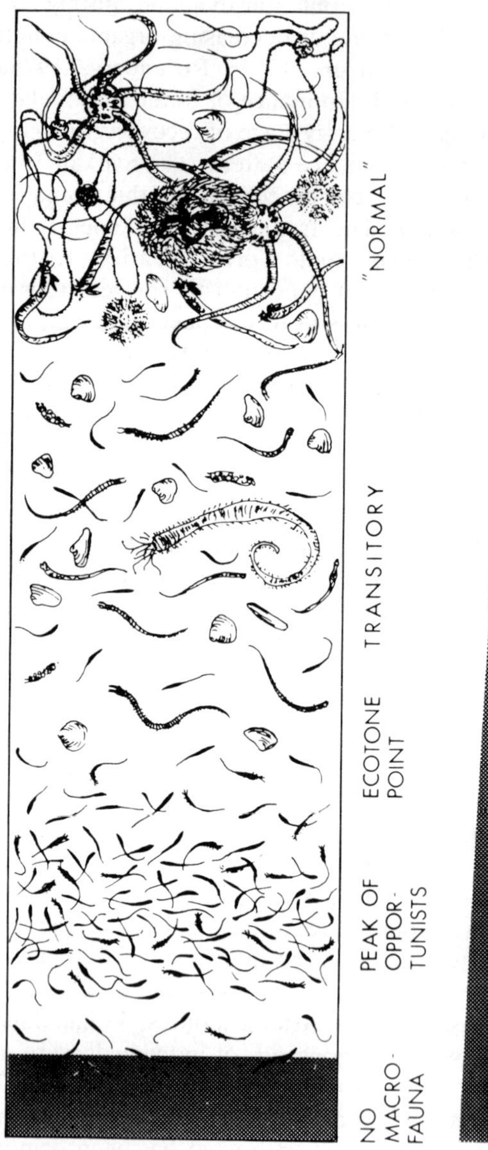

Figure 12. Diagrammatic representation of changes in abundance and species types occurring along a generalized organic enrichment gradient (From Pearson and Rosenberg, 1978). (Reproduced by permission of *Ambio*.)

to about 50 m need approximately 5 to 10 years to recover, whereas at greater depths the time of recovery is more extended. These suggested times seem to be realistic estimations in well-flushed organically polluted estuaries in boreal, temperature, and tropical regions.

4 REFERENCES

Ahmed, M., and Focht, D. D. (1973). Degradation of polychlorinated bifenyl by two species of *Achromobacter*. *Can. J. Microbiol.*, **19**, 48–52.

Arlt, G. (1975). Remarks on indicator organisms (meiofauna) in the coastal waters of the GDR. *Merentutkimuslait. julk.* **Skr. 239,** 272–279.

Aston, S. R., Bruty, D., Chester, R., and Padgham, R. C. (1973). Mercury in lake sediments: A possible indicator of technological growth. *Nat.*, **241**, 450–451.

Augustin, A., and Anger K. (1974). Experimente über Substratpräferenzen von *Capitella capitata* (Fabricus) *Kieler Meeresforsch.* **30**, 28–36.

Bagge, P. (1969). The succession of the bottom fauna communities in polluted estuarine habitats in the Baltic–Skagerak region. *Merentutkimuslait. julk*; **Skr. 228,** 119–130.

Banat, K., Förstner, U., and Müller, G. (1972). Schwermetalle in Sedimenten von Donau, Rhein, Ems, Weser, und Elbe in Bereich der Bundesrepublik Deutschland. *Naturwiss.*, **59**, 525–528.

Berner, R. A. (1971). *Principles of Chemical Sedimentology*, 240 pp. McGraw-Hill Book Co., New York.

Boesch, D. F., Diaz, R. J., and Virnstein, R. W. (1976). Effects of tropical storm Agnes on soft-bottom macrobenthic communities of the James and York estuaries and the lower Chesapeake Bay. *Chesapeake Sci.*, **17**, 246–259.

Borchert, H. (1965). Formation of marine sedimentary iron ores. In Riley, J. P., and Skirrow, G. (eds.), *Chemical Oceanography*, Vol. 2, 159–204, Academic Press, London, New York and San Francisco.

Bowden, K. F. (1967). Circulation and diffusion. In Lauff, G. H. (ed.), Estuaries, *Am. Assoc. Adv. Sci. Publ.*, **83**, 15–35, Washington D.C.

Bradford, W. L. (1972). A study on the chemical behaviour of Zn in Chesapeake Bay water using anodic shipping voltammetry. *Techn. Rep. Chesapeake Bay Inst.*, **76**, 103 pp.

Braman, R. S., and Foreback, C. G. (1973). Methylated forms of arsenic in the environment. *Sci.*, **182**, 1247–1249.

Brunswig. D., Arntz, W. E. and Rumohr, H. (1976). A tentative field experiment on population dynamics of macrobenthos in the Western Baltic. *Kieler Meeresforsch.*, *Sonderheft Nr.* 3, 49–59.

Buchanan, J. B. (1964). A comparative study of some features of the biology of *Amphiura filiformis* and *Amphiura chiajei* (Ophiuroidea) considered in relation to their distribution. *J. mar. biol. Ass. U.K.*, **44**, 565–576.

Cato, I. (1977). Recent sedimentological and geochemical conditions and pollution problems in two marine areas in south-western Sweden. *Striae* 6, 158 pp.

Cato, I., Olsson, I., and Rosenberg, R. (1975). Sedimentologiska och bottenfaunistiska undersökningar i Välen—ett område att restaurera. Göteborgs Naturhistoriska Museum, Årstryck, 1975, 13–35.

Cederwall, K., and Svensson. T. (1975). Sediment flushing after dredging in tidal bays. *Medd. 84, Hydraul. Div. Chalmers Inst. Technol. Göteborg.*, 25 pp.

Chamroux, S., Boucher, G., and Bodin, P. (1977). Etude expérimentale d'un écosystème sableux II Evolution des populations de bactéries et de méiofaune *Helgoländer wiss. Meeresunters.*, **30**, 163–177.

Cline, J. T., and Upchurch, G. B. (1973). Mode of heavy metal migration in the upper strata of lake sediments. *Proc. 16th cong. Great Lakes Res.*, 1973, 349–356. Am. Arbor. Mich.

Dahlgren, L. (1962). *Allogromia crystallifera* n. sp., a monothalamous foraminifer. *Zool. Bidr.* Uppsala, **35**, 451–455.

Dauer, D. M., and Simon, J. L. (1976). Habitat expansion among polychaetous annelids repopulating a defaunated marine habitat. *Mar. Biol.*, **37**, 169–177.

Dean, D., and Haskin, H. H. (1964). Benthic repopulation of the Raritan River estuary following pollution abatement. *Limnol. Oceanogr.*, **9**, 551–563.

Driscoll, E. G. (1975). Sediment–animal–water interaction, Buzzards Bay, Massachusetts. *J. Mar. Res.*, **33(3)**, 275–302.

Duinker, J. C., and Nolting, R. F. (1976). Distribution model for particulate trace metals in the Rhine Estuary, Southern Bight and Dutch Wadden Sea. *Neth. J. Sea Res.* 10(1), 71–102, Texel.

Eisma, D. (1973). Sediment distribution in the North Sea in relation to marine pollution. In Goldberg, E. D., (ed.), *North Sea Science*, 131–150. Massachusetts Institute of Technology Press, Boston.

Elderfield, H., and Hepworth, A. (1975). Diagenesis, metals and pollution in estuaries. *Mar. Pollut. Bull.*, **6**, 85–87.

Elliott, J. M. (1971). Some methods for the statistical analysis of samples of benthic invertebrates. *Freshwater Biological Association, Scientific Publications*, **25**, 144 pp. Titus Wilson & Son Ltd., Kendal.

Elmgren, R. (1975). Benthic meiofauna as indicator of oxygen conditions in the northern Baltic proper. *Merentutkimuslait. julk. Skr.*, **239**, 265–271.

Elmgren, R., Rosenberg. R., Andersin, A.-B., Evans, S., Kangas, P., Lassig, J., Leppäkoski, E., and Varmo, R. (1979). Benthic macro- and meiofauna in the Gulf of Bothnia (Northern Baltic). *Pr. morsk. Inst. ryb. Gdyni* (in press.)

Evans, D. W., and Cutshall, N. H. (1973). Effects of ocean water on the soluble-suspended distribution of Columbia River radionuclides. In *Radioactive contamination of the marine environment, IAEA*, Vienna, 125–140.

Fagerström, T., and Jernelöv, A. (1971). Formation of methyl mercury from pure mercuric sulphide in aerobic organic sediment. *Water Res.* 5, 121–122.

Fagerström, T., and Åsell, B. (1973). Methyl-mercury accumulation in aquatic food chain. A model and some implications for research planning. *Ambio*, **2**, 164–171.

Feick, G., Horne, R. A., and Yeaple, D. (1972). Release of mercury from contaminated freshwater sediments by the runoff of road deicing salt. *Science*, **175**, 1142–1143.

Fenchel, T. M. (1970). Studies on the decomposition of organic detritus derived from the turtle grass *Thalassia testudinum*. *Limnol. Oceanogr.*, **15**, 14–20,

Fenchel, T. M., and Riedl, R. J. (1970). The sulfide system: a new biotic community underneath the oxidized layer of marine sand bottoms. *Mar. Biol.*, **7**, 255–268.

Fischer, R. A., Corbett, A. S., and Williams, C. B. (1943). The relationship between the number of species and the number of individuals in a random sample of an animal population. *J. Anim. Ecol.*, **12**, 42–58.

Förstner, U. (1976). Lake sediments as indicators of heavy-metal pollution. *Naturwiss.*, **63**, 465–470.

Förstner, U., and Müller, G. (1974). Schwermetallanreicherungen in datierten Sedimentkernen aus dem Bodensee und aus dem Tegernsee. *Tschermaks Mineral.*

Petrol. Mitt., **21**, 145–163.

Förstner, U., and Müller G. (1976). Heavy-metal pollution monitoring by river sediments. *Fortschr. Miner.*, **52**, 271–288.

Förstner, U., and Patchineelam, S. R., (1976). Bindung und Mobilisation von Schwermetallen in fluviatilen Sedimenten. Chem. Ztg., 100 (2), 49–57.

Gerlach, S. A. (1971). On the importance of marine meiofauna for benthos communities. *Oecologia*, (Berl.), **6**, 176–190.

Gerlach, S. A., (1977). Means of meiofauna dispersal. *Mikrofauna Meeresboden*, **61**, 89–103

Gerlach, S. A., (1978). Food chain relationships in subtidal silty sand marine sediments and the role of meiofauna in stimulating bacterial productivity. *Oecologia* (Berl.), **33**, 55–69.

Goldberg, E. D. (1976). The health of the oceans. *The UNESCO press*, Paris 172 pp.

Grassle, J. F. (1977). Slow recolonisation of deep-sea sediment. *Nature*, Lond. **265**, 618–619.

Grassle, J. F., and Grassle, J. P. (1974). Opportunistic life histories and genetic systems in marine benthic polychaetes. *J. mar. Res.*, **32**, 253–284.

Gray, J. S. (1977). The stability of benthic ecosystems. *Helgoländer wiss. Meeresunters.*, **30**, 427–444.

Gray, J. S., and Johnson, R. M. (1970). The bacteria of a sandy beach as an ecological factor affecting the interstitial gastrotrich *Turbanella hyalina* Schultze. *J. exp. mar. Biol. Ecol.*, **4**, 119–133.

Groot, A. J. de (1973). Occurrence and behavior of heavy metals in river deltas, with special reference to the rivers Rhine and Ems. *North Sea Sci.*, 308–325. Massachusetts Institute of Technology Press, Boston.

Groot, A. J. de, Geoij, J. J. M. de, and Zegers, C. (1971). Contents and behaviour of mercury, as compared with other heavy metals in sediments from the rivers Rhine and Ems. *Geol. Mijnbouwkd.* **50**, 393–398.

Håkanson, L. (1973). Kvicksilver i några svenska sjöars sediment—möjligheter till tillfriskning. *IVL. Publ. a* 92, 171–186, Stockholm.

Hallberg, R. O. (1973). Paleoredox conditions in the eastern Gotland Basin during the last 400 years. *Contrib. Askö Lab. Univ. Stockholm* 2:1, 89–117 Stockholm.

Hanks, R. W. (1968). Benthic community formation in a 'new' marine environment. *Chesapeake Sci.*, **9**, 163–172.

Holland, A. F., Zingmark, R. G., and Dean, J. M. (1974). Quantitative evidence concerning the stabilization of sediments by marine benthic diatoms. *Mar. Biol.*, **27**, 191–196.

Holmes, C. W., Slade, E. A., and McLerran, C. J. (1974). Migration and redistribution of Zinc and Cadmium in marine estuarine system. *Environ. Sci Technol.*, **8**, 255–259.

Hunter, J., and Arthur, D. R. (1978). Some aspects of the ecology of *Peloscolex benedeni* Udekem (Oligochaeta: Tubificidae) in the Thames estuary. *Estuar. cstl mar. Sci.*, **6**, 197–208.

Jensen, S., and Jernelöv, A., 1969: Biological methylation of mercury in aquatic organisms. *Nature*, **223**, 753–754.

Jernelöv, A. (1970). Release of methyl mercury from sediments with layers containing inorganic mercury at different depth. *Limnol. Oceanogr.*, **15**, 958–960.

Jernelöv, A. (1971). Återställande av förgiftade vatten. In *Praktisk miljökunskap-miljögifter*, 281–289.

Jernelöv, A. (1972). Mercury—A case study of marine pollution. In Dyrssen, D., and Jagner, D. (eds): *The Changing Chemistry of the Oceans. Nobel Symp.*, **20**,

161–169. Almqvist and Wiksell, Uppsala.

Jernelöv, A., and Rosenberg, R. (1976). Stress tolerance of ecosystems. *Environ. Cons.*, **3**, 43–46.

Johnston, R. (1970). The decomposition of crude oil residues in sand colon. *J. Mar. Biol. Ass.*, **50**, 925–937.

Kharker, D. P., Turekian, K. K., and Bertine, K. K. (1968). Stream supply of dissolved Ag, Mo, Sb, Se, Cr, Co, Rb, and Cs to the oceans. *Geochim. Cosmoch. Acta*, **32**, 285–298.

Kuijpers, A. (1974). Trace Elements at the depositional interface and in sediments of the outer parts of the Eckernförder Bucht, western Baltic. *Meyniana*, **26**, 23–28.

Lee, F. G. (1970). Factors affecting the transfer of materials between water and sediments. *Eutrophication Inf. Program Water Resour. Cent. Univ. Wisconsin*, 50 pp.

Leppäkoski, E. (1975). Assessment of degree of pollution on the basis of macrozoobenthos in marine and brackish water environments. *Acta Academiae Aboensis, Ser. B*, **35(2)**, 1–90.

Loring, D. H. (1975). Mercury in the sediments of the Gulf of St. Lawrence, *Can. J. Earth Sci.*, **12**, 1219–1237.

Margalef, R. (1958). Temporal succession and spatial heterogeneity in natural phytoplankton. In *Perspectives in Marine Biology*, 323–349. Univ. of California Press, Berkeley and Los Angeles.

McBride, B. C., and Wolfe, R. C. (1971). Biosynthesis of dimethylarsine by methanobacterium. *Biochem. J.*, **10**, 4312–4317.

McCall, P. L. (1977). Community patterns and adaptive strategies of the infaunal benthos of Long Island Sound. *J. mar. Res.*, **35**, 221–266.

McEnery, M., and Lee J. J. (1976). *Allogromia laticollaris*: A foraminiferan with an unusual apogamic metagenic lifecycle. *J. Protozool*, **23(1)**, 94–108.

Meade, R. H. (1969). Landward transport of bottom sediments in estuaries of the Atlantic Coastal Plain. 3. *Sediment Petrology*, **39**, 222–234.

Meade, R. H. (1972). Transport and deposition of sediments in estuaries. In Nelson, B. W. (ed.), Environmental Framework of Coastal Plain Estuaries. *Mem. Geol. Soc. Amer.*, 133, 91–120.

Mileikovsky, S. A. (1970). The influence of pollution on pelagic larvae of bottom invertebrates in marine nearshore and estuarine waters. *Mar. Biol.*, **6**, 350–356.

Müller, G., and Förstner, U. (1975). Heavy metals in sediments of the Rhine and Elbe Estuaries: Mobilization or mixing effect? *Environ. Geol.* **1**, 33–39, New York.

Nyholm, K.-G. (1957). Orientation and binding power of recent monothalamous Foraminifera in soft sediments. *Micropaleontology*, **3**, 75–76.

Nyholm, K.-G., Olsson, I., and Andrén, L. (1977). Quantitative investigations on the macro- and meiobenthic fauna in the Göta River estuary. *Zoon*, **5**, 15–28.

Olausson, E. (1972). Water-sediment exchange and recycling of pollutants through biogeochemical processes. In Ruivo, M. (ed.), *Marine Pollution and Sea Life*. 158–161. FAO, Fishing News Books Ltd., London.

Olausson, E. (1974). Sedimentation, water-sediment exchange: Marine sediments and their use in environmental studies. In First FAO/SIDA training course on marine pollution in relation to protection of living resources. *FAO/SIDA/TF 95—Suppl.*, **1**, 1825. Rome.

Ott, J., and Fedra, K. (1977). Stabilizing properties of a high-biomass benthic community in a fluctuating ecosystem. *Helgoländer wiss. Meeresunters.*, **30**, 485–494.

Pearson, T. H. (1975). The benthic ecology of Loch Linnhe and Loch Eil, a sea-loch system on the west coast of Scotland. IV. Changes in the fauna attributable to organic enrichment. *J. exp. mar. Biol. Ecol.*, **20**, 1–41.

Pearson, T. H., and Rosenberg, R., (1978). Macrobenthic succession in relation to organic enrichment and pollution of the marine environment. *Oceanogr. Mar. Biol. Ann. Rev.*, **16**, 229–311.

Pearson, T. H., and Rosenberg, R., (1978). Macrobenthic succession in relation to organic enrichment and pollution of the marine environment. *Oceanogr. Mar. Biol. Ann. Rev.*, **16**, 299–311.

Presley, B. J., Kolodny, Y., Nissenbaum, A., and Kaplan, I. R. (1972). Early diagenesis in a reducing fjord, Saanich Inlet. British Columbia. II. Trace-element distribution in interstitial water and sediment. *Geochim. Cosmochim. Acta*, **36**, 1073–1090. Oxford.

Pritchard, D. W. (1967). Observations of circulations in coastal plain estuaries. In Lauff, G. H. (ed.), Estuaries. *Am. Assoc. Adv. Sci. Publ.*, **82**, 37–44. Washington D.C.

Rashid, M. A., and Leonard, J. D. (1973). Modification in the solubility and precipitation behaviour of various trace metals as a result of their interaction with sedimentary humic acid. *Chem. Geol.*, **11**, 89–97.

Reinhard, D., and Förstner, U. (1976). Metallanreicherungen in Sedimentkernen aus Stauhaltungen des mittleren Neckars. *N. Jb. Geol. Paläontol. Mh. H.*, **5**, 301–320.

Reish, D. J. (1971). Effect of pollution abatement in Los Angeles harbours. *Mar. Pollut. Bull.*, **2**, 71–74.

Renfro, W. C. (1973). Transfer of ^{65}Zn from sediments by marine polychaete worms. *Mar. Biol.*, **21**, 305–316.

Rhoads, D. C., Aller, R. C., and Goldhaber, M. B. (1977). The influence of colonizing benthos on physical properties and chemical diagenesis of the estuarine seafloor. In *Ecology of Marine Benthos*, ed. B. C. Coull, Univ. South Carolina Press, Columbia, pp. 113–138.

Riemann, F., and Schrage, M. (1978). The mucustrap hypothesis on feeding of aquatic nematodes and implications for biodegradation and sediment texture. *Oecologia* (Berl.), **34**, 75–88.

Rohatgi, N., and Chen, K. Y. (1975). Transport of trace metals by suspended particulates on mixing with seawater. *J. Water/Pollut. Con. Fed.*, **47**, 2298–2316.

Rosenberg, R. (1971). Recovery of the littoral fauna in Saltkällefjord subsequent to discontinued operations of a sulphite pulp mill. *Thalassia jugosl.*, **7**, 341–351.

Rosenberg, R., (1972). Benthic faunal recovery in a Swedish fjord following the closure of a sulphite pulp mill. *Oikos*, **23**, 92–108.

Rosenberg, R. (1973). Succession in benthic macrofauna in a Swedish fjord subsequent to the closure of a sulphite pulp mill. *Oikos*, **24**, 244–258.

Rosenberg, R. (1976a). Benthic faunal dynamics during succession following pollution abatement in a Swedish estuary. *Oikos*, **27**, 414–427.

Rosenberg, R. (1976b). The relation of treatment and ecological effects in brackish water regions. *Pure Appl. Chem.*, **45**, 199–203.

Rosenberg, R. (1977). Effects of dredging operations on estuarine benthic macrofauna. *Mar. Pollut. Bull.*, **8**, 102–104.

Rosenberg, R., Olsson, I., and Ölundh, E. (1977). Energy flow model of an oxygen-deficient estuary on the Swedish west coast. *Mar. Biol.*, **42**, 99–107.

Rust, B. R., and Waslenchuk, D. G. (1976). Mercury and bed sediment in the Ottawa River, Canada. *J. Sediment. Petrol.*, **46**, 563–578.

Sanders, H. L., Grassle, J. F., and Hampson, G. R. (1972). The West Fallmouth oil spill. *Woods Hole Oceanogr. Inst. Technical Report*, I. Biology (unpublished manuscript.).

Schafer, C. T. (1976). Transport and recolonization of Foraminifera. From: *Report of activities, Part C*; *Geol. Surv. Can.*, Paper 76–1 C, 27–32.

Schafer, C. T., and Young, J. A. (1977). Experiments on mobility and transportability of some nearshore benthonic Foraminifera species. From: *Report of Activities, Part C, Geol. Surv. Can.*, Paper 77-1 C, 27–31.

Scheibel, W. (1974). Submarine experiments on benthic colonization of sediments in the western Baltic Sea. II. Meiofauna. *Mar. Biol.*, **28**, 165–168.

Scheibel, W., (1976). Quantitative Untersuchungen am Meiobenthos lines Profils unterschiedlicher Sedimente in der westlichen Ostsee. *Helgoländer wiss. Meeresunters.* **28**, 31–42.

Shephard, F. P. (1963). *Submarine geology*: Harper and Row, 557.

Summer, A. O., and Silver, S. (1978). Microbial transformations of metals. *Ann. Rev. Microbiol.*, **32**, 637–672.

Teal, J. M., Burns, K., and Ferrington, J. (1978). Analyses of aromatic hydrocarbons in intertidal sediments from two oil spills of No. 2 fuel oil in Buzzards Bay, Massachusetts. *J. Fish. Res. B. Can.*, **35**, 510–520.

Tenore, K. R. (1975). Detrital utilization by the polychaete, *Capitella capitata*. *J. Mar. Res.*, **33 (3)**, 261–274.

Turekian, K. K. (1977). The fate of metals in oceans. *Geochim. Cosmochim Acta*, **41**, 1139–1144.

Wefer, G., and Richter, W. (1976). Colonization of artificial substrates by Foraminifera. *Kieler Meeresforsch.*, Sonderheft 3, 72–75.

Winston, J. E., and Anderson, F. E. (1971). Bioturbation of sediments in a northern temperate estuary. *Marine Geol.*, 10, 39–49.

Wolff, W. J. (1974). Benthic diversity in the Rheine-Meuse estuary. *Hydrobiol. Bull.*, 8, 242–252.

Wong, P. T. S., Chau, Y. K., and Luxon, P. L. (1975). Methylation of lead in the environment. *Nat.*, **253**, 263–264.

Subject Index

Systematic Index

Index of Estuaries, Rivers, and Bays